Exploration
Magnetics

This book is dedicated to the memory of the late Professor Don Emerson, AM, PhD, who as a founding member of the Australian Society of Exploration Geophysicists, ASEG Past President, ASEG Gold Medal recipient, Co-Convenor of the ASEG's 1st Conference and Exhibition and former Editor of the ASEG's scholarly journal, **Exploration Geophysics,** *made extraordinary contributions to the profession of exploration geophysics in Australia over a career spanning six decades. For 28 years, Don was one of Australia's most eminent university teachers and researchers in exploration geophysics, mentoring many explorationists and researchers, as Head of Geophysics and finally as Head of Department of Geology and Geophysics, University of Sydney. After his university career, his passionate interest in petrophysical studies led to the establishment of Systems Exploration, which developed a comprehensive rock-properties testing facility, servicing the exploration and engineering industries. The Systems Exploration laboratory had extensive collaborations with CSIRO, particularly in the areas of palaeomagnetism, magnetic petrophysics and potential field methods.*

Exploration Magnetics
Theory and Practice

Editors: Phil Schmidt, James Austin, David Clark, Keith Leslie, Mark Lackie and Clive Foss

CSIRO

PUBLISHING

A catalogue record for this book is available from the National Library of Australia

ISBN: 9781486315574 (pbk)
ISBN: 9781486315581 (epdf)
ISBN: 9781486315598 (epub)

How to cite:

Schmidt P, Austin J, Clark D, Leslie K, Lackie M, Foss C (Eds) (2025) *Exploration Magnetics: Theory and Practice*. CSIRO Publishing, Melbourne.

Published by:

CSIRO Publishing
36 Gardiner Road, Clayton VIC 3168
Private Bag 10, Clayton South VIC 3169
Australia

Telephone: +61 3 9545 8400
Email: publishing.sales@csiro.au
Website: www.publish.csiro.au
Sign up to our email alerts: publish.csiro.au/earlyalert

Front cover: Vertical derivative of gravity image and magnetic (TMI) contours over Gairdner Dykes in the Gawler Craton of South Australia. Image by Clive Foss using data from the Geological Survey of South Australia (GSSA).

Cover design by Cath Pirret
Typeset by Envisage Information Technology
Index by Max McMaster
Printed in Australia by Jossimo Print

CSIRO Publishing is a business unit within the Commonwealth Scientific and Industrial Research Organisation (CSIRO) and publishes and distributes scientific, technical and health science books, magazines and journals from Australia to a worldwide audience and conducts these activities autonomously from the research activities of CSIRO. The views expressed in this publication are those of the author(s) and editor(s) and do not necessarily represent those of, and should not be attributed to, CSIRO. While all appropriate care has been taken to ensure the accuracy of the content at the time of publication, CSIRO, the author(s) and any editors or contributors shall not be liable to any person for any errors, omissions or inaccuracies herein. The reader/user accepts all risk and responsibility for losses, damages, costs and other consequences resulting directly or indirectly from using this information.

CSIRO acknowledges the Traditional Owners of the lands that we live and work on across Australia and pays its respect to Elders past and present. CSIRO recognises that Aboriginal and Torres Strait Islander peoples have made and will continue to make extraordinary contributions to all aspects of Australian life including culture, economy and science. The use of Western science in this publication should not be interpreted as diminishing the knowledge of plants, animals and environment from Indigenous ecological knowledge systems.

Foreword

It is indeed a privilege to introduce this latest earth science work of research and expertise from the Commonwealth Scientific and Industrial Research Organisation (CSIRO), Australia's national science agency, with a global record of discoveries, innovations, inventions and leading-practice science in the mining and resources industries.

There is a clear line of sight between the economic importance of our nation's mineral resources industry and the opportunity that CSIRO has given to fostering a very deep dive into this practical collected research in exploration magnetics, both comprehensively underpinned by our national discovery-rich precompetitive magnetic datasets.

In 2024, the Australian Bureau of Statistics reported $3.95B of expenditure on mineral exploration across Australia. Successful exploration, leading to economic discovery, is fundamental to the nation's economy and the sustainability of Australia's mining industry, reported to have contributed a record $455B in export revenue in the 2022-23FY.

One of the key impediments to successful mineral exploration is the weathered cover sequence of rocks and sediments that conceals around 80% of Australia's prospective basement rocks. Using multiple geophysical methods along with geology and geochemistry has become fundamental in the undercover search for critical mineral resources, essential for Australia's industries and future economic prosperity.

Australia's world-leading precompetitive aeromagnetic coverage is a primary foundation dataset for the exploration industry in the analysis of prospectivity, optimum search parameters and project specific targeting. The challenge for magnetic exploration undercover is to extract the fullest understanding of the magnetic subsurface from the aerial survey measurement of total magnetic intensity.

As background for non-specialists and mineral explorers in international jurisdictions, it's relevant to reflect on the path of Australia as an early innovator and world leader in precompetitive geoscience and aerial magnetic surveying.

I had the pleasure in 2020 of working closely with author Doug Morrison on the production and publication of his book *Measuring Terrestrial Magnetism*[*], which recounts how the measurement of terrestrial magnetism has influenced the history of the world up until 1950.

In the 1940s there were major advances in magnetometry, and by the end of World War 2 the sensitivity of military magnetometers was rapidly improved. The mineral exploration industry applied these magnetometers to find new economic deposits of magnetic mineral ores. Countries including Australia, Canada and the United States directed their national geological survey departments to establish programs of major aerial magnetic surveying and mapping in the search for minerals and energy.

In 1945, Harold Raggatt, Director of the Australian Government Mineral Resources Survey, later Bureau of Mineral Resources (BMR), and his Chief Geophysicist, Jack Rayner, visited the American and Canadian geological surveys to investigate the potential for magnetic mapping of Australia from the air.

In the final chapter of his book, Morrison recounts the story of the BMR purchase in the late 1940s of two airborne magnetometers from the Royal Navy, which were significantly modified in Australia and installed in a DC3 aircraft in preparation for the first government magnetic aerial survey programs in the 1950s. In the following two decades, BMR first flew aeromagnetic and radiometric regional surveys across vast regions of northern Australia, and in collaboration with state and territory geological surveys began the systematic government regional magnetic mapping of known mineral domains and sedimentary basins with petroleum potential. In 1949, the first industry exploration aeromagnetic surveys were undertaken in Australia by major mining companies, including The Zinc Corporation, Broken Hill Proprietary Company Limited and Western Mining Corporation.

In the past six decades, the positive geoscience and funding collaboration of state, territory and federal geoscience agencies through successive government-funded exploration initiatives and joint national geophysical

---

[*] Morrison WD (2020) *Measuring Terrestrial Magnetism: The Evolution of the Airborne Magnetometer and the First Anti-submarine and Aeromagnetic Survey Operations.* Australian Society of Exploration Geophysicists, Sydney.

mapping programs, covering every corner of Australia, has produced the world's best precompetitive high-resolution aeromagnetic datasets. Indeed, The Fraser Institute Annual Survey of Mining Companies over the past decade has regularly ranked Australia's prospectivity, underpinned by our free online national geoscience data, in the top five of all global jurisdictions.

Turning to *Exploration Magnetics*, this is a unique and complete advanced reference work for mineral explorers and specialist exploration geophysicists on all aspects of the theory and contemporary practice of magnetics. It covers the importance of analysing and understanding rock magnetic properties at the micro-scale through to the inversion/interpretation at exploration scale of magnetic source depth and configuration, induced and remanent magnetisation direction, continuously mapping subsurface magnetisation and contrasts in magnetisation properties.

Geoscientists with non-specialist experience in geophysics will nevertheless find some real exploration gems of wisdom in this work, including the extensive case study applications.

For the past four decades, Australia's aerial geophysical survey practitioners and contractors have continued to innovate, invent and refine their magnetic survey data acquisition systems, the majority delivering increasingly sensitive and ever lower noise levels in the final primary product – Total Magnetic Field maps and datasets.

We have known theoretically for many years that there is much more three-dimensional information that can be directly measured in the field, including total field gradiometry and even magnetic tensor gradiometry.

I refer to an excellent journal paper very much ahead of its time on magnetic interpretation by two authors of this book (Schmidt and Clark 2000[**]). This paper outlines compelling advantages for magnetic tensor gradiometry, noting the technical and instrumentational challenges at that time of operating SQUID technology in field environments.

In 2025, SQUID and field practical instrumentation has moved on, such that there are magnetic tensor gradiometry fixed wing and helicopter systems operating in North America and Southern Africa, but yet to be contracted for Australian survey operations. Sometime in the next few years, I hope that we will see the application of magnetic tensor gradiometry to our challenging and complex magnetic terrains under cover.

As a measure of the future-proof science in this book, I was delighted to see that estimation of magnetisation direction of a dipole source is covered in Chapter 7 for axial ratios of magnetic field components AND gradient tensor elements!

I am privileged to have had an association with all of the editors of this exceptional publication, each one recognised internationally and awarded as experts in this highly specialist area of exploration geophysics. Phil Schmidt and David Clark are both recipients of The Gold Medal of the Australian Society of Exploration Geophysicists (ASEG), the society's highest award for excellence in exploration geophysics. James Austin and his team are internationally recognised through their publications for the work on CSIRO's world-leading petrophysical rock property laboratory. Keith Leslie is also one of CSIRO's acknowledged inventors and experts in developing new geophysical systems, including magnetic surveying using the latest SQUID technology, an essential technology for the emerging area of magnetic tensor gradiometry. Clive Foss and Mark Lackie are widely recognised through their eminent publications as world experts in the practical analysis, interpretation and inversion of large minerals and environmental exploration magnetic datasets.

It is my view that this is an outstanding publication on the latest leading practice application of magnetisation and magnetic field studies to mineral exploration. This book is a must have for all practising exploration geophysicists, government geological surveys and other diverse users of geophysics for environmental, forensic and military applications and archaeology, as well as university earth science departments.

Congratulations to the editors and authors on your grand plan for producing this far-sighted work!

There is nothing else like this book. In the right hands, *Exploration Magnetics: Theory and Practice* will accelerate real breakthroughs in the search for Australia's next generation of critical mineral discoveries under cover.

Very Highly Commended,

[**] Schmidt PW, Clark DA (2000) Advantages of measuring the magnetic gradient tensor. ASEG *Preview*, April, pp. 20–26.

Ted Tyne
Former Director, Geological Survey NSW & Exploration NSW, Govt of New South Wales
Former Executive Director, Mineral Resources SA & PACE – Plan for Accelerating Exploration, Govt of South Australia
Founding Member, Past President & Honorary Member, Australian Society of Exploration Geophysicists (ASEG)

Dr Tyne has led a distinguished career in public geoscience programs in the NSW and SA Governments and the international airborne survey industry as well as in AMIRA geophysics research and teaching. He is known for his leadership of government-funded exploration initiatives, particularly high-resolution geophysical mapping surveys designed to stimulate exploration and bring forward new mineral discoveries. Ted has been directly responsible for initiating geoscience strategies and economic policies and delivering airborne geophysical projects, totalling more than 3.5 million line-kilometres of state and national precompetitive geophysical survey mapping, in partnership with Geoscience Australia, CSIRO and other state and territory geological surveys. In retirement, Ted continues to contribute to ASEG publications, symposia and conventions. (https://www.aseg.org.au/public/200/files/Ted%20Tyne%20Nov%202022a.pdf)

Contents

Section 2: Estimation of magnetisation direction from magnetic field analysis and inversion

Preface

Geology is in large part a model-building science attempting to make the most of limited information to infer and extrapolate across large volumes of a sparsely or completely unsampled subsurface. This book investigates how subsurface models of magnetisation properties and distribution can be derived from magnetic field data.

An incorrect but nevertheless instructive statement is that 'the best geologist is the one who has seen the most rocks'. The basis for this statement is that no two examples of geological systems are identical and that geology is poorly predictable. The equivalent statement that 'the best magnetic field geophysicist is the one who has seen the most magnetic fields' is similarly untrue but conveys the same advantage of experience gained in every magnetic field investigation. Few published methods for magnetic field interpretation address irregular geological distribution of magnetisation or imperfections in magnetic field data. In consequence (and with notable exceptions) many academic publications have little meaningful impact in applied magnetic field studies. In this book we focus on the practical solution of problems and present extensive case study applications.

It is no accident that this book has originated from Australia. Australia has long been a leading supplier of metals to the world, and this forms a major pillar of the national economy. Discovery of future ore deposits beneath cover depends on remote sensing for which geophysics plays the key role. Australia has major advantages for mineral exploration. Necessarily, it has the geological endowment of resources to be discovered. It also has an excellent national coverage of FAIR (findable, accessible, interoperable and reusable) regional aeromagnetic data curated by Geoscience Australia and distributed by their GADDS (geophysical archive data delivery system) web utility. This regional data acquired mostly by Federal and State and Territory Governments, often as part of 'exploration initiative' programs, plays a critical role in the primary selection of areas for mineral exploration. Free availability of magnetic field data has been a key component in attracting mineral exploration to Australia and provides the data used throughout this book. CSIRO's palaeomagnetic and rock magnetic studies over many years are crucial in augmenting this magnetic field data to complete the circle of relationships between geology, mineralisation and the magnetic field.

Rocks are not expressly designed to optimise recording of the Earth's magnetic field or of processes such as mineralisation that they may have been subject to (the challenges of engineering a material to achieve that would be considerable). Ferromagnetism of rocks is of intricate complexity and is highly idiosyncratic, particularly for rocks within multi-phase mineral systems. The magnetisation of a rock is completely controlled by minerals that in most cases constitute only a few per cent or less of its volume, and the magnetisation of those minerals is in turn critically dependant on minor differences in chemistry, oxidation state and the size, shape and crystallographic imperfections of the mineral grains. Furthermore, recent advances in imaging ferromagnetism at ever smaller scales has revealed that fine-scale intergrowths play a more substantial role in rock magnetism than had formerly been widely appreciated. The palaeomagnetic and rock magnetism laboratory established by CSIRO in the 1970s has conducted leading research into the magnetisation of mineralised systems in all of Australia's major metallogenic provinces. This research has played a critical role in the exploration and development of Australia's largest mines and has promoted understanding of the considerable complexity of the magnetisation of mineral systems. Historical reports are available for download from the CSIRO potential fields team site (https://research.csiro.au/potential-fields).

The understandings gained from rock magnetic studies of known mineral systems across Australia are of great value in maximising recognition of similar as-yet-undiscovered systems beneath cover. For buried systems we lose the advantage of directly measured magnetisations and until samples for direct measurement of magnetisation are recovered from drilling we can only conduct poorly constrained inversion of magnetic field data. Substantial challenges of non-uniqueness in magnetic field inversion include lack of reliable analytic or statistical estimates of how distant a proposed solution may be from the truth. At present these limitations are insufficiently recognised and there is poorly justified acceptance of continuous, three-dimensional (voxel) inversion models of the subsurface. Only for 'brownfields' inversion in immediate proximity to existing mines and supported by extensive drilling constraints can space-filling magnetic field inversion models be developed with reasonable levels of confidence. A major suggestion we make repeatedly through this book, but that is only occasionally raised in published literature or scientific debate, is that the magnetic field is only informative at specific locations (that we term 'sweet-spots') determined by the subsurface geology and the magnetic field measurements. This message seems negative, but it is not. By isolating the most reliable information from a magnetic field inversion, that information and its higher reliability is highlighted and is not compromised by dilution with apparent information of little or no reliability.

Sweet-spots are predominantly due to the shallower magnetisations that produce field variations with sharp changes of curvature. The over-printed magnetic fields of deeper magnetisations are much poorer in this relatively diagnostic information. Space-filling voxel models distribute deeper magnetisations according to a pre-determined expectation of 'depth functions' in the inversion algorithms. Many users of these models assume that the models are discoveries revealed by the inversion, but the depth distribution of magnetisation is as significantly determined by the inversion algorithm as it is by the ground. Alternative parametric models that we use extensively in this book are subject to equal if not greater discreditation if they are proposed to directly represent subsurface magnetisations. However, we use these models to recover their key statistics (parameters) that can be selectively presented as the most reliable information available from magnetic field inversion. Consider, for instance, a compact magnetisation for which the magnetic field cannot be distinguished from that of a homogeneous spherical magnetisation. Voxel inversions distribute the magnetisation across a cluster of neighbouring prisms, reporting a specific magnetisation in each voxel. A parametric model of homogeneous spherical magnetisation reports the horizontal and vertical coordinates of the centre of magnetisation, its mean strength and direction, and the radius of the sphere. The intensity of magnetisation and volume of the sphere can be multiplied to give the magnetic moment, and this should be considered the key estimate quantifying the magnetisation. Without independent knowledge of either the volume or intensity of magnetisation, no model specifying either value is justified. All acceptable models should, however, have a narrow range of magnetic moment values, common centres of magnetisation and common mean magnetisation directions. Inversions

are better reported as a population of these reliable statistics, rather than as magnetisation or susceptibility values. Despite this, magnetic moments are rarely reported for models and few geoscientists are familiar with the units used or the values to expect.

The second section of this book (Chapters 5 to 15) focusses on recovery of source magnetisation direction from magnetic field data – a field to which CSIRO has made considerable contributions, including the PhD thesis and multiple publications of David Clark, and through industry collaborations (particularly with Encom Technology and more recently Tensor Research). For many years, magnetic field inversion was performed with a 'head in the sand' neglect of remanent magnetisation because of conceived difficulties in addressing the issue of an unknown magnetisation direction. Obvious expression of remanent magnetisation in magnetic field data is only recognised in specific cases where it has a similar or greater amplitude than the induced magnetisation and a significantly different direction. Nevertheless, remanent magnetisation is an intrinsic characteristic of ferromagnetism and is a significant contribution to all measured magnetic field variations sourced in crustal geology. The traditional measure of the relative strength of remanent to induced magnetisation (the Koenigsberger ratio or Q factor) is insufficient to determine recognition of remanent magnetisation in magnetic field data. Other factors that are significant are the departure of the remanent magnetisation direction from the local geomagnetic field direction, the spatial distribution of magnetisation, overlap of magnetic fields due to adjacent magnetisations and the measurement sampling of the field. The external static magnetic field expression of a magnetisation reveals only the total magnetisation that is the vector resultant of its induced and remanent components. Since approximately the start of this century, there has been a substantial transition from inversions that neglected remanent magnetisation to inversions that report supposed continuous subsurface distributions of magnetisation of varying direction. These complex models face the standard restriction of inversion, that matching the input field does not in itself justify the model. Geologists like and often ascribe significance to irregularities in such models because these irregularities are consistent with the expected geology, but it is specifically those details that are of lowest reliability. Wherever a simple, homogeneous model satisfactorily explains a set of magnetic field measurements, considerable independent justification is required to propose a more complex, inhomogeneous model. All models should, if possible, be reduced to their minimum key statistics.

A further factor repeatedly raised in the book (see Chapter 4 in particular) is the resurrection of what have become mostly archaic terms of 'proximal' and 'distal' to describe the declining information content of a magnetic field as it is measured further from a source magnetisation.

The concept of continuously mapping subsurface magnetisation follows from the apparently continuous mapping of the magnetic field imaged by magnetic field grids. These grid images are derived from a finite set of samples of the magnetic field and are not the continuous coverage they appear to be, with many grids much poorer representations of the true magnetic field than is commonly realised. Magnetic field grids generated from measurements at closer line spacing or at lower elevation can substantially revise perception of the underlying distribution of magnetisation. There is a theoretical ability to 'downward continue' a magnetic field grid from the measurement elevation to a lower elevation but this is rarely feasible because of sampling limitations, and the process does not create new information. Grids also support enhancement of data, such as computation of gradients (as widely used throughout this book).

However, any use of enhancements must carefully consider limitations of the primary data because they are distorted by any insufficiencies in both the primary measurements and the gridding algorithm used. Quantitative analyses of the magnetic field should where possible be restricted to primary data measurements. As discussed in Chapters 2 and 3, estimation of depth to the top of magnetisation from grid data is particularly prone to error.

We take full responsibility for the work and views reported in this book but happily acknowledge input from interactions with colleagues both within the vibrant community of exploration geophysicists across Australia, well represented by the Australian Society of Exploration Geophysicists (ASEG), and internationally. We particularly wish to acknowledge the key roles played by the late Peter Milligan and Richard Lane of Geoscience Australia in enriching development of potential field geophysics in Australia.

Finally we would also like to acknowledge CSIRO's Deep Earth Imaging Future Science Platform for supporting the publication of this book and the encouragement of Tim Munday to achieve this.

**James Austin, David Clark, Clive Foss,
Mark Lackie, Keith Leslie, Phillip Schmidt**

List of contributors

James Austin
CSIRO Mineral Resources

David Clark
Integrated Magnetics LLC (and formerly CSIRO)

Clive Foss
CSIRO Mineral Resources

Gunther Kletetschka
Charles University, Prague

Mark Lackie
Formerly Macquarie University and CSIRO

Keith Leslie
CSIRO Manufacturing

Blair McKenzie
Tensor Research

Sarath Patabendigedara
CSIRO Mineral Resources

Phillip Schmidt
Magnetic Earth (and formerly CSIRO)

Marian Takáč
Charles University, Prague

Peter Warren
CSIRO Mineral Resources

Map index

Australian study areas are shown on the map below, along with a table (opposite) listing corresponding chapter sections that include those studies.

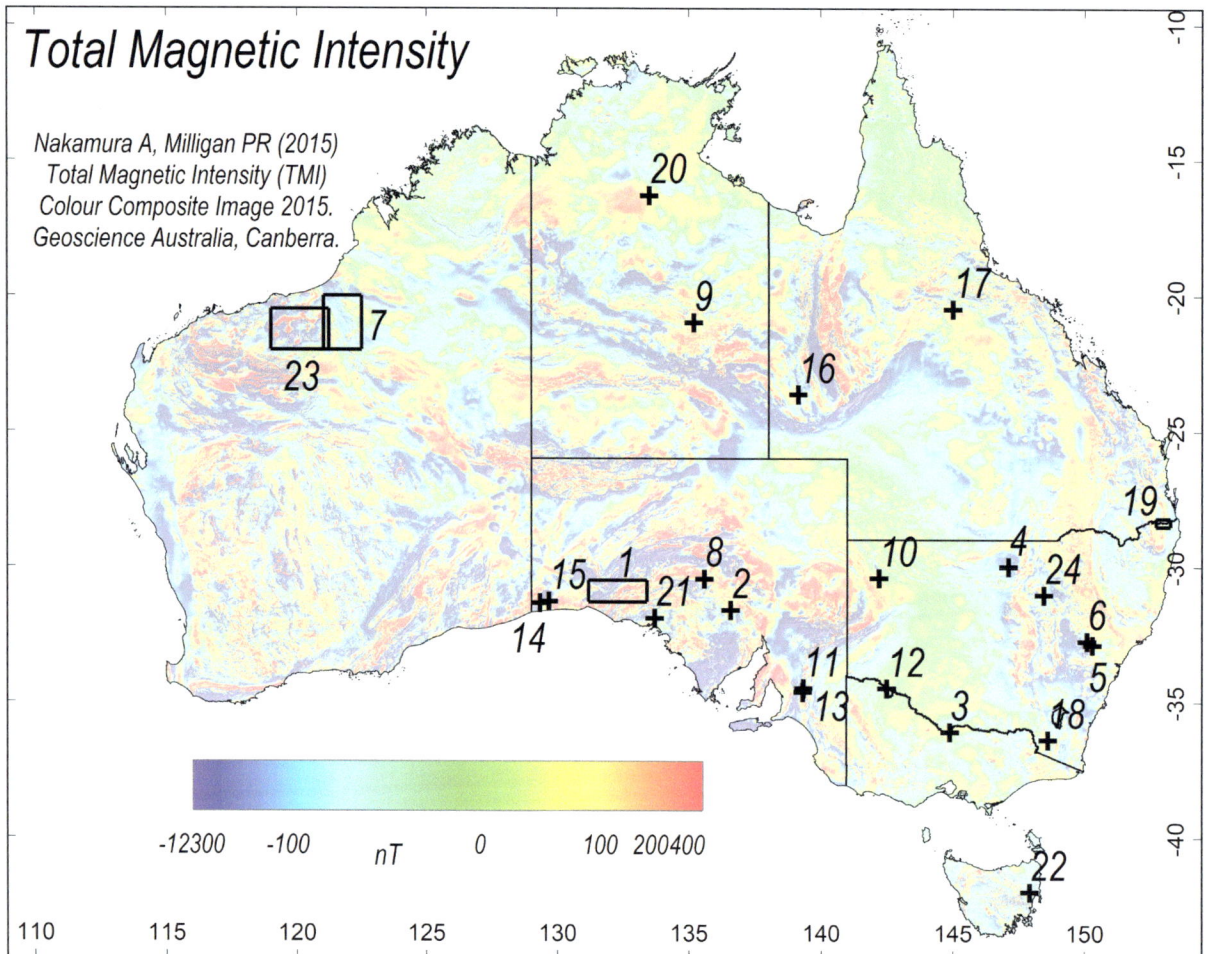

Total Magnetic Intensity

Nakamura A, Milligan PR (2015)
Total Magnetic Intensity (TMI)
Colour Composite Image 2015.
Geoscience Australia, Canberra.

-12300   -100   nT   0   100   200 400

Index	State	Chapter section	Figures	Name	ARAD entry	Description
1	SA	1.3.8	1.6, 1.7	Barton		Matching of regional gravity and magnetic fields
2	SA	1.3.8	1.8, 1.9	Torrens		Matching TMI and AGG data
3	Vic	2.4	2.1 to 2.5	Echuca		Image enhancement to resolve the shape of magnetisation
4	NSW	2.5	2.6 to 2.10	Brewarrina		Resolution of overlapping magnetic anomalies
5	NSW	2.6	2.11 to 2.16	Rylstone	319, 320	Magnetic anomaly separation from a regional field
6	NSW	6.8	6.24 to 6.47	Rylstone	267	Detailed inversion of magnetisation position and shape
7	WA	2.7	2.17 to 2.27	Waukarlycarly		Gravity and magnetic inversions over a graben
8	SA	2.8	2.28 to 2.32	Mount Vivian		Inversion of magnetic anomalies over Gairdner dykes
9	NT	2.9	2.33 to 2.38	Elkedra		Mapping dips of a magnetic sheet
10	NSW	2.10	2.39 to 2.42	Cobham Lake		Selection of features for inversion in regional studies
11	SA	3.4	3.18 to 3.34	Sedan		Magnetisation estimates from regional and detailed surveys
12	NSW	4.11	4.17 to 4.27	Kemendok Park	355, 356	Investigation of magnetic anomalies over buried sources
13	SA	5.6	5.10 to 5.12	Black Hill Norite	001	Influence of remanence on the RTP transform
13	SA	8.4	8.8 to 8.12	Black Hill Norite	001	Symmetry analysis of a magnetic field anomaly
14	SA	5.8	5.25 to 5.27	Coompana	275	Drill-target magnetic field interpretation
15	SA	5.9	5.28 to 5.31	Coompana	015	Inversion of complex magnetic anomalies
16	QLD	7.4	7.12, 7.13	Ethabuka	137	Magnetisation analysis of an isolated magnetic anomaly
17	QLD	9.4	9.12 to 9.18	White Mountains Park	327–330	Estimation of magnetisation direction of volcanic plugs
18	NSW	10.2	10.7 to 10.24	Jindabyne	309, 314–316	Aeromagnetic and drone anomalies over volcanic bodies
19	NSW	11.2	11.1 to 11.11	Tenterfield		Normal and reverse terrain-generated magnetic anomalies
20	NT	12.2	12.3 to 12.7	Daly Waters		Magnetisation deficit anomalies from holes in a basalt sheet
21	SA	12.3	12.8 to 12.15	Ceduna		Magnetisation deficit anomaly over a buried granite
22	Tas	13.2	13.2 to 13.8	Lost Falls Forest		Valley-excavation magnetic anomaly over a basalt sheet
23	WA	15.6.2	15.21 to 15.23	Pilbara	325, 326, 352–354, 357	Analysis of anomalies in aeromagnetic and EMAG2 grids
24	NSW	1.5.1	1.11 to 1.15	Combara	310	Estimation of magnetic moment for a compact source

1
Introduction

C.A. Foss

ABSTRACT

This book presents studies from geological mapping and mineral exploration projects for which delivery of practical solutions was paramount. Individual chapters are self-contained, with Chapters 2 to 4 largely focusing on estimation of depth to magnetisation and Chapters 5 to 15 investigating issues related to magnetic field expression of remanent magnetisation. The qualitative interpretation of geology from magnetic field imagery is well covered by Isles and Rankin (2013) and we do not address that topic here. Fundamental mathematics supporting magnetic field interpretation is explained in 'Potential theory in gravity and magnetic applications' (Blakely 1995) and additional background information can be found in 'Aeromagnetic Surveys: principles, practice and interpretation' (Reeves 2005) and 'Gravity and magnetic exploration: principles, practices, and applications' (Hinze *et al.* 2013). Rock magnetism and palaeomagnetism are well described in the textbook 'Palaeomagnetism: magnetic domains to geological terranes' (Butler 1992) that is freely available on the internet and Clark *et al.* (2003) review the geological controls on rock magnetism and their significance in mineral exploration. The role of magnetic field methods in mineral exploration projects where they are combined with other geophysical techniques is described in 'Geophysics for the mineral exploration geoscientist' by Dentith and Mudge (2014).

At the time of writing this book there is a pressing need for discovery of new metal deposits to replace those we are exhausting, particularly in support of the transition to a low-carbon economy. Conferences and publications by geophysical societies across the world bring together government, company and academic geoscientists and play a key role in enabling the discovery of new mineral deposits. These discoveries require drilling and drilling requires models. The current expectation of many geophysicists in mineral exploration is not (as it was previously) that they understand the complexities of relationships between geology and gravity and magnetic fields, but that they can manage the mechanics of running software programs, with responsibility for problem solving progressively ceded to the computer. Furthermore, creation of geological models from geophysical data is currently in danger of being switched to 'automatic' mode by adoption of artificial intelligence (AI) that places more import on its own rules than in the laws of physics. Hopefully this book, in focussing on practical aspects of the application of magnetisation and magnetic field studies in mineral exploration will support more interpretively guided mineral discoveries.

The value of magnetic field data is not as a commodity but in the information it contains. The passage from magnetic field data to an understanding of geology and mineralisation is complicated. The study of geology is largely based on the concept of uniformitarianism – that geology results from processes consistent across the globe and across time. However, each geological system is unique with long and complex sequences of events influencing and modifying each other. Furthermore, in many cases only a small part of a geological system is known from exposure or drill intersection. The search for ore deposits is especially difficult as these are, almost by definition, unusual systems. Throughout this book we repeatedly affirm that quantitative information (such as estimates of depth or magnetisation direction) is only recovered with confidence from magnetic field data in favourable locations where compact magnetisation generates distinctive local curvature in the magnetic field. We define these locations, the data they are characterised by and the magnetisations that give rise to them as 'sweet-spots'. This terminology has not been widely used previously because limits on computational size and speed restricted inversions to only these selected features. With massive advances in computational power it is now feasible to generate large models of the complete subsurface, but as a consequence we now have unfortunate and incorrect acceptance of these models as continuous and reliable three-dimensional maps of physical property values. Geophysicists would prefer this to be true, as would exploration managers, all other consumers of the models and geophysical software developers. It is an 'inconvenient truth' and is not a universally welcomed message that only selective parts of the magnetic field support meaningful inversion. Acceptance of these models also introduces the associated implication that magnetic fields continuously propagate information about their source magnetisation to whatever elevation they are measured at or upward continued to. Previously this was well understood to not be the case, with references to 'distal' and 'proximal' fields as a classification of the level of information in magnetic field data. Unfortunately those terms have fallen out of regular use.

Image processing and enhancement of medium- to high-resolution aeromagnetic surveys has been remarkably successful in support of qualitative geological interpretation and mapping (Isles and Rankin 2013). This does not, however, mean that those same geological features can necessarily be reliably recovered from a three-dimensional inversion model. The success across Australia of geological mapping from magnetic field imagery is in part because of focus on the shallowest substantial magnetisations that are typically at or just beneath the basement unconformity. At greater depths the reliability of magnetic models rapidly dissipates, except in those few windows where there are no significant overlying magnetisations. The central and base sections of a magnetisation are difficult regions to investigate because they lie beneath shallower sections of magnetisation. Obscuring by shallower magnetisation causes much greater loss of reliability for the deeper sections of a magnetisation model than the (already considerable) attenuation of magnetic field signal with depth.

1.1 MAGNETISATION STUDIES IN MINERAL EXPLORATION

The two key applications of rock magnetic and palaeomagnetic studies to mineral exploration are as a tool to investigate the process of mineralisation and to help constrain magnetic field interpretation.

In Australia drill core obtained in mineral exploration programs constitutes a large proportion of the material available for petrophysical studies. Until recently, core from boreholes drilled in short-lived exploration programs was commonly discarded together with its metadata at the end of unsuccessful exploration projects. Fortunately, there is now systematic retention and documenting of core by State and Territory Geological Surveys with internet access to the metadata. However, there is still a backlog of petrophysics measurements on the cores, many of which are unoriented. Systematic growth of data will progressively increase regional knowledge of petrophysical properties, including magnetic susceptibility and remanent magnetisation, and better reveal linkages between those properties and geology, mineralogy and mineralisation. Measured magnetisation values help to constrain magnetic field inversions that extend mapping away from the boreholes and the linkages of magnetisation to mineralogy, alteration and mineralisation support more confident transformation of magnetisation models to geological models.

1.1.1 Ferromagnetism

Magnetisation is a vector field description of the density of magnetic dipole moments in a material. Ferromagnetic magnetisation is characterised by its high strength and a behaviour known as hysteresis, in which self-sustaining magnetisation persists after an applied magnetic field is removed. In consequence, ferromagnetic minerals carry both a component of magnetisation induced in the presently applied field and a 'remanent' component acquired in their previous history. Ferromagnetism is an ordering effect and temperature disorders, preventing magnetisation above a critical temperature for each ferromagnetic mineral, known as its Curie point (580°C for magnetite). Basic igneous rocks are intruded at 1,200° to 1,400° and solidify above their Curie point. As an igneous rock cools through its Curie point it becomes ferromagnetic, but each grain according to its size, shape, chemistry and crystallography may have a short relaxation time within which magnetisation resets. At some temperature below the Curie point, referred to as its blocking temperature, the relaxation time of each grain increases abruptly to periods of geological extent. The population of blocking temperatures in a rock determines the history with which it acquires thermoremanent magnetisation (TRM) on cooling and has it reset by any subsequent reheating, and the vector sum of remanent magnetisations of all ferromagnetic grains in a rock determines its bulk remanent magnetisation (natural remanent magnetisation or NRM).

Within a ferromagnetic mineral grain the space across which a consistently directed magnetic ordering extends is known as a domain. Large ferromagnetic grains optimise their magnetisation energy by arrangement into domains separated from each other by walls (in some cases pinned at crystallographic imperfections) and these grains are known as multi-domain (MD). Magnetisations of MD grains are relatively easily reset by movement of the domain walls and in many cases carry 'viscous' remanent magnetisation (VRM) acquired at low temperatures in the recent geomagnetic field. Grains too small to support an internal domain wall are known as single-domain (s.d.). s.d. grains carry extremely stable remanent magnetisations that can persist since the original formation of the rock provided they are not reheated above their blocking temperatures. However, these grains have small volumes and do not generate a strong NRM unless they are present in large numbers. Fortunately, the range of stable magnetisation is considerably extended by elongate pseudo-single domain (PSD) grains that have only a few internal domain walls with limited options for repositioning. In some igneous or metamorphic rocks, magnetite needles exsolved in the cleavage traces of amphiboles and pyroxenes carry stable ('hard') remanent magnetisation while the larger, more visually obvious blobby magnetite grains predominantly carry viscous ('soft') remanent magnetisation. The contribution of different mineral grains to the magnetisation of a rock was investigated in studies such as reported by Dunlop and Özdemir (1997) at a time when magnetic remanence carriers were

the key to digital data recording. Newly developed methods of probing magnetisation of individual mineral grains include scanning magnetic microscopy (e.g. Pastore *et al.* 2019) and micromagnetic tomography (Cortés-Ortuño *et al.* 2022).

1.1.2 Dominant ferromagnetic minerals

Magnetic field variations measured in aeromagnetic surveys do not carry diagnostic evidence of the specific ferromagnetic minerals that generate them; however, only a small number of minerals are significant sources of measured magnetic field variations. Magnetite (ferrous) and hematite (ferric) oxides of the titanomagnetite and titanohematite series respectively as shown in Fig. 1.1 carry many rock magnetisations. These minerals are solid solutions at high temperatures, but at lower temperatures exsolution occurs together with oxidation of titanomagnetites, creating complex magnetisations (Robinson *et al.* 2002; Kasama *et al.* 2004). In a study of the Iron Knob magnetic field anomaly in the Gawler Craton of South Australia extreme remanent magnetisations are associated with fine intergrowths of magnetite and maghemite in hematite that may be an analogue for strong magnetisations in Martian rocks (Schmidt *et al.* 2007).

The titanomagnetite series is a widely distributed common constituent of many igneous, metamorphic and sedimentary rocks. Titanohematites are also widely distributed (in some cases from oxidation of titanomagnetite) but typically have weaker magnetisations. In its monoclinic form the iron sulphide pyrrhotite is also strongly magnetic (with magnetisation intensities similar to magnetite) but pyrrhotite is only weakly magnetic or non-magnetic in its hexagonal form. Pyrrhotite is also

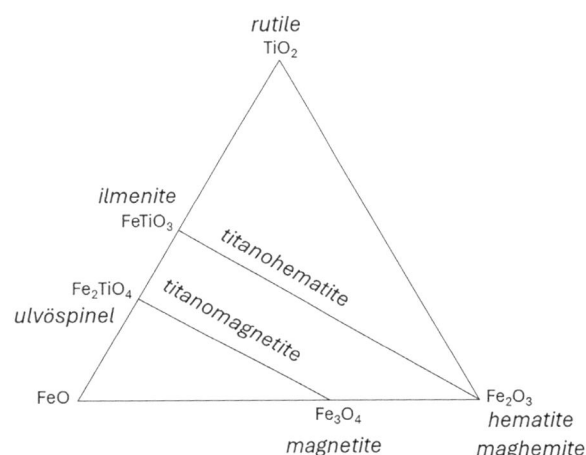

Fig. 1.1. Major iron and iron-titanium oxide ferromagnetic minerals.

more restricted in its occurrence than magnetite, mostly in basic or ultrabasic rocks (in some cases associated with sulphide mineralisation), in contact metamorphic aureoles or in shales created in reducing conditions.

Maghemite is a strongly magnetic iron oxide, in many cases produced at surface, such as in grass fires (Löhr *et al.* 2017). In Australia there are many arid or semi-arid regions where maghemite is widespread in the regolith and gives rise to extensive, sharp variations of tens to hundreds of nanoTeslas (nT) in aeromagnetic data, including dendritic patterns over drainage and paleo-drainage systems.

1.1.3 Induced and remanent magnetisations

All ferromagnetic materials carry both induced and remanent magnetisations, but their relative strengths as defined by the Koenigsberger ratio (the ratio of remanent to induced magnetisation) are highly variable and poorly predictable. The Koenigsberger ratio cannot be determined from magnetic field data alone and this scalar measure is insufficient to describe the vector relationship between induced and remanent magnetisation that determines the magnetic field expression of magnetisation (see Chapter 5). For a Koenigsberger ratio of 1, remanent magnetisation parallel to the induced magnetisation doubles the resultant magnetisation, while remanent magnetisation anti-parallel to the induced magnetisation gives a resultant magnetisation of zero intensity. As shown in Fig. 1.2A, the influence of remanent magnetisation differently directed to the

geomagnetic field is more fully specified if the Koenigsberger ratio is supplemented with the apparent rotation angle (ARRA) that measures the difference between induced and resultant magnetisation directions (Foss 2017). In favourable situations this angle can be estimated directly from magnetic field data. Figure 1.2B shows the stereographic projection (a projection from unit vectors on a sphere to a horizontal plane) of induced, remanent and resultant magnetisation. The great circle plots the plane in which these three vectors lie, with the intermediate position of the resultant direction determined by the Koenigsberger ratio.

Remanent magnetisation is often and unjustifiably neglected as a contribution to magnetic field anomalies. All ferromagnetism includes remanence contributions, and magnetic field inversions using only magnetic susceptibility values are invalid unless specifically justified by extensive measurement of low Koenigsberger ratios. For inversions performed using only magnetic susceptibility, recovered values should be referred to as apparent susceptibility, allowing that part of that magnetisation may be remanent rather than induced. Furthermore, susceptibility and magnetisation values derived from magnetic field inversion are strictly contrasts rather than absolute values. They are also subject to uncertainty linked with corresponding uncertainty of the magnetisation volumes. Estimates of total susceptibility or magnetisation (magnetic moment) contrasts derived from magnetic field inversion are more reliable than individual estimates of susceptibility, magnetisation or volume.

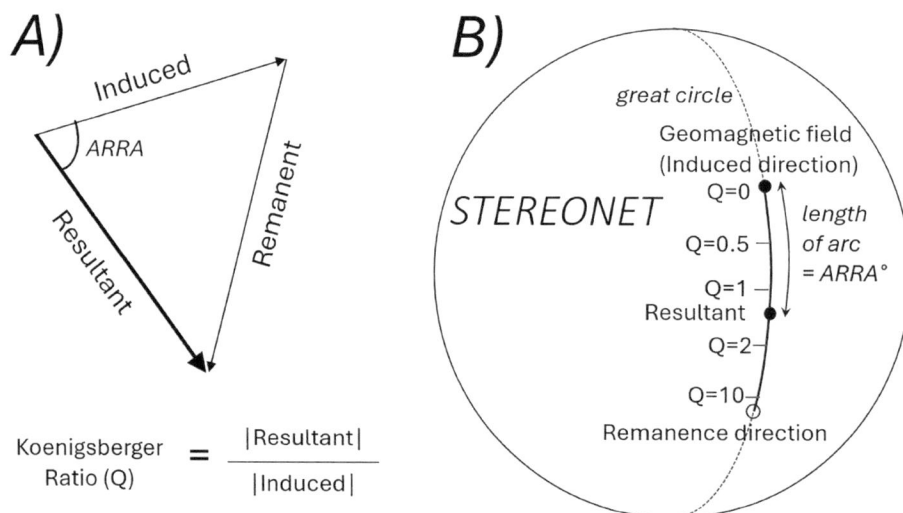

Fig. 1.2. A) coplanar induced, remanent and resultant magnetisation vectors with the Apparent Resultant Rotation Angle, B) Stereonet projection of the geomagnetic field, remanence and resultant magnetisation directions.

Magnetic field anomalies can be inverted with a free magnetisation direction to provide simultaneous estimates of the spatial distribution and resultant magnetisation, but inversion cannot resolve resultant magnetisation into its component parts without independent knowledge of any one of: magnetic susceptibility, Koenigsberger ratio or remanent magnetisation direction. The results of any inversion that limits magnetisation to the present field direction should be confirmed by inversion allowing the magnetisation to have a free direction. If assumption of magnetisation parallel to the field is correct then inversion allowing a free magnetisation direction should confirm that.

For a compact magnetic field variation, inversion provides a reliable estimate of only the mean magnetisation direction. Various inversion methodologies force assumption of a homogeneous magnetisation or allow spatial variation of magnetisation direction, but there is little or no justification to assert that inversion supports either internal distribution. Provided a magnetic field variation is reliably separated from overlapping fields, it is commonly possible to recover estimates of the (contrast) resultant magnetisation direction more reliably than the (contrast) resultant magnetisation intensity.

A major reason why many magnetic field interpretations are made with use of magnetic susceptibility measurements only is the difficulty and expense of remanent magnetisation measurements. Rock magnetic and palaeomagnetic measurements are generally made in specially equipped petrophysical laboratories and the results may not be available at the time they would be needed for input to an active exploration program. An exception is the portable Qmeter (Schmidt and Lackie 2014; Schmidt 2015) that can measure induced and remanent magnetisations in a core shed or at a drill rig for rapid turn-around in magnetic field modelling or inversion. A comprehensive review of the wide range of methods available to determine magnetisation direction is provided by Clark (2014).

1.1.4 Self-demagnetisation

A magnetisation does not sit directly in the background field but in a modification of that field by its own internal field. This influence is termed 'self-demagnetisation' because the internal magnetic field is reverse to the applied field and acts to weaken it and reduce the effective induced magnetisation. At low magnetic susceptibilities the internal field is weak compared to the primary field and can be ignored. At magnetic susceptibilities of ~0.1 SI self-demagnetisation effects become significant, and at susceptibilities greater than 0.4 SI it is essential to incorporate self-demagnetisation effects in any magnetic field modelling. Self-demagnetisation acts to rotate the induced magnetisation direction towards the plane of elongation of a body. At high magnetic susceptibilities, independent constraints become critical to modelling and inversion because the relationship between magnetic susceptibility and the external magnetic anomaly amplitude is highly non-linear. Analytic solution for self-demagnetisation only exists for spherical and ellipsoidal bodies. For sources with sharp edges, corners or internal inhomogeneity there are only iterative approximations that become less reliable with increasing magnetic susceptibility. Iterative computation using equivalent source arrays of dipoles (Purss and Cull 2005) supports computation of self-demagnetisation effects both within a magnetised body and within assemblages of closely grouped bodies.

1.1.5 Anisotropy of magnetic susceptibility

Anisotropy of magnetic susceptibility (AMS) in ferromagnetic materials arises from a combination of crystallographic and shape effects. AMS measurements of oriented samples provide rock fabric information that reveal details of the deposition, intrusion or deformation of those rocks (Hrouda 1982, 2007). AMS also influences the external magnetic field of the induced magnetisation (Biedermann and McEnroe 2017). AMS is described by a symmetric, traceless 3x3 tensor. In extreme cases (such as some banded iron formations) what would otherwise be consistent magnetisation within a folded sheet-like unit gives rise to considerable along-strike variation in its magnetic field expression due to its variable orientation with respect to the inducing field. The differential ease of induction of AMS causes rotation of the induced magnetisation away from the external field direction and towards the 'easy' magnetisation direction. AMS properties known from direct measurement can be incorporated in forward modelling of induced magnetisation but (without considerable independent constraints on other parameters) it is not feasible to ask inversion of magnetic field data to find unknown AMS properties.

1.1.6 Spurious magnetisations

One of the objectives of measuring induced and remanent magnetisations is to help constrain magnetic field interpretation. However, some measured rock magnetisations can give misleading results. Measurement of induced or remanent magnetisation of outcropping rocks may record unrepresentative results from a veneer of weathered surface material even where that weathering is not obvious visually. Conversely, rocks at the site of a lightning strike can display sharp and irregular increase in intensity of remanent magnetisation by several orders of magnitude through acquisition of isothermal remanent magnetisation (IRM). Drill-core can carry magnetisations induced by the drilling and IRM acquired when geologists use a pencil magnet to search for magnetic minerals. Some rocks have particularly soft magnetisations, with VRM components acquired over periods as short as days or hours. This can be detected by storing the samples in a known orientation for a short time before remeasuring their NRM to detect if it has changed. Other samples with even less stable remanent magnetisation may be super-paramagnetic and remagnetise during measurement, but this is readily detected by monitoring the measurement statistics.

Another form of spurious magnetisation arises from misreported units, especially of susceptibility. Magnetic susceptibility is unfortunately defined differently in the cgs and SI systems, giving rise to a difference of 4π (just over 12) between values even though in both cases the units are dimensionless. Magnetic susceptibility is the physical property most measured by geoscientists in the field, using convenient hand-held meters. To accommodate users working in either system, many meters have switches between cgs and SI measurement (with a physical switch on older meters and a software selection on newer meters). Digital output of the newer meters records the metadata of which scale is in use and this information must be retained when transferring data between spreadsheets. Values measured with older meters are of uncertain value unless they have been carefully documented. Neglect of remanent magnetisation in specifying magnetisation in forward or inverse magnetic field modelling is also a form of spurious magnetisation as magnetic susceptibility values alone misrepresent the true magnetisation, even if that magnetisation is parallel to the local field.

1.2 MAGNETIC FIELDS

Magnetic fields are explained by potential field theory as described in many publications (e.g. MacMillan 1930; Kellogg 1967; Blakely 1995). Magnetic potential is the integral of work done to bring a reference pole from infinity to a point and is independent of the track taken. Points of equipotential define a surface and the magnetic field vector at a point is perpendicular to that surface with strength proportional to the potential gradient. The magnetic field can be described as a vector of specified strength and direction (with direction generally measured as a declination angle in the horizontal plane and an inclination angle in the vertical plane) or alternatively as three orthogonal vector components (e.g. east, north and vertical). The magnetic field is piecewise-continuous and smoothly differentiable. In exploration geophysics many measurements of the magnetic field quote only its strength (the total magnetic intensity or TMI) but this is the amplitude of a vector rather than a scalar property.

Some magnetic field studies focus on gradients of the field, either measured or derived by differentiation of field data. For magnetic fields defined by high precision measurements close to each other and of wide extent, a single description of the field carries all the information in the field and any one description is sufficient to derive another (except for indefinite integrals in transforming from gradients to fields). Conversely, if the field is undersampled (e.g. along a single flightline or in a borehole) multi-channel measurements such as different components, tensor gradients or field component and gradient combinations carry more information than single-channel measurements and partially compensate for that restricted sampling. Gradients are an enhancement or 'sharpening' transform relative to the field description and preferentially express nearby sources but they also have more demanding sampling requirements than for the field (Reid 1980).

The three orthogonal gradients of the three orthogonal field components define the gradient tensor. This tensor is symmetric and traceless and is completely defined by one set of off-diagonal terms and two of the three principal elements. The gradient tensor can be directly measured, for instance by low temperature SQUIDs (Schiffler et al. 2014), high temperature SQUIDs (Schmidt et al. 2004) or diamond nitrogen vacancy sensors (Kuwahata et al. 2020). The gradient tensor can also be derived by combination of three orthogonal phase transforms of a TMI grid and then orthogonal gradient transforms of those field components. The relative advantages of direct gradient measurement or transform of TMI data depends on fidelities of both gradient and field measurements, survey line spacing and the proximity and distribution of the magnetisations.

Magnetic fields are additive. The field of a complex distribution of magnetisation can be approximated by summing fields of a suitable array of magnetisation elements – either dipoles or rectangular voxels, or of parametric model bodies such as rectangular prisms, sheets or pipes. In an array of sources, each sits in the external field of the others. However, this consideration is almost never required as the external field of adjacent magnetisations is in almost all cases substantially weaker than the Earth's field. Therefore, it is generally sufficient to sum the independent fields of each source.

1.2.1 The magnetic field of the Earth

Magnetic fields investigated for geological mapping or mineral exploration are samples of the Earth's magnetic field. There are three major contributions to this field; fields generated by current flow in the fluid outer core, ferromagnetic magnetisations in the Earth's crust where temperatures are below Curie point isotherms and fields generated by movement of charged particles in the free space above the Earth. The geomagnetic dynamo is the most prominent contribution and gives the field its major form of an axial dipole approximately aligned with the Earth's rotation axis. Local irregularities in the core field are believed to arise from vortices below the core/mantle boundary. These irregularities grow, die and migrate over periods of hundreds to thousands of years. A key factor of the geomagnetic dynamo field is that it has reversed polarity on multiple occasions, possibly following instances of high symmetry in current flow within the core that reduces the amplitude of the dominant dipole component. The time-average approximation of the geomagnetic field to that of an axial dipole is the basis for latitude estimation in paleomagnetic studies. Across the Earth there is a network of geomagnetic observatories, some of which have provided continuous records over centuries and from which long-term (in the human reference frame) field variations can be tracked. The level of detailed and reliable magnetic field data acquisition is expanding across the Earth to meet increasing needs for geological mapping, mineral exploration, directional drilling and military applications. The growing resource of aeromagnetic data worldwide is also supplemented with satellite data, particularly from the Eusropean Space Agency (ESA) SWARM mission.

The International Geomagnetic Reference Field (IGRF) is a time-varying description of the Earth's magnetic field developed by the International Association of Geomagnetism and Aeronomy (IAGA). Several online tools are available to compute the strength, components, direction and rate of change of the field at any point on the Earth for a specified date and elevation. The 13th generation field is described and defined by Alken et al. (2021). The difference in amplitude from over 60,000 nT at the poles to less than 30,000 nT in some equatorial regions (Fig. 1.3A) is broadly consistent with the factor of 2 variation between Gauss-A and Gauss-B positions of a dipole field. Steepening of inclination towards the poles (Fig. 1.3B) that is highly significant for geophysical exploration projects at different latitudes is also broadly consistent with an axial dipole model.

The principal consequence of making TMI measurements in the Earth's magnetic field are that the expression of secondary anomalous fields is strongly influenced by their vector addition with the dominant background field. In the steep geomagnetic inclinations at high latitudes the vertical geomagnetic field component dominates and orthogonal horizontal components in the secondary field contribute little to TMI. In the low geomagnetic inclinations of equatorial latitudes the horizontal northerly-directed geomagnetic field component dominates and only secondary field components parallel or anti-parallel to that contribute significantly to TMI. A further inclination-dependant difference is that the secondary field above an induced magnetisation is parallel to the geomagnetic field in a vertical inclination field, giving rise to predominantly positive TMI anomalies, but is anti-parallel to it in a horizontal inclination field, giving rise to predominantly negative TMI anomalies, as shown in Fig. 1.4. The reduction to pole (RTP) transform seeks to reduce the complexity in relative positioning of a magnetisation and the peak of its TMI anomaly in non-vertical geomagnetic inclinations by a phase transform of both the magnetic field component and the magnetisation. The standard RTP transform assumes that magnetisation is parallel to the geomagnetic field and this transform is invalidated if the magnetisation is differently directed (see Chapter 5). The RTP transform is also unstable if applied to data measured in particularly low geomagnetic inclinations. The phase transform of TMI to vertical component (B_z) data also suffers instabilities in low inclination fields but nevertheless B_z and its vertical derivative $B_{z,z}$ provide considerable advantage in magnetic field analysis (see Chapter 9) without the dependence on magnetisation direction of the RTP transform.

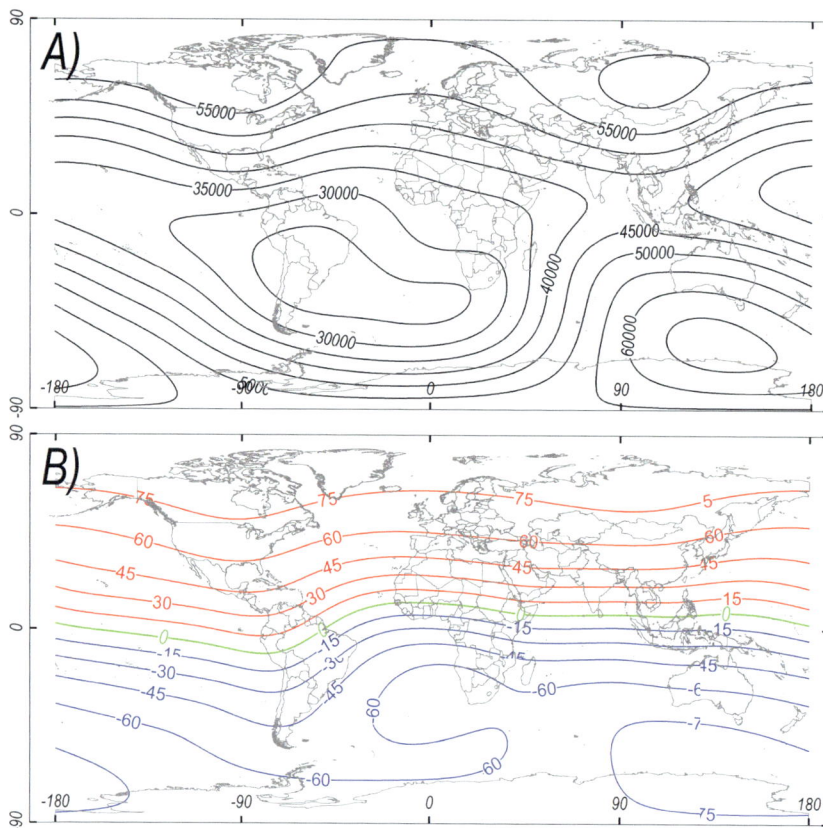

Fig. 1.3. A) Contours of IGRF intensity (nT) and B) contours of inclination of the IGRF (degrees).

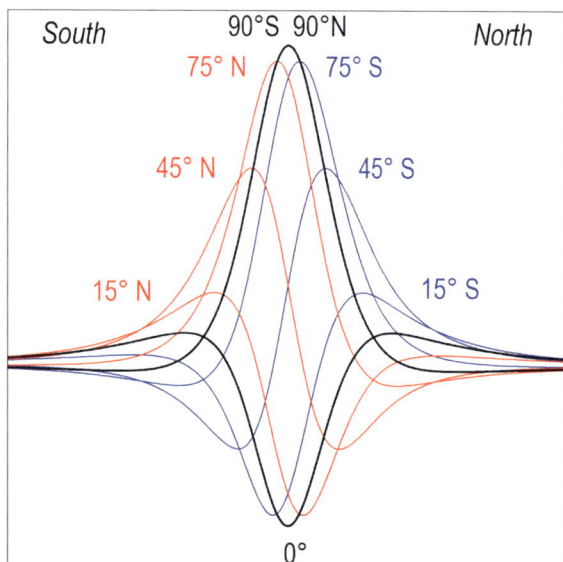

Fig. 1.4. South to north TMI profiles at different geomagnetic inclinations over dipoles with field-parallel magnetisation.

The vertical component of the geomagnetic field is directed downwards (by convention positive) in the northern hemisphere and upwards (negative) in the southern hemisphere. The considerable latitude variation of TMI anomalies generated by field-parallel magnetisations is highlighted in Fig. 1.4. Plotted in black, anomalies in both vertical (± 90°) and horizontal (0°) inclinations are symmetric and almost single-signed. The horizontal-inclination TMI curve is predominantly negative and of lower amplitude than the predominantly positive vertical-inclination TMI curve. At intermediate inclinations the anomaly curves are of dual polarity. In the southern hemisphere (the blue curves in Fig. 1.4) the anomaly peak is to the north of magnetisation with a trough to the south, and in the northern hemisphere (the red curves in Fig. 1.3) the anomaly peaks are to the south of the magnetisation with a trough to the north.

1.3 MAGNETIC FIELD DATA

1.3.1 Sufficiency of sampling the field

We only have imperfect samples of the magnetic field and this invariably imposes limitations in the analyses and interpretations we can make according to data resolution, reliability and distribution. In analysis of magnetic field data it is important to always remain alert to this limitation because as data are progressively processed, enhanced, modelled and inverted the outputs become further removed from the original measurements and our understanding of their information limits. Any processing or enhancement only distils information – it does not create it, whereas poorly considered processing can destroy or distort information.

Most magnetic field data acquired for geological mapping and mineral exploration is measured on aeromagnetic surveys. Measurements are much closer along the flightlines than between them and spatial resolution of the data is mostly determined by line spacing. There is also greater continuity in sensor orientation, measurement elevation and subtraction of diurnal variation along lines than across them. Rule-of-thumb sufficiency estimates can be derived from fast Fourier transform (FFT) analysis of the data sampling in both across-line and along-line directions as discussed in Reid's (1980) seminal paper. Importantly, Reid's analysis emphasises the different data requirements for quantitative map interpretation, working with gradient enhancements and modelling or inversion of the data. In steep geomagnetic inclinations there is obvious advantage in setting the flightline direction perpendicular to geological strike and elongation of magnetic features. Low geomagnetic inclinations introduce the complication of strong horizontal directionality of the geomagnetic field and favours a north–south flight line orientation. Mapping magnetic fields of north–south elongated magnetisations at low geomagnetic inclinations is fundamentally problematic. There are polarisation anomalies at the northern and southern terminations of the magnetisations and only weak magnetic field expression across their central section. This problem is only partially alleviated by choice of measurement or transform strategies.

Bifold classification of an area as having or not having aeromagnetic data available is of restricted meaning. Mineral exploration programs are exercises in problem solving and the critical question of any existing aeromagnetic data is whether it is sufficient to contribute to solving the specific challenges of an exploration program. Existing data over an area should always be examined, but if the line spacing is too wide its main value may be to provide input to design of a more detailed survey. Many of the datasets presented in this book are from regional surveys flown at 200 or 400 m line spacing and typically with 8 m along-line spacing. Chapter 2 includes investigation of resolution differences between a survey flown at 400 m and 100 m line spacing.

1.3.2 Magnetic field grids

It is commonly assumed that a magnetic field grid at all points represents the true magnetic field. However, most grid nodes lie off the survey flightlines and their values are substantially influenced by the gridding algorithm that is used. Gridding is not a 'passive' operation – it creates apparent data where there was none initially. Grids are commonly generated at cell sizes of one-quarter or one-fifth of the flightline spacing. This is a compromise between de-sampling of the line data (for a line spacing of 400 m a cell spacing of 80 m is about 10 times the normal measurement spacing) and suppression of short-wavelength artefacts between lines. As shown in Chapter 2, the magnetic field is severely misrepresented by gridding where flightlines are sub-parallel to the field gradients. At present, grid interpolation is generally performed with minimum curvature algorithms. The limitations of grid interpolation are most evident in regions of particularly sharp field curvature that give rise to 'bicycle-chain' or 'string-of-pearls' artefacts. For data with sharp curvature in one direction, anisotropic diffusion algorithms (Naprstek and Smith 2019; Davis 2022) reduce these problems (as discussed in Chapter 2).

1.3.3 Separation of magnetic fields

Magnetic fields are additive. It is simple to combine fields of different sources (as is regularly performed in forward modelling calls of magnetic field modelling or inversion) but there is no reliable method to perform the inverse operation and achieve unique and correct separation of overlapping fields from different sources. In mineral exploration no magnetic field is measured in complete isolation. More distant or much deeper sources add magnetic field variations to that from any shallower magnetisation of interest. However, it may be possible to separate these fields quite effectively in a process termed 'regional-residual separation' utilising their different curvatures. There are various analytic methods to perform this process but regional-residual separation is fundamentally interpretive. Removal of unwanted shorter wavelength field variations due to shallower magnetisations is also based on differences in field curvature and is also interpretive. The most problematic isolation of magnetic fields is between fields of similar curvature from adjacent magnetisations at similar depth. Field separation is also problematic if fields are insufficiently sampled or if the measurements are of insufficient extent. In these cases field separation contributes substantially to errors and uncertainties in modelling, inversion and interpretation.

Vertical derivative anomalies are more compact than field anomalies and have greater range of curvature with source depth. Both these factors assist anomaly separation but place greater requirements on the spacing and resolution of data. Removal of short wavelength field variations can be best achieved from the original line data because in gridding those variations are partially redistributed and their characteristics become less diagnostic. Line-based filters act on the primary data but only access data on the same line. For instance, the vertical derivative computed as an along-line filter is made on the assumption that the horizontal crossline gradient is zero. Fortunately, this is not a limitation as line-based filters can be applied with advantage in inversion provided they are applied identically to both measured and model-computed data channels (as discussed in Chapter 2).

Anomaly separation is particularly important in estimation of magnetisation direction. The symmetry/asymmetry of a magnetic anomaly is a critical expression of the source magnetisation direction that is also influenced by any elongation of body shape or by superposition of other fields across the anomaly or some part of it.

1.3.4 The vector nature of TMI data

Almost all magnetic field surveys for geological mapping and resource exploration measure TMI. As shown in Fig. 1.3A, the IGRF amplitude varies from ~30,000 nT near the equator to more than 60,000 nT close to the poles. Few measured local magnetic field variations have amplitudes greater than 1,000 nT and therefore they do not significantly deflect the local geomagnetic field direction. However, at locations of particularly strong magnetisation (including commercial magnetite deposits) magnetic field variations can be of much higher amplitude, resulting in significant rotation of TMI within those areas. In such cases, measured TMI must be corrected to a consistently directed vector using an iterative scheme (Lourenço and Morrison 1973; Clark 2013) before it can be treated as a potential field. This step is not required for modelling or inversion where TMI is computed at each point from its three individual field components.

In the Earth's magnetic field it is not feasible to accurately measure individual magnetic field vector components that are penalised by their slightest sensor misorientation in the background field. Chapter 11 considers the quite different challenge of making magnetic field measurements in weak magnetic fields away from the Earth. In weak fields, orientation penalties are substantially reduced and conversely there is a problem with TMI measurements because the field orientation is highly variable and poorly predictable. The problem of orientation sensitivity in the strong background field of the Earth is substantially reduced by working with gradients. Background gradients of the Earth's magnetic field are typically much weaker than those of magnetisations of interest sourced at shallow to intermediate depths. Horizontal along-line TMI gradients can be reasonably recovered from aeromagnetic survey data by differencing sequential along-line measurements, and in some surveys horizontal crossline gradients are measured using a pair of wingtip magnetometers. To date, application of these data has been mostly restricted to improved processing of TMI grids with few efforts to exploit the data for modelling and inversion studies. Vertical gradients have also been measured by mounting a pair of vertically displaced TMI sensors on the aircraft tail, but this has had limited success due to the short baseline. Recently, measurements have been extended to the gradient tensor using SQUID detectors (Schmidt *et al.* 2004; Schiffler *et al.* 2014) and diamond nitrogen-vacancy magnetometers (Kuwahata *et al.* 2020). Tensor gradiometer surveys have yet to be adopted in standard aeromagnetic surveying and have mostly found application in the search for small and shallow kimberlite targets in diamond exploration or for unexploded ordnance. For these anomalies defined on only a few flightlines there is significant advantage in multi-channel data, provided the gradients can be measured with sufficient precision. A major challenge to any gradient data, whether derived from direct measurement or transform of field component measurements, is presence of short-wavelength noise from data imperfections and/or platform or near-surface geological magnetisations.

1.3.5 Magnetic field expression of the centres and edges of magnetised bodies

A magnetic field anomaly due to a magnetisation parallel to the inducing field can be simplified with a standard RTP transform (Baranov and Naudy 1964). Using a modified transform, RTP can also be performed for another, known magnetisation direction. Correctly applied RTP transforms peak over or close to the horizontal centre of magnetisation (unless displaced by the disproportionate influence of the shallowest sections of a plunging magnetisation). An alternative approach to locating the centre of magnetisation is to apply a transform such as the total gradient or 'analytic signal' (Nabighian 1972) that approximately locates the centre of a compact magnetisation with reduced influence of its magnetisation direction. The normalised source strength (NSS) transform (Beiki *et al.* 2012) mostly locates the centre of magnetisation more closely but at the cost of greater computational complexity. Both the analytic signal and NSS transforms enhance the field contributions of the shallowest magnetisations.

In the case that a body of magnetisation is wide compared to the elevation at which the field is measured, magnetic fields and their associated gradients change more rapidly towards the edges of magnetisation rather than towards its centre. In a non-vertical geomagnetic field, with magnetisations parallel to the field, there is polarisation over the northern and southern margins with quite different field variations over those two margins and with lower amplitude expression of the eastern and western margins. In a steep geomagnetic field shallow wide sheets of small depth extent with induced magnetisation generate a continuous or near-continuous anomaly peak just within the margin of the magnetisation and a flanking trough just outside it. Steeply dipping edges of wide magnetisations can be approximated located using the peak horizontal or peak total gradient of the RTP field (Blakely and Simpson 1986; Grauch and Cordell 1987; Fedi and Florio 2001). Roest *et al.* (1992) and Wijns *et al.* (2005) developed gradient-based analyses to enhance the expression of magnetisation edges in magnetic field data.

If an upward continuation is applied to magnetic field data before resolving gradients, the gradient peaks migrate in the direction of dip (Archibald *et al.* 1999; Hornby *et al.* 1999) because the upward continuation more heavily weights contributions from the deeper zones of magnetisation relative to the shallower ones. Assistance in analysis of complex gravity and magnetic field variations is provided by an automated vector analysis applied to the gradient maxima, termed 'multiscale edge analysis' or 'worming' (Boschetti *et al.* 2001). The success of this method depends on the nature of the physical property contrasts in the ground and the sufficiency of data coverage. The horizontal gradient transform applied to gravity data or appropriately transformed RTP magnetic data highlights shallow density or magnetisation contrasts respectively. Applying an upward continuation

to the gravity or magnetic field data before the horizontal gradient transform preferentially attenuates field variations from the shallowest property contrasts and places greater focus on deeper property contrasts. A series of analyses with increasing upward continuations produces a series of vectors with progressively increased weighting of deeper property variations (although it is not feasible to link analysis at any one continuation height to property contrast at any specific depth).

1.3.6 Spurious magnetic field variations

A sound approach to inspection of new data is to be suspicious of everything and expect errors. Magnetic field data are geolocated and one source of error in the data that might not be obvious in its initial inspection is that it might be incorrectly located. At present, with GPS positioning and use of standard projections there should be little scope for incorrect location, but in Australia historic aeromagnetic data are still being used that was acquired when two projection systems were in common use (AGD66 and WGS84) with differences of ~200 m. This unfortunate situation resulted in frequent mislocation of data and it is by no means certain that all data of that vintage has a correctly attributed projection.

Magnetic field expressions of man-made objects in aeromagnetic data are commonly referred to as 'cultural' anomalies. In some cases these have diagnostic characteristics that reveal their origin but in a survey with many shallow-sourced anomalies there may be several of uncertain origin. We now have unprecedented coverage of surface imagery (including Google Earth) on which a magnetic field image can be overlaid to investigate whether an anomaly of uncertain origin coincides with a man-made feature, although this check does not detect temporary features that may have been present when the aeromagnetic survey was flown but not when the surface imagery was acquired. If a survey is to be flown over a developed area where cultural anomalies are anticipated, surface imagery can be acquired with the survey to evaluate possible cultural anomalies (as is conveniently done for small area drone surveys).

Most evaluation and interpretation of magnetic field data is performed using grid imagery. One of the more common artefacts encountered in grid data are flightline level shifts that appear as linear features in the flightline direction. Micro-levelling attenuates these features but may leave subtle, longer wavelength features that can be confused with geological signal. Survey grids can also include point anomalies due to gridding of any unremoved spikes in the line data (particularly at line terminations). Grids should be tightly clipped to the extent of input data to avoid spurious apparent coverage of unsurveyed areas where the grid is an extrapolation of the gridding algorithm rather than interpolation of measurements. Magnetic field analysis and interpretation is generally best optimised on application to individual surveys, but sometimes data are available only as composite grids of multiple surveys and there may be spurious level shifts between surveys or differences in line spacing, line orientation and/or flying height that confuse evaluation of features in the different survey areas. Grid stitching that is performed where there is overlap between adjacent survey grids acts to disguise differences between those surveys, effectively by distorting the data in the area of overlap where the resulting grid may be incompatible with both input grids. The result may be cosmetically more acceptable and support enhancement of the combined grid that would otherwise be distracted by the local disconformities at grids margins, but a mosaic of grids can mislead interpretation of the data.

1.3.7 Static and dynamic magnetic fields

Conventional magnetic field surveys are classified as 'static' because they aim to define the magnetic field at one instant of time through elimination of time variations of the field over the duration of the survey (using diurnal corrections made at a fixed base station).

'Dynamic' surveys with a time series of measurements at each survey station should provide capabilities to map spatial variation in time variations of the field (Goldstein and Ward 1966; Clark *et al.* 1998). These field variations due only to changes in induced magnetisation can be analysed or inverted with the advantage of the known magnetisation direction. Dynamic survey procedures are challenging because the amplitude of 'quiet day' diurnal variation that induces the dynamic anomaly is only of the order of 1,000th of the total geomagnetic field that induces the conventional static anomaly, and because each station requires multiple measurements in highly repeatable position and orientation at different times. However, if a reliable model can be generated from inversion of the dynamic anomaly then the static magnetic anomaly can be subsequently resolved into induced and remanent magnetisation components using that spatial model. We now have SQUID magnetometers and gradiometers of sufficient precision to attempt these surveys on the ground (Foss *et al.* 2019). The time required for multiple repeat measurements is likely to limit surveys to a small number of stations. However, the method may be useful for investigation of elongate anomalies over sheet magnetisations for which separation of structural dip and inclination of magnetisation is particularly uncertain in conventional magnetic field analysis. Potentially, high-resolution inversion of anomalies measured on conventional aeromagnetic surveys flown decades apart might support a dynamic field analysis but this would require

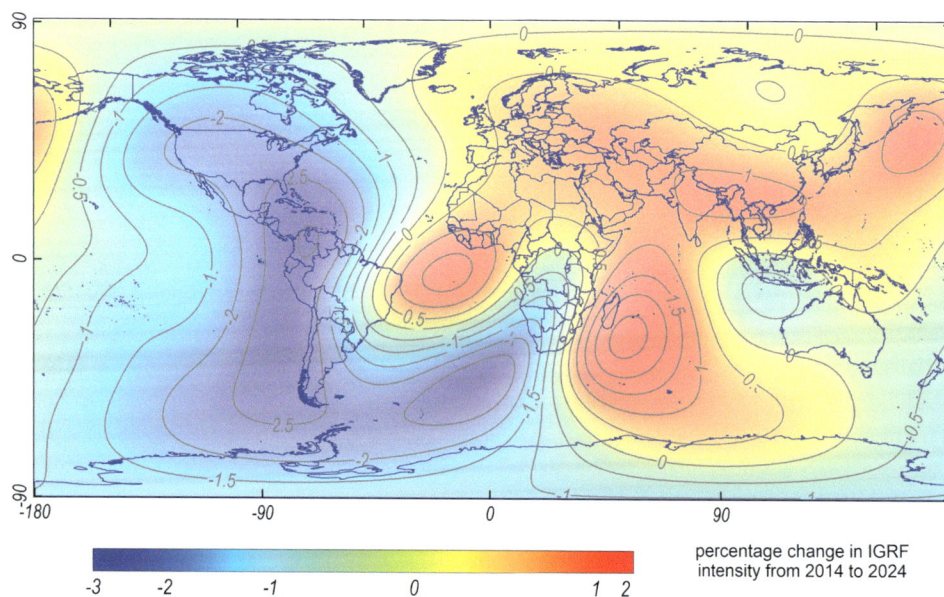

Fig. 1.5. Percentage change in IGRF Intensity over the period 2014.5 to 2024.5.

extreme measurement precision and substantial differences in the geomagnetic field between the different survey dates. Figure 1.5 shows that high-amplitude decade-long variations in the field occur in a few (mostly offshore) regions across the globe, with onshore TMI intensity variations over the last 10 years mostly less than 1% of the total field strength (increase or decrease in the field strength are equally suitable for dynamic field analysis). At present, the only convincing detection of time differences in naturally sourced magnetic anomalies are from surveys flown above active volcanoes (Koyama *et al.* 2013) where magnetisation differences may be due to physical movement of magnetisation and/or abrupt temperature variations.

1.3.8 Relationships between magnetic and gravity fields

Rock magnetism is dominated by ferromagnetic minerals that typically constitute only a few per cent of a rock or less. In contrast, the density of a rock is determined by all its components, including fluid-filled pore spaces. Many observed variations in the gravity field are generated by density contrasts of the order of 10% of total density and the issue of density contrast is foremost in consideration for interpretation of gravity data. For many interpreted magnetic field variations the magnetisation contrasts are almost the same as the absolute magnetisation values and the concept of contrast is commonly neglected. Furthermore, magnetisation contrasts are vector rather than just scalar contrasts.

Gravity and magnetic fields are governed by common mathematical laws of potential field theory (Blakely 1995) with the dipole magnetic field an order of curvature higher than the monopole gravity field. As determined by Poisson's theorem, for a constant relationship between density and magnetisation, the vertical gradient of gravity $g_{z,z}$ is directly proportional to the vertical magnetic field due to vertical magnetisation (the RTP of TMI). With an assumed or known magnetisation direction, gravity fields can be expressed as equivalent magnetic fields using the pseudomagnetic transform (Blakely 1995). Conversely, magnetic fields can be expressed as equivalent gravity fields using the pseudogravity transform (Baranov 1957). Complex density and magnetisation variations invalidate the basic assumptions of these transforms and only in quite rare circumstances do pseudogravity or pseudomagnetic transforms approximate to the true gravity and magnetic fields. Interpretation from visual evaluation of large gravity

and magnetic images is challenging because of the complexity and different dynamic ranges of the individual datasets. Figure 1.6 shows Bouguer gravity and TMI images of the Barton area of the southern Gawler Craton in South Australia (Foss *et al.* 2019a). The Bouguer gravity grid is derived from station spacings of 8 km over the west and centre of the area and 2 km over eastern part of the area, and the magnetic field is measured on east-west flightlines at 200 m spacing across the complete area. There are clear associations between the TMI and Bouguer gravity grids but direct comparison is challenging because of the complexity of each image.

Figure 1.7 shows vectors developed from multi-scale edge analysis (see section 1.3.5) of both the gravity and magnetic fields. These are independent fields mapped by completely independent data. The gravity data are measured at ground stations and the aeromagnetic field on flightlines at a nominal 80 m ground clearance, yet there are considerable similarities between the vectors derived from the two datasets. Each vector set is extracted from a series of upward continuations. For the lower-level continuations in each set (with signal predominantly sourced in the shallow subsurface) there is a larger number of vectors, including many short vectors. For the higher-level continuations (with signal more strongly sourced from greater depths) only the longer vectors persist. In Fig. 1.7 I have selected gravity vectors from one continuation level (1286 m) and magnetic vectors from a deeper level (1970 m) to best highlight the correspondence between the two fields. There are many features with near-identical traces in the two data vector sets, revealing that they are generated by geological contrasts with both density and magnetisation expression. Some features have combined expression along segments of their lengths and only gravity or only magnetic expression along other segments. There are also vectors almost perpendicular to each other, where a vector from one set terminates against a vector from the other. There is clearly additional information relating to the geology available from combination of the two geophysical methods.

Combined airborne gravity gradiometer (AGG) and aeromagnetic surveys are flown in mineral exploration programs, but unfortunately cost has to date restricted widespread application of AGG for regional geological mapping. AGG surveys are more appropriate for revealing relationships between gravity and magnetic fields than are aeromagnetic and ground gravity surveys because the measurements are co-located and are of

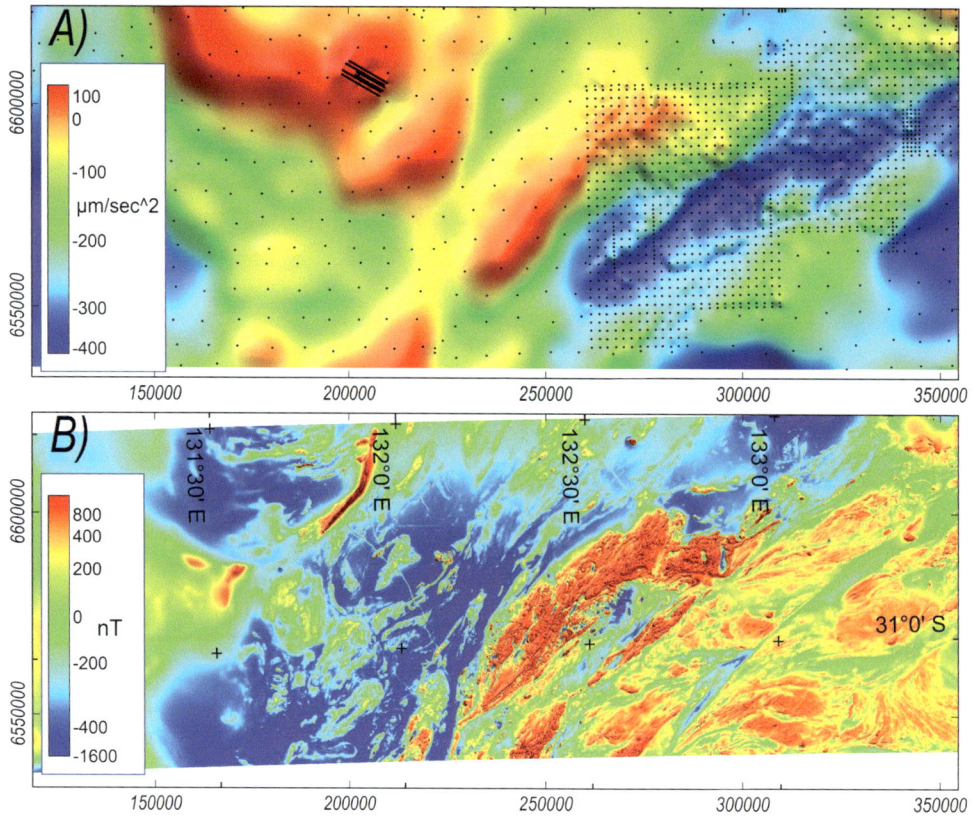

Fig. 1.6. A) Bouguer gravity (reduction density 2670 kg/m^3) and B) RTP of TMI over the Gawler Craton, Barton 4A survey area.

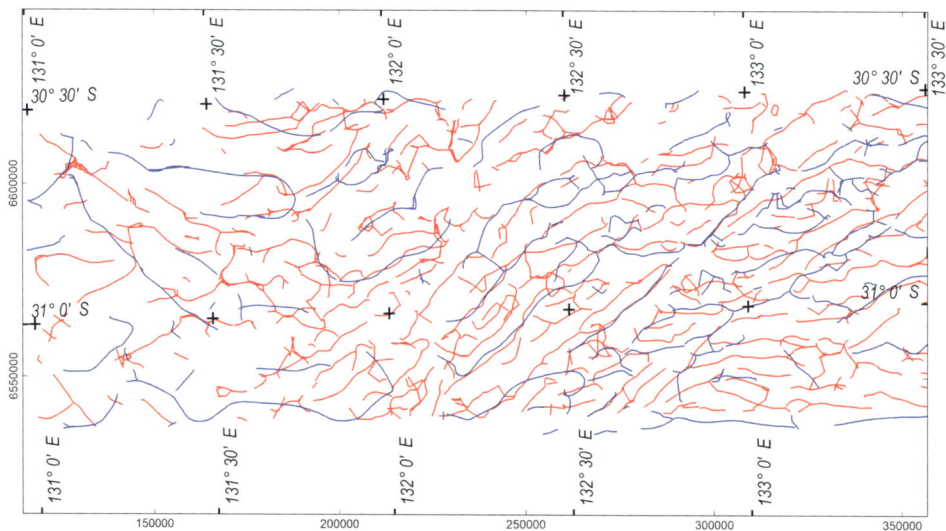

Fig. 1.7. Multiscale edges (horizontal gradient peaks) of the Bouguer gravity imaged in Fig. 1.6A (in blue) and of the RTP of TMI imaged in Fig. 1.6B (in red).

matching curvature expressions of the fields (gravity field gradients and the magnetic field). However, even for combined AGG and aeromagnetic surveys the two datasets in most cases highlight different aspects of the geology. In some areas the most prominent magnetic field variations are of small but shallow magnetisations of only small anomalous mass that may not be detectable or are only weakly evident in the AGG data. Conversely, many bodies large enough to have recognisable expression in the AGG data may either be magnetically bland with little expression in the magnetic field data or have substantial internal magnetisation variations and appear as multiple smaller magnetic sources.

An unusually clear correlation between ground gravity and aeromagnetic anomalies is shown in Fig. 1.8 over a segment of the Gairdner dyke swarm in the Torrens 3B Gawler Craton Airborne Survey (Foss *et al.* 2018) in South Australia. The gravity field is mapped by a 17 km × 18 km survey at a station spacing of 400 m (approximately 2,000 stations). Figure 1.8 shows RTP of TMI contours at a 50 nT interval over an image of the vertical gradient of Bouguer gravity generated by FFT. The strong correlation is because the gravity field is mapped to an unusual degree of detail by the 400 m spaced stations, because the dykes are unusually wide and have large volume, and because the dykes intrude through the crystalline basement into the lower-density overlying siliclastic Pandurra Formation. The density contrast for this shallow section of the dykes may be as high as 500 kg/m^3. Figure 1.9 shows a section of an east–west profile constructed from the 400 m spaced gravity stations. The individual

dykes are scarcely recognisable in the gravity (g_z) profile (Track A in Fig. 1.9) although they can be traced in a contoured map view because of their continuity between successive profiles. The dykes are much more easily recognised in the vertical gradient of gravity derived from a grid FFT and sampled back onto the line (Track B in Fig. 1.9). Three prominent anomalies in centre of this track are labelled 'a', 'b' and 'c' in Fig. 1.9B. Track C shows a profile of TMI sampled from the grid at each of the gravity stations. The TMI anomalies are strongly undersampled at the 400 m gravity station spacing, generating sharp, almost single-station anomalies for the western and eastern gravity anomalies 'a' and 'c'. There is only weak TMI expression for the central gravity anomaly 'b'. Track D in Fig. 1.9 shows the TMI profile along the nearest, almost co-located flightline. From the enhanced gravity map image in Fig. 1.8 (and the corresponding but higher-resolution TMI map not shown here) the central gravity anomaly 'b' is two weaker magnetic anomalies (as seen in Fig. 1.9 track D) with magnetic field expression poorly sampled at the gravity stations but with a

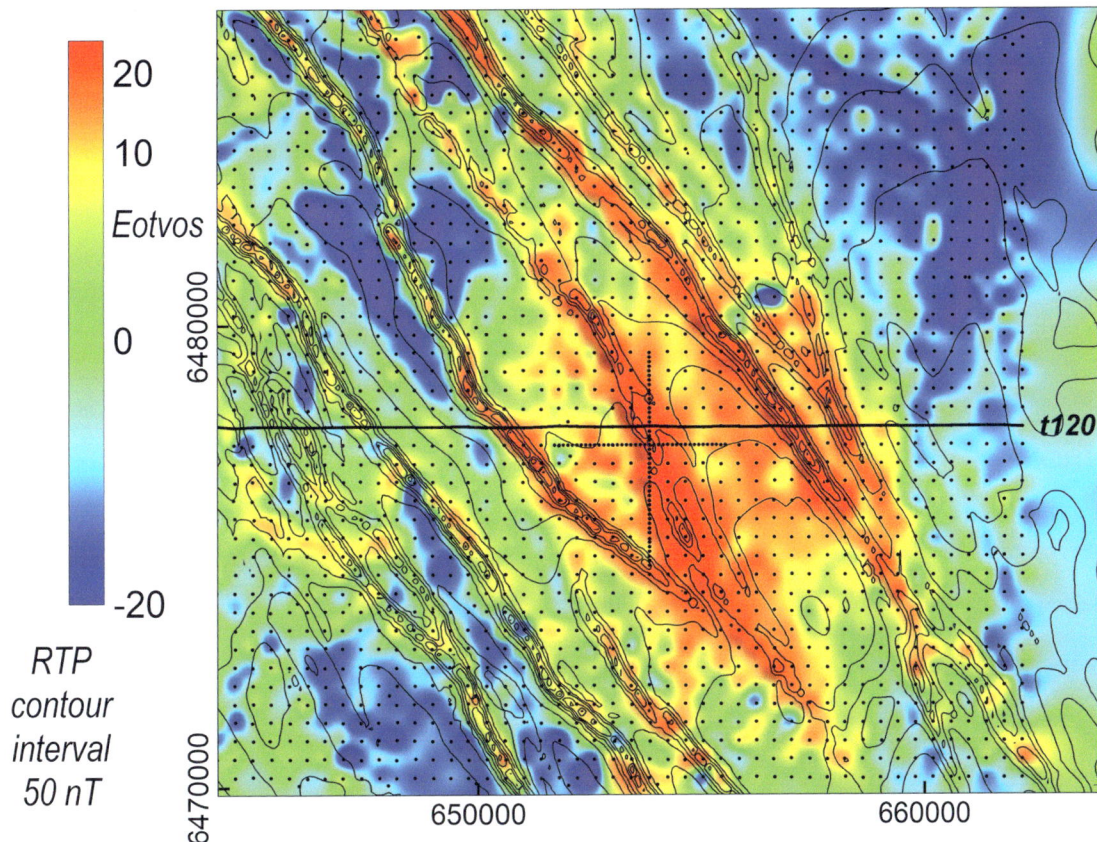

Fig. 1.8. Vertical derivative enhancement of Bouguer gravity as an image and aeromagnetic TMI contours over a part of the Torrens Area 3B of the Gawler Craton Airborne Survey in South Australia. Gravity stations are at 400 m spacing. T120 is an aeromagnetic survey flightline imaged in Fig. 1.9.

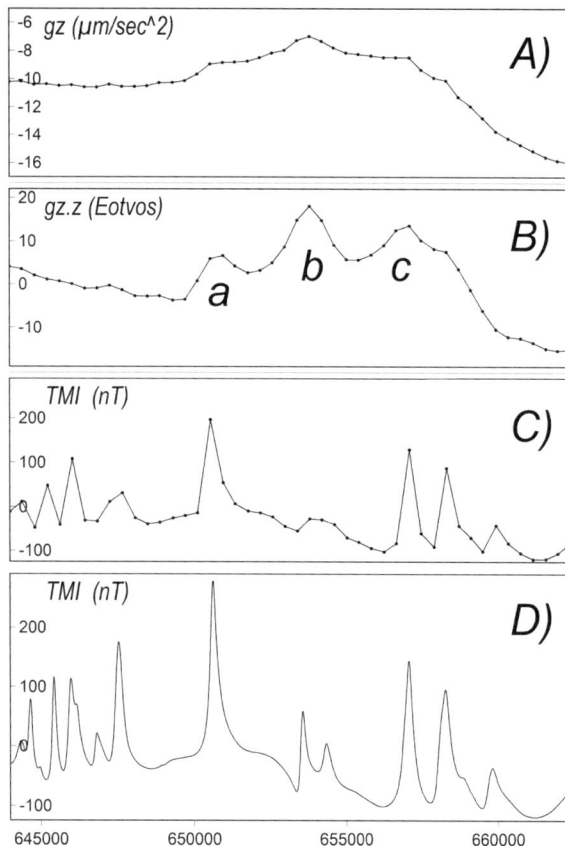

Fig. 1.9. A) ground Bouguer gravity, B) vertical derivative of Bouguer gravity (from a grid FFT), C) TMI sampled at the gravity stations and D) measured TMI along the aeromagnetic flightline T120.

strong combined expression. The considerable difference in resolution between the magnetic field tracks C and D in Fig. 1.9 reveals the major advantage of aeromagnetic data with close point spacing (in this case < 7 m) compared to the wider gravity station spacing.

1.4 MAGNETIC FIELD INVERSION

The objective of magnetic field inversion is to investigate and evaluate the information content of the input data. The lesser achievement of finding a model or a set of models that can match the data can easily be misleading. Most published inversion studies are prefaced with an acknowledgement of the limitations of non-uniqueness but still provide and draw conclusions from models that inadequately represent the range of possible solutions. Non-uniqueness arises from the ability of variation in one or more details of a magnetisation to effectively compensate for variation in others. The fundamental restriction of non-uniqueness is that we cannot know whether an inversion model is a reasonable representation of the

true magnetisation in the ground however well it matches the input data. It is therefore not feasible to assign meaningful conventional uncertainty measures to inversion models (if we cannot know the fundamental validity of a model we also cannot constrain it with uncertainty bounds).

Many different challenges arise in day-to-day magnetic field inversion, and the optimised solution of each problem requires a versatile library of different inversion utilities. The key determinants in making the choice about how best to perform an inversion are:

1) what question is to be answered?
2) what is already known about the geology?
3) what data are available?
4) how much time is available to derive a solution?
5) how much exploration of alternatives is required or justified?
6) what inversion resources are available and how can they be optimised for this task?

Non-uniqueness of the inverse problem opens the possibility to such a wide range of solutions that it is invaluable to have a basis of geological knowledge and understanding of the area to be able to hypothesise about the source of a magnetic anomaly and place some limits on the expected distribution of magnetisation. A physicist with no knowledge or understanding of geology can efficiently invert a magnetic field variation to obtain a source magnetisation model, but unless there is basic geological guidance it is unlikely that the model obtained will be geologically feasible. The two most common choices in style of inversion are voxel inversion (the most popular) and parametric inversion. This classification is based on the nature of the models used and the mechanics of how the inversion is performed. Voxel inversion provides much more apparent detail than parametric inversion of the same data but that additional apparent detail is mostly unjustified.

1.4.1 Voxel inversion

The factors to be addressed in voxel inversion are primarily in design of a model objective function that oversees the underdetermined problem of assigning values to a large number of cells (voxets) as well reviewed by Li and Oldenburg (1996). Of particular importance is a depth weighting function to supervise distribution of magnetisation with depth for which the magnetic field provides little definitive assistance. Depth weighting is empirically applied to encourage extension to depth of magnetisations that would otherwise occupy only the shallowest parts of the model. The depth weighting is also generally combined with guidance that variations in magnetisation should maximise smoothness or compactness of magnetisation. These are sound principles for selection of models (just as acceptable as the homogeneous magnetisation assumptions of most parametric modelling) but, as with the parametric model assumptions, there is no *a priori* reason that a geological magnetisation will conform to these assumptions.

Once a voxel inversion scheme has been designed, it is quite straightforward for a user to perform an inversion by following a routine process of defining the architecture of the model (its three-dimensional extents and cell sizes) and submitting the data in an appropriate format. This involves few critical decisions and does not require an expert understanding of the inversion process. Ease of use is a key attraction for voxel inversions. The major step of isolating the field variation to be explained by the model can also be automated and included seamlessly in the inversion process, but field separation is fundamentally non-unique and interpretive. Li and Oldenburg (1998) provide a robust separation method based on prediction from the surrounding field variation that is well suited to automation and simplicity of use. In optimal cases of well-isolated anomalies, most regional-residual separation methodologies should provide very similar output. However, an interpreter should carefully inspect the field data to make an optimum selection for inversion and to develop a strategy to best address issues of anomaly separation. Decisions and findings from that process should be described in any subsequent review and use of the resulting model (possibly including alternative separations). A drawback of a fully automated process is that users of the inversion results are unaware of any issues that may have influenced field separation and thereby the output model.

Each individual rectangular prism of the subsurface magnetisation is specified by its three-dimensional address in the space-filling voxel array, and the summed magnetisations of that array generate the model-computed magnetic field. The prisms may have uniform dimensions, or to lessen computation load, the size and depth-extent of the voxels may increase with depth. A voxel model has the attraction to its users of representing magnetisation throughout the complete subsurface volume, although it is misleading to report values for many parts of the model in which there is no justified confidence.

There are two common representations of voxel models: thresholding the model by displaying only those voxels of magnetisation greater than a selected value, or an isosurface (or isosurfaces) from three-dimensional gridding of the inverted magnetisation. Note that neither of these options clearly displays the magnetisation distribution that matched the anomaly. Both exclude representation of magnetisation below the threshold or isosurface value, and both fail to represent internal inhomogeneity within the displayed magnetisation. Alternative displays are colour-coded model slices along appropriate lines of section that better reveal details of the variation in magnetisation.

Lelièvre and Oldenburg (2009) developed algorithms to allow for inversion of data where magnetisation direction is unknown. Throughout this book many examples are shown to illustrate that magnetisation direction for a compact magnetisation is a bulk property, and that provided the field of a magnetisation can be reasonably separated from other fields, the mean magnetisation direction can be recovered with reasonable confidence (this can be achieved by voxel or parametric inversion); however, there is no analytic justification that spatially variation in magnetisation direction can be reliably recovered from magnetic field data. The concern is not that a model cannot be found but that any model that is found might not be justified. Both voxel and parametric estimates of magnetisation direction are therefore more reliably constrained as single estimates of mean magnetisation direction for each spatially separate feature in the magnetic field (each discrete anomaly or sweet-spot).

1.4.2 Parametric inversion

Parametric inversion has a quite different approach to voxel inversion. With parametric inversion a model is constructed based on interpretive decisions about the feasible distributions of magnetisation required to explain it. A fully space-filling model identical to a voxel inversion could be constructed (with difficulty) by parametric inversion but inclusion of components that are essentially unconstrained by the magnetic field variation is of no advantage. Minimalist models constructed with one source assigned to each discrete magnetic field variation are more easily justified. The forward modelling codes used in parametric inversion are derived for ideal geometries (polyhedral bodies, ellipsoids etc.) of homogeneous magnetisation, but models need not be accepted at this exact level of representation and their magnetisations should only be interpreted as mean values without resolution of internal distribution. Few geological bodies, except for some basalt or dolerite sheets such as dykes, have both homogeneous properties and near-ideal geometries. The most acceptable representation of parametric inversion models is as sets of key statistics, including centre points, approximate horizontal and vertical extents and orientations and total contrast magnetisations (rather than individual magnetisation and volume values).

Parametric inversion is well suited to one-to-one matching between sweet-spots and inversion model bodies. Inversions can address multiple sweet-spots using multiple bodies with advantage if their magnetic fields overlap. However, inversions generally benefit from as tight a focus as feasible. Another aspect of parametric inversion well matched to the sweet-spot approach to magnetic field modelling and interpretation is the strength of parametric modelling in support of model testing and sensitivity studies. Parametric sensitivity tests are performed by offsetting an individual parameter from its value in the global minimum misfit set and repeating inversion with that parameter or parameters held at their offset values to investigate what increase of the misfit statistic this causes, and what adjustments this evokes from the other parameters. Compensation can also be restricted to variation of one other single parameter to investigate more specific cross-correlation between parameters. These tests are most conveniently performed with parametric inversion, but their findings illustrate fundamental aspects of inversion that apply to understanding results of any inversion methodology.

1.5 RELATIONSHIPS BETWEEN MAGNETISATION AND THE MAGNETIC FIELD

The relationship between magnetisation and the magnetic field at first appears a straightforward causative relationship as depicted in the left-hand vertical side of the schematic in Fig. 1.10. It might be thought that this cause-and-effect relationship should be simple to establish. However, this is a complex, non-unique task. The intermediate step in linking an unknown subsurface magnetisation to its remotely sensed expression in the magnetic field is creation of a model or models of the magnetisation (as shown in the other two sides of the triangle in Fig. 1.10). The models may be simple conceptual ideas of what or where the magnetisation is, or specific definitions of magnetisation values and distributions. Until a magnetisation is extensively intersected by drilling or is excavated, we only know it by its model representation. If we already had the model we could verify it by forward computing its magnetic field expression and measuring the field data to test it, but we almost invariably have to work in the opposite direction of first measuring field data and then inverting that data to find a model. From the magnetic field to a model is a one-to-many relationship and ambiguity is widely perceived to be the source of all problems in magnetic field inversion. However, potential field theory places constraints on what we can know from the magnetic field data in much the same way that quantum

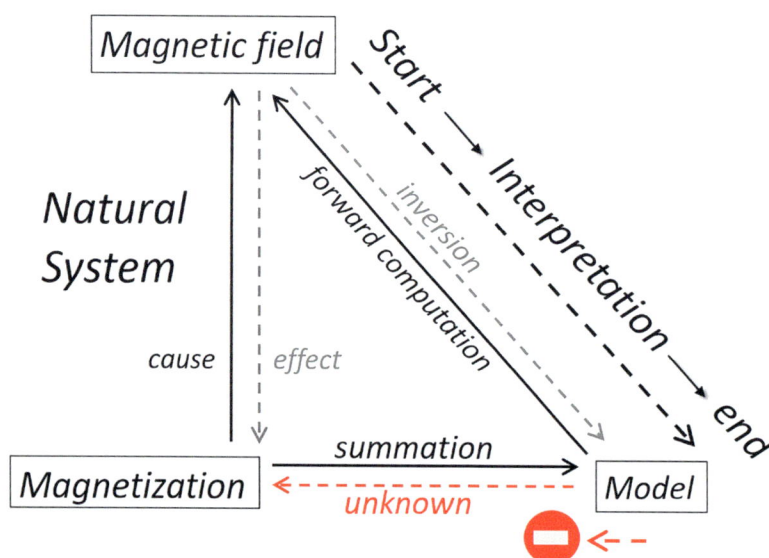

Fig. 1.10. A schematic representation of the relationships between magnetisation, the magnetic field and models.

uncertainty prevents simultaneous knowledge of the position and momentum of a subatomic particle. We should consider that a model obtained by inversion of magnetic field data is a speculation rather than a direct representation of the magnetisation and reclassify the problem to lie mostly between the model and magnetisation as shown in Fig. 1.10 rather than between the model and magnetic field.

Many of the challenges of relating measured magnetic field variations to measured magnetisations depend on scale. Most magnetic susceptibility measurements are made with hand-held susceptibility-meter measurements that sample only a few cubic centimetres of sample and remanent magnetisation is generally known from measurements of a small set of cylindrical plugs of 2.5 cm diameter and 2.5 cm height. Conversely, many magnetic field variations are measured at elevations tens of metres above the Earth's surface and have horizontal extents of tens of metres to kilometres. For even the smallest and shallowest-sourced anomalies we never have direct sampling of more than a very small fraction of the magnetisation that causes the anomaly. This problem is compounded by variability of magnetisation at all scales, within borehole cores and undoubtedly throughout the remainder of the source body that is not tested by drilling. Extensive programs of drilling and measurement of magnetic susceptibility and remanent magnetisation to directly relate geology or mineralisation with magnetisation are rarely feasible. The challenge can be divided into two parts: the challenge of representing the true magnetisation distribution with a model and the challenge of resolving a model from the magnetic field data.

1.5.1 The relationship between magnetisations and their models

Subsurface magnetisation can be envisaged as the true (voxel) distribution that is a digitisation of the magnetisation at regular horizontal and vertical spacing. The challenge of either voxel or parametric models is to best and most reliably represent this distribution of magnetisation. There are different geologies that variously favour representation by sharp discontinuities in magnetisation across individual body margins of parametric models or by the smooth variations imposed on most voxel models. For estimation of the detail of depth to the top of magnetisation, specific shape assumptions of parametric bodies with horizontal tops and sharp edges are required (or at a minimum these can be substituted by shape-related 'structural indices'). If a magnetisation

is diffuse and of gradual spatial variation (as assumed or allowed in many voxel inversions) it is not possible to reliably invert for depth to its top. Voxel models are favoured by geologists in part because of their irregular shape and variation in intensity of magnetisation that appears to mimic variation expected of most geological systems. However, it is specifically this level of detail that is the least reliable aspect of the models and is an embellishment of the information that can reliably be recovered from a model. Neither can the implied homogeneity of parametric models be justified. A model of subsurface magnetisation is best envisaged as a set of statistics that summarise a distribution that cannot be known in detail from its magnetic field expression. The set of statistics is necessarily sparser for deeper parts of the model or for models derived from measurements at higher elevation.

To highlight the range of models that can explain a magnetisation I show multiple inversions of a sweet-spot measured in the Coonabarabran Survey (P1290) in northwest New South Wales. Figure 1.11A shows the TMI image of a section of the survey that was flown for the Geological Survey of New South Wales in 2017, on east–west flightlines at 250 m spacing and 60 m ground clearance. Figure 1.11B shows the detail of a discrete anomaly recorded in the survey. This concentration of magnetisation and the aeromagnetic measurements that define its magnetic field expression constitute a sweet-spot from which magnetisation details can be recovered with relative reliability. Inversion with ellipsoid- and elliptic-section pipe models recovered consistent and stable magnetisation direction estimates of inclination −67°, declination 0° in a geomagnetic field of inclination −62°, declination 11°. The departure of the resultant magnetisation from the geomagnetic field (ARRA) of 12° is sufficient to suggest that the magnetisation is not completely induced, but small enough that the centre of magnetisation should be reasonably estimated by modelling with an assumed induced magnetisation only.

Figure 1.12 shows three alternative source models derived by different inversions. Figure 1.12A shows thresholded and isosurface displays of a voxel inversion assuming induced magnetisation performed with the UBC (University of British Columbia) 3D inversion code. As described previously, neither the threshold or isosurface displays faithfully represent the magnetisation that causes the anomaly. Figure 1.12B shows two parametric models: an ellipsoid and a plunging elliptic-section pipe, both of homogeneous magnetisation that is directed

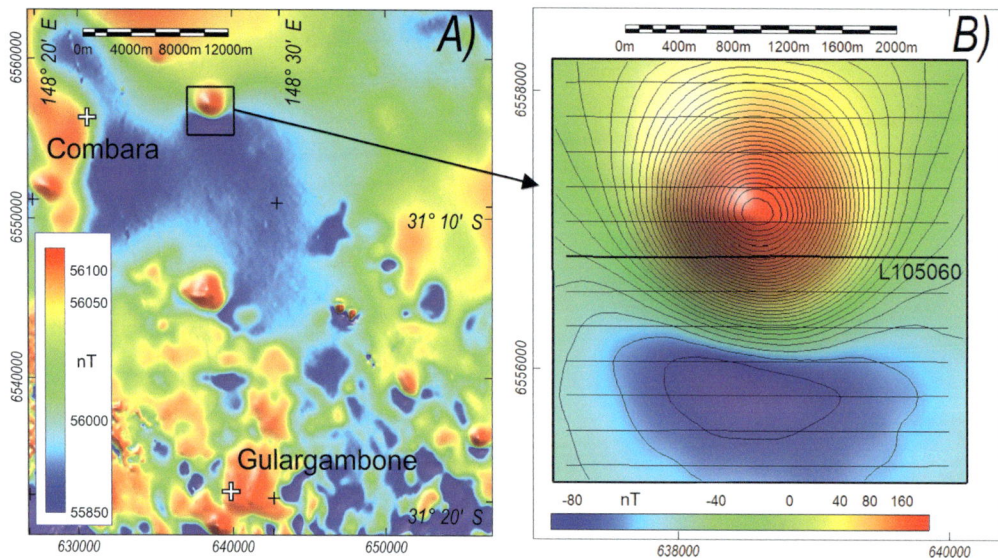

Fig. 1.11. A) part of the TMI map from a regional aeromagnetic survey in New South Wales, Australia and B) detail of a selected anomaly.

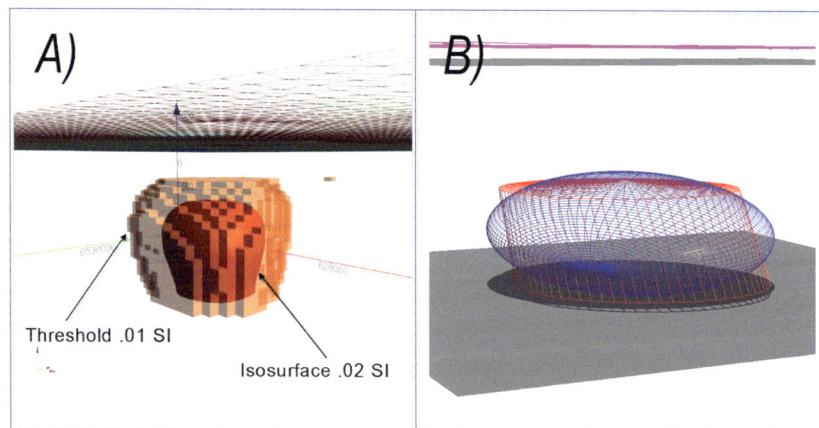

Fig. 1.12. A) 3D Isosurface and threshold representations of a voxel model inversion of the anomaly in Fig. 1.11B, and B) alternative ellipsoid (blue) and elliptic-section pipe (red) models.

slightly away from the local geomagnetic field. However, difference in magnetisation direction plays only a small role in difference between the voxel and parametric models. Figure 1.13 shows the very close fits to the measured field of the computed fields from both parametric models in Fig. 1.12B. No significance can be ascribed to the different post-inversion residual data misfits of the models and it is not feasible to discriminate between the model fields according to their success in matching the measured field. The magnetisation directions and centres of magnetisation of the two parametric models are almost identical. There is a difference of 13% in depth below surface to the top of the two models due to the

details of their differences in shape. Only the horizontal-topped pipe model is suitable for depth estimation (the voxel model is also excluded from depth analysis). Shape and depth to the top of a magnetisation are details that are poorly constrained by inversion. The direction of magnetisation and its effective horizontal centre are more reliably determined for this and most compact magnetisations.

Figure 1.14A shows cross-sections on Flightline 105060 through the centre of the anomaly (for location see Fig. 1.11) for the best-fit ellipsoid model (in green) with a magnetisation of 3.8 A/m, and for models with magnetisations double that (in red) and half

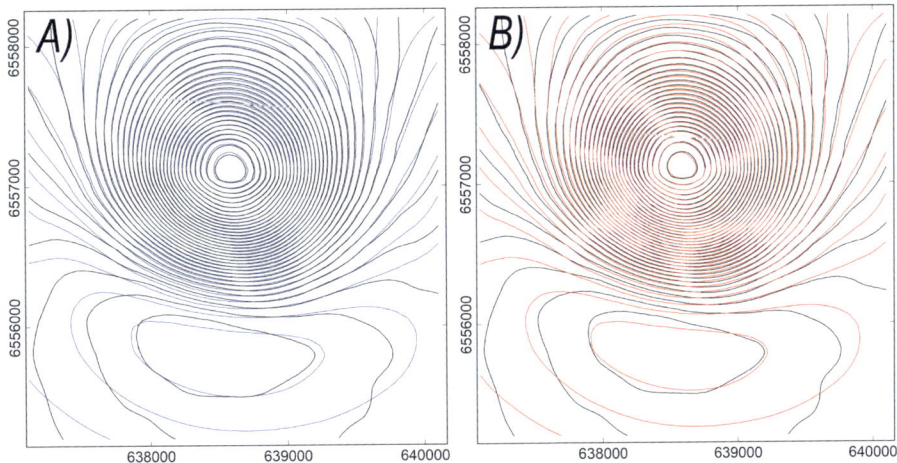

Fig. 1.13. TMI contours: (black) measured and A) (blue) computed from the best ellipsoid inversion model and B) (red) computed from the best elliptic pipe inversion model.

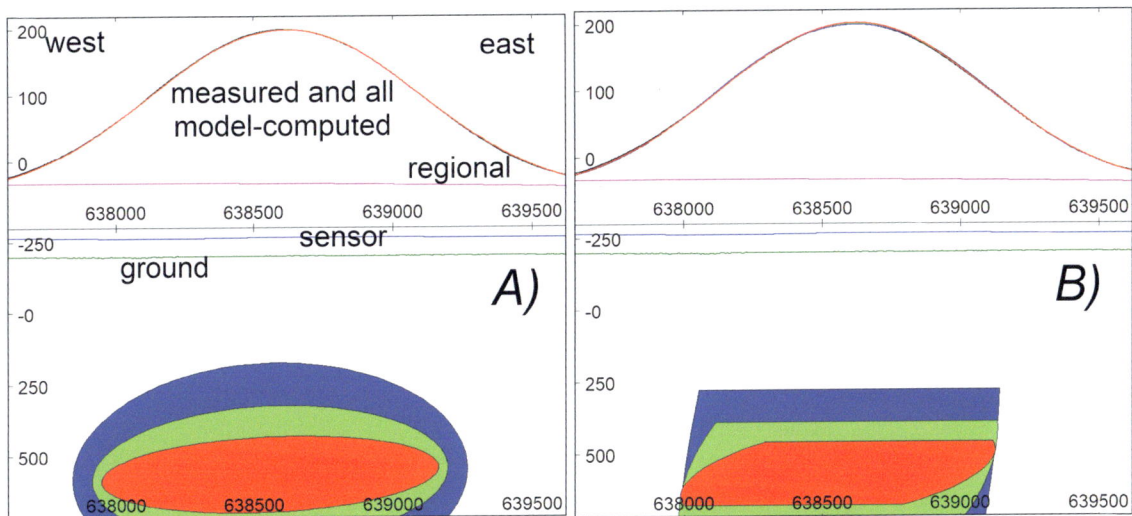

Fig. 1.14. Model sections through flightline 105060 (for location see Fig. 1.11B). A) the best ellipsoid model (red), best-fit ellipsoid of half the susceptibility (blue) and twice the susceptibility (green), and B) the matching set of elliptic pipe models.

of that (in blue). As shown, the computed fields all almost completely match the measured field. Figure 1.14B shows the sections and data-fits for the corresponding pipe models (the best pipe model has a magnetisation of 5.0 A/m. It is not feasible to discriminate between these models based on their goodness of fit to the measured field. Note that this indeterminacy in intensity of magnetisation applies equally to voxel models and discredits any claim that those inversions map detail of variation in magnetisation from voxel to voxel.

Figure 1.15 shows a cross-plot of intensity of magnetisation against volume for a series of ellipsoid and elliptic-section pipe models. The range of the figure spans more than an order of magnitude in both volume and intensity of magnetisation, with very little ground for confidently selecting a value for either parameter. However, the two curves plotted of constant magnetic moment (best estimated as 15.6 A/m^2 for the ellipsoid models and 16.3 A/m^2 for the elliptic-section pipe models show that all bodies have a magnetic moment very close to 16 A/m^2, regardless of their type or individual magnetisation and volume values. This compound parameter of magnetic moment is therefore much more reliably determined from the inversions than the individual parameter values and is a more suitable statistic than intensity of magnetisation or apparent magnetic susceptibility to summarise the source magnetisation for the anomaly.

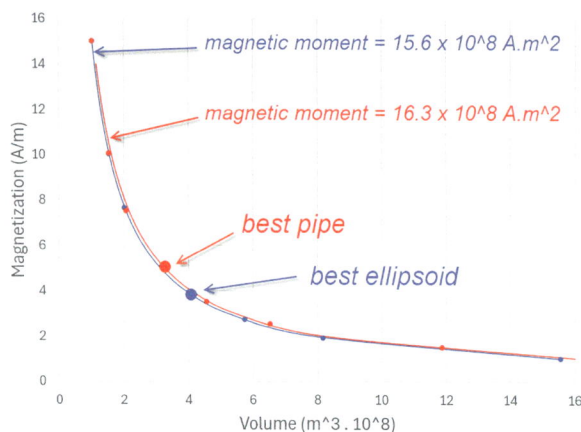

Fig. 1.15. Cross plot of intensity of magnetisation against volume for pipe and ellipsoid models.

1.5.2 The relationship between models and magnetic field anomalies

As has been extensively discussed above, the relationship between models and magnetic fields is extremely anisotropic with an exact relationship in the 'forward' direction and substantial non-uniqueness in the 'reverse' direction. The challenge of the inverse relationship that we mostly deal with is of optimising information recovery of the most reliable model statistics that are made on the basis of justified assumptions and of attributing those recovered statistics with realistic indications of their reliability. The statistics and their reliability estimates are fundamentally constrained by the geology and magnetic field measurements. The apparent detail of parametric and voxel models are very different but if models from well-conducted parametric and voxel inversions are distilled to key statistics, those statistics should be similar.

1.6 MAGNETIC FIELD STUDIES IN MINERAL EXPLORATION

Many mineral deposits have a magnetic field expression, but with exceptions of extreme amplitude anomalies such as those associated with magnetite deposits, few magnetic field features are diagnostic of mineralisation. A review of the magnetic field expression of different styles of mineralisation is given by Clark *et al.* (2003) and by Dentith and Mudge (2014). Dentith *et al.* (1994) and Dentith (2003) report the geophysical expression of various mineral deposits in Western Australia and South Australia respectively. For several forms of mineralisation (e.g. the search for kimberlites in diamond

exploration) magnetic field surveys may be a primary exploration method, but in many cases aeromagnetic surveys are used to map the major distribution of geology and structures in an area and provide context to optimise location of exploration methods such as detailed gravity surveys, induced polarisation (IP) and electromagnetics that are more expensive but may also be more diagnostic of mineralisation. Many magnetic field expressions associated with mineralisation are indirect and interpretive, for instance the presence of demagnetised zones caused by alteration, recognition of magnetic field patterns characteristic of deposits, or tracing of the intersection of geological units with structural features. Some magnetic field expressions of mineralisation might be overlooked (e.g. a low amplitude anomaly due to a large body of haematite where the expectation was for a magnetite body) and some forms of mineralisation have no associated magnetisation contrasts and therefore no magnetic field expression. The most common disappointments in application of magnetic field studies to mineral exploration occur in 'bump hunting' where a magnetic field anomaly may be targeted because it is the most prominent anomalous feature in an area. This approach to mineral exploration is most likely to be successful if proven nearby mineralisation is associated with similar magnetic anomalies. The most sophisticated and generally most successful application of magnetic field interpretation in mineral exploration is in construction of geological maps in association with all other available information, supplemented with quantitative modelling or inversion studies where appropriate. In an increasing number of countries around the world, government- or state-funded regional aeromagnetic surveys are flown to provide data sufficient to encourage and enable mineral exploration companies to investigate large areas within which they can selectively commit to exploration tenements where they will perform exploration activities, possibly including more detailed magnetic field surveys to better constrain features of interest.

1.6.1 Aeromagnetic surveys

Most of the magnetic field data used in mineral exploration are measured on aeromagnetic surveys. Application of these data in exploration geophysics started in the late 1940s after wartime development of the technique for submarine detection. Early surveys used vertical component magnetometers but these were soon replaced with total field instruments. Methods of aeromagnetic surveying are described by Gunn (1997) and Reeves (2005) and details of the current technologies can be found from inspection of acquisition reports of recent surveys. Magnetometers have developed considerably over the last 50 years but the most influential advance was the introduction of differential global satellite positioning (DGPS) in the mid-1990s. Before that, positioning was largely by flight recovery from photographs or installing beacons for triangulation. It was only with advent of DGPS that mapping magnetic fields across large areas at high resolution became feasible. Regional surveys that had been flown at 1 km or 1 mile line spacing or greater were reduced to 200 or 400 m line spacing and could subsequently be infilled with detailed surveys at 50 m spacing or closer if required. Many early surveys used magnetometers deployed in a 'bird' towed behind a plane or beneath a helicopter to reduce the magnetic noise associated with the aircraft. However, with much improved magnetic cleaning of aircraft and compensation for their fields, most aeromagnetic data is now acquired with magnetometers in 'stingers' attached to the nose or rear of an aircraft or to the skids of a helicopter. Compensation for aircraft orientation and manoeuvre noise is generally performed using a three-dimensional set of fluxgate vector sensors that are highly sensitive to orientation. This system is calibrated before a survey using a 'comp-box' of data for aircraft roll, pitch and yaw flown at high elevation in the principal survey line directions. Acquisition of this data is highly specialised and is mostly performed by contract companies that maintain specialist aircraft and instruments and employ pilots experienced in flying these surveys, instrument engineers and data processors. On larger surveys client companies employ independent consultants to oversee data acquisition and ensure that all data are of required standard before the contractor leaves the survey area.

Fixed-wing aircraft are most economic for large areas of modest topography. Data are typically acquired at 10 or 20 hertz that equates to spacings of 5 to 10 m according to flight speed. Helicopters are used for smaller surveys or in areas of more rugged terrain. Surveys by either fixed-wing or helicopter are generally specified in terms of constant ground clearance (except where legally mandated over residences or roads) but if there is substantial terrain there are generally advantages in reducing abrupt changes from line to line by flying a 'smooth drape' rather than maintaining constant ground clearance. This includes consistency in flying in the up and down directions over major terrain features that requires considerable skill of the pilot. To reduce abruptness of turn, surveys are not generally flown sequentially along adjacent lines but in a 'racetrack' pattern, with turns between lines performed outside the area of required coverage.

The magnetic field varies in the time that it takes to fly a survey. These time variations are referred to as diurnal variation. On 'quiet' days they follow a common pattern according to the survey latitude, proximity to the ocean and regional variations in crustal and mantle conductivity. However, the variations are not predictable in detail and must be measured at a fixed base station so that they can be subtracted from the same variations experienced by the survey magnetometer. Typical 'quiet day' diurnal variations have ranges of 50 to 100 nT but amplitude and details vary from day to day. There are also days with magnetic storms that occur when flares from sunspot activity and coronal mass ejections impact the Earth. These introduce streams of charged particles that disturb the balance in the Earth's magnetosphere and cause high-amplitude, sharp magnetic field variations that are most severe towards the poles where associated visual effects are known as aurora. It is not feasible to reliably remove these extreme variations from the magnetic field data and surveys must be suspended over these periods. An aeromagnetic survey contract specifies how the level of activity of the field is quantified on the base station records and at what value of a specified variation statistic flying should be suspended. Almost all onshore aeromagnetic surveys are flown in conjunction with acquisition of radiometric data, and following periods of rain when the radiometric signal is supressed there may also be contractual requirements to suspend flying.

REFERENCES

Alken P, Thébault E, Beggan CD, et al. (2021) International Geomagnetic Reference Field: the thirteenth generation. *Earth, Planets, and Space* **73**, 49. doi:10.1186/s40623-020-01288-x

Archibald N, Gow P, Boschetti F (1999) Multiscale edge analysis of potential field data. *Exploration Geophysics* **30**, 38–44. doi:10.1071/EG999038

Baranov V (1957) A new method for interpretation of aeromagnetic maps: pseudo-gravimetric anomalies. *Geophysics* **22**, 359–383. doi:10.1190/1.1438369.

Baranov V, Naudy H (1964) Numerical calculation of the formula of reduction to the magnetic pole. *Geophysics* **29**, 67–79. doi:10.1190/1.1439334

Beiki M, Clark DA, Austin JR, Foss CA (2012) Estimating source location using normalized magnetic source strength calculated from magnetic gradient tensor data. *Geophysics* **77**, J23–J37. doi:10.1190/geo2011-0437.1

Biedermann AR, McEnroe SA (2017) Effects of magnetic anisotropy on total magnetic field anomalies. *Journal of Geophysical Research. Solid Earth* **122**, 8628–8644. doi:10.1002/2017JB014647

Blakely RJ (1995) 'Potential theory in gravity and magnetic applications'. Cambridge University Press, pp. 441.

Blakely RJ, Simpson RW (1986) Approximating edges of source bodies from magnetic or gravity anomalies. *Geophysics* **51**, 1494–1498. doi:10.1190/1.1442197

Boschetti F, Hornby P, Horowitz FG (2001) Wavelet based inversion of gravity data. *Exploration Geophysics* **32**, 48–55. doi:10.1071/EG01048

Butler RF (1992) 'Palaeomagnetism: magnetic domains to geologic terrains'. Blackwell Scientific Publications, pp. 238.

Clark DA (2013) New methods for interpretation of magnetic vector and gradient tensor data II: Application to the Mount Leyshon Anomaly, Queensland. *Exploration Geophysics* **44**, 114–127. doi:10.1071/EG12066

Clark DA (2014) Methods for determining remanent and total magnetisations of magnetic sources - a review. *Exploration Geophysics* **45**, 271–304. doi:10.1071/EG14013

Clark DA, Schmidt PW, Coward DA, Huddleston MP (1998) Remote determination of magnetic properties and improved drill targeting of magnetic anomaly sources by Differential Vector Magnetometry (DVM). *Exploration Geophysics* **29**, 312–319. doi:10.1071/EG998312

Clark DA, Genua S, Schmidt PW (2003) Predictive exploration models for porphyry, epithermal and iron-oxide copper-gold deposits: implications for exploration. *AMIRA International Exploration and Mining Report* **1073R**, 398.

Cortés-Ortuño D, Fabian KV, de Groot L (2022) Mapping magnetic signals of individual magnetite grains to their internal magnetic configurations using micromagnetic models. *Journal of Geophysical Research, Solid Earth* **127**, e2022JB024234. doi:10.1029/2022JB024234

Davis A (2022) Nested anisotropic geostatistical gridding of airborne geophysical data. *Geophysics* **87**, E1–E12. doi:10.1190/geo2021-0169.1

Dentith MC (Ed.) (2003) 'Geophysical Signatures of South Australian Mineral Deposits'. ASEG Special Publication 12.

Dentith MC, Mudge ST (2014) 'Geophysics for the mineral exploration geoscientist'. Cambridge University Press, pp. 438.

Dentith MC, Frankcombe KF, Trench A (1994) Geophysical signatures of Western Australian mineral deposits: an overview. *ASEG Extended Abstracts* **1994**(1), 29–54. doi:10.1071/ASEGSpec07_03

Dunlop DJ, Özdemir Ö (1997) 'Rock magnetism: fundamentals and frontiers (Cambridge studies in Magnetism). Cambridge University Press, pp. 596.

Fedi M, Florio G (2001) Detection of potential fields source boundaries by enhanced horizontal derivative method. *Geophysical Prospecting* **49**, 40–58. doi:10.1046/j.1365-2478.2001.00235.x

Foss CA (2017) 'Resultant-magnetization based magnetic field interpretation'. In *Proceedings of Exploration 17: Sixth Decennial International Conference on Mineral Exploration*. (Eds V Tschirhart and MD Thomas) pp. 637–648.

Foss CA, Gouthas G, Fabris A, Werner M, Katona LF, Hutchens M, Reed G (2018) 'Gawler Craton Airborne Geophysical Survey Region 3b, Torrens – Enhanced geophysical imagery and magnetic source depth models'. Report Book 2018/00038. Department for Energy and Mining. South Australia, Adelaide.

Foss CA, Clark DA, Keenan S, Leslie K (2019) Constraining structural dip and magnetization direction of a sheet from its static and dynamic magnetic anomalies. *ASEG Extended Abstracts* **2019**, 1–5. doi:10.1080/22020586.2019.12073127

Foss CA, Gouthas G, Katona LF, Wise TW, Pawley MJ (2019a) 'Gawler Craton Airborne Geophysical Survey Region 4a, Barton – Enhanced geophysical imagery and magnetic source depth models'. Report Book 2019/00012. Department for Energy and Mining. South Australia, Adelaide.

Goldstein NE, Ward SH (1966) The separation of remanent from induced magnetization in situ. *Geophysics* **31**, 779–796. doi:10.1190/1.1439810

Grauch VJS, Cordell L (1987) Limitations of determining density or magnetic boundaries from the horizontal gradient of gravity or pseudogravity data. *Geophysics* **52**, 118–121. doi:10.1190/1.1442236

Gunn PJ (Ed.) (1997) 'Airborne magnetic and radiometric surveys'. *AGSO Journal of Geology and Geophysics* **17**(2).

Hinze W, Von Frese R, Saad A (2013) 'Gravity and Magnetic Exploration: Principles, Practices, and Applications. Cambridge: Cambridge University Press. doi:10.1017/CBO9780511843129

Hornby P, Boschetti F, Horowitz FG (1999) Analysis of potential field data in the wavelet domain. *Geophysical Journal International* **137**, 175–196. doi:10.1046/j.1365-246x.1999.00788.x

Hrouda F (1982) Magnetic anisotropy of rocks and its application in geology and geophysics. *Geophysical Surveys* **5**, 37–82. doi:10.1007/BF01450244

Hrouda F (2007) 'Magnetic susceptibility, anisotropy'. In *Encyclopedia of Geomagnetism and Paleomagnetism*. (Eds D Gubbins, E Herrero-Bervera) pp. 546–560.

Isles DJ, Rankin LR (2013) 'Geological interpretation of aeromagnetic data'. Australian Society of Exploration Geophysicists, pp. 365.

Kasama T, McEnroe SA, Ozaki N, Kogure T, Putis A (2004) Effects of nanoscale exsolution in hematite-ilmenite on the acquisition of stable remanent magnetization. *Earth and Planetary Science Letters* **224**, 461–475. doi:10.1016/j.epsl.2004.05.027

Kellogg OD (1967) 'Foundations of potential theory'. Springer-Verlag, pp. 383.

Koyama T, Kaneko T, Ohminato T, Yanagisawa T, Watanabe A, Takeo M (2013) An aeromagnetic survey of Shinmoe-dak volcano, Kirishima, Japa, after the 2011 eruption using an inmanned autonomous helicopter. *Earth, Planets, and Space* **65**, 657–666. doi:10.5047/eps.2013.03.005

Kuwahata A, Kitaizumi T, Saichi K, *et al.* (2020) Magnetometer with nitrogen-vacancy center in a bulk diamond for detecting magnetic nanoparticles in biomedical applications. *Scientific Reports* **10**, 2483. doi:10.1038/s41598-020-59064-6

Lelièvre PG, Oldenburg DW (2009) A 3D total magnetization inversion applicable when significant, complicated remanence is present. *Geophysics* **74**, L21–L30. doi:10.1190/1.3103249

Li Y, Oldenburg DW (1996) 3D inversion of magnetic data. *Geophysics* **61**, 394–408. doi:10.1190/1.1443968

Li Y, Oldenburg DW (1998) Separation of regional and residual magnetic field data. *Geophysics* **63**, 431–439. doi:10.1190/1.1444343

Löhr SC, Murphy DT, Nothdurft LD, Bolhar R, Piazolo S, Siegel C (2017) Maghemite soil nodules reveal the impact of fire on mineralogical and geochemical differentiation at the earth's surface. *Geochimica et Cosmochimica Acta* **200**, 25–41. doi:10.1016/j.gca.2016.12.011

Lourenço JS, Morrison HF (1973) Vector magnetic anomalies derived from measurements of a single component of the field. *Geophysics* **38**, 359–368. doi:10.1190/1.1440346

MacMillan WD (1930) 'The theory of the potential'. Dover Publications, New York, pp. 324.

Nabighian MN (1972) The analytic signal of two-dimensional magnetic bodies with polygonal cross-section: its properties and use for automated anomaly interpretation. *Geophysics* **37**, 505–517. doi:10.1190/1.1440276

Naprstek T, Smith RS (2019) A new method for interpolating linear features in aeromagnetic data. *Geophysics* **84**, JM15–JM24. doi:10.1190/geo2018-0156.1

Pastore Z, Church NS, McEnroe SA (2019) Multistep parametric inversion of scanning magnetic microscopy data for modelling magnetization of multidomain magnetite. *Geochemistry, Geophysics, Geosystems* **20**, 5334–5351. doi:10.1029/2019GC008542

Purss MBJ, Cull JP (2005) A new iterative method for computing the magnetic field at high magnetic

susceptibilities. *Geophysics* **70**, L53–L62. doi:10.1190/1.2052469

Reeves C (2005) 'Aeromagnetic surveys: Principles, practice and interpretation'. Earth-works, Washington DC, 155 p.

Reid AB (1980) Aeromagnetic survey design. *Geophysics* **45**, 895–982. doi:10.1190/1.1441102

Robinson P, Harrison RJ, McEnroe SA, Hargraves RB (2002) Lamellar magnetism in the hematite-ilmenite series as an explanation for strong remanent magnetization. *Nature* **418**, 517–520. doi:10.1038/nature00942

Roest WR, Verhoef J, Pilkington M (1992) Magnetic interpretation using the 3-D analytic signal. *Geophysics* **57**, 116–125. doi:10.1190/1.1443174

Schiffler M, Queitsch M, Stoltz R, Chwala A, Krech W, Meyer HG, Kukowski N (2014) Calibration of SQUID vector magnetometers in full tensor gradiometry systems. *Geophysical Journal International* **198**, 954–964. doi:10.1093/gji/ggu173

Schmidt PW (2015) The Qmeter - a portable tool for remanence and susceptibility. *ASEG Extended Abstracts* **2015**(1), 1–3. doi:10.1071/ASEG2015ab235

Schmidt PW, Lackie MA (2014) Practical considerations: making measurements of susceptibility, remanence and Q in the field. *Exploration Geophysics* **45**, 305–313. doi:10.1071/EG14019

Schmidt PW, Clark DA, Leslie KE, Bick M, Tilbrook D, Foley C (2004) GETMAG – a SQUID magnetic tensor gradiometer for mineral and oil exploration. *Exploration Geophysics* **35**, 297–305. doi:10.1071/EG04297

Schmidt PW, McEnroe SA, Clark DA, Robinson D (2007) Magnetic properties and potential field modelling of the Peculiar Knob metamorphosed iron formation, South Australia: an analogue for the source of the intense Martian magnetic anomalies? *Journal of Geophysical Research* **112**, B03102. doi:10.1029/2006JB004495

Wijns C, Perez C, Kowlczyk P (2005) Thetamap: edge detection in magnetic data. *Geophysics* **70**, L39–L43. doi:10.1190/1.1988184

SECTION 1:
LOCATION OF MAGNETISATION BY MAGNETIC FIELD INVERSION

2

Data selection and optimisation for magnetic field inversion

C.A. Foss

ABSTRACT

Magnetic field inversion is data-centric. In this chapter I explain how the capabilities and limitations of inversion depend on the input data, including data acquisition and processing. We do not invert the magnetic field – only samples of it, and those samples are subject to distortion as well as misrepresentation by sampling bias and limitation. Gridding in particular, is a major transformation of the primary measurements, generating apparent data where no measurements have been made. Inversion of gridded data should be performed only where necessary or with a thorough evaluation and justification.

The Earth's magnetic field combines the fields of multiple crustal magnetisations together with core and ionospheric contributions into a single field. One of the most challenging tasks of inversion is to ascribe measured magnetic field variations to selected magnetisations so that inversion can be performed on appropriate data. Field separation is interpretive, and we should not rely on inversion itself to perform that separation without careful supervision. It is only those parts of the magnetic field that dominate its local curvature that can be reliably inverted. These are generally field contributions from the shallower magnetisations.

2.1 INTRODUCTION

This chapter presents case studies to illustrate unconstrained magnetic field inversion. Reading through these examples will provide an understanding of how to select data appropriate for magnetic field inversions and how to design inversions to achieve a required output. For successful inversion the magnetic field must include distinct features that can be attributed to distinct magnetisations. This restriction is not because less distinct field variations cannot be inverted but because without distinctive features ambiguity severely restricts the value of any inversion results and those results might well be misleading.

A common misconception is that inversion is an output model-focussed process in which data play only a passive role. A more holistic view is that inversion is data-centric with output models as advanced analytic data products. Clearly inversion can only be expected to return a justified model of subsurface property distribution if the input data are valid and sufficient. Input of reliable data to an inversion requires attention throughout acquisition and processing and those operations should be documented with comprehensive metadata. Incorrect assumptions about data units, geolocation and what processing has been applied to data are common and avoidable causes of invalid inversion results. Statistical checks should be performed to ensure that all data values are in expected ranges and those numerical error traps should be complemented with visual inspection of the data to rapidly reveal data flaws that if overlooked might invalidate inversion results. An inversion which reveals a flaw in the input data is wasted time and effort.

2.2 SURVEY PLANNING

Exploration geophysics is task oriented. Surveys are undertaken as a cost-effective means to support decision-making in a project. The decision to run a survey and the design of that survey are informed by consideration of the project objectives. Important survey design decisions cover survey extent, instrumentation, measurement spacings and, in the case of airborne surveys, line direction and terrain clearance. A key paper by Reid (1980) provides rules of thumb about line spacings required for subsequent analysis of magnetic field data and a free E-book by Colin Reeves (2005) provides an excellent review of the aeromagnetic method. Synthetic data modelling has a role to play in survey design such as optimisation of flightline spacing and/or flying height. Decisions made in survey planning, data acquisition and processing determine the capabilities and limitations of subsequent inversions and possibly the success of the project.

2.3 NETWORK ADJUSTMENT OF MAGNETIC SURVEY DATA

A common process applied in airborne surveys is to fly tie-lines perpendicular to the flightlines at a spacing of five to ten times the flightline spacing. A network adjustment is then applied to minimise residual differences at each tie-line and flightline intersection. This empirical process is critical to ensure data quality but there should be concern if tie-line levelling significantly changes the data. Data commonly available from aeromagnetic surveys may include 'raw' magnetics without diurnal correction, and the subsequent processing product of 'tie-line-levelled' magnetics includes diurnal correction, removal of heading errors and network adjustment. Neither data channel is ideal for inversion. 'Raw magnetics' includes unwanted noise and for the tie-line-levelled data there is no longer an appropriate elevation channel because network adjustment includes adjustments to accommodate elevation differences between flightlines and tie-lines without providing a new elevation reference for the magnetic data. Nevertheless, tie-line-levelled data is generally the preferrable data channel for inversion. Elevation inconsistencies between flightlines and tie-lines and between adjacent flightlines are most problematic in areas of rugged terrain. These problems should be anticipated in survey planning and reduced by flying a smooth drape survey focussed primarily on flying lines within a smooth envelope above terrain rather than honouring a constant ground clearance.

Tie-line levelling is commonly followed by micro-levelling to attenuate residual artefacts which would otherwise have prominent expression in subsequent grid enhancements such as vertical gradient computation. Micro-levelling uses a combination of directional and wavenumber filtering designed to preferentially attenuate field variations parallel to the flight direction and related to the line spacing (Minty 1991). Micro-levelling, however, has low discrimination and reduces the amplitude of many true field variations. Micro-levelling significantly modifies more problematic data. Remedial reprocessing designed to correct those problems as far as possible is a better option. Micro-levelled grids are less suitable than tie-line-levelled grids for quantitative applications such as inversion. It is also a common practice to resample the micro-levelled grid back onto line data, but again it is preferable to invert the tie-line levelled channel in which levelling artefacts may be evident, rather than the micro-levelled channel where they have been cosmetically supressed possibly modifying the signal of interest.

2.4 IMPORTANCE OF DATA VISUALISATION

Isles and Rankin (2013) review the expression of geological features and structure in magnetic field imagery and this geological basis is crucial for informed inversion of magnetic field variations due to geology. Data visualisation is of critical importance in magnetic field inversion and supports the understanding required to make decisions about its inversion and interpretation. Images should convey the true amplitude of features. These amplitudes are obscured in histogram equalisation but can be restored by overlaying the images with contours. Addition of shadows and highlights in image sun-shading emphasises the expression of features which may be valid and of interpretational importance but which may not have sufficient amplitude for meaningful inversion. All inversions should be preceded by careful visual inspection of the input data, and no inversion results should be accepted until the computed field of the output model has been visually inspected.

Figure 2.1 shows a TMI image over a region near Echuca in Victoria derived from a survey with east–west flightlines at 400 m spacing and nominal terrain clearance of 100 m. The image shows a string of discrete, well separated and slightly elongate anomalies. On cursory

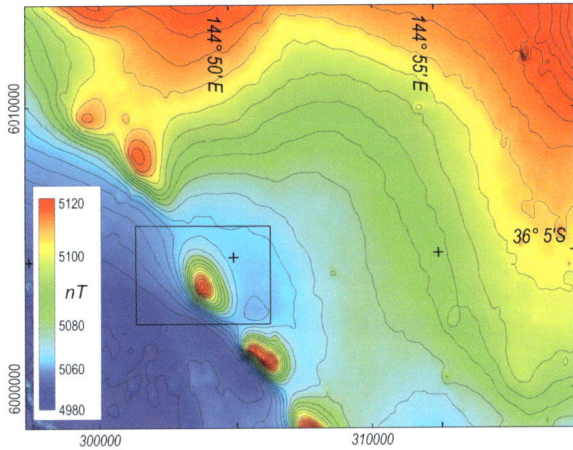

Fig. 2.1. TMI image over a string of magnetic bodies near Echuca, Victoria.

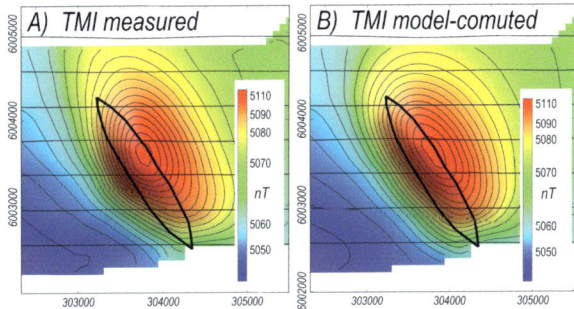

Fig. 2.2. A) measured and B) model computed TMI over the anomaly highlighted in Fig. 2.3. The outline of the top surface of the inversion body is plotted over the images.

inspection of the anomaly in the inset area of Fig. 2.1 it seems that a simple model may be most appropriate for its inversion. The source magnetisation is interpreted to be a steeply plunging pod (possibly associated with a sub-vertical fault) truncated at the top of basement beneath a younger cover. This anomaly is shown in more detail in Fig. 2.2. An elliptic-section horizontal-top cylinder seems suitable to represent the magnetisation with sufficient versatility and an appropriate level of detail. The background field on which the anomaly is superimposed has a gentle gradient from north-west to south-east and for the modelling is represented by an almost planar surface of that trend. The external magnetic field of a geological body is an expression of its resultant magnetisation – the vector sum of its induced and remanent magnetisations. For anomalies such as those shown in Fig. 2.1 which by visual inspection seem consistent with magnetisation in the geomagnetic field direction, source magnetisation can initially be assigned to be induced only. Alternatively, these well-separated, well-defined anomalies also allow the magnetisation to be assigned as including remanence of unknown direction. For simplicity in this study I use an induced-only magnetisation assumption to generate one possible model of the magnetisation. I positioned a pipe body with elliptical section and horizontal-top surface beneath the anomaly and gave it an initial test susceptibility. I then adjusted susceptibility and top depth of the model using a few trial computations to produce a suitable starting model for inversion. I ran the inversion with free horizontal position, depth to top, elliptic axial section radii and azimuth, depth extent and magnetic susceptibility. The resulting best-fit inversion model has a depth below surface of 720 m, length 2130 m,

maximum width 340 m, and magnetic susceptibility 0.024 SI. Images of the input measured and inversion-model computed fields are shown in Fig. 2.2. The model field is the combination of the proposed regional field and the proposed anomalous model field. The two images in Figs 2.2A and 2.2B show a strong but imperfect match, suggesting that the model is a candidate representation of the magnetisation.

Figure 2.3A shows stacked profiles of the measured, model-computed and regional fields. This display provides greater discrimination of the match between observed and computed fields and reveals systematic line-to-line displacements between them suggesting that the model would benefit from additional complexity to reduce this misfit. I replaced the initial elliptic section pipe with a polygonal section pipe which I subsequently inverted allowing individual horizontal displacement of the vertices. The other model parameters were also free to adjust their best values to optimise for this new model type. A plan view of the top of this more complex inversion model with its stacked profiles is shown in Fig. 2.3B. The data misfit shown by the stacked

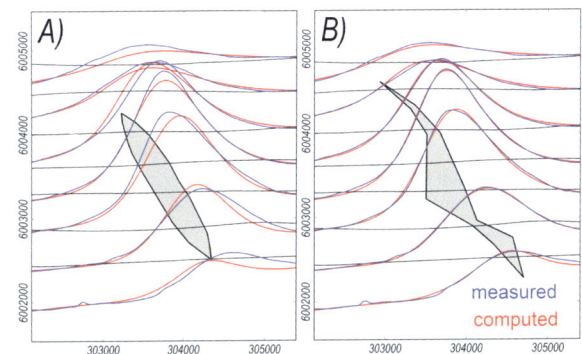

Fig. 2.3. Stacked profiles of measured and model computed TMI over A) the elliptic section pipe model and B) the polygonal section pipe model.

profiles in Fig. 2.3A is considerably reduced by the more complex model that has a depth below surface of 520 m, length 2950 m, width 450 m, and magnetic susceptibility 0.024 SI. These statistics are not substantially different to those of the elliptic pipe model, but due to improved fit to the data the more complex polygon section model shown in perspective view in Fig. 2.4 is (quite subjectively) considered a superior representation of the subsurface magnetisation.

Figure 2.5 shows a plot of the top of the model outline over an image of the tilt of TMI. The tilt is the ratio of vertical and horizontal derivatives of the field (Miller and Singh 1994) with units of degrees. With only indirect expression of field amplitude this enhancement is not well suited as input to inversion but the match between

the image and the model section provides visual support for the additional model complexity. Data visualisation played a significant role in the development and acceptance of this inversion model.

2.5 DATA FOCUS AND SEPARATION OF FIELDS TO BE INVERTED

In inversion it is important to optimise the expression and detection of differences between measured and model computed fields and thereby narrow the range of acceptable models. With increasing computer power and speed ever-larger inversion models are being run – necessarily diluting their focus. A superior strategy is 'divide and conquer' by isolating individual features in the magnetic field data and restricting the data used to be only that required to invert for each feature separately. In many areas only a small proportion of the magnetic field carries source-diagnostic information and little is lost in excluding the remaining data. Figure 2.6 shows an image of TMI over an area near Brewarrina in New South Wales. In this image all the information that can support meaningful inversion of the subsurface magnetisation occurs within less than 10% of the area. The most effective inversion of the data is to focus on each anomaly in turn using data clips as the one shown in Fig. 2.6 that isolates data for two immediately adjacent anomalies. The overlap between these anomalies prevents their reliable individual inversion. The anomalies are defined by survey data on east–west flightlines at 400 m spacing and nominal terrain clearance 80 m flown for the New South Wales Geological Survey to promote mineral exploration. Another five or

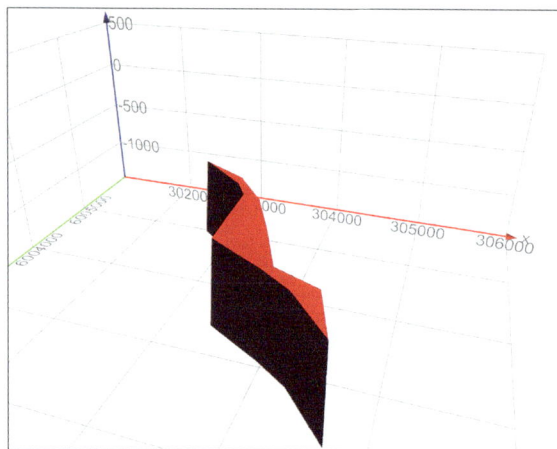

Fig. 2.4. Perspective view of the polygonal section pipe model.

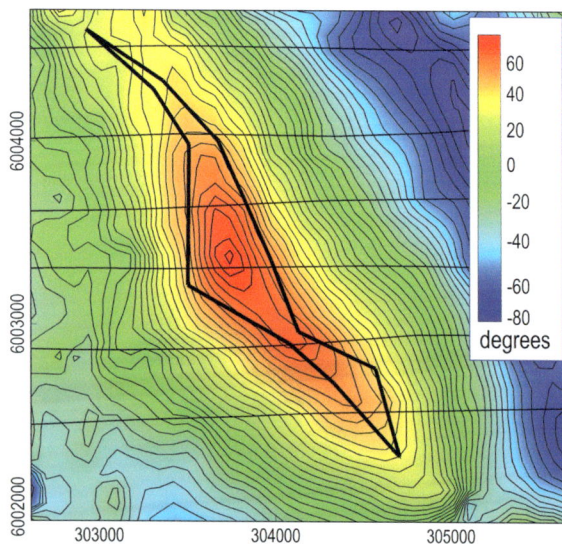

Fig. 2.5. Plan of the top surface of the polygonal pipe model over TMI tilt image.

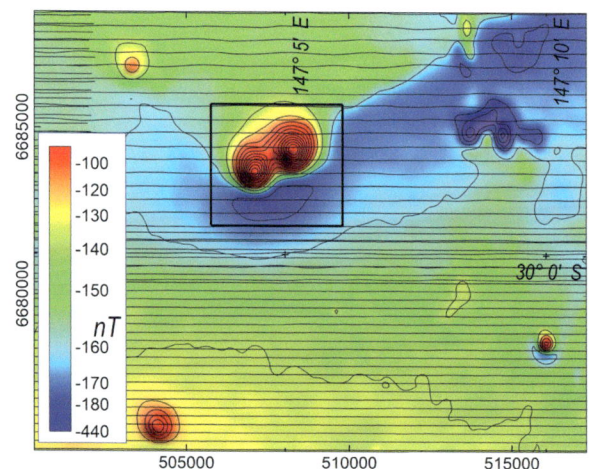

Fig. 2.6. TMI image of two overlapping magnetic pipe anomalies near Brewarrina, New South Wales.

six data clips could be applied to the data imaged in Fig. 2.6 to investigate other discrete anomalies.

Once data selection is optimised the next task is to separate that part of the field variation that is to be explained by the inversion model. This separation is best performed before inversion so that inversion begins with a suitable starting model, thereby improving chances of convergence to an acceptable solution. There are many proposed methods of regional-residual separation to isolate an anomaly from its background field but no method is automatically justified. Regional and residual fields need not conform to any analytic definition and are not better qualified if they do. Spectral separation of fields rarely achieves the degree of separation ideal for inversion because regional and residual field spectra overlap. A method of regional-residual separation by Li and Oldenburg (1998) performs an inversion across a wider zone around the area of interest to define the regional field. This method may work well for some anomalies but lacks the flexibility to be optimised by human guidance.

Regional-residual separation is so critical to an inversion that no automated method should be trusted to make this separation unsupervised. Field separation is fundamentally interpretive and visual inspection of both the regional and residual fields is an important quality control operation before submitting the residual data for inversion. As an inversion proceeds it may become evident that a different separation might be more appropriate, in which case it is quite acceptable to interrupt the inversion and adjust the separation. Adjustments of the anomaly separation are most commonly required for inversion of deep-going magnetisations for which fields of the modelled magnetisation extend considerable distances with only gradual transition to the background field. For bodies of shallow depth extent the transition from the anomaly to the surrounding background field is generally sharper and closer to the margin of the magnetisation. If it is not possible to reliably separate background and residual fields before an inversion is started then it is unlikely that the inversion itself will supply a meaningful solution, although inversion can be used to make modest adjustments to the regional. Multiple inversions using different regionals may be required for particularly problematic regional-residual separations to test the consequences of that uncertainty.

The overlapping anomalies in the data-clip area of Fig. 2.6 are proposed to be due to adjacent intrusive pipes truncated at the top of a weakly magnetised crystalline

basement beneath an unknown thickness of non-magnetic cover. An appropriate model geometry for this interpretation is a pair of vertical or steeply plunging circular or elliptic section cylinders with horizontal top surfaces (similar to the initial model used in the previous Echuca case study). Anomaly separation was straightforward in the Echuca case because of the very different characteristics of the regional and residual fields. That is also the case for the isolated anomalies in the Brewarrina area as shown in Fig. 2.6 but anomaly separation between the fields of the two pipes themselves is more problematic and neither anomaly can be reliably inverted alone without resolving the overlapping field of the other. The degree of overlap is highlighted in the three-dimensional display of the TMI anomalies shown in Fig. 2.7. The anomalies in a two-dimensional FFT vertical derivative transform of the TMI data also shown in Fig. 2.7 have significantly reduced overlap and independent inversion of each anomaly would be assisted by inversion of the derivative rather than the TMI data. However, I decided to simultaneously invert the two TMI anomalies using a two-body model.

As shown in Fig. 2.6, the anomalies are predominantly positive with a weak negative to the south as generally expected for bodies of induced magnetisation at this geomagnetic inclination. However, the two anomalies to the south-west in Fig. 2.6 are both positive-only and nearby there are similar anomalies of negative polarity, suggesting that remanent magnetisation of alternate polarities is significant for this population of bodies. For these compact, sufficiently sampled anomalies there is little reason other than their overlap not to invert for both magnetisation direction and the spatial distribution of magnetisation and I chose this option. If the magnetisation is parallel to the local geomagnetic field then the magnetisation direction recovered by the inversion should be that of the local geomagnetic field, with the advantage that magnetisation direction has

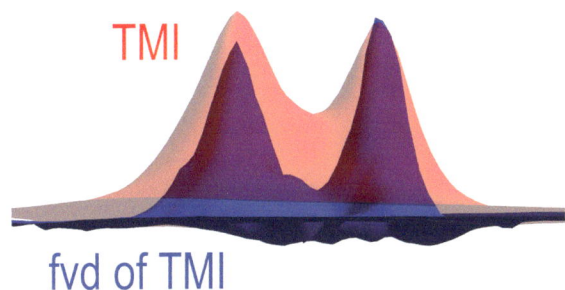

Fig. 2.7. Three-dimensional display of overlapping anomalies of TMI (red) and vertical derivative pf TMI (blue).

been tested rather than assumed. Without independent information only the resultant magnetisation can be recovered from a static magnetic field anomaly. In this study I arbitrarily assigned all magnetisation to be remanent but gave it an initial direction parallel to the geomagnetic field.

I generated a starting mode for inversion by placing a vertical cylinder with magnetisation in the local geomagnetic field direction directly beneath each of the two anomalies and adjusted their depth and magnetisation intensity to approximately match the measured anomalies. From this starting model I ran an inversion with free horizontal position, depth to top, depth extent, cross-section radii and azimuth, axial plunge and plunge azimuth, and magnetisation intensity and direction for each body. A further advantage of working with a small data subset cut to just beyond the extent of

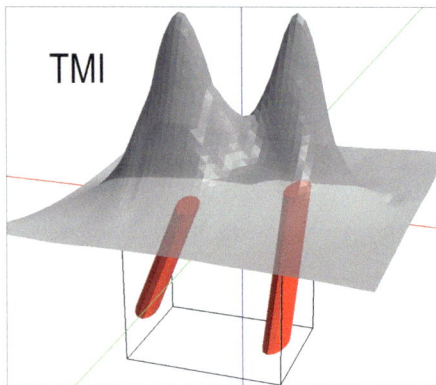

Fig. 2.9. Flightlines and stacked profiles of the measured (blue), regional (magenta), and model-computed (red) fields due to the model of two elliptic pipes.

the anomalies is that the inversions proceeded rapidly, even with many iterations to investigate the 24-dimensional model space. Success of the inversions was encouraged by having a reasonable starting model and small bounds ranges forcing multiple iterations in small steps within the parameter space rather than large and possibly unstable jumps. Figure 2.8 shows the post-inversion bodies beneath the TMI anomalies and Fig. 2.9 shows stacked profiles of the input measured field, regional field, and output computed field.

The concern of simultaneously inverting for two source bodies is that each body may in part explain field variation due to the other. Both non-vertical plunge and magnetisation direction of these deep-going bodies can

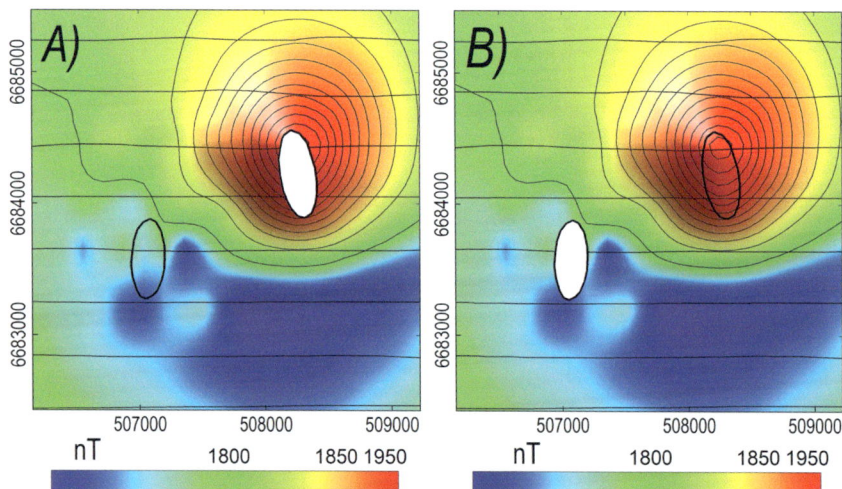

Fig. 2.8. 3D display of the TMI anomalies over the inversion source model.

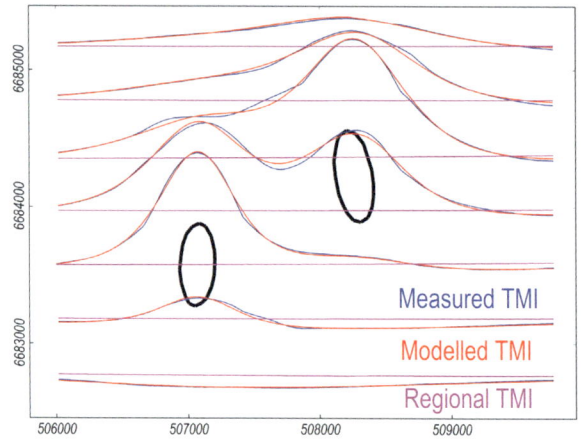

Fig. 2.10. A) and B) Images of fields generated by subtracting from measured TMI the fields computed from the individual inversion model bodies. In the combined inversion each body performs the task of explaining a single anomaly.

adjust to explain asymmetry in their anomalies and this provides opportunity for either body to incorrectly explain a part of the adjacent anomaly. However, an appropriate starting model reduces this opportunity. To illustrate how effectively each body explains only its own anomaly despite the opportunity to do otherwise, Fig. 2.10 shows the fields computed by subtraction of the individual computed body fields from the measured TMI. In each case there is no residual indication of the anomaly due to the subtracted body as would be seen for cross-contribution between the model bodies and anomalies. The model also has acceptable internal consistency, with an angular difference of only 6° both between the plunge of the two bodies and also between their magnetisation directions.

This study highlights the advantage of inversions designed to link specific field variations to their interpreted sources. The decisions to be made require understanding the geology and recognition of any diagnostic characteristics of the field, generally from visual inspection of the data. Without this guidance it is less likely that an inversion will produce a sensible result. It is reassuring that from this starting model the guided inversion resulted in each model body performing its allocated task.

2.6 SEPARATION OF ANOMALIES FROM A STEEP REGIONAL GRADIENT

In each of the previous case studies the regional gradient across the extent of the anomalies causes much smaller field variation than the residual anomaly to be inverted. Figure 2.11 shows examples where anomalies are superimposed on steep background gradients that cause field variations with significant amplitude compared to the residual anomalous field. In these cases correct estimation and treatment of the regional field is particularly critical to the success of the inversion.

Figure 2.11 is an image from a heli-mag survey over the Rylstone area of New South Wales, flown on east–west flightlines with a nominal terrain clearance of 50 m. The Rylstone area contains many intrusive and

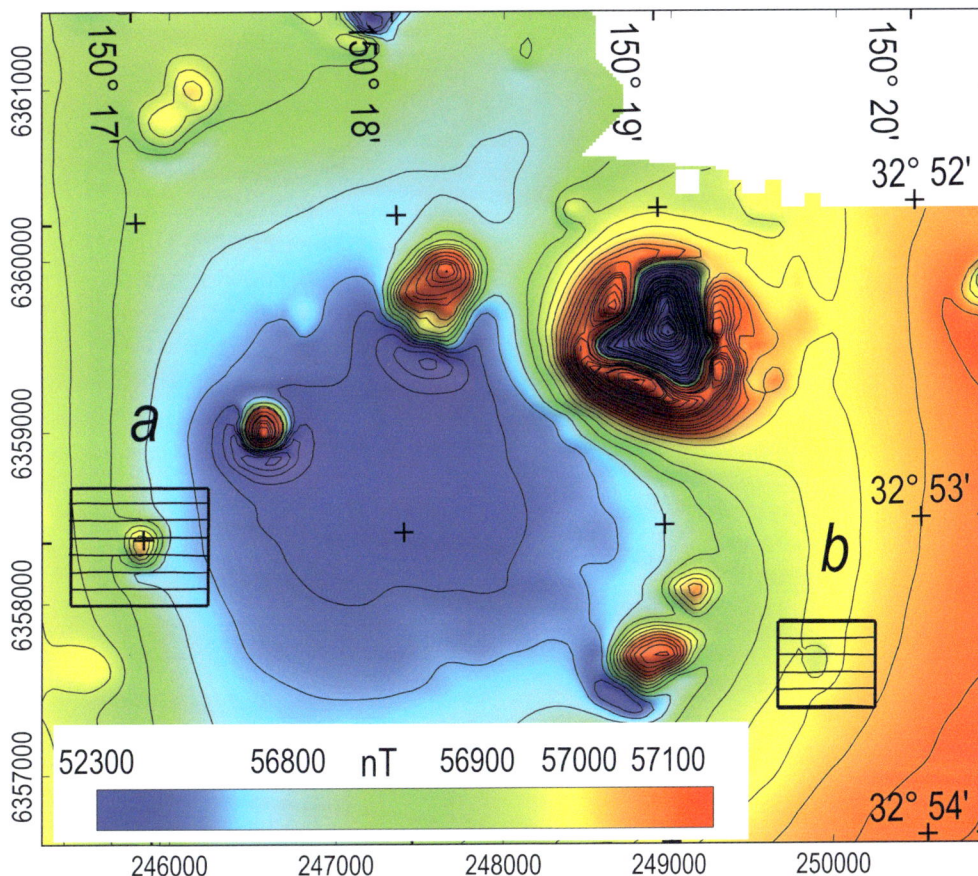

Fig. 2.11. TMI of an area near Rylstone with two anomalies 'a' and 'b' superimposed on steep background gradients.

volcanic bodies of different ages, many with prominent remanent magnetisation. The anomalies 'a' and 'b' in Fig. 2.5 (Australian Remanent Anomalies Database anomalies 319 and 320) are believed to be narrow Tertiary-age pipes with dominant normal and reverse remanent magnetisation respectively. Figures 2.12 and 2.13 show model sections on flightlines through the centres of the anomalies (each is the centre line from a seven-line inversion). For the predominantly positive anomaly 'a' the regional field variation and the peak residual anomaly amplitude are both ~120 nT. For anomaly 'b' the regional field variation is of the order of 80 nT and the central anomaly trough amplitude is 50 nT. The model sections show that on the centre lines for both anomalies the regional field is well represented as a near-linear gradient. The anomalies are both well matched by addition to the regional field of anomalous

fields derived by inversion of horizontal-top, elliptic-section, remanently magnetised cylindrical bodies.

The positive anomaly 'a' is imaged in Fig. 2.14A. The anomaly is strongly asymmetric in the direction of the regional gradient with the contours compressed to the side of the anomaly where regional and anomaly gradients reinforce and are drawn apart on the side where the two gradients are in opposition. The regional field is computed as a second-order polynomial from all model lines but is almost planar as imaged in Fig. 2.15 A. The residual field derived by grid subtraction of the regional field from the measured TMI is imaged in Fig. 2.15B. There is no indication in either image of contamination by the other field component, suggesting that the regional-residual separation is highly effective with the appropriate residual field submitted to the inversion. The negative anomaly 'b' as shown in Fig. 2.14B has a smaller amplitude in comparison to the regional gradient across it and is expressed primarily as a deflection of the regional field contours. The regional and residual separations for this anomaly are shown in Fig. 2.16 A and B respectively, again with very effective separation between the fields.

These results establish the feasibility of performing inversion in the presence of a dominant regional field gradient. However, these are not the most challenging cases of regional-residual separation. Although the regional gradients are strong they are almost planar, easily predictable from the surrounding data, and with a substantial contrast in curvature between the regional and residual fields. More challenging separations arise when these conditions are not met and a residual field to be inverted has to be separated from a regional field of similar curvature, or where the anomaly is close to the edge of a survey where some of the data required to define the regional field are unavailable.

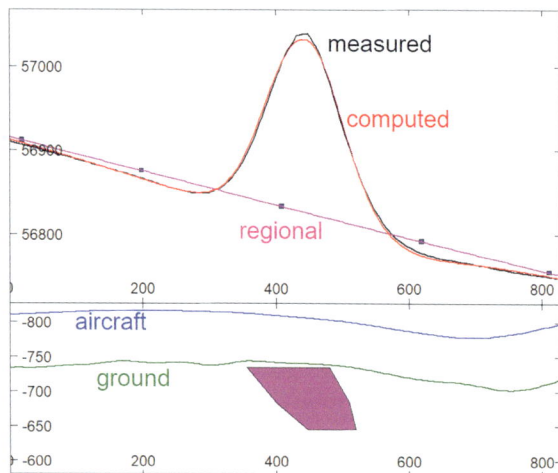

Fig. 2.12. Central flightline section for anomaly 'a'.

2.7 THE INFLUENCE OF DATA SPACING

Input data to an inversion must be not only valid but also of sufficient precision, sample spacing and extent to reliably represent spatial variations in the magnetic field. Using spectral analysis, Reid (1980) suggested data spacing limits beyond which the magnetic field is unacceptably aliased and these have long been adopted as guidelines for survey planning. Reid's recommendations were that TMI surveys should use a flightline spacing no wider than twice the height above the shallowest magnetisation, that gradient surveys require a line spacing at one-half of that

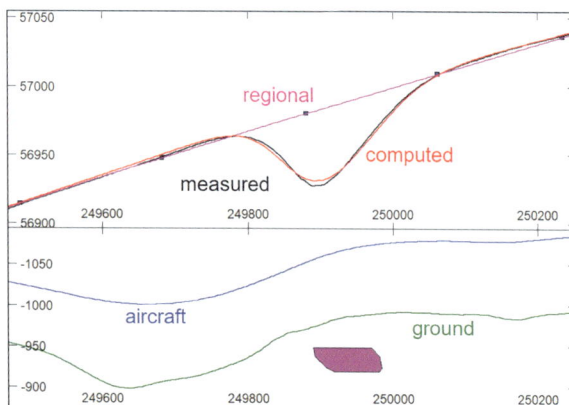

Fig. 2.13. Central flightline section for anomaly 'b'.

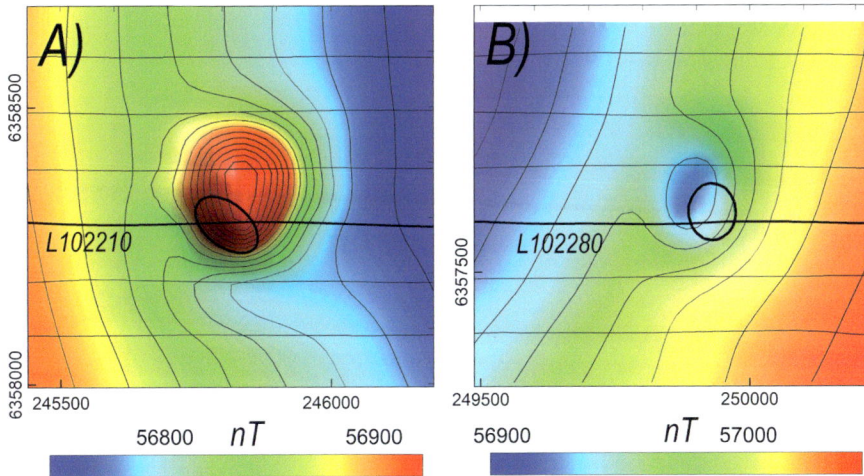

Fig. 2.14. Measured TMI anomalies 'a' and 'b' with outlines of the model top surfaces.

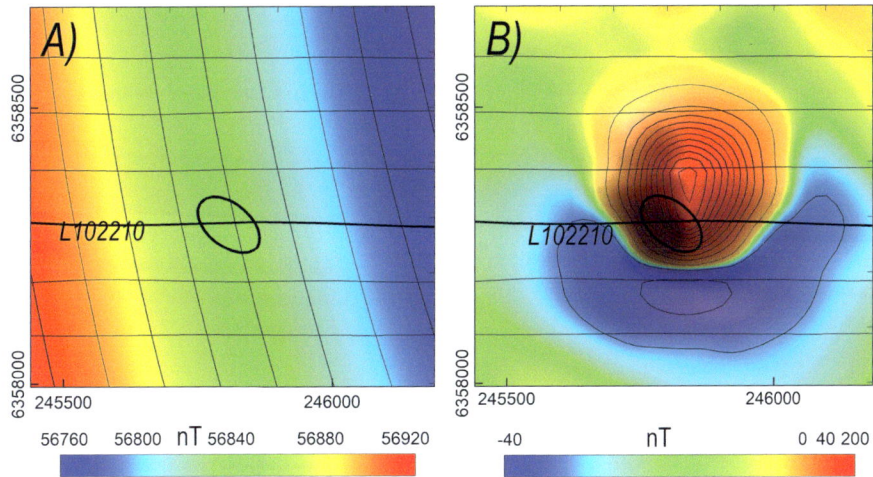

Fig. 2.15. A) regional and B) residual TMI separations for anomaly 'a'.

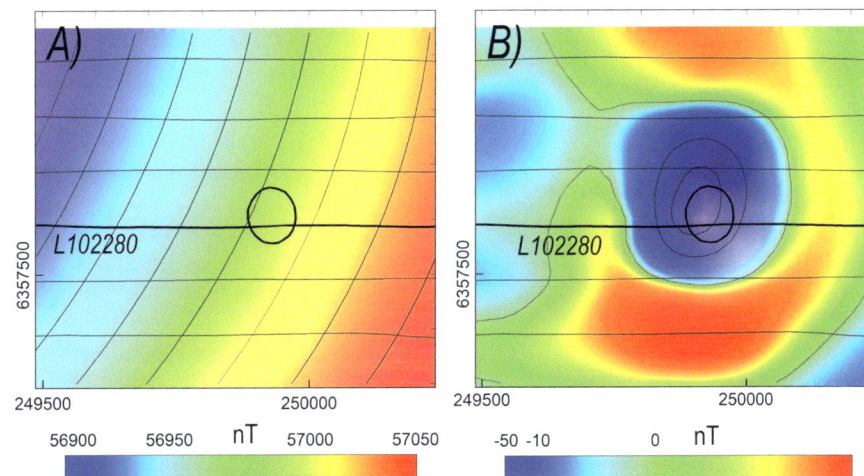

Fig. 2.16. A) regional and B) residual TMI separations for anomaly 'b'.

limit and that surveys designed to collect data for modelling or inversion should use a line spacing no wider than half the height above the top of magnetisation. These limits avoid substantial aliasing but it may well be of advantage to further increase sample density and provide additional resolution and fidelity of the mapped field and its gradients beyond minimum requirements. Consistent with Reid's analysis, mapping of short wavelength field variations near a source requires closer sample spacing and higher spatial precision than measurement of the longer-wavelength distant parts of the field.

Inversion optimises a model not to matching of the magnetic field but to matching the specific sample of the field that is submitted to it. Even with high-quality data, if those data do not sufficiently sample the field, do not extend sufficiently, or are at too high an elevation above the magnetisation there can be substantial challenges in mapping the magnetisation.

To illustrate linkage between data sampling and inversion results I present an example of gravity and magnetic surveys over the Waukarlycarly Graben and surrounding shallow Proterozoic basement of the Paterson region on the western edge of the Canning Basin in Western Australia as located in Fig. 2.17. The geology of the area is described by Alavi (2013) and the data inversions are from Foss and Purcell (2006). The gravity low over the area is interpreted as due to an accumulation of sediments, interpreted to have started with a rifting episode in the early Cambrian and to have been reactivated through to the Permian. A single seismic line traverses the main gravity anomaly with a reflection interpreted as acoustic basement at 1.7 to 1.8 s two-way time (2.3 to 2.5 km depth depending on the velocity estimates). There are several shallower

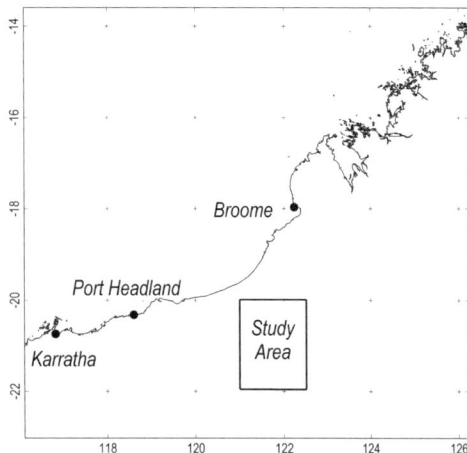

Fig. 2.17. Location of the Waukarlycarly study area.

sub-horizontal and fault-disrupted events but with no borehole control these are not confidently attributed to specific stratigraphic surfaces.

2.7.1 Gravity modelling and inversion

Figure 2.18A shows the gravity image of the area from previous sparse measurements, predominantly from stations at an 11 km spacing. The rules from Reid (1980) are not directly relevant for these data where the density contrasts extend almost to surface but the general principles of the need for sufficient data coverage still apply. A likely explanation of the observed elongate gravity low of up to 400 μm/sec^2 is variation in depth to the Proterozoic basement although the sparse data could also be matched by an intra-basement density low. Figure 2.18B shows a gravity image from a survey with a station spacing of 2.5 km, which is an increase in station density by a factor of almost 20. The sharp linear and parallel gradients of the gravity low defined by these higher-resolution data clearly suggest that the gravity low is due to a graben with sub-vertical faulted margins because these sharp gradients cannot be matched with a deep-going intra-basement density contrast. To enhance the gravity data I applied a two-dimensional FFT vertical derivative filter to the Bouguer grids to give the images shown in Fig. 2.19. Minimum curvature gridding of such sparse data introduces image artefacts in the form of 'pimples' at the location of each station. To supress these artefacts I preconditioned the data with a 500 m upward continuation before applying the gradient filter. This preconditioning has the advantage that it can be directly addressed in subsequent modelling and inversion by a corresponding increase in elevation of the computations. The vertical gradient filter accentuates expression of the graben that is the shallowest density contrast, but as noted by Reid (1980) a closer data spacing is required not just for gradient surveys but also to be able to effectively image gradients from transforms of field data. Gradient enhancement of the previous data (Fig. 2.19A) reveals the shape and extent of the gravity low more clearly than does the Bouguer gravity image but it is not clear that it is due to a fault-bounded graben. Gradient enhancement of the new data with the same 500 m upward continuation preconditioning shown in Fig. 2.19B is much more revealing, with substantial benefit from the increased data density. Enhancement of the higher resolution data provides convincing support for the gravity anomaly being due to a graben, with the sharp gravity variation over the fault margins tightly constrained by the data. Detailed inspection of the gravity and gravity gradient images of the

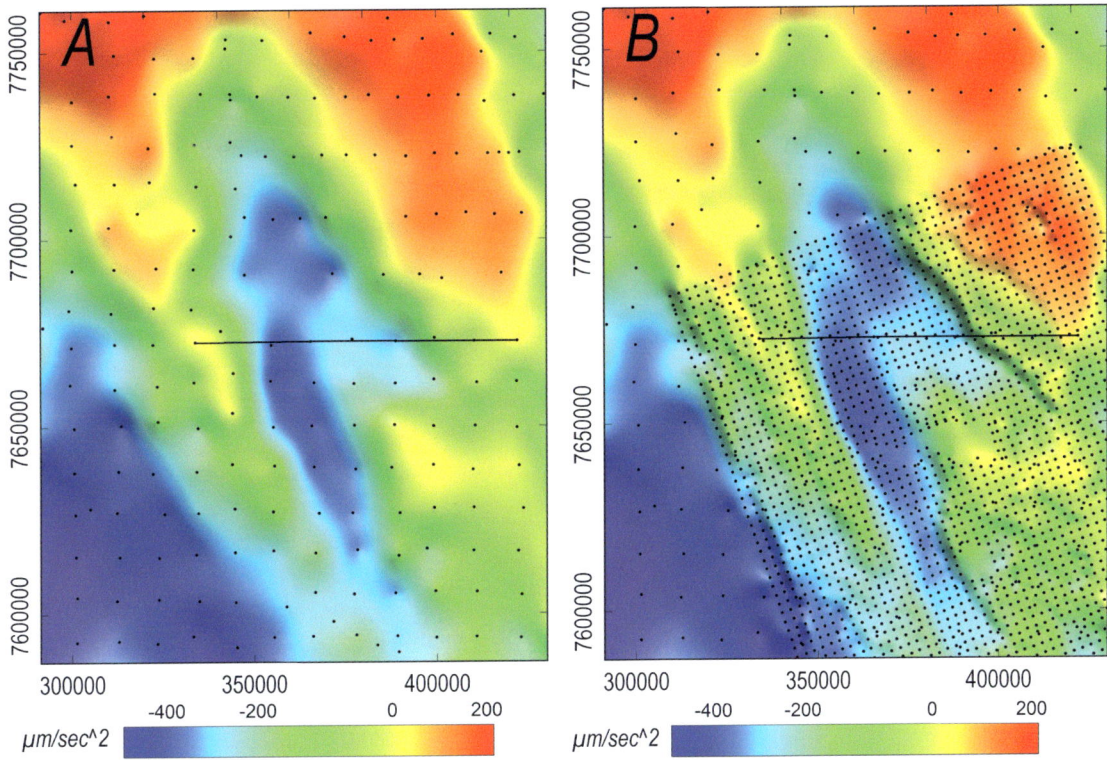

Fig. 2.18. Bouguer gravity images from gridding of A) previous sparse data and B) new closer-spaced data.

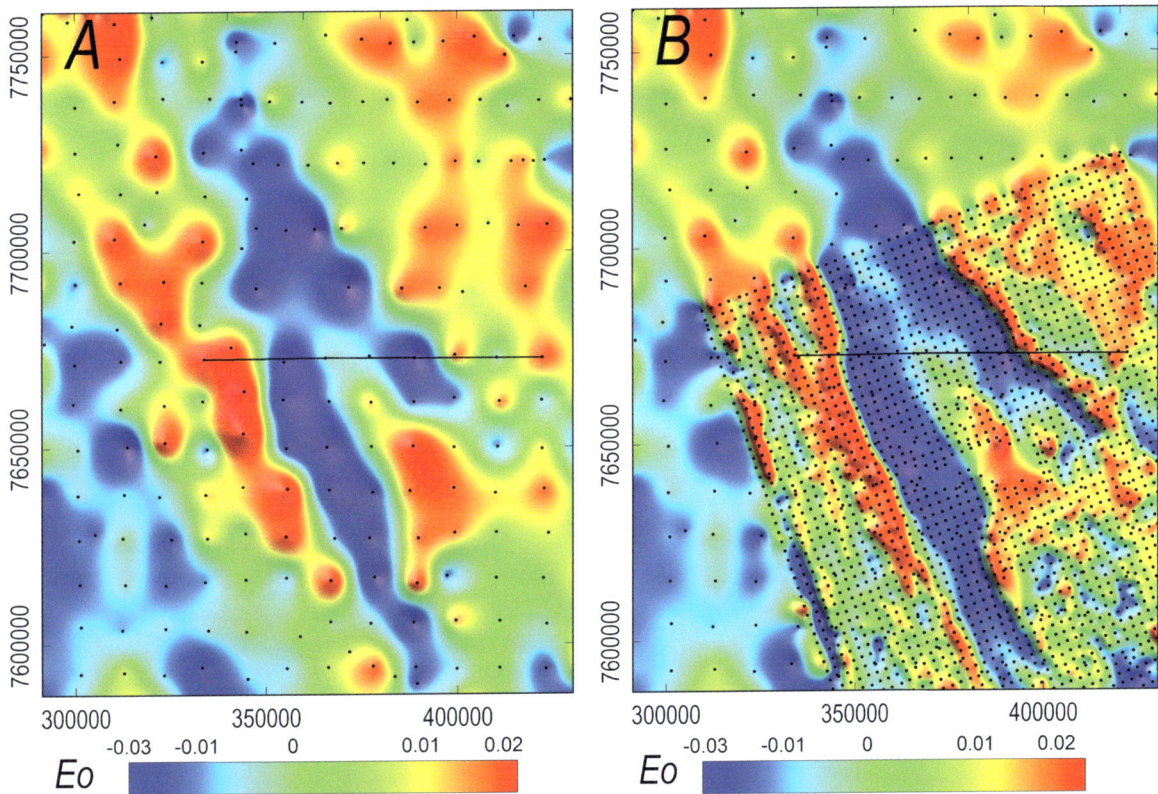

Fig. 2.19. Vertical derivative of Bouguer gravity from gridding of A) previous sparse data and B) new closer-spaced data.

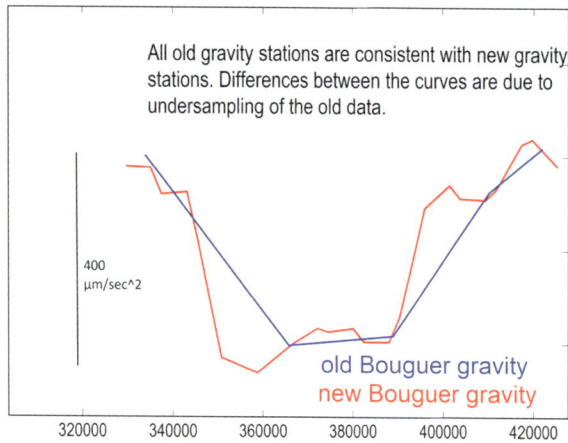

Fig. 2.20. Co-located profiles of previous (blue) and new (red) survey Bouguer gravity.

higher-resolution data in Figs 2.18B and 2.19B suggest that the graben is in fact two opposite and outward-facing half-grabens with an intervening central basement high (see also Alavi 2013).

A traverse across the basin is shown in Fig. 2.20. The previous data traverse contains only five stations and provides an insufficient mapping of the gravity field variation to confidently discriminate between different source models. The new data traverse contains 20 stations and localises the abrupt changes in gravity that are distinctive of the bounding sub-vertical faulted margins of the graben. The previous gravity data are of low quality by today's standards with pre-GPS positioning provided by feature recognition on aerial photography and elevation control by barometric altimeter. Nevertheless, these previous gravity stations are individually consistent with the new data. The weakness of the previous gravity coverage is the small number of stations and their wide spacing.

Once the initial interpretation has been made that the negative gravity anomaly is due to a graben, the gravity modelling and inversion depends substantially on the density contrast applied between the basement and graben fill. There are low amplitude circular and elliptic negative gravity anomalies over granites or granitoids in nearby shallow basement and using these as a best density estimate (because granites are composed mostly of minerals with similar densities to produce a reasonably predictable rock density) I selected a density of 2700 kg/m^3 to represent the basement. For the Palaeozoic graben sediments I selected a density of 2400 kg/m3 to give a density contrast of 300 kg/m3. The basin fill is expected to consist mostly of clastic sediments and even if these are uniform, their density will vary with depth due to compaction. Substantial density variations within the basin can also occur at major unconformities between different age sediments or if there are thick units of limestone, dolomite or salt (none of which are known to be present in this case). The underlying and surrounding basement is also inhomogeneous and may have density variations across large volumes. Despite this, there is little or no justification for a density model more sophisticated than a uniform contrast between basement and basin infill. Because steep density interfaces extend almost to surface it is not possible to match the measured gravity variation using a density contrast that is significantly smaller than the true contrast, but models can always be produced with too high a density contrast. The density contrast value was calibrated by matching the base depth of gravity models using different density contrast values to the acoustic basement depth from seismic and to the tops of magnetic model bodies assumed to represent magnetisations truncated at the basement unconformity. These tests support the initial density contrast estimate of 300 kg/m^3.

For convenience in constructing and manipulating the gravity model I decided to use grid traverses perpendicular to the trend of the gravity field rather than profiles of gravity stations. This option was feasible because the field variations of interest are well sampled by the 2.5 km station spaced survey and are reliably represented in the gravity grid derived from that data. I inverted the terrain-corrected Bouguer gravity data. The data already correct for small gravity variations due to terrain and allow use of a model with a horizontal top. I generated a set of 31 parallel traverses at a spacing of 6250 m and on those traverses defined a regional field that is slightly higher than the Bouguer gravity values beyond the basin edges. This regional field is high to the north through to low in the south, representing a broad field variation not necessarily related to the basin. This regional field variation could have been included as a wider and much deeper density contrast surface in the model but that would require additional work which would not have added value to the modelling objective. Once the regional gravity field is defined the task of the gravity model is to explain the ('residual') difference between the measured and regional fields. The difference is considered the anomalous field to be explained by the anomalous negative density contrast of the basin sediments against the basement of reference density. This approach conveniently avoids the need for a model body to represent the basement.

I constructed a starting inversion model using a polygonal-section horizontal prism with a strike length of 190 km to match the strike length of the basin and used this model to test the density contrast value. The horizontal top surface of the model is defined by only the two end vertices and the bottom surface is defined by vertices positioned beneath local gravity minima and maxima. After manual adjustment of this model quickly achieved an approximate match to the measured gravity data I reduced the strike length of the body to be the same as the data traverse spacing of 6250 m and duplicated it beneath each grid traverse to generate a continuous model of 31 adjacent strips, initially identical to the single body of the same cross-section and similar strike length. To invert this new composite body I first inverted the central strip of the model allowing only vertical adjustment of the bottom surface vertices to best fit the data on the central traverse immediately above it (all the model strips contributed to the forward field computations). Subsequently I similarly inverted the model strips to either side of the central strip to best fit only those traverses. These adjustments slightly reduced the data-fit previously achieved on the central traverse. However, because the density contrast is only shallow (a maximum depth of 3 to 4 km compared to a strip width of 6.25 km) gravity variation along each traverse can be matched with just a few iterations of individual or adjacent model strips or with one simultaneous inversion of all strips beneath the activated profiles. In this iterative fashion I worked out to both ends of the model, at which stage the combined fields of the complete set of model strips representing the basin matched the field on the complete set of traverses representing the anomaly.

The stepwise discontinuities between adjacent model strips occur halfway between the traverses where they are least significant to the measurements that are all computed on the central axes of the strips. Figure 2.21A shows an example model strip and the field computed from the complete basin model on the traverse above it. This model section clearly shows the two half-grabens with steep external basin margin faults and more gentle basement slopes to a central intra-basement high. Figure 2.21B shows the segmented model that reproduces the complete basin anomaly. The green-coloured segments in the south are controlled by the new higher-resolution data and the yellow segments to the north by the older, lower-resolution data. This model should not be envisaged as a final product but as an interim model best justified by currently available data and suitable for

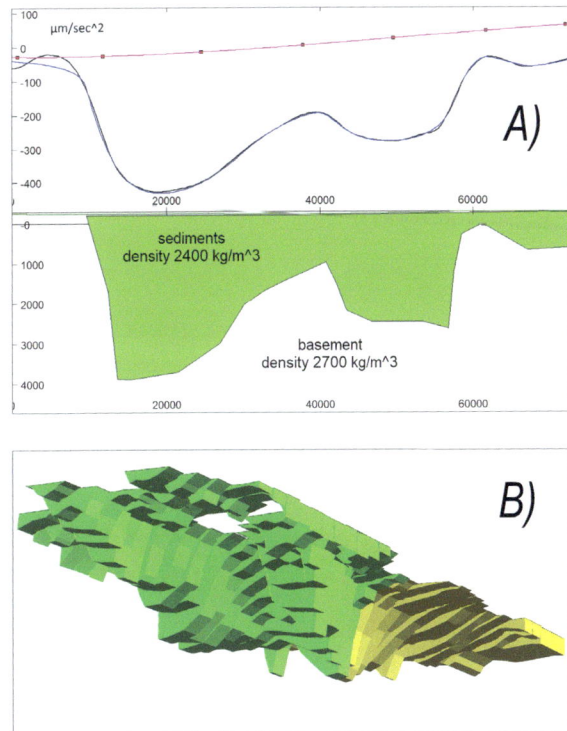

Fig. 2.21. A) example model section slice and B) perspective view of the multi-slice model. The green section is developed from the new high-resolution data and the yellow section from the older lower-resolution data.

designing next-stage exploration of the basin. When any new data becomes available this model can be updated as required. Any additional information from closer-spaced gravity or airborne gravity gradient (AGG) data would mostly be significant in the narrow regions of abrupt gravity variation over the bounding faults.

2.7.2 Magnetic field modelling and inversion

The previous magnetic field coverage of the area was from surveys flown on east–west flightlines at 1,600 m spacing and ground clearances of between 60 and 100 m. The reduced to pole (RTP) field computed from this data is shown in Fig. 2.22A. Flightlines are not shown because at this scale they would obscure the image. Figure 2.22B shows an equivalent image from more recent magnetic field surveys with line spacings between 150 and 400 m and ground clearance of 60 to 80 m. The major features evident in the recent survey image of Fig. 2.22B are also evident in the older survey image of Fig. 2.22A. This is in part because many features are elongate perpendicular to the flightline direction and even in the wider line-spaced survey they have multiple flightline intersections.

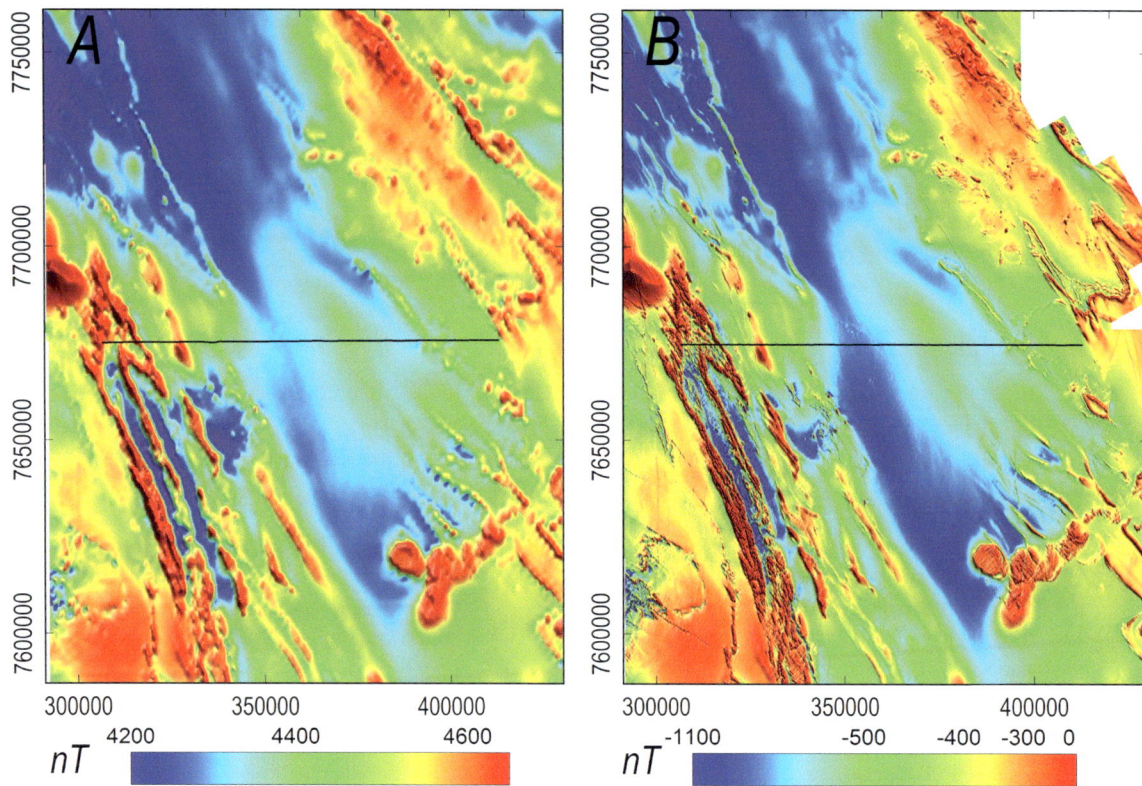

Fig. 2.22. TMI from A) previous 1,600 m line-spaced data and B) new 400 m line-spaced data.

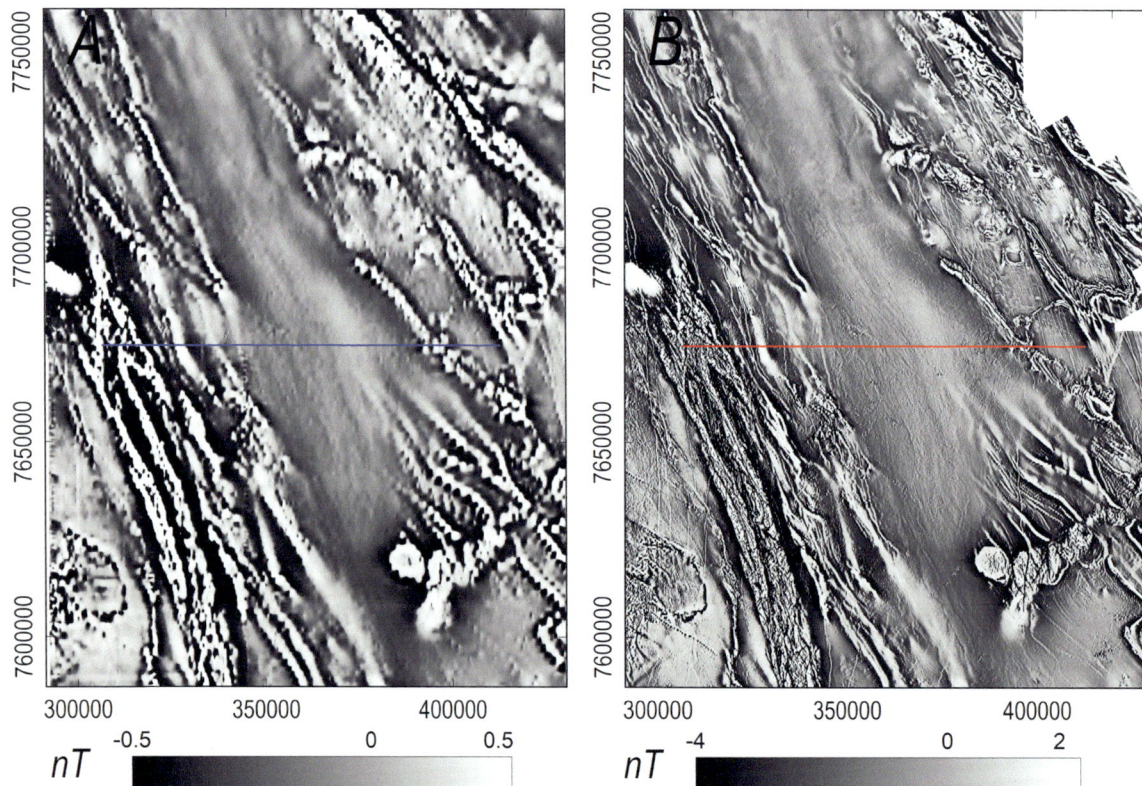

Fig. 2.23. The vertical derivative of TMI from A) previous 1,600 m line-spaced data and B) new 400 m line-spaced data.

However, all features are more sharply resolved by the closer line-spaced survey. Figure 2.23 shows corresponding vertical derivatives of the RTP. As for the gravity field, the gradient enhancement of the magnetic field benefits more from the closer line spacing than the field data do. I did not need to pre-condition this magnetic field data with an upward continuation before computing the vertical derivative. However, enhancement of the older, more widely spaced line data has significant artefacts in the form of 'bicycle-chain' or 'string-of-pearls' patterns where the minimum curvature griding used cannot interpolate sharp gradients from one profile intersection to the next. This is consistent with the case made by Reid (1980) that a closer line-spaced survey is required if gradients are to be derived from the measured field data. The gridding artefacts are not evident in the vertical derivative of the closer line-spaced survey data in Fig. 2.23B and in consequence geological inferences about the source magnetisations can be made more reliably from this data (see for instance Alavi 2013).

Figure 2.24 shows a comparison of two almost coincident flightlines from the two surveys. Differences between the flightlines are inconsistent, in part possibly because the individual lines have local vertical and horizontal departures. Improvements in magnetic field measurement and processing have been incremental but cumulative between the acquisition of these two surveys but more significantly there has been a major improvement in navigation and positioning, which went through a revolutionary change in the early to mid-1990s with the advent of GPS and differential GPS positioning that are critical to support the close line spacing of high-resolution aeromagnetic surveys. As with the gravity

data, the differences between the two vintages of magnetic survey are due much more to the difference in data density than in data quality.

Gravity and magnetic field characteristics are consistent with the same potential field theory, with reduced to pole (RTP) magnetic fields due to a transformed vertically directed magnetisation in a transformed vertical magnetic field equivalent to the vertical derivative of the gravity field (Blakely 1995). However, the vertical gradient of gravity images in Fig. 2.19 recognisable as enhancements of the gravity field images in Fig. 2.18 are very different to the RTP magnetic images in Fig. 2.22. This difference arises predominantly from differences in distribution of contrasts in the causative properties of density and magnetisation. As discussed above, the gravity field variation is primarily determined by the bulk difference in property of the complete basin against the complete basement and this determines the approach to the gravity inversion. That approach is not feasible and would not be productive for inversion of the magnetic field data. Much of the basement has magnetisation similarly low or not much greater to the magnetisation of the basin sediments, and across the basement unconformity surface there are only small areas of useful magnetisation contrast. Instead, the magnetic field variation is dominated by fields of strong magnetisations of minor, shallow parts of the basement. The consequence of this is that a combined inversion of gravity and magnetic fields at this scale is unlikely to be productive. Instead, the undoubted benefits to be gained from investigating the two fields due to different properties is best exploited by ensuring compatibility between the output models of the two independent inversions, or else by introducing the results of one inversion as constraints in the other. The magnetic field expression of the basin is most obvious as the absence of sharp anomalies due to local, strong magnetisations present in the surrounding areas of shallow basement. This absence of anomalies is not readily exploited in an inversion. There is a weak magnetic field expression of the two basin deeps evident in the gravity images and resolved in the gravity inversion. In the magnetic field image these appear as subtle magnetic lows separated by a weak magnetic high over the central basement and gravity high. This demonstrates a weak contrast in magnetisation between the bulk basement and the basin sediments, but the magnetic field variation (evident in the images because of histogram equalisation) is of such low amplitude that a combined gravity and magnetic

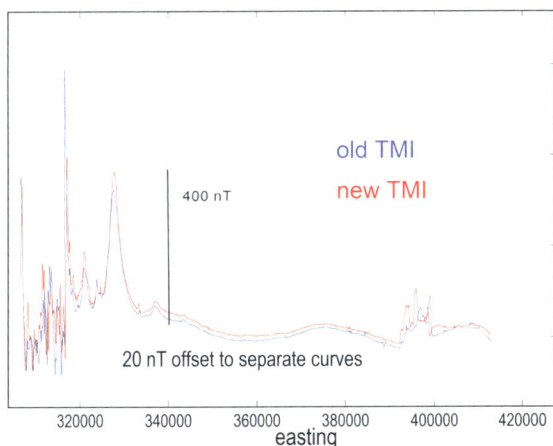

Fig. 2.24. Co-located profiles of previous and new survey TMI data.

inversion would be unlikely to be of advantage. Instead, I inverted the magnetic field data just as in the previous magnetic field studies using the sweet-spot approach of focussing individual inversions on small individual packets of information in the magnetic field data.

Figure 2.25 shows an example sweet-spot magnetic field inversion of two source bodies. Their source depths can be estimated in the following steps:

1) draw a grid traverse through the anomalies
2) extrapolate magnetic field and ground surface grid data onto the traverse and add the nominal flying height of 60 m to the ground elevation to represent the sensor height
3) choose an appropriate source model (for these bodies which I assume to be truncated at an unconformity I selected tabular prisms with horizontal top and bottom surfaces)
4) place the bodies under each anomaly and in map view adjust their strike length and azimuth to approximately match the anomalies
5) assign a background or regional field to which the body fields will be added to match the anomalies

6) adjust the depth and magnetic susceptibility of the bodies to produce an approximate starting model for the inversion
7) select appropriate free parameters for the inversion – in this case depth, depth extent, width, dip and magnetic susceptibility of each body. Note that their azimuth and strike extent should not be changed as there is almost no sensitivity to these parameters on the single data traverse
8) if the inversion converges acceptably it provides estimates of each body parameter.

In areas of the shallowest sources I modified the above procedure to use measured flightline data and avoid concerns of the magnetic field gradients being modified in the process of gridding and subsequent interpolation. This sweet-spot method utilises only the most appropriate data for source depth estimation. The location where depth estimates are derived and the spatial density of the estimates is controlled by the magnetic field data. In some areas there are many suitable anomalies and a selection of those anomalies was made for analysis, in other areas the analysis had to be applied to

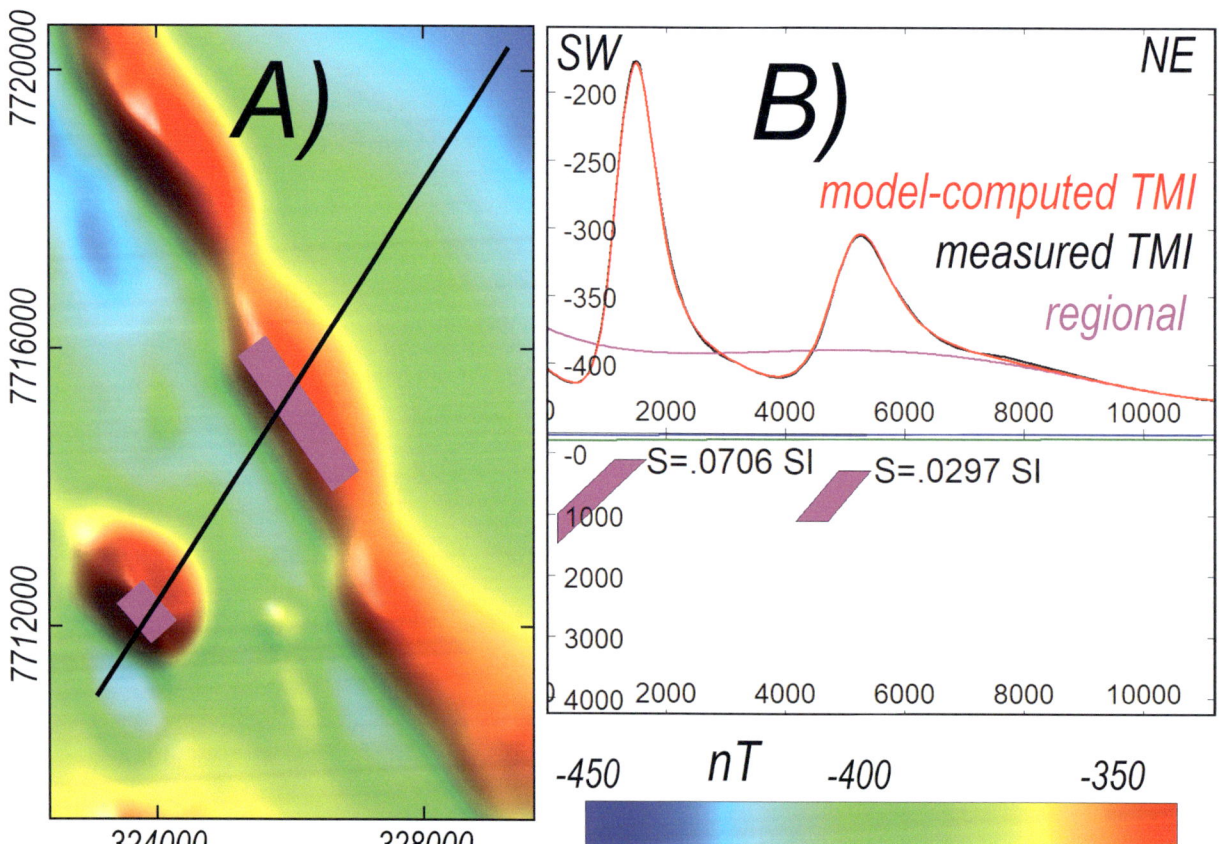

Fig. 2.25. Magnetic depth traverse: Top) cross-section view and Bottom) plan of the traverse over an image of TMI.

less suitable anomalies and in some places there were no suitable anomalies from which meaningful depth estimates could be derived. There are many suitable anomalies over shallow basement where small magnetised bodies can be detected. Where basement is more deeply buried there are fewer suitable anomalies.

Full three-dimensional inversion of the magnetic bodies can give a more reliable representation of the magnetisation but provided the strike and azimuth of the model bodies are appropriate to match the anomalies, a superior estimate of depth to the top of magnetisation is derived by closely fitting the field gradients on well-selected individual profiles. The depth estimates are to the top of magnetisation, but these magnetisations are not stratigraphically attributed. To investigate the Palaeozoic Waukarlycarly Graben all magnetisations truncated at the base of the basin fill can be classified as Proterozoic basement, but for mineral exploration within the Palaeozoic basement it is of advantage to discriminate between magnetisations sourced in separate units.

The spot depth estimates are the most reliable depth information recovered from the magnetic field analysis. In many cases a depth surface is required (in this case the elevation of Proterozoic basement) and these surfaces must be interpolated between the relevant depth points. Geological surfaces can be complex and include abrupt discontinuities so any interpolation is necessarily speculative and interpretive. Any interpreted depth surface should always be accompanied with the depth points from which it has been constructed.

Figure 2.26A shows the ensemble of magnetic source depth models. Normally a depth surface would be constructed by gridding of the top surface depth points or by drawing the surface in a three-dimensional GIS environment (including any boreholes or depth-converted seismic sections) but in this case bodies were imported to provide guidance in the gravity modelling and inversion where the base of the basin infill gravity model is expected to coincide with tops of the intra-basement magnetic models. This correspondence was very strong as shown

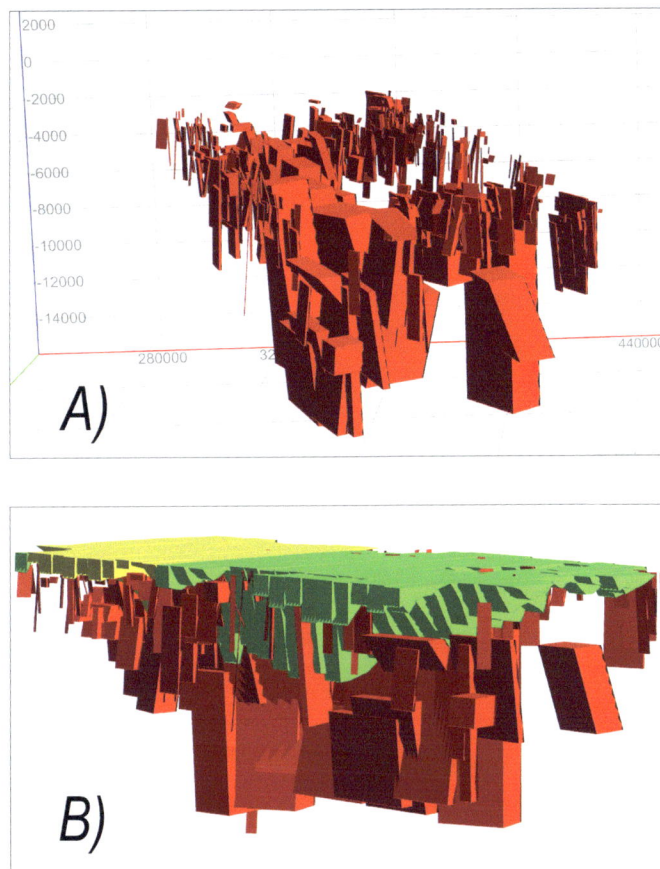

Fig. 2.26. A) Ensemble of magnetic depth models and B) combination of the gravity basin model (yellow and green) and intra-basement magnetic models (red).

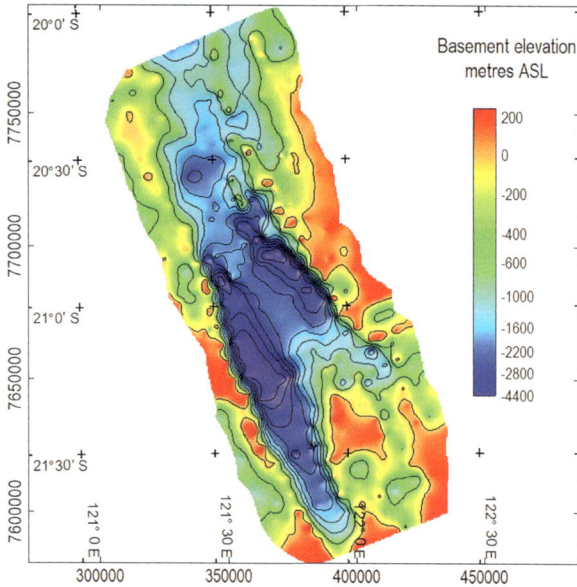

Fig. 2.27. Top of basement elevation map.

by the view of the combined gravity and magnetic inversion models in Fig. 2.26B. The final Palaeozoic basement surface imaged in Fig. 2.27 was derived from gridding of the vertices defining the base of the gravity model but a near-identical surface would be derived from generating a surface through the tops of the magnetic models.

The major conclusion I draw from the Waukarlycarly case study is the need for careful and detailed measurement of the fields and understanding of the relationships between geology, physical properties, and the gravity and magnetic fields. Information can then be efficiently recovered with focus on suitable features in the fields.

2.8 THE INFLUENCE OF GRIDDING ON INVERSION

Gridding is often considered a passive repositioning of data from measurement points to nodes of a regular mesh. However, grid data is a new data type significantly different to the primary data and this has significant consequences. Grid data are convenient to use but unquestioned acceptance of grids as valid representation of a gravity or magnetic field is a common and serious problem in gravity and magnetic studies. The influence of gridding is least significant where measurements have been acquired on a regular mesh which requires little or no adjustment of the primary data values to minimally displaced grid cell positions. Gridding is most significant where data have variable spacing or (as for aeromagnetic data) with different spacings in along-line

and cross-line directions. Over the last few decades many magnetic field grids have been generated with algorithms based on the minimum curvature method of Briggs (1974) which assigns grid characteristics broadly consistent with expected potential field variations. A grid cell size of one-quarter or one-fifth of the line spacing is commonly chosen as a compromise between retaining as much detail of the line data as possible and restricting distortion in interpolation between lines. One of the most distinctive gridding artefacts of profile data is known as a 'string-of-pearls' or 'bicycle-chain' pattern where a linear sharp gradient in the field breaks down to a series of individual features at each profile intersection. This artefact is visible in the south-eastern segment of the Waukarlycarly wide line-spaced magnetic field data in Fig. 2.22A and because the vertical gradient has sharper curvature than the field itself the artefact is more strongly emphasised in the corresponding vertical gradient image in Fig. 2.23A. In this study I investigate limitations in inversion of grid data with these artefacts to recover estimates of depth to the top of magnetisation. The artefacts represent only a small proportion of more widespread and cryptic problems with grid data. Recently new and more sophisticated methods of anisotropic gridding have been published (Naprstek and Smith 2019; Davis 2022). However, no gridding methodology completely compensates for insufficiencies in distribution of the primary data.

Figure 2.28 shows a section of the south-east-north-west trending swathe of prominent Gairdner Dyke magnetic field anomalies near Mount Vivian in South Australia (Foss *et al.* 2019; Pawley *et al.* 2021). The survey was flown on north–south flightlines at a spacing of 200 m and nominal terrain clearance of 60 m as part of the Gawler Craton Airborne Survey (GCAS) project (Katona *et al.* 2021) and the grid was well produced within the capabilities of the minimum-curvature method used. The dyke anomalies have subtle expression of 'string-of-pearls' or 'bicycle-chain' artefacts at each profile inter-section, mostly evident in sun-shading of the image. Figure 2.29 shows a contour map of the single anomaly inset in Fig. 2.28. For this anomaly tight contour closures mark each profile intersection with the anomaly. The contour interval is 10 nT and the amplitudes of the artefacts are up to 35 nT on an anomaly of peak to trough amplitude 240 nT. Figure 2.30 shows profile modelling of a north–south flightline. The anomaly was inverted with an assumption that magnetisation is due only to induced magnetisation. An identical inversion was

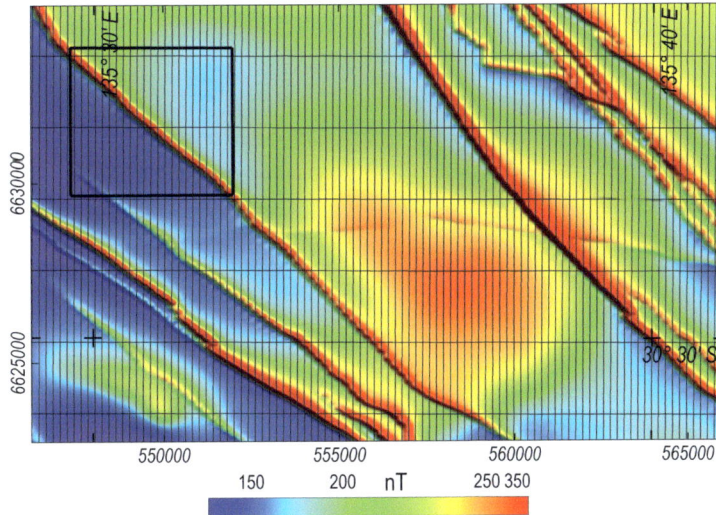

Fig. 2.28. TMI image of a section of the Gairdner Dyke Swarm, South Australia with inset of the single-dyke study area.

performed of an east–west tie-line that intersects the anomaly at the same location (see Fig. 2.29). Inversion of these independent lines of data produces two consistent models with key statistics listed in Table 2.1. The difference in depth to their tops is only 5 m in an average depth below sensor of 108 m. For this study the mean of these estimates is accepted as the best available depth for the top of the dyke. There is a difference of less than 4° between the apparent dip estimates of the dyke models derived on the two orthogonal lines.

Figure 2.31 shows a contour plot from an alternative gridding of the data using the anisotropic method developed by Naprstek and Smith (2019) and implemented in the Oasis Montaj software package (grid provided by

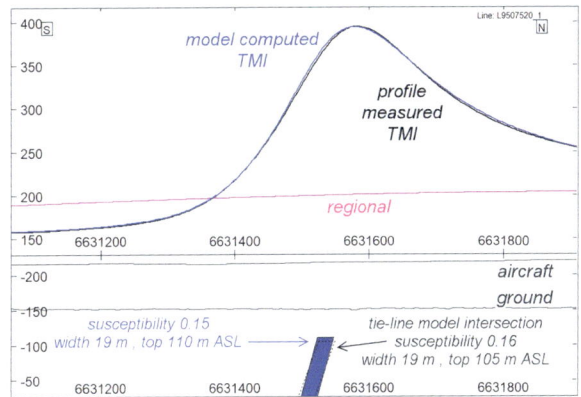

Fig. 2.30. A model section on north–south flightline L9507520 (for location see Fig. 2.29).

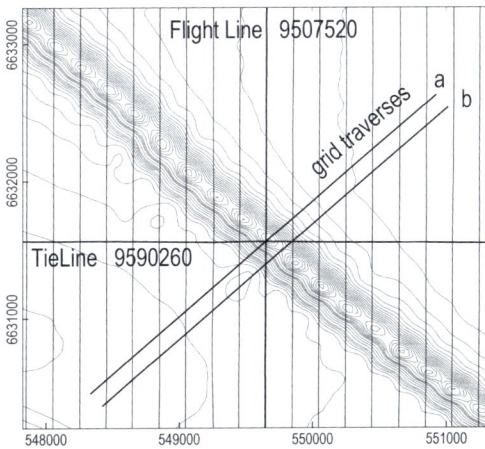

Fig. 2.29. Segment of a Gairdner dyke magnetic anomaly defined by a minimum-curvature TMI grid. The contour interval is 10 nT. The 'string-of-pearls' artefacts have amplitudes of c. 35nT and the anomaly amplitude is c. 240 nT.

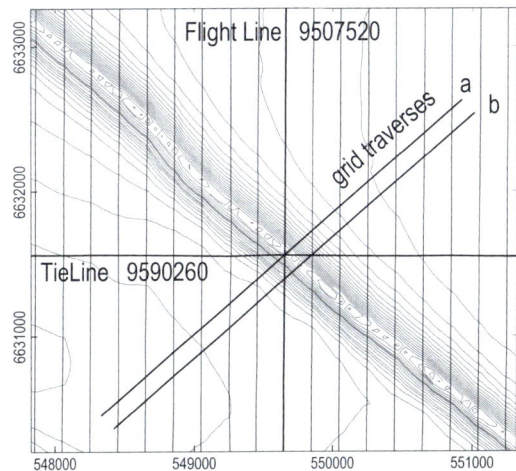

Fig. 2.31. Flightlines and grid traverses over anisotropic gridded TMI contour map. Contour interval 10 nT and anomaly amplitude c. 240 nT.

Table 2.1. Dyke inversion model statistics.

Grid	Line	Susc SI	Elev m ASL	Width m	Dip	Difference in depth below sensor	Difference in depth below surface
–	9507520	0.145	–110	19	75°		
–	9590260	0.157	–105	19	78°		
Min curv	9507520	0.145	–111	20	75°	3% (shallower)	7%
Min curv	west 100	0.0112	–154	227	81°	42% (shallower)	100%+
Min curv	a	0.016	–130	180	75°	20% (shallower)	51%
Min curv	b	0.013	–133	197	83°	23% (shallower)	58%
Anisotropic	9507520	0.137	–108	22	72°	0	1%
Anisotropic	west 100	0.131	–113	23	73°	5% (deeper)	12%
Anisotropic	a	0.146	–102	21	74°	5% (deeper)	14%
Anisotropic	b	0.137	–107	22	75°	1% (deeper)	2%

Cericia Martinez). The individual contour closures at profile intersections prominent in the minimum curvature gridding of Fig. 2.29 are removed and it is not feasible to locate line intersections from the gridded data. To investigate the consequence of inverting grid data I interpolated data channels from both the minimum curvature and anisotropic grids onto flightline L9507520 and inverted those channels. The data are modified not just in the process of creating the grids but also in the interpolation used to resample them onto the profile. The minimum curvature grid cell size is 40 m, the anisotropic gridding cell size is 50 m, and the average along line sample spacing is 3.5 m. Single profile inversions of both grid-interpolated datasets provide acceptable models, with depths consistent with the flightline model depths to within 5%. There is a slight northward shift of the interpolated anisotropic grid data results in a corresponding 25 m northward shift of the inversion model (one-half of the grid cell width).

Depth below sensor is an appropriate statistic to evaluate sensitivity of depth estimation but for geological exploration programs a more relevant measure is depth below the ground surface. The relationship between the percentage errors on these two depths depends on the ratio of depth to magnetisation below the ground and flying height above it, with proportional errors and uncertainties rising steeply for shallow magnetisations.

Inversions of the L9507520 flightline grid interpolations only test the consequences of inverting interpolated grid data in immediate proximity to the measurements. I also inverted data onto a grid traverse 100 m to the west, halfway to the adjacent flightline and

intersecting the grid anomaly between the pearls of the minimum curvature gridding. Continuity of the anomaly on multiple flightlines indicates that interpolations along traverses between the flightlines should be the same as interpolations of traverses coincident with the flightlines. However, interpolation from the minimum curvature grid onto the intermediate traverse is quite different to the primary profile data on the flanking flightlines. The consequence of inverting the minimum curvature grid traverse away from the flightline is a significant increase in apparent thickness, decrease in apparent susceptibility by an order of magnitude and a decrease in depth to top of over 40% in depth below sensor and 100%+ in depth below the ground surface (see Table 2.1). This depth estimate would be useless and even misleading in an exploration program. In marked contrast, interpolation of the anisotropic grid on this traverse produces an anomaly similar to the directly measured profiles on the adjacent anomalies and its inversion recovers a model very similar to the profile data inversions with increases in depth of only 5% and 12% below sensor and below ground respectively.

A major advantage of using grid data for inversion is the freedom in position and orientation of traverses, which for elongate anomalies such as this Gairdner dyke anomaly would generally be selected perpendicular to the feature. Figure 2.32 shows two grid traverses perpendicular to the dyke (traverses 'a' and 'b' in Fig. 2.29); the first passing through a dyke and flightline intersection (through a minimum curvature gridding pearl) and the second midway between two adjacent pearls. These traverses do not have primary profile data to compare with the grid interpolations but we can confidently emulate

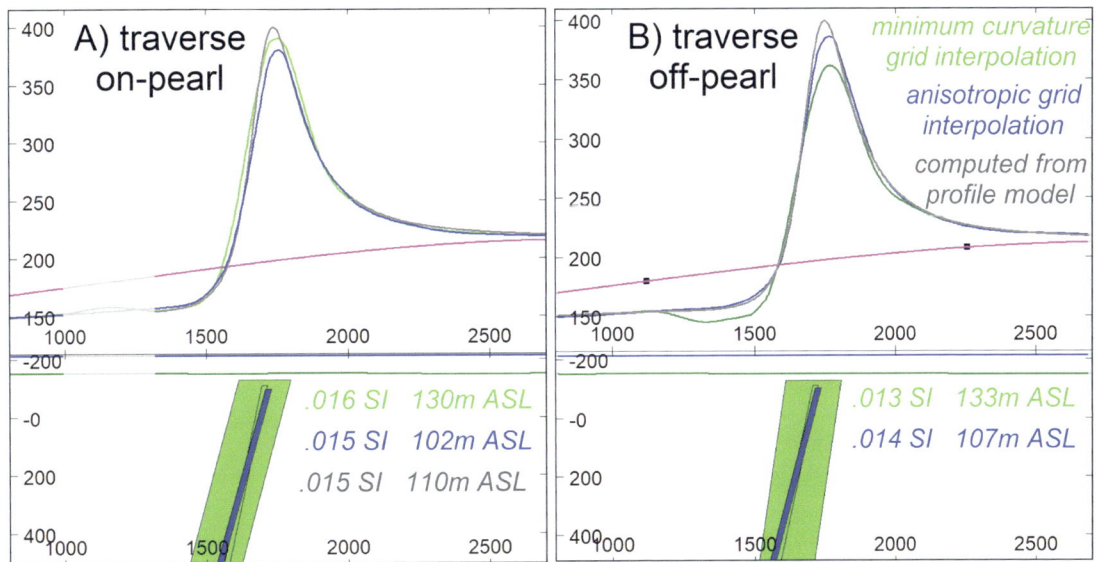

Fig. 2.32. Line 9507520 minimum curvature (mc) and anisotropic (an) grid interpolations and inversion model bodies.

that data using the already derived inversion model which closely matches the primary flightline data. On the interpolated curves along both traverses plotted in Fig. 2.32 the broader curvature of the minimum curvature grid data is obvious. The amplitude of the minimum curvature interpolation through the pearl (grid traverse 'a') is higher than on the grid traverse between the pearls (grid traverse 'b') but the erroneous curvature is similar on both traverses. Because of the erroneous curvature of these data the dyke models derived from its inversion are also in error. The models are substantially wider and with lower apparent susceptibilities than the profile data inversion models and are shallower by over 20% of depth below sensor and over 50% of depth below ground (see Table 2.1).

The data interpolated onto the traverses from the anisotropic grids have curvature more consistent with the field forward computed along the traverses from the profile data inversion model. Inversions of the anisotropic grid data also produces models more consistent with the profile data inversion model (see Table 2.1). The models from inversions of both lines have widths and susceptibilities similar to the profile data inversion models and increases in depth of 5% below sensor and just over 10% below ground. These imperfections in inverting grid data increase for shallower magnetisations with sharper gradients and for surveys of wider line spacings (inversion model imperfections are also greater for anomalies over bodies with short strike length that give rise to sharp field variations in all horizontal directions).

This Gairdner Dolerite dyke study clearly shows that for depth to top of magnetisation one-half of the line spacing and with 'string of pearls' artefacts weakly evident, minimum curvature gridded data misrepresents the magnetic field to cause over 20% difference in estimates of depth to the top of magnetisation compared to estimates made directly from the profile data. Anisotropic gridding reduced those differences to 5%. Further systematic studies are required to map these inadequacies in grid data as a function of line spacing, depth to magnetisation and magnetisation body geometry. For a single survey dataset there are only minor overheads in using profile data for inversion and this should be the preferred option. If depths to top of magnetisation are less than one-half of the line spacing, use of grid data becomes less acceptable and gives increasingly misleading results. Unfortunately, this problem is unlikely to be detected if only grid data are inverted because inversions still match that data very closely. Use of micro-levelled grids is a further concern because micro-levelling also distorts representation of the magnetic field with little evidence from which that distortion can be recognised or corrected. Unfortunately, some clients of aeromagnetic surveys focus only on the grid deliverables from the contractor and the line data may be unavailable for depth estimation and inversion. The grids may also become separated from the processing report and metadata which records the operations and settings applied to generate them. It is safer to invert profile data unless there is a pressing reason otherwise.

2.9 THE USE OF VERTICAL GRADIENT ENHANCEMENTS IN INVERSION

The previous study of the Gairdner dykes dealt with issues related to horizontal gradients of the magnetic field for a case that they vary in a single horizontal direction. Cross-line horizontal gradients can be measured with advantage in aeromagnetic surveys with wing-tip mounted sensors and along-line horizontal gradients can be derived from the closely spaced sequential measurements. The vertical gradient of TMI has been measured on aeromagnetic surveys using vertically displaced sensors on the tail fin of the aircraft (Hood and Teskey 1989) but this application faces challenges of low signal to noise ratios. Fortunately, the vertical and total horizontal gradients form a Hilbert pair (Blakely 1995) and the vertical gradient can be derived by a two-dimensional FFT of horizontal gradients estimated from closely spaced TMI measurements in the horizontal plane. The two-dimensional FFT filter to derive a map of the vertical gradient of TMI from a horizontal survey of TMI data requires construction of grids with the corresponding limitations discussed in the previous section, and with the increased resolution requirements of gradient data as discussed by Reid (1980). Gradient measurement of the three orthogonal field components to give the second-rank gradient tensor (Schmidt *et al.* 2004; Chwala *et al.* 2012) has considerable advantage in detailing small and shallow magnetisations but the increased cost of acquisition of these data has to present restricted its use.

Figure 2.33A shows an image of a minimum-curvature TMI grid of the Elkedra Area in the southeast Northern Territory derived from a survey flown on north–south flightlines at 400 m line spacing and a nominal 60 m terrain clearance. The TMI variation is dominated by a series of curved, elongate anomalies of 100 to 200 nT amplitude. The near-surface geology is shallow marine to fluvial sediments and interbedded felsic and mafic volcanics of the Proterozoic Davenport Province Hatches Creek Group (Blake and Horsfall 1986). The prominent anomaly imaged in Fig. 2.33A is sourced in a recessive volcanic unit near the base of the Alinjabon Sandstone in the Errolola Syncline (Fig. 2.33B). Figure 2.34A shows the minimum curvature

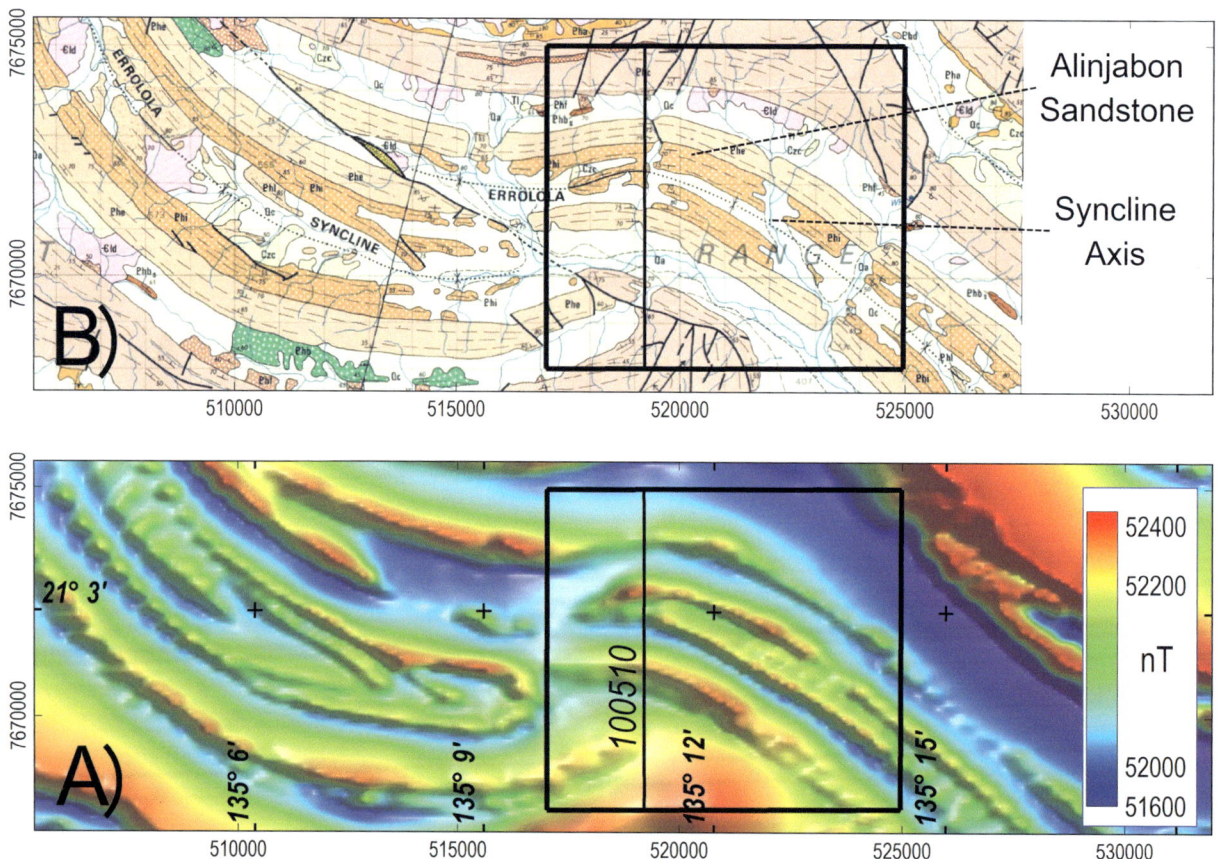

Fig. 2.33. A) TMI and B) geology (Blake and Horsfall 1986) over the Elkedra area in the Northern Territory.

grid in the inset area of Fig. 2.33. The bicycle-chain or string-of-pearls artefacts in this image are much more pronounced than those over the Gairdner dykes in the previous section. The magnetisations in the two areas are of similar depth, width and elongation but the line spacing in the Elkedra area is twice that of the Gairdner area. Re-gridding of the data with the anisotropic method of Davis (2022) created the grid imaged in Fig. 2.34B (grid supplied by Aaron Davis) which significantly reduces these artefacts and offers identical advantages

for inversion as established in the previous study by the anisotropic gridding of Naprstek and Smith (2019). Gradient filters preferentially enhance the shortest wavelength features in data, which in the case of the minimum-curvature grid are gridding artefacts. The advantage of anisotropic gridding shown by comparison of the TMI grids in Fig. 2.34 is amplified in mapping the vertical derivative as shown in Fig. 2.35. Despite this significant improvement we will use the line data in investigation of these shallow magnetisations,

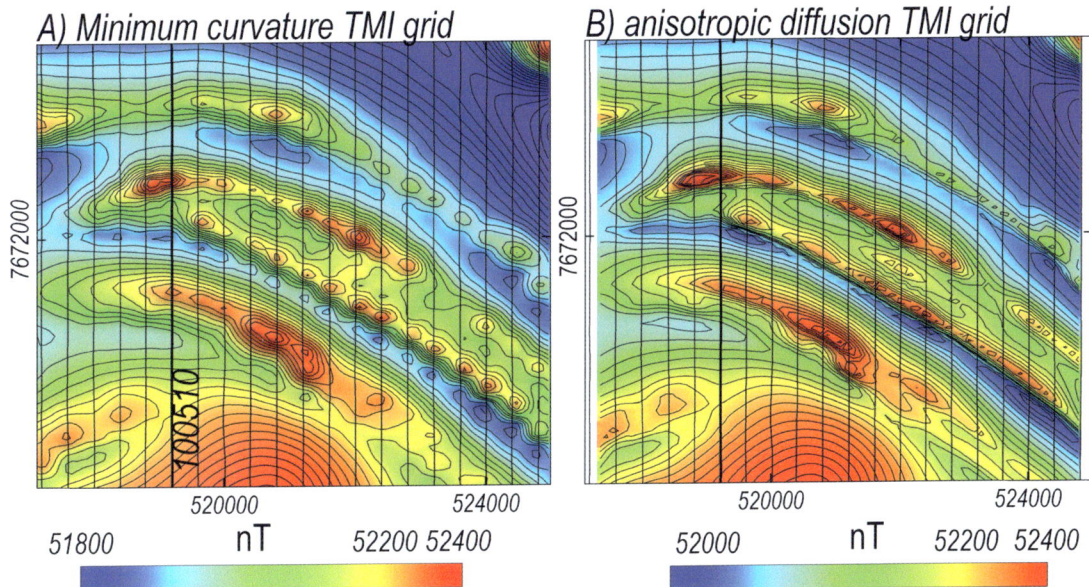

Fig. 2.34. Vertical derivative of TMI from A) minimum curvature gridding and B) anisotropic gridding (Davis 2022) for the inset area in Fig. 2.33.

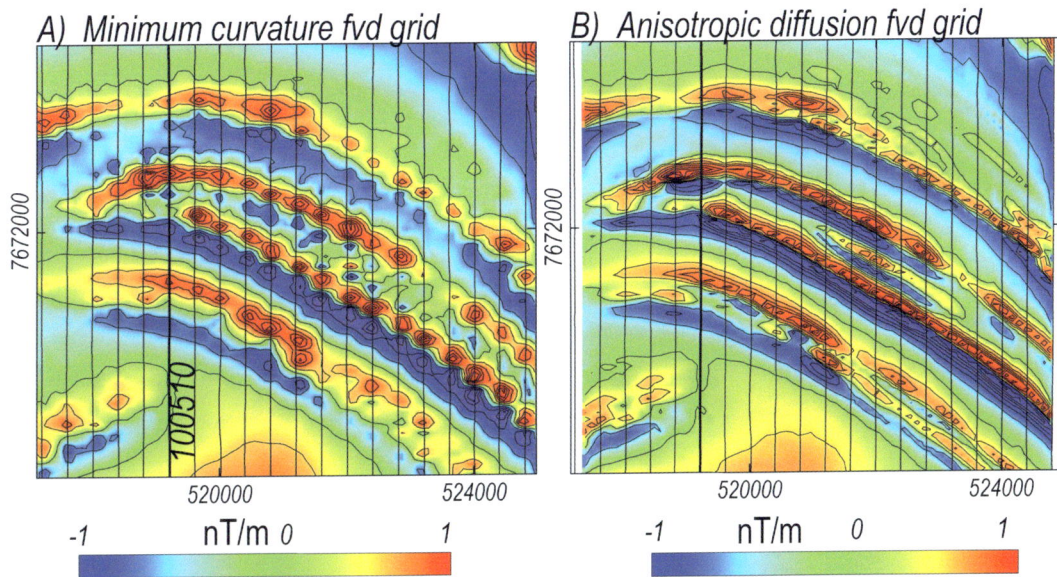

Fig. 2.35. Vertical derivative of TMI from A) minimum curvature gridding and B) anisotropic gridding (Davis 2022) for the inset area in Fig. 2.33.

particularly in consideration of applying gradient enhancements.

Figure 2.36 shows a model section along the north–south flightline 100510 located in Figs 2.33 to 2.35. The bottom section shows model bodies semi-automatically generated by the ModelVision implementation of AutoMag, a magnetic source depth estimator based on the method of Naudy (1971) and further developed by Shi (1991) and Shi and Boyd (1993) to analyse field gradients. The Naudy method nominates a target source body – in this case a dyke of specified width and top depth ('dyke' is used as a geometric term for a thin sheet). The TMI expression of the target body in a nominated window width is split into symmetric and asymmetric parts, moved along the measured profile and cross-correlated with a similar splitting of the measured field (in this case the splitting is applied to vertical derivative filters of those fields). Where the correlation coefficient triggers a solution a more advanced analysis is performed, followed by a simple table-lookup inversion. In this method the solutions contain all the information required to convert them to model bodies that can be tested against the primary data by forward modelling.

To match the measured and model-computed TMI shown in the centre track of Fig. 2.36 clearly requires addition of a long wavelength background field gradient. The vertical gradient filters of measured and computed TMI shown in the top track of Fig. 2.36 match much more closely than the equivalent field curves in the TMI track below because the gradients are only weakly sensitive to the regional field and because analysis was tuned to the gradient filter. The vertical derivative of TMI has the advantage that it is more sensitive to the shallowest magnetisation, that it has only low sensitivity to the regional gradient, and that the anomalies are compressed which reduces overlap of adjacent body fields. In using the profile data to retain the advantage of its close sample spacing and avoid griding artefacts we do not have access to the cross-line gradients. The initial Automag analysis is performed on a two-dimensional assumption that source magnetisations are perpendicular to the line and extend large distances to either side, in which case there is no horizontal cross-line gradient. With only the along-line component of the horizontal gradient available the vertical gradient filter may not provide a valid estimate of the true vertical gradient of

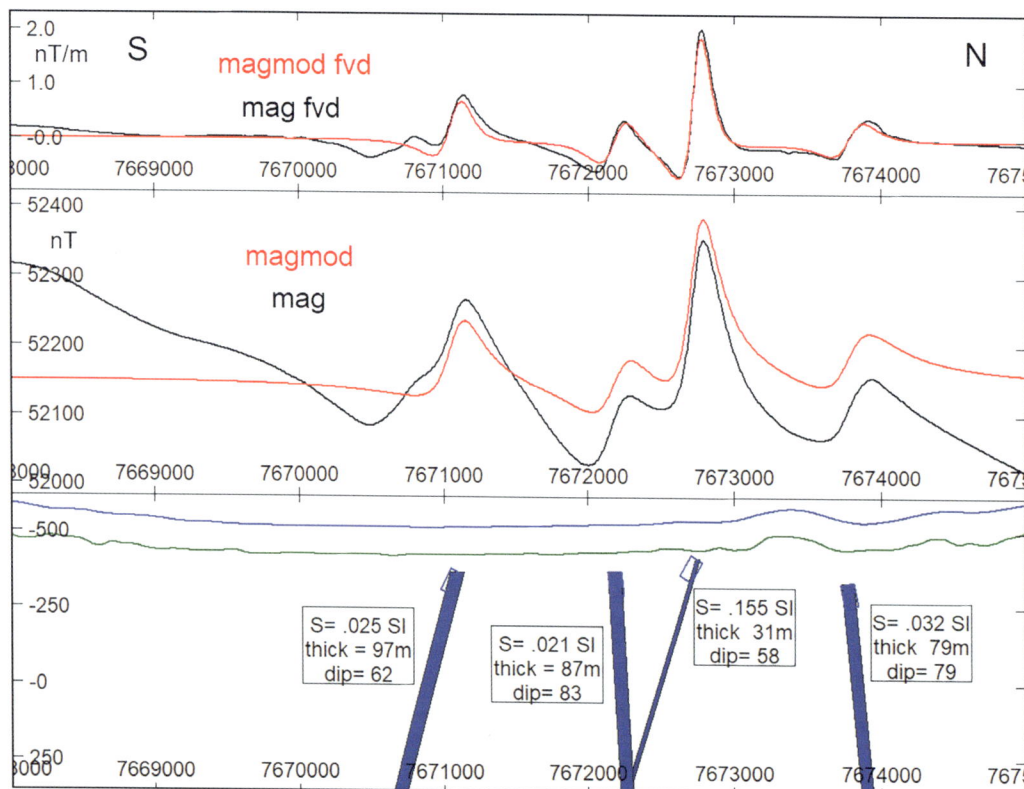

Fig. 2.36. South to North flightline 100510 with (bottom) source bodies generated from Automag solutions, (middle) measured and model-computed TMI, and (top) linear filter vertical derivatives of measured and model-computed TMI.

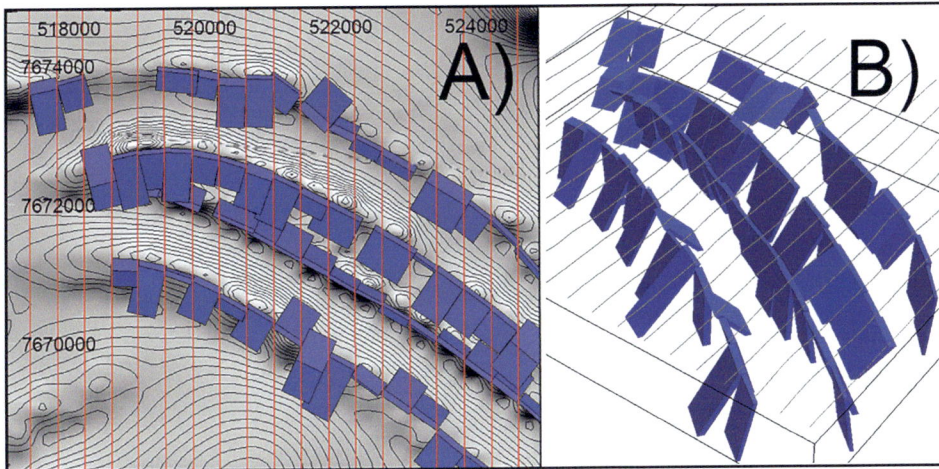

Fig. 2.37. A) plan view of strike-adjusted Automag-derived source models over an image of TMI, and B) perspective view of the source models.

the field. However, comparison between the vertical gradient filter of the data and the identical vertical gradient filter of the model-computed field remains valid. Even if the filter is not the true vertical gradient it provides similar advantages.

I finetuned the first phase analysis on the selected flightline to optimise solutions from the anomalies of interest while reducing inappropriate or unwanted solutions, and then generated solutions in batch mode on all the lines. I then converted those solutions with a second phase analysis to source model bodies as shown in map view in Fig. 2.37A. This operation requires rotation of the solutions to their true orientation with associated adjustment of their parameter values. Rotation can be performed using the local field trend interpolated from a suitable grid (ideally based on anisotropic gridding of Naprstek and Smith 2019 or Davis 2022). In this case, I selected sequential solutions along individual anomalies which the analysis reorients to align the bodies 'head to tail'. As this realignment is performed the parameter values are automatically adjusted for rotation to their new strike direction. Figure 2.38B shows a perspective view of the rotated solutions which clearly represent the Errolola syncline and provide an excellent starting model for inversion. This model with a body at each profile and anomaly intersection was more complex than required and I thinned it to fewer, longer bodies before proceeding to a user-guided inversion. Figure 2.42 shows the post-inversion flightline model of Fig. 2.40. Just as with Automag, I tuned inversion to match the line-based vertical derivative filter. Vertical gradient inversion should not modify the regional field or the depth extent of bodies

because of lack of sensitivity to those parameters, but if required those values can be adjusted with interspaced TMI inversions. For instance, I adjusted the centre body depth extents to restrict 'scissoring' (see Fig. 2.40) to which there is little model sensitivity even in the TMI inversions. Having produced a good general fit to the data, any subsequent improvement of individual bodies is best achieved by inverting only that body or a combination of immediately adjacent bodies to best fit local selections of the data. Any nearby bodies with overlapping fields must be included in the forward modelling calls of inversion even if those bodies are not inverted themselves.

The Elkedra study illustrates the considerable power of tuning inversions using gradient data of the primary, closely spaced flightline measurements with suitable

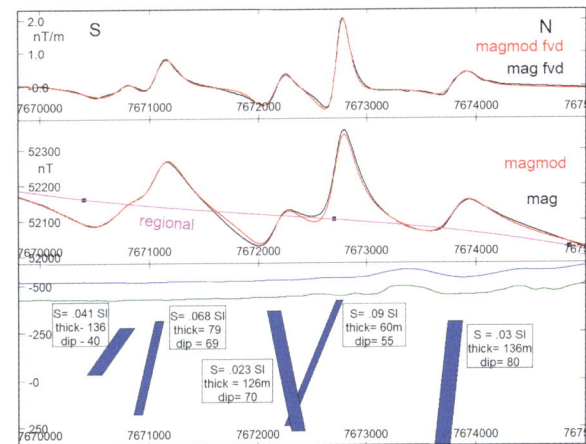

Fig. 2.38. South to North flightline 100510 with (bottom) source bodies from vertical derivative inversion, (middle) measured, regional and model-computed TMI, (top) linear filter vertical derivatives of measured and model-computed TMI.

strategies to compensate for loss of cross-line horizontal gradients.

2.10 USE OF GRADIENT ENHANCEMENTS TO SEPARATE FIELDS THAT CAN BE MEANINGFULLY INVERTED FROM THOSE THAT CANNOT

Many consumers and some authors of inversion models assume or accept that inversion software has sophistication buried within it to produce definitive models. Unfortunately, this is not true. Inversions produce models which should be (at least mathematically) possible explanations of the measured field. However, without input of considerable independent information inversion cannot resolve fundamental non-uniqueness and models are not definitive because they have been developed by an inversion. In the previous Elkedra case study I illustrated how gradient data can be used to improve recovery of shallow magnetisation models. In this case study I will show that for a model to be meaningfully derived by inversion (still not for it to be uniquely defined) it must primarily focus on explaining data features that dominate curvature of the measured magnetic field.

Figure 2.39 shows a TMI image of the Cobham Lake area in north-west New South Wales. The survey was flown as part of a regional geological mapping program on east–west flightlines at 250 m spacing and a nominal 60 m terrain clearance. The area has extensive regolith cover, sparse basement outcrop and few boreholes. The geological map in Fig. 2.40 (Hegarty 2017) was developed

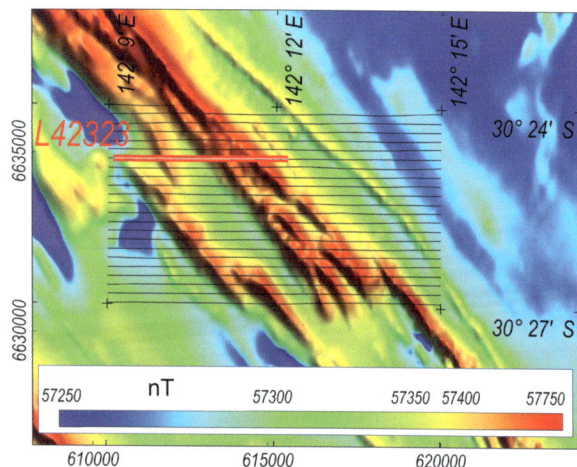

Fig. 2.40. Location of the sections in Figs 2.41 and 2.42 over the Cobham Lake geological map (Hegarty 2017).

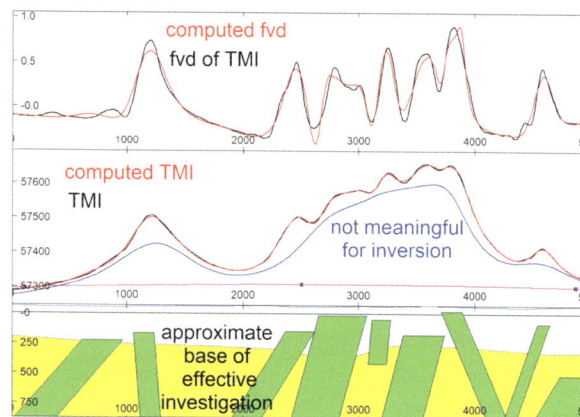

Fig. 2.41. Bottom) Model X-section from TMI inversion. The top of the yellow zone is the approximate limit below which magnetisation does not dominate curvature of the field and therefore is not resolved by inversion. Centre) measured and computed TMI, Top) vertical derivative of measured and computed TMI.

with considerable input from the magnetic field data in Fig. 2.39 but is a simplified synthesis restricted by the limited availability of information from boreholes and outcrop. Multiple, overlapping field variations along the section of flightline modelled in Fig. 2.41 are separated into only the two geological units of the Cambrian Ponto Group and Wonnaminta Formation. Both are described as packages of predominantly quartzo-feldspathic metasandstones and phyllites, differentiated predominantly by the stronger magnetic signature of the Ponto Group compared to the Wonnaminta Formation. Variability in the TMI image suggests that these packages have significant internal petrophysical variation but it is not feasible to map those variations as specific geological units. The magnetic field data can, however, map

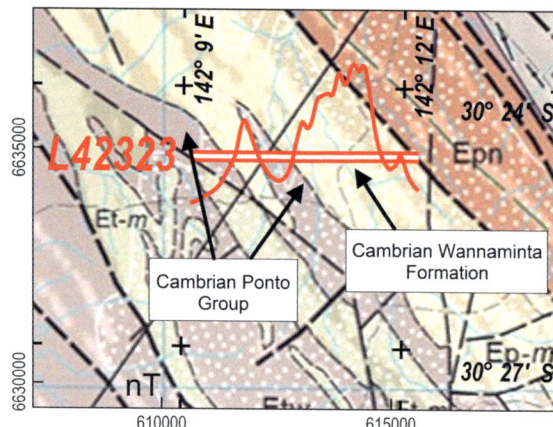

Fig. 2.39. TMI image and 250 m spaced east–west flightlines near Cobham Lake, New South Wales.

contacts across which there is variation in magnetisation and support inversion to estimate apparent susceptibility contrasts, dips of contacts, and at suitable locations, depth to the top of magnetisation.

Figure 2.41 shows a model section of the profile located in Figs 2.39 and 2.40. The objective of this study was to map depth to the top of magnetisation and recover statistics about its distribution such as thicknesses and dips of prominent magnetic units, rather than to build a complete three-dimensional model for which most of the volume is not meaningfully constrained. I positioned nine bodies to match prominent features along the TMI profile and adjusted their position, strike extent and azimuth in map view to match the anomaly extents and orientations. I assumed a horizontal background field, with the bodies explaining all the field variation about an initially unknown base level. Explaining these poorly separated field variations with the minimum number of discrete sources is justified but because the anomalies overlap so strongly any space-filling model of continuous distribution of magnetisation is highly non-unique. The model shown in Fig. 2.41 is also non-unique, but the non-uniqueness has been minimised by focusing on the more reliable features in the field.

The inversion to create the model shown in Fig. 2.41 was permitted to change all model parameters other than northing, strike extent and azimuth, to which there is no sensitivity on the individual profile. The close match of measured and post-inversion model-computed TMI is shown in the centre panel in Fig. 2.41. Typical of least-squares fits to data the most noticeable misfits are rounding of the local maxima and minima. These misfits represent a very small proportion of the data, but unfortunately the part most critical and informative. The top track of Fig. 2.41 shows line-based vertical gradient filters of the measured and model-computed fields which more clearly reveal failings of the model in reproducing the sharpest measured TMI variations.

The sum of the magnetisation in the model matches the measured field variation but there is considerable variation in sensitivity to different parts of the model. The tops of the shallowest bodies are less than 200 m below the measurements and bodies appear to extend more than a kilometre beneath that. The deeper parts of the magnetisation produce small contributions to the magnetic field variations and do not contribute at all to the local field gradients which carry the most diagnostic information. In Fig. 2.41 I have drawn an approximate and indicative surface below which there is little

justification to delineate magnetisation. In the model of Fig. 2.41 the deeper magnetisation is produced by the lower segments of the bodies for which the tops at shallow depth match the partially separated sharp field variations. Alternatively, the deeper magnetisations could be replaced across a horizontal discontinuity by any number of different deep magnetisation distributions. In Fig. 2.41 I have also shown the approximate corresponding split between magnetic field contributions from shallow, partially resolved magnetisations and deeper unresolved magnetisations. This twofold classification is a considerable simplification but it is a great improvement on supplying the model of Fig. 2.41 without discrimination between aspects which are reasonably justified and aspects which are not justified at all. Failure to provide at least an indication of this separation both devalues those parts of the model which have some reliability and includes features which might be interpreted but which have no justification.

The upper gradient filter track of Fig. 2.42 highlights the curvature features in the lower TMI track that have been designated as those parts of the field with significance for analysis and interpretation. All the information about magnetisation that can be meaningfully recovered from the inversion has expression in the gradient data, and the gradient data consists mostly of this separation of the field. The bottom track of Fig. 2.42 shows an alternative model from inversion of the gradient filter. Small features are more prominent in the gradient filters than in the field data and I used 14 bodies to match these gradient anomalies compared to nine for the TMI inversion.

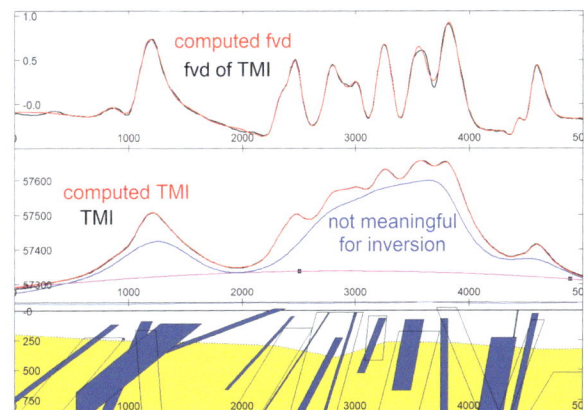

Fig. 2.42. Bottom) Model X-section from inversion of the vertical derivative filter of TMI. The top of the yellow zone is the approximate limit below which magnetisation does not dominate curvature of the field and therefore is not resolved by inversion. Centre) measured and computed TMI; Top) vertical derivative of measured and computed TMI.

Most of the misfit for this vertical derivative inversion (just as for the previous TMI inversion) is in rounding local maxima and minima. However, because the inversion focuses on matching this small part of the field, it matches those gradients much more closely than is achieved by the TMI inversion of Fig. 2.41. For the more sharply defined anomalies depths to the top of magnetisation are reasonably consistent between the TMI and vertical gradient inversions. Focus on the short wavelength magnetic variations using the gradient data necessarily reduces sensitivity to the longer wavelength variations. However, having matched the gradient filters in the top track of Fig. 2.36, that model can be subsequently inverted to best fit the TMI data by changing only the depth extent of the bodies and the regional field. This second-stage inversion does not significantly disrupt the already achieved fit of the gradients.

There is a further selective optimisation in terms of results obtained for effort spent. In this study I have run a single profile inversion. The model is simple (nine individual and independent bodies for the TMI inversion and 14 for the vertical gradient inversion) and the number of data points is small (600) so that inversions run very quickly even with many iterations. The final models are easily summarised in a set of data points which contain statistics for each model body such as depth to top, apparent susceptibility, width, azimuth and dip. Inclusion of the vertical gradient inversion increases the time spent but improves efficiency in recovering reliable results. A full three-dimensional volume inversion would take longer, be less interactive, and would not necessarily provide any additional reliable information. Information from the few-body profile inversion is concisely condensed and easily imported to GIS packages for integration with other information. The location of the profile can be selected from inspection of the field data to ensure that the most significant field variations are sampled and having completed one profile inversion, the decision can be made about how far to step to the next one. This process gives a versatile focus on the most informative field variations across the area or those most relevant to specific exploration objectives.

REFERENCES

Alavi SN (2013) 'Structure, stratigraphy, and petroleum prospectivity of the Waukarlycarly Embayment, Canning Basin, Western Australia'. Record 2013/10. Geological Survey of Western Australia, Perth.

Blake DH, Horsfall CL (1986) 'Elkedra Region, Northern Territory (First Edition). 1:100 000 geological map commentary, parts of 5955, 5855 and 6055'. Bureau of Mineral Resources, Australia, Canberra.

Blakely RJ (1995) 'Potential theory in gravity and magnetic applications'. (Cambridge University Press)

Briggs IC (1974) Machine contouring using minimum curvature. *Geophysics* 39, 39–48. doi:10.1190/1.1440410

Chwala A, Stolz R, Zakosarenko V, Fritzsch L, Schulz M, Rompel A, Polome L, Meyer M, Meyer HD (2012) Full Tensor SQUID Gradiometer for airborne exploration. *ASEG Extended Abstracts* 2012(1), 1–4. doi:10.1071/ASEG2012ab296

Davis A (2022) Nested anisotropic geostatistical gridding of airborne geophysical data. *Geophysics* 87, E1–E12. doi:10.1190/geo2021-0169.1

Foss CA, Purcell PG (2006) Structure and hydrocarbon prospectivity of the Waukarlycarly graben, West Australia in 'Abstracts: American Association of Petroleum Geologists; AAPG International Conference and Exhibition'. (AAPG: Perth, Western Australia).

Foss CA, Gouthas G, Wilson TC, Katona LF, Heath P (2019) Gawler Craton Airborne Geophysical Survey, Region 9A, Childara – enhanced geophysical imagery and magnetic source depth models Report Book 2019/00008. Department for Energy and Mining, South Australia, Adelaide.

Hegarty RA (2017) 'Cobham Lake 1: 250 000 Geophysical-Geological Interpretation Map'. Sheet SH/54–11. (Geological Survey of New South Wales, Maitland)

Hood PJ, Teskey DJ (1989) Aeromagnetic gradiometer program of the Geological Survey of Canada. *Geophysics* 54, 1012–1022. doi:10.1190/1.1442726

Isles DJ, Rankin LR (2013) 'Geological Interpretation of Geological Data.' (Australian Society of Exploration Geophysics)

Katona LF, Reed GD, Heath PJ (2021) 'The Gawler Craton Airborne Survey, 2017 – 2021: Final Report, Report Book 2021/00017.' Department for Energy and Mining, South Australia, Adelaide.

Li Y, Oldenburg DW (1998) Separation of regional and residual magnetic field data. *Geophysics* 63, 431–439. doi:10.1190/1.1444343

Miller HG, Singh V (1994) Potential field tilt - a new concept for location of potential field sources. *Journal of Applied Geophysics* 32, 213–217. doi:10.1016/0926-9851(94)90022-1

Minty BRS (1991) Simple micro-levelling for aeromagnetic data. *Exploration Geophysics* 22, 591–592. doi:10.1071/EG991591

Naprstek T, Smith RS (2019) A new method for interpolating linear features in aeromagnetic data. *Geophysics* 84, JM15–JM24. doi:10.1190/geo2018-0156.1

Naudy H (1971) Automatic determination of depth on aeromagnetic profiles. *Geophysics* 36, 717–772. doi:10.1190/1.1440207

Pawley M, Irvine J, Melville A, Krapf C, Thiel S, Gonzale-Alvarez I, Kelka U, Martinez C (2021) Automated lineament analysis of the Gairdner Dolerite dyke swarm of the Gawler Craton. *MESA Journal* 95, 30–40.

Reeves C (2005) Aeromagnetic surveys: Principles, practice and interpretation. Earth-works, Washington DC, 155 p.

Reid AB (1980) Aeromagnetic survey design. *Geophysics* **45**, 895–982. doi:10.1190/1.1441102

Schmidt PH, Clark DA, Leslie KE, Bick M, Tilbrook D, Foley C (2004) GETMAG – a SQUID magnetic tensor gradiometer for mineral and oil exploration. *Exploration Geophysics* **35**, 297–305. doi:10.1071/EG04297

Shi Z (1991) An improved Naudy-based technique for estimating depth from magnetic profiles. *Exploration Geophysics* **22**, 357–362. doi:10.1071/EG991357

Shi Z, Boyd D (1993) AUTOMAG - An automatic method to estimate thickness of overburden from aeromagnetic profiles. *Exploration Geophysics* **24**, 789–794. doi:10.1071/EG993789

3

Sweet-spot estimation of depth to the top of magnetisation

C.A. Foss

ABSTRACT

It is commonly assumed that any segment of magnetic field data can automatically be inverted to create a space-filling model of subsurface magnetisation. This is true, but unless the magnetic field data carries information required to reasonably constrain magnetisation models there will be no value to any models generated from it. Any degree of confidence in even the most reliable space-filling models applies only to a small proportion of the model volume. I refer to the locations where the magnetic field carries significant source information, the measurements at those locations and the subsurface magnetisation contrasts giving rise to those field variations collectively as 'sweet-spots'. Because of non-uniqueness, no magnetic field data is truly diagnostic of its source magnetisation. However, suitable segments of data provide reliable source models with only simple and generally reasonable assumptions. Away from these favourable data segments, confidence in recovered models falls abruptly giving an almost binary classification of sweet-spots where information can be reliably recovered and 'other' areas where the magnetic field is only weakly informative about subsurface magnetisation. A necessary condition of a sweet-spot is that the magnetic field contains a discrete feature for which the curvature can be ascribed to a single discrete magnetisation. These are generally the shallower magnetisations in an area.

Depth to the top of a magnetisation is a detail of its distribution and can only be recovered for stated model assumptions, generally that the magnetisation has a horizontal top and sharp edges. As an alternative to specific source geometry, structural indices can also be used to summarise the distribution of a magnetisation. I provide an analysis of the sensitivity with which we can hope to recover estimates of depth to magnetisation from magnetic field inversion and illustrate this with a case study from the Sedan area of South Australia.

3.1 INTRODUCTION

A common reason to conduct a magnetic field survey is to map depth to the top of magnetisation. There is a comprehensive literature of methods to do this (e.g. Werner 1953; McGrath and Hood 1970; Naudy 1971; Nabighian 1972; Thompson 1982; Ku and Sharp 1983; Reid *et al.* 1990; Almond and Fitzgerald 1998; Silva and Barbosa 2003; Vallée *et al.* 2004; Hansen 2005). Different methods use derivations based on potential field theory and all methods necessarily include some characterisation of the the distribution of magnetisation (either as a specific shape or as a structural index). Some analyses include their own estimation of source shape from second or higher order derivatives of the field, but as outlined by Reid (1980) use of derivatives places greater requirement on the sampling and precision with which the field is measured. In practice, measurement imperfections restrict analysis of higher-order derivatives from magnetic field measurements.

Potential field theory allows that on any surface a film of virtual magnetisation can replicate any magnetisations beneath that surface as the source of a magnetic field measured at any elevation above it (even though this requires infinitesimally detailed variation of unacceptably extreme magnetisation intensities). As impractical as this solution may be, it highlights the challenge of establishing a minimum depth-to-magnetisation. It is commonly assumed that depth to its top is one of the most pronounced and easily recovered statistics of a distribution of magnetisation, but for compact sources this is not true. Figure 3.1 shows a magnetic profile over concentric sources with top depths of between 20 and 250 m below the measurement surface. For concentric spherical sources of equal total magnetisation the magnetic anomalies are identical and there is no sensitivity to depth to the top (other than that it is shallower than depth to the centre). The co-centred cube with side length 100 m and depth to the top 250 m has an anomaly with only minor difference to those of the spherical sources, and the cubes with side length 200 and 300 m (depths to top 200 and 150 m respectively) also have quite similar anomalies.

As shown in Fig. 3.2, differences between the magnetic fields of these sources are also difficult to discern in map view, even over the centres of the anomalies where those differences are largest. Interpretation of the significance of subtle differences between model-computed and measured fields requires close-spaced, high-resolution measurements. For the anomalies shown in Figs 3.1 and 3.2 with an amplitude range almost 200 nT the maximum differences between the fields are 15 nT for the 300 m side-length cube and other bodies, and only 5 nT between those other bodies. Figure 3.3 shows near-identical north–south magnetic field profiles over the centre of a cube of side length 200 m and depth to top 200 m and two bodies of polygonal section, homogeneous magnetisation and identical strike length and azimuth. The shallowest of the bodies has a depth to top only 50% of the cube. Equalities are not limited to such '2.5d' bodies (bodies with constant cross-section along their strike axis) but exist for any magnetised body. For inversion of measured magnetic field data, causative geological bodies are not known to be of simple geometry. However, if an anomaly can be reasonably matched using a model of simple geometry then introduction of any additional complexity must be justified. If the magnetic profile shown in Fig. 3.3 is inverted on the assumption that the source is a simple sharp-edged, horizontal-top body it should return a depth estimate close to 200 m, although either of the

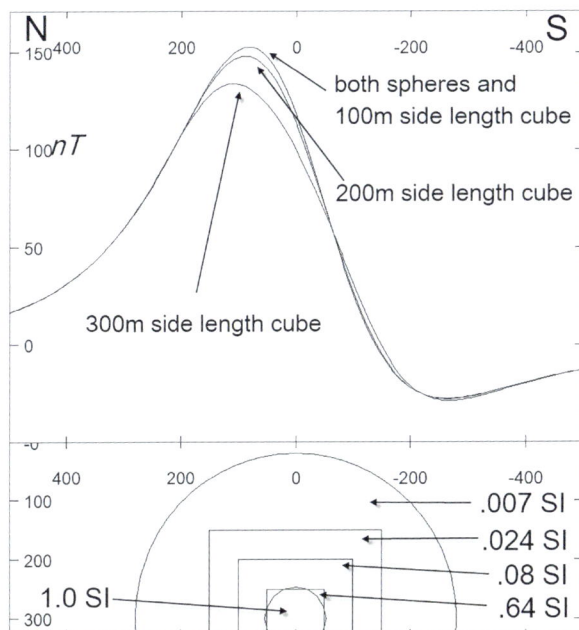

Fig. 3.1. Central north–south sections over concentric spheres and cubes of equal total magnetisation.

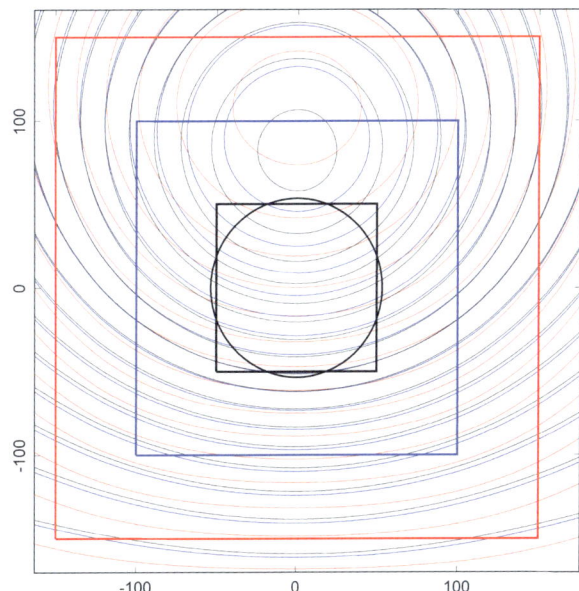

Fig. 3.2. Contour displays of the magnetic anomalies shown in Fig. 3.1. Red – 300 m side-length cube and contours. Blue – 200 m side-length cube and contours. Black – spheres, 100 m cube and their field contours. All contour intervals are 10 nT.

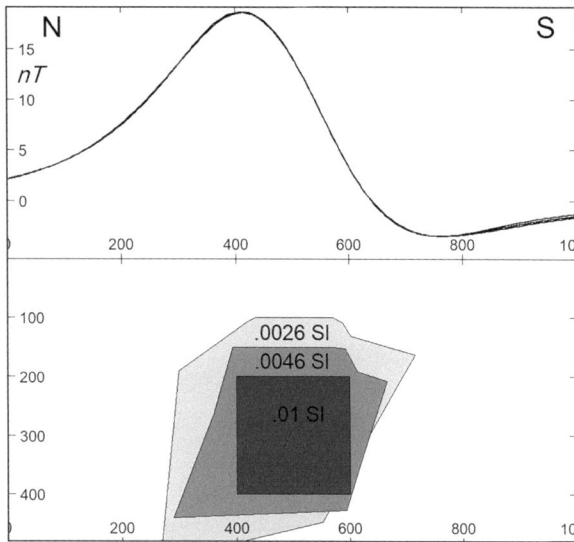

Fig. 3.3. Bodies with near-identical magnetic fields but quite different depths to top.

two shallower alternative models might represent the true magnetisation. Examples such as this reinforce the known requirement that a specified geometry is necessary rather than optional for source depth estimation. Assumption of shape is particularly important for the top surface of a model. If the assumption of a horizontal top surface is relaxed then estimation of depth to the top of magnetisation is unsupported. Fortunately, specification of a horizontal top surface is in many geological cases a reasonable assumption. For instance, in Australia many magnetic field features are due to magnetisations terminated beneath a sub-horizontal basement unconformity that is broadly consistent with this model assumption.

The most relevant statistic for evaluation of magnetic depth estimation methods is their predictive success, but unfortunately there is no definitive database recording this statistic, and almost all claims of precision in estimating depth to magnetisation have been made using synthetic data or already-known drilling results. If exploration geophysics had a regulatory authority then such a database could be created and maintained from pre-drilling predictions using block-chain authentication. Authenticated empirical results are particularly important because non-uniqueness precludes a strictly analytic evaluation of different methods and obscures the issues considerably. In this chapter I outline the method of 'sweet-spot' magnetic depth estimation. This method is subject to the same non-uniqueness limitations as for all other methods but it has a key advantage that solutions are directly tested against the field variations they are

proposed to explain. Although I specifically present the sweet-spot method, the issues I discuss in this chapter apply to all depth estimation methodologies.

Sweet-spot depth estimation combines three concepts from which different implementations can be developed. These concepts are:

1) Source magnetisation information is only selectively available from magnetic field data at locations of suitable subsurface magnetisation contrasts. If the magnetic field measurements above these locations are of high quality, the location and magnetic field data define as a sweet-spot.
2) The small data package defining a sweet-spot is optimum for intensive, focussed and individually tuned analyses or inversions.
3) Magnetic depth estimates do not have conventional uncertainty statistics but can be characterised by model sensitivities with an accompanying caveat that model assumptions are valid.

Magnetic field surveys typically provide uniform data coverage. Geology, however, is not uniform. The location and characteristics of abrupt and substantial contrasts in subsurface magnetisation control the overlying magnetic field variations and determine the feasibility of recovering information from analysis of magnetic field data. Any magnetic source depth estimator should be able to return reasonable results where magnetisation distributions are supportive. If the geological distribution of magnetisation is not supportive then no method will provide reliable results.

An ideal sweet-spot is a simple magnetic field variation with amplitude well above measurement noise and superimposed field variations. The measurements should extend across several flightlines and be well separated from field variations due to other magnetisations. In Chapter 2 (section 2.8) I presented a study of sweet-spots over the Gairdner dykes of South Australia that are particularly well suited to source depth estimation. Other field variations may be less appropriate because they are too complex, overlap, or are defined by insufficient measurements. In other areas there may be no sweet-spots at all. In consequence, many depth estimation studies are a compromise between restricting the number of solutions to retain reliability of results and expanding the number of solutions to provide more information, albeit at lower reliability.

The initial recognition of a sweet-spot is as a discrete and well-sampled magnetic field anomaly. As the different models in Fig. 3.3 illustrate, even an apparently suitable field variation can return misleading depth estimation results. However, this can only be known from subsequent testing with independent information. In consequence I classify all field variations apparently supportive of depth estimation as sweet-spots and allow that results may include some solutions seemingly well justified by the available data but which may later be shown to be misleading of the true magnetisation distribution.

3.1.1 Sweet-spot data packages

The three models shown in Fig. 3.3 produce similar but not identical fields (difference between the fields can be further reduced to any degree by adding more model complexity). To discriminate between source models from analysis of their magnetic fields requires confidence in minor features of the data. To focus on such small field differences, inversions should be optimised. Ongoing advances in computing speed and memory size permit generation of extremely large models from inversion of extremely large datasets. Small increments of magnetisation in these models contribute meaningfully to only small sections of the measured field and no part of the resulting model is optimised as well as can be achieved by isolation of smaller, local models with smaller, local data packages. Where possible, magnetic source depth estimations should ensure that results are compatible with surrounding data and models. Small models and datasets also encourage application of sensitivity tests by scanning ranges of model parameter values to investigate rates of divergence of data-fit away from the optimum solution.

After the extent of a dataset for inversion is specified, the next step is to split that data into anomalous (residual) and background (regional) parts. This operation is interpretive and generally requires inclusion of data at or beyond the margin of the anomaly. However, the data used in an inversion should not extend far beyond the anomaly as that dilutes focus on the critical data. Unfortunately, the only section of the background field that requires definition is the part overlapped by the anomaly itself, where it cannot be directly measured or easily estimated. Widening the area across which the background field is estimated does not in all cases improve fidelity of anomaly separation.

It is essential to establish the authenticity and suitability of any data to be used for magnetic source depth estimation. For methods using gridded data this includes the fidelity with which the grid represents the magnetic field. This chapter includes evaluation of source depth estimates derived from gridded data.

3.1.2 Forward modelling tests of depth estimates

Inability to prove correctness of magnetic source depth solutions due to non-uniqueness can discourage what should be best efforts to optimise depth estimation results. All depth estimate solutions should be tested by forward modelling to establish that they explain the field variation from which they are derived. This does not establish that they are correct but can readily detect solutions that are obviously inappropriate. Automation of magnetic source depth estimation generally involves tuning to generate either a small number of solutions at high discrimination, or many solutions at low discrimination. Depth estimation methods that are unable to directly interrogate solutions with forward modelling apply two styles of confidence or quality tests: (i) solutions are plotted over magnetic field images and those associated with the most distinct features in the magnetic field are attributed higher reliability, or (ii) clusters of adjacent solutions with similar value are accepted to be of higher reliability. Neither of these evaluations is a direct test of the validity of solutions other than allowing rejection of inconsistent results.

3.2 KEY PRACTICAL CHALLENGES OF MAGNETIC SOURCE DEPTH ESTIMATION

Major and overlapping sources of error in a model to explain a measured magnetic field are:

1) fundamental non-uniqueness and model insensitivity
2) imperfection or insufficiency of the magnetic field, elevation or topography data
3) incorrect separation of the magnetic field due to the magnetisation
4) inappropriate model specification
5) incorrect geological attribution of the magnetisation.

I illustrate some of these issues with synthetic data and in a case study over the Sedan area of South Australia.

3.2.1 Fundamental insensitivity of magnetic field inversion

In favourable cases of suitable geology and high-quality magnetic field data, the fundamental limitations of non-uniqueness account for a substantial part of the uncertainty in magnetic field source depth estimates as I will show in the following investigations. In Figs 3.1 to 3.3 I illustrated how different source magnetisations can generate similar magnetic fields. I now quantify those differences and relate statistics of difference between two magnetic fields to variations in depth to the top of their source magnetisations. Figure 3.4 shows the central north–south magnetic field profiles over regular circular section cylinders, and equal volume, thickness and depth extent rectangular prisms in a field of strength 50,000 nT, declination 0° and inclination −60°. Depths to the tops of the magnetisations are 320, 400 and 480 m (400 m +/− 20%). For each source depth the magnetic field profiles of the cylinders and prisms are almost identical. For the shallower bodies there is a slight increase in amplitude for the rectangular prisms but this can be readily modified by minor adjustment of thickness, strike extent or magnetic susceptibility. The decrease and increase in depth by 20% of the central depth of 400 m causes an increase and decrease in peak

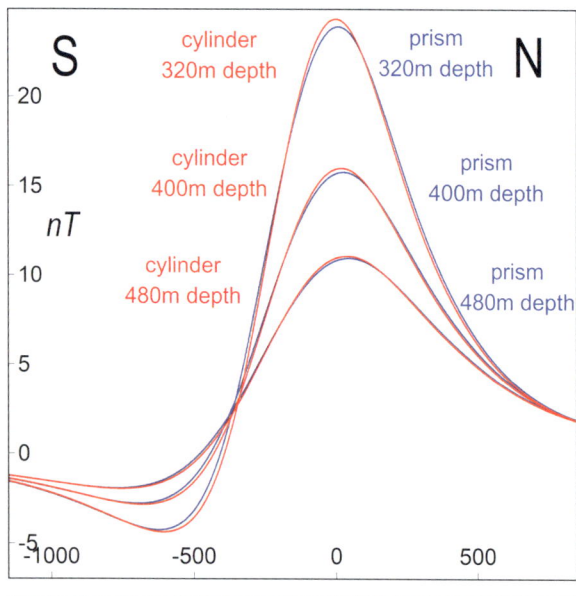

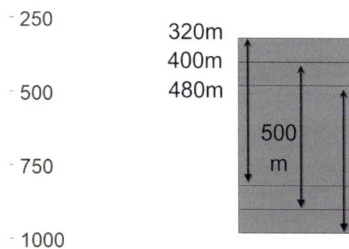

Fig. 3.5. Magnetic profiles over depth-offset prisms (blue) and cylinders (red) to best fit the profile of a cylindrical magnetisation (green) at the reference depth. The inset shows magnification of the almost imperceptible differences at the anomaly peaks.

amplitude for the anomalies of either body by +52% and −31% respectively. This compares to differences of only 1% to 2% between peak amplitudes of the two body-type anomalies at each depth. From the magnetic field profiles of Fig. 3.4 there appears to be high sensitivity to source depth for both model geometries. Unfortunately, that sensitivity is significantly reduced by compensating variation of any unknown parameters, particularly magnetic susceptibility, thickness and depth extent. This is illustrated in Fig. 3.5, in which the magnetic fields of the bodies at 320 and 480 m depth are best-fitted to the field of the cylindrical body at 400 m by variation of those other parameters. The reduced differences between field curves from Fig. 3.4 to Fig. 3.5 clearly establishes that uncertainty of the values of other parameters is the primary cause of low sensitivity to source depth. Further reduction of differences between fields from magnetisations at different depths (further increase in the scope of non-uniqueness) is possible if the specification of body geometry is relaxed.

The difference between two magnetic fields can be quantified by their percentage root-mean-square (rms) misfit. This is the standard deviation of difference between the fields normalised to the standard deviation

Fig. 3.4. Central magnetic field profiles over 400 m wide dodecagonal cylinders (red) and rectangular prisms (blue) at depths of 320, 400 and 480 m.

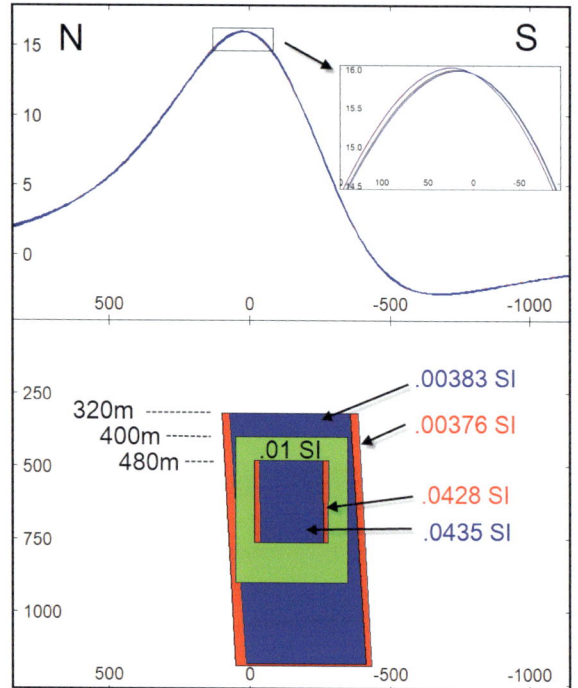

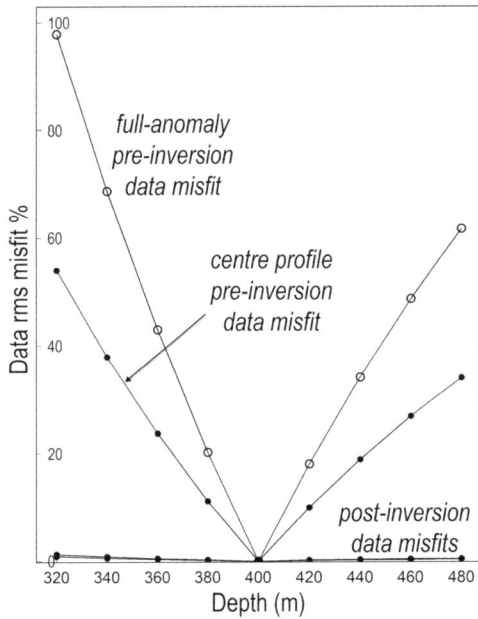

Fig. 3.6. Pre- and post-inversion misfit rms statistics for the full anomaly and centre profile of a circular cylinder at 400 m depth and best-fit cylinders at depth offsets of up to +/− 20%.

of the reference field and multiplied by 100. The statistic is directly related to objective terms minimised in inversion algorithms, including that of the Levenburg–Marquadt ridge-regression method (Marquardt 1970) implementation in the ModelVision software used in this book (Pratt *et al.* 2020). Unfortunately, it is not possible to assign an interpretational threshold to this or any equivalent statistic because their significance is subjective. The pre- and post-inversion rms misfit values both between the central profiles plotted in Figs 3.4 and 3.5 (for a single profile inversion) and between the complete anomalies (for full anomaly inversions) are plotted in Fig. 3.6. The increase by a factor of less than 2 from single-profile to complete-anomaly misfit values is of little significance because it is overwhelmed by differences in anomaly separation and weighting of model parameters between the two inversions. The similar pattern of these curves, and particularly their identical minima, supports application of central-profile analysis in source depth estimation. The key feature of Fig. 3.6 (the most informative figure in this chapter) is the 99% loss of sensitivity to source depth between the pre-inversion and post-inversion misfit values arising from uncertainty in other source parameter values. Parametric inversion is ideal to investigate these relationships but the findings are equally relevant to all other inversion methods, even if the output of those inversions are not expressed in terms of individual source parameters.

The role of individual parameters in compensating for error in estimation of depth-to-magnetisation is highlighted in Fig. 3.7. This figure plots parameter values for each of the best-fit models found by inversion at different offset depths. Figure 3.7A plots the magnetic susceptibility values. Unknown or incorrect magnetic susceptibility is a major contribution to uncertainty in source depth estimation. Increase in estimated source depth is facilitated by higher magnetic susceptibility values, and conversely decrease in estimated source depth is facilitated by lower estimated magnetic susceptibility values. Figure 3.7A plots the fivefold increase and decrease respectively in apparent magnetic susceptibility values associated with a 20% increase and decrease in estimated depth. With this ratio, the challenge of estimating subsurface magnetic susceptibility values is clearly more substantial than the challenge in estimating depth to the top of that magnetisation. There would be major advantage in constraining magnetic source depth estimates with true magnetic susceptibility values, but because of geological variability across many scales this is rarely feasible to better than an order of magnitude, even where some susceptibility measurements are available. The analysis presented here finds the single apparent magnetic susceptibility values of the best-fit models at each depth offset. Most of the variation in depth to the top of magnetisation related to incorrect magnetic susceptibility values arises from magnetic susceptibility variations within a factor of up to 3 or 4. Larger departures of the magnetic susceptibility values accommodate only slight additional depth changes.

Figure 3.7B plots the thickness of the best-fit models for each depth offset. Variation in source thickness is less substantial than the associated variation in magnetic susceptibility, with a proportional variation in thickness of approximately two to three times the proportional variation in depth. Deeper bodies have reduced thickness to compensate for the reduced anomaly sharpness arising from the increased source depth.

Depth extent plotted in Fig. 3.7C also shows an inverse relationship with depth, in this case with a scaling factor of three to four times the change in depth. The change in depth extent compensates for change in sharpness of anomalies by adding or removing long wavelength field variations from the base of the model to contribute towards adjustment for changes in depth to its top.

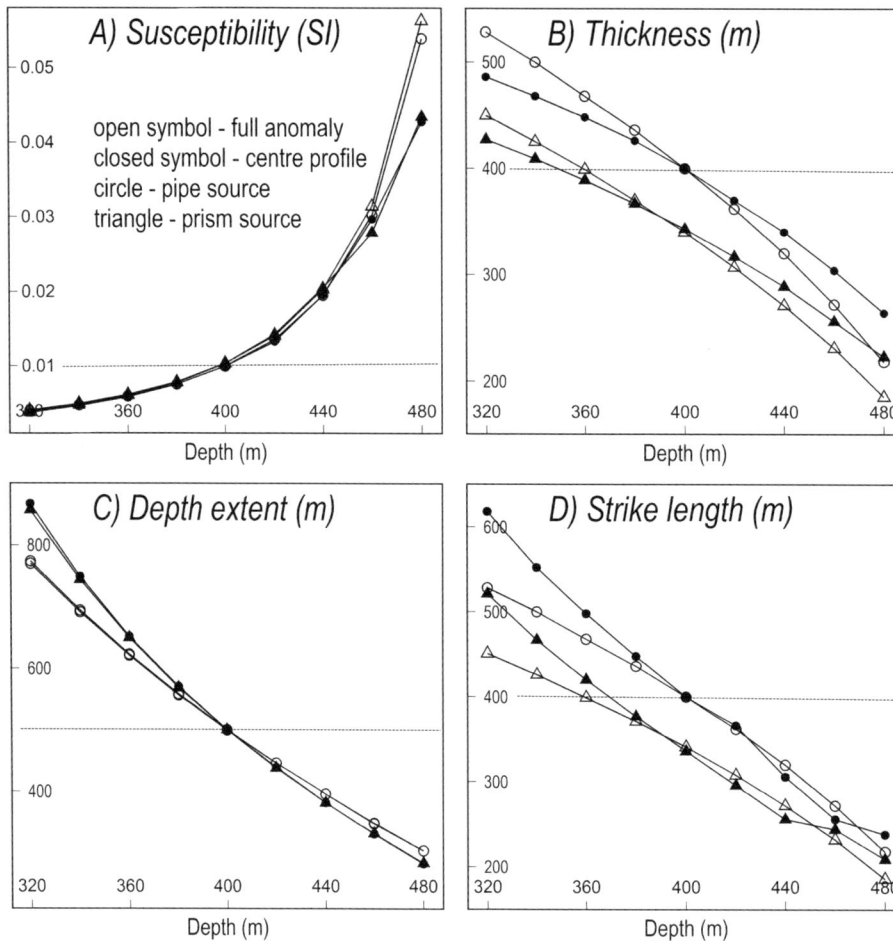

Fig. 3.7. Mapping of parameter values for the best-fit inversions at different depth offsets. Open symbols are for full anomaly inversions and closed symbols for centre profile inversions. Circle symbols are for cylinder models and triangles for prism models.

Figure 3.7D shows a near-linear inverse variation in strike length by a factor of 2 for the 20% increase and decrease in apparent depth. There is an offset between plots for the rectangular and elliptic section bodies, with larger strike lengths of the elliptic section bodies compensating for taper in width along their strike axes. The strike length of a magnetic body is almost unconstrained by inversion of data on only the central profile of an anomaly, but assigning a strike length (together with azimuth of strike) to the magnetisation based on inspection of the field variation in map view is generally sufficient to specify this parameter.

Variation in shape between the bodies used in this study is of low significance. The curves plotted in Fig. 3.7 are for both cylindrical and rectangular prism models. The parallel behaviour of each individual parameter for these two bodies supports substitution of either body type for the other in inversions to estimate depth

to the top of magnetisation. The similar values and patterns of variation of the post-inversion difference statistics plotted in Fig. 3.8 (which is a vertical magnification of the post-inversion curves plotted in Fig. 3.6) also support this conclusion, with no significant offset in best-estimated depth between the two model types.

3.2.2 Data selection – single profile analysis

In the previous section I have shown that horizontal-top models of simple cross-section can be interchanged in estimation of source depth. Simple cross-section shape and homogeneous magnetisation are in most cases justified, with the exception of magnetisations at shallow depth for which there may be substantial expression of any shallow inhomogeneities. In this section I investigate analysis of individual profiles rather than complete

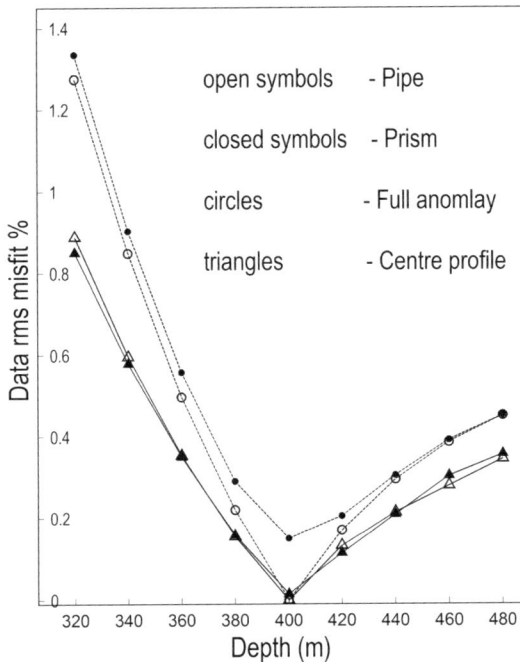

Fig. 3.8. Minimum (post-inversion) data misfit at each depth offset for full anomaly and central profile analyses of cylinder and prism models (expansion of the post-inversion curves in Fig. 3.6).

anomalies. Analysis and interpretation of measured anomalies face problems in separation of overlapping fields, particularly towards their margins. A central profile generally provides more reliable data and in some cases may be the only feasible option if other parts of the anomaly are inaccessible or are overprinted with fields of adjacent magnetisations.

Matching of a complete anomaly can provide the best-qualified bulk estimates of the magnetisation, such as total magnetisation, average magnetic susceptibility or total magnetisation and volume. Depth to its top is, however, a detail of a magnetisation, and its estimation is optimised differently. The most informative aspect of the magnetic field regarding depth to top of magnetisation is field curvature. Matching a well-positioned profile rather than the complete anomaly allows the most critical region of field curvature to be better honoured. Matching a single profile does, however, raise the additional challenge of estimating strike length and azimuth. For highly elongate anomalies, such as those due to dykes, this is not a problem as strike azimuth is well estimated from multiple profile intersections and there is little sensitivity to variation of large strike extent. For more equidimensional magnetisations, the almost linear inverse relationship

between strike length and estimated depth to the top of magnetisation plotted in Fig. 3.7D reveals that an overestimation of strike length produces a tendency towards underestimation of depth and conversely, underestimation of strike length encourages overestimation of depth.

For inversion of measured field anomalies a suitable approach to address strike length is to set strike length and azimuth from inspection of the magnetic field variation in map view and to leave them fixed in the initial inversion stages. Once inversion has reduced the major data misfit, in subsequent inversion runs strike length can optionally be enabled as a free parameter. Figure 3.9 plots post-inversion data misfits for multiple cylindrical-section and rectangular-section models with fixed but incorrect strike length. These inversions allow variation of magnetic susceptibility, thickness, depth extent and dip in attempts to compensate for the incorrect strike length. For both body types, the global data misfit minima are at the true depth of 400 m. Consistent with results plotted in Fig. 3.7D, the minima of curves for bodies of erroneously large strike length provide underestimates of depth-to-magnetisation, and for bodies of erroneously short strike length the minima provide overestimates of depth-to-magnetisation. For body types of both cross-section geometries, a misrepresentation of strike length by 25% (+/− 100 m for the models shown) is

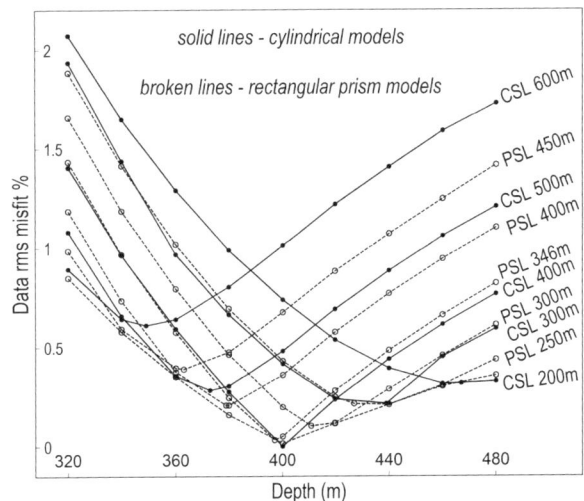

Fig. 3.9. Post-inversion data misfit curves for models of different strike length (CSL – cylinder strike length and PSL – prism strike length). The global minimum misfit at the true 400 m depth is given by the 400 m strike length cylinder and 346 m strike length prism models.

associated with an error of less than 10% in the estimated depth to the top of magnetisation.

Single profile analysis is a powerful (but not essential)
aspect of sweet-spot depth estimation. The major
remaining issue impacting on estimation of depth to the
top of magnetisation is data sampling that I investigate
in the following study.

3.3 A SYNTHETIC-DATA STUDY OF DEPTH TO MAGNETISATION ESTIMATION

Having established the fundamental limitations in estimating depth to the top of magnetisations we can now
evaluate the additional uncertainty arising from overlapping fields and incomplete data sampling. For aeromagnetic surveys conducted on flightlines, inadequacy
in sampling the field generally involves flightline spacing. For helicopters and fixed-wing aircraft with measurements made at 10 or 20 Hz, the along-profile spacing
is typically between 1 and 10 m whereas survey line
spacings are typically in the range of 20 to 400 m. In
many cases this gives between-line to within-line measurement spacing ratios greater than 10. To investigate
the effect of line spacing I again first use synthetic,
noise-free data.

Figure 3.10 shows north–south synthetic flightlines
at 50 and 400 m spacing over a set of 31 synthetic models with horizontal top surfaces, vertical plunge and

different size, shape and orientation. The bodies have a
range of susceptibilities from 0.05 SI for the smaller
bodies to 0.003 SI for the larger bodies so that a range of
body sizes contribute significantly to the magnetic
field. The tops of the bodies are all at a common elevation and there is no superimposed background field.
Nevertheless, even this ideal model presents considerable challenges in solution of the inverse problem to
estimate depth to those magnetisations from fields at
moderate elevations above them or with insufficient
sampling. The field is computed for induced magnetisation in a geomagnetic inclination -60° and declination 0°. In this steep but non-vertical field each body
produces a dipole anomaly with a dominant peak
slightly to the north of the centre of magnetisation and
a weaker negative further to the south. The magnetic
field can be simplified by application of a reduced to
pole (RTP) transform, but this introduces artefacts for
insufficiently sampled data as well as dependence on
magnetisation direction. I use TMI computed from the
model in Fig. 3.10 to illustrate limitations in mapping
depth to the top of magnetisation according to elevation of the measurement surface above the magnetisations, and analysis of multiple and single flightlines
and grid traverses.

3.3.1 Influence of measurement elevation

Figure 3.11A is a contour map of TMI computed at 12.5 m
cell spacing and an elevation of 100 m above the top of
magnetisation. This high-resolution proximal field provides rich information about the distribution of magnetisation. All anomalies would qualify as sweet-spots
suitable for source depth estimation within the limitations of non-uniqueness previously discussed. The clusters of small adjacent bodies give rise to elevated
background values where the anomalies overlap, but
otherwise each of the 31 magnetisations is marked by an
individual and reasonably separated anomaly that supports a depth estimate.

Figure 3.11B is a contour map of the field computed
over the same magnetisation model at the same horizontal locations but at a higher elevation of 400 m above the
top of magnetisation. In this case the 31 magnetisations
give rise to only 14 closure peaks in the magnetic field.
The clusters of smaller magnetisations each give rise to
single anomalies, and the fields of several adjacent magnetisations merge into combined anomalies from which
contributions from the individual sources cannot be
reliably separated.

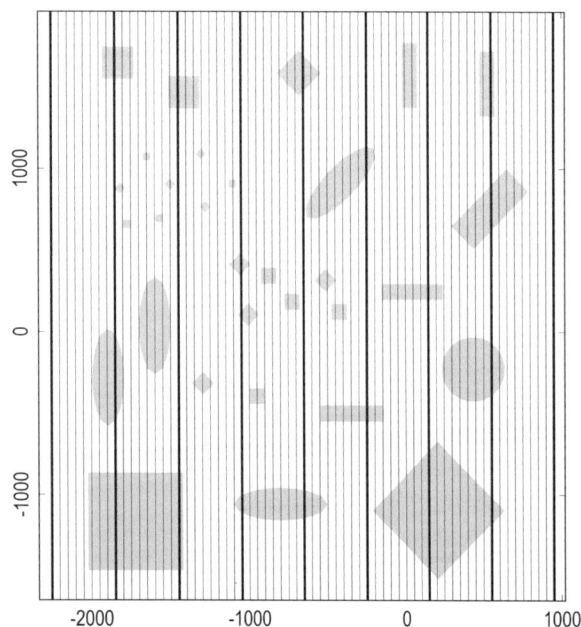

Fig. 3.10. Plan view of a magnetisation model with synthetic
flightlines at 50 m (feint) and 400 m (bold) spacings.

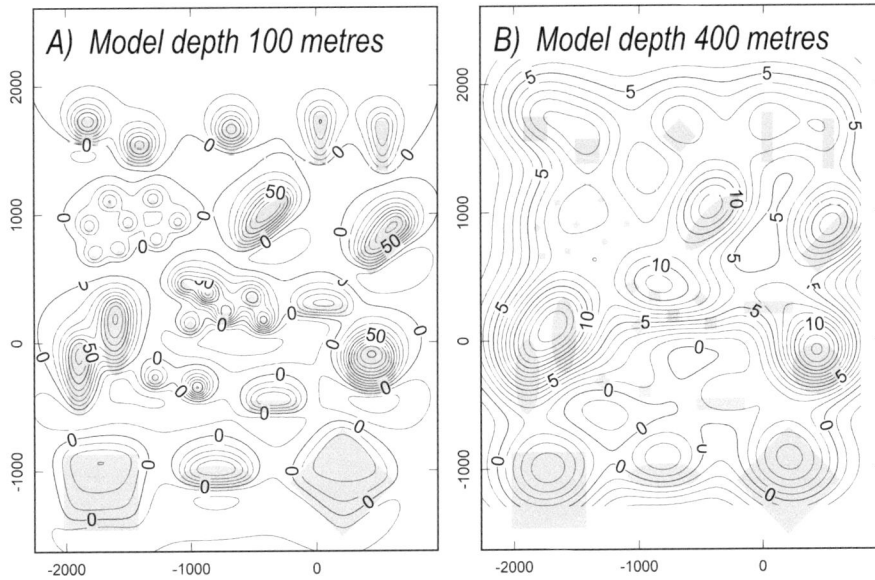

Fig. 3.11. A) TMI computed at 100 m elevation above magnetisation (contours at 5nT interval) and B) TMI computed at 400 m elevation (contours at 1 nT interval).

3.3.2 Complete survey multi-line inversion

Figure 3.12 shows the depth to the top of magnetisation values derived from inversion of the 400 m elevation, 100 m spaced flightline dataset after creating a starting model with individual bodies assigned to explain each of the recognised anomalies. The bodies are flat-topped (as were the bodies of the input model), have elliptic cross-section and are allowed to plunge (the input model bodies were all vertical). The inversion also introduces the freedom of a planar background field to represent

uncertainty in making a regional separation (as encountered when working with field data). At most locations the field computed from the inversion model matches the input field very closely. Residual mismatches could be further reduced by introducing additional bodies or allowing additional complexity of the existing bodies but it is by no means clear than additional improvement in fitting the data would improve the validity in representing the input magnetisation. A statistical summary of the model is given in the top row (row 'A') of Table 3.1.

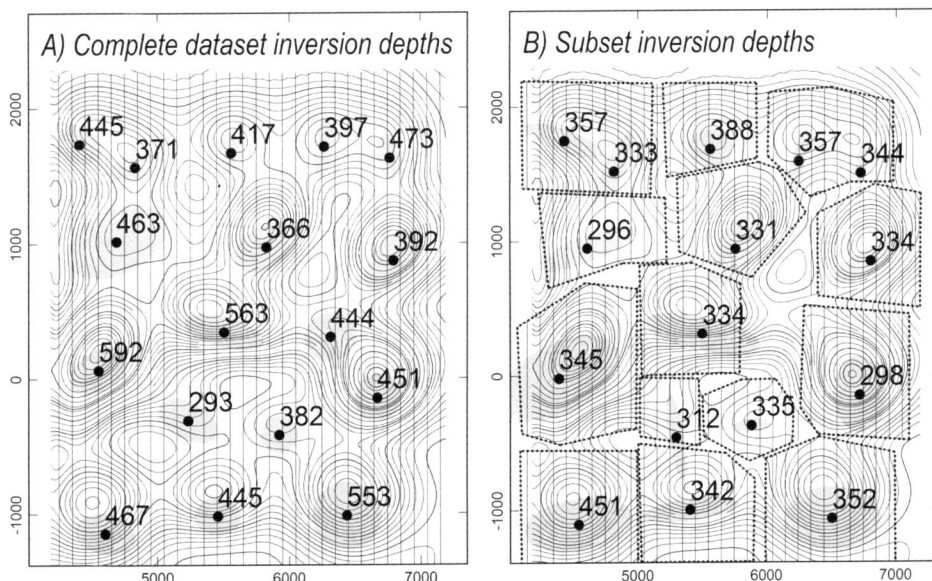

Fig. 3.12. Contours of TMI gridded from 100 m spaced flightlines (black) and of output inversion model field (dotted) with plan of the tops of the magnetisation models annotated with depth in metres. All input model depths are 400 m.

Table 3.1. Model depth-to-magnetisation estimates.

Inversion data	Spacing (m)	Number	Mean depth (m)	Std dev (m)	Min depth (m)	Max depth (m)
A) Complete lines	100	17	442 (+11%)	76 (19%)	293	592
B) Multi-lines	100	16	344 (–14%)	36 (9%)	296	451
C) Single lines	100	15	373 (–7%)	39 (10%)	306	456
D) Computed traverses	–	14	400 (< 1%)	31 (8%)	351	473
E) Grid traverses	100	14	402 (+1%)	44 (11%)	336	480
F) Complete lines	400 odd	13	464 (+16%)	54 (14%)	370	550
G) Complete lines	400 even	14	454 (+14%)	119 (29%)	229	608
H) Multi-lines	400	11	416 (+4%)	113 (28%)	241	594
I) Single lines	400 odd	20	359 (–10%)	47 (12%)	284	456
J) Single lines	400 even	20	368 (–8%)	87 (22%)	243	572
K) Grid traverses	400 odd	12	374 (–7%)	68 (17%)	341	464
L) Grid traverses	400 even	13	316 (–21%)	63 (16%)	238	465

The mean depth to magnetisation is over-estimated by 11% and there is a standard deviation of almost 20% and maximum errors of -27% and +48% in the estimated depths. The details of this inversion model are poorly repeatable because there is trade-off between individual body parameters and between adjacent bodies, but the mean and standard deviation of the population of depth values remains consistent between repeat inversions. The smooth envelope of the least-squares best-fit field provides a general overestimation of depth, although there are also bodies of underestimated depth.

3.3.3 Multi-line single anomaly inversions

The inversion model shown in Fig. 3.12A is an attempt to best-fit the fields of all the magnetisations simultaneously, with weak focus on each and cross-compensation of errors between adjacent magnetisations. I also separately performed multi-line inversions of the individual anomalies. This method requires more work but better focuses the inversions by making only local adjustments of the magnetisation models to improve the local fit to the field above them. Where an anomaly appears to be due to multiple sources so close together that their fields cannot be separated, a compromise is made of performing a multi-body inversion for that complex anomaly. Results are shown in Fig. 3.12B.

As a starting model for each of these inversions I create an elliptic-section body with horizontal top and bottom faces and vertical sides. I assign a test magnetic susceptibility to the body and adjust the starting regional field to bring the sum of the regional and anomalous fields to an approximate match with the input field. A

modest fit between the input field and that of the starting model is generally sufficient to ensure stable convergence of inversion. If the inversion does not proceed as intended it can be undone and the initial model, regional field or selection of free parameters can be adjusted to encourage different behaviour. I find this user guidance preferrable to 'hands-free' inversions that follow their own path, unbiased though that is. For these relatively simple inversions the only intervention required was to define the regional field and adjust it if required as the inversion progresses. The process of inversion, including any interpretive guidance or intervention, must be clearly documented so that the results can be fairly evaluated. The free model parameters in the inversion are the unknown values of magnetic susceptibility, easting and northing, depth, depth extent, axes lengths and dip. Adjustment of the regional can optionally be included in the inversion, or if preferred it can be kept fixed or can be adjusted by hand.

Statistics of the separate multi-line inversions are listed in the second row (row 'B') of Table 3.1. The main difference from the single inversion of the complete dataset is that depths are in most cases underestimated. Compared to the complete model inversion results the individual anomaly inversion depths have a reduced range of values and a standard deviation only one-half of that for the complete model inversion.

The inversion results listed in the two top rows of Table 3.1 suggest that the previously estimated uncertainty in recovered depth estimates of isolated simple geometry bodies of 5% to 10% due to fundamental non-uniqueness, increases to 10% to 15% if those magnetisations are close to other surrounding unknown magnetisations. This loss of confidence in the depth values increases gradually from widely separated anomalies to anomalies with extensive overlap.

3.3.4 Single profile inversions

In section 3.2.2 I investigated depth estimation from single data profiles. Figure 3.13A shows the distribution of segments of flightlines selected for single-profile inversion of each anomaly. The main difference from multi-line inversions is that a single line inversion cannot obtain estimates of the strike length and azimuth of the magnetisation (because on a single line there is little sensitivity to those parameters). I first adjust the strike length and azimuth of the model in a map window to match the trend and strike extent of the anomaly and retain those settings throughout the inversion. Also, for inversion of a single profile it is generally more convenient to use a body of constant rectangular section unless there is a geological reason to prefer some other shape. Multi-line inversions allow freedom of movement of the horizontal reference coordinates of the magnetisation (most conveniently the centre of the top face). For single profile inversion there is little sensitivity to transition of the magnetisation in the across-profile direction. Therefore, horizontal displacement of the body is restricted to movement backwards or forwards along the profile (as if

the body is moving along a track). The reference point need not be on the data profile if inspection of the grid suggests that the profile used is not centred over the magnetisation. For this example of closely spaced flightlines, displacements of the model centres from the selected lines are mostly small.

Single profile inversion simplifies data selection, reduces issues of complexity resulting from the horizontal shape and homogeneity of the magnetisation and focuses inversion on the more reliably defined, high-amplitude central sections of the anomalies. The single flightline inversion results are summarised in row C of Table 3.1. The range and standard deviation of the depth estimates are similar to the multi-profile inversion results but the error in mean depth is halved to 7% of total depth. Depth values are again generally underestimated.

3.3.5 Grid-traverse inversions

Another example of single profile inversion is the inversion of traverses through grids. Selection of grid traverses allows data to be extracted passing through anomaly maxima and minima parallel to the local field gradient. The quality of a grid in representing the magnetic field depends on the quality and distribution of the primary data, the variation patterns of the true magnetic field and the interpolation algorithm used. There is also degradation of data in resampling from the grid to a

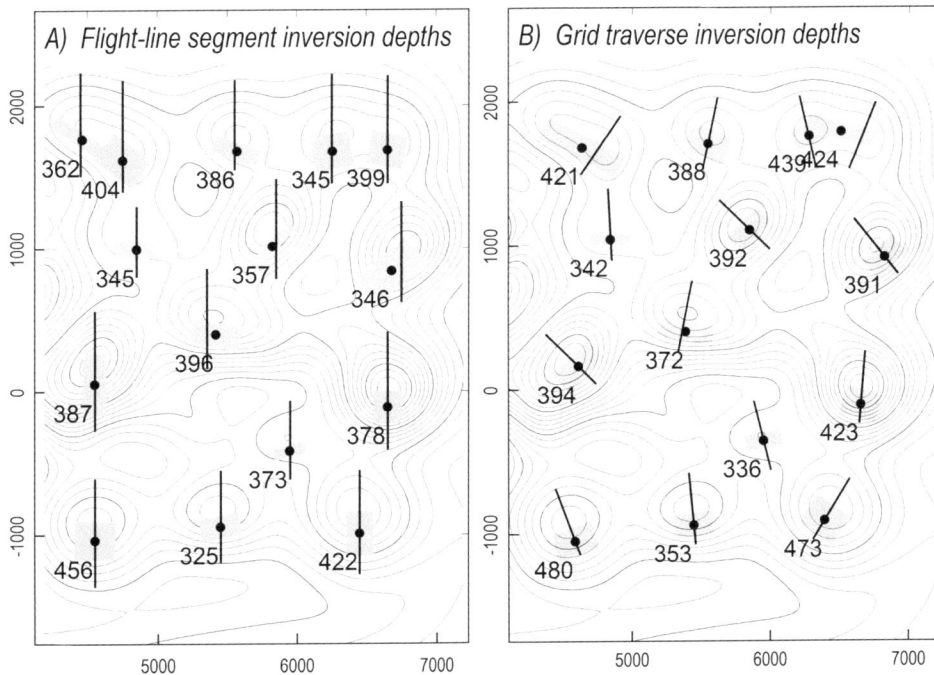

Fig. 3.13. Magnetisation depth 400 m, line spacing 100 m. TMI contours with A) inverted flightline segments and B) inverted grid traverses. Annotations are depths to top of magnetisation in metres.

traverse through it, unless those traverses are along grid rows or columns. To objectively investigate inversion of grid data I used an automated algorithm to draw traverses through each grid peak in the direction of maximum gradient. These grid traverses are shown in Fig. 3.13B. To provide a reference against which to evaluate the influence of gridding I also directly computed the magnetic field along the traverses. Statistics of the depth estimates from inversions of these directly computed traverses are listed in row D of Table 3.1. There is a low error in the mean depth value and the standard deviation and range of the depth solutions are similar to those for the single flightline inversions (as should be expected because both datasets are directly computed from the same input model). The models and annotated depth values shown in Fig. 3.13B are from individual inversions of the gridded data interpolated along the traverses. The results are listed in row E of Table 3.1. For this grid created from adequate sampling of close line-spaced data there is only a 1% error in the mean depth estimate. The standard deviation and range of the solutions are 11% and 36% respectively. These values are similar to those for solutions from inversion of the directly computed data traverses (we will see that this is in contrast to traverses through grids generated from an insufficient line spacing).

3.3.6 Under sampling of the magnetic field

In this section I investigate how inversion of multiple flightlines, single flightlines and grid traverses performs for an insufficient sampling of a magnetic field defined from flightlines at 400 m spacing (equal to the depth to the top of magnetisation). Each of the 400 m spaced flight lines is identical to a line in the 100 m spaced set, the difference is the absence of three-quarters of the lines. Following the rules of Reid (1980) this is an inadequate representation of the magnetic field for inversion or depth estimation. The line data are valid but at the wide line spacing do not reliably map the magnetic field. Some anomalies are not sampled by this wider line spacing and some separations between adjacent anomalies are not sampled, creating fewer, larger anomalies. Under-sampling of the field in some cases misses areas of sharp gradient that are most diagnostic of how deep a source is, encouraging overestimation of depth. However, in some cases under-sampling also causes misrepresentations that give rise to underestimation of depth.

Where significant differences between grids derived from subsets of survey lines (e.g. between sets of alternate odd- and even-numbered survey flightlines) reveals that those subsets insufficiently sample the magnetic field there is concern that their combination may also under sample the field. Figures 3.14A and 3.14B show

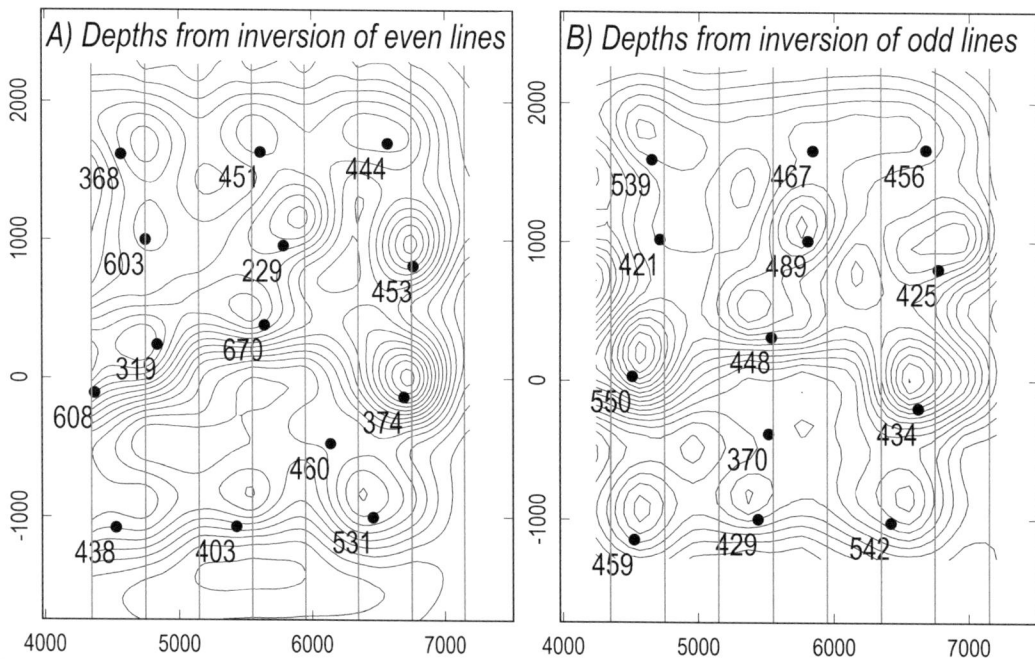

Fig. 3.14. TMI contours and annotated depths (in metres) to magnetisation derived from inversion of two 400 m spaced line sets A) and B) with a 200 m displacement between them.

contours of grids produced from two sets of 400 m spaced flightlines displaced from each other by 200 m, together with the annotated depth estimates from inversion of each subset of lines (these values can be compared to inversion of the complete 100 m spaced line set shown in Fig. 3.12A). Inversion produces 13 source models from inversion of one dataset and 14 from the other. There are 11 pairs of solutions for the same (but differently sampled) anomalies. Of those 11 pairs, seven have depth values within 10% of each other but there are also other pairs with differences of over 50%. There are also some substantial horizontal displacements between the centres of bodies derived from the two inversions. Statistics of the depth values are listed in rows F and G of Table 3.1. The average depth values for the two inversions are overestimated by 15% and the mean standard deviation is 22%.

3.3.7 Under-sampled survey multi-line anomaly inversions

For inversion of the 100 m line-spaced data there was an improvement of the depth estimates derived from single-anomaly, multi-line inversions rather than from simultaneous inversion of the complete dataset. For the 400 m line-spaced data there are few anomalies suitably sampled for individual multi-line inversion. In a compromise the data were divided into subsets of overlapping anomalies. The inversion depths are listed in row H of Table 3.1.

The mean depth error is significantly reduced from simultaneous inversion of the complete dataset but the standard deviation and range of values is similar.

3.3.8 Under-sampled survey single profile inversions

Single flightline inversions also suffer from under-sampling of the magnetic field because critical segments of the field that would have been more suitable for depth estimation may not be measured and because the horizontal centre location and horizontal extents of the magnetisations are misrepresented by the under-sampling. Flightline segments for single-line inversion were selected from both sets of 400 m spaced lines. Figures 3.15A and 3.15B show segments of the flightlines selected for individual inversion of each anomaly, together with centre points of the inversion models and their annotated depth values. Many of the anomalies are centred on profiles, because at a wide line spacing individual anomalies may have strong expression on only one profile. Just as for the complete dataset inversions, there is reasonable agreement between many of the pairs of depths from the even and odd alternate line samplings. Statistics of the depth values are listed in rows I and J of Table 3.1. There is a mean underestimation of depth by ~10% for each of the sets of depths.

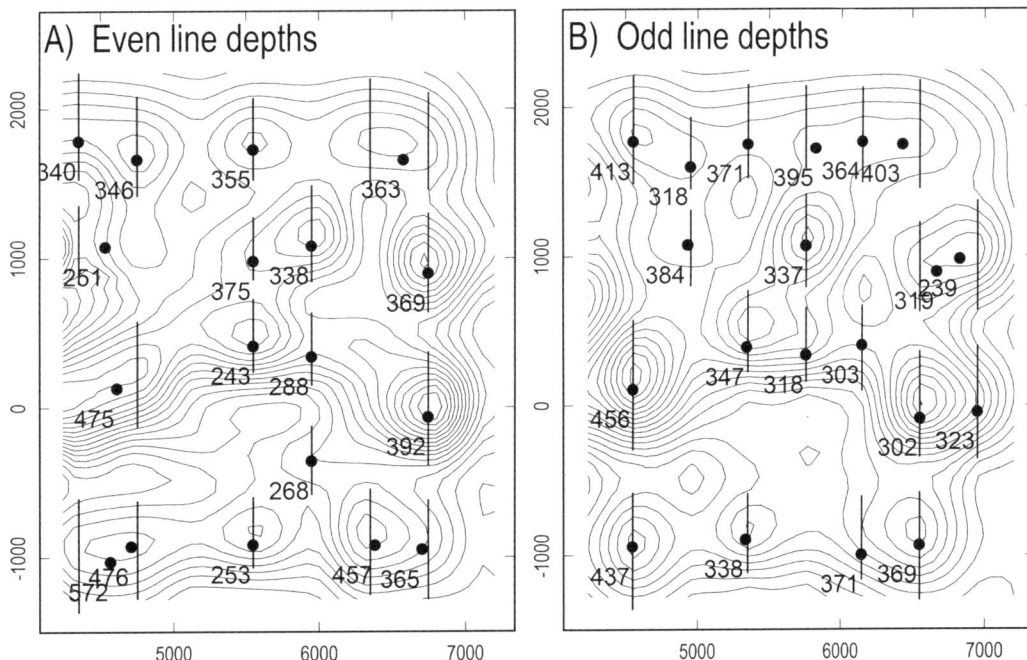

Fig. 3.15. TMI Contours and annotated model depths from inversion of segments of the 200 m displaced sets A) and B) of 400 m spaced flightlines with the annotated depths in metres.

3.3.9 Under-sampled survey grid-traverse inversions

Grid traverses derived from the 100 m and the 400 m line-spaced data are quite different to each other. For the 100 m spaced line data the gridded TMI interpolated onto traverses through the grid maxima is similar to TMI directly computed along those lines from the input model. However, on traverses through grids of the 400 m spaced flightlines there are substantial differences between the directly computed data and data interpolated from the grid, as shown in Fig. 3.16. This reveals substantial inadequacy in gridding of widely spaced data to represent the true magnetic field variation. Significant horizontal shifts between peaks of the directly computed and grid-interpolated channels show that in many cases the TMI grid mis-locates the magnetisations, and significant differences in curvature of the channels gives rise to errors in estimation of magnetisation depths.

Figure 3.17 shows the tracks of grid traverses automatically generated through maxima of the two grids of 400 m line-spaced data with annotated depth estimates from inversion of the grid data interpolated onto those traverses. Statistics of the depth values are listed in rows K and L of Table 3.1. Few of the depth values are correctly centred on the input model magnetisations and more plot off the magnetisation bodies than over them. Just as for the individual flightline segment inversions, the solutions are mostly underestimates of depth.

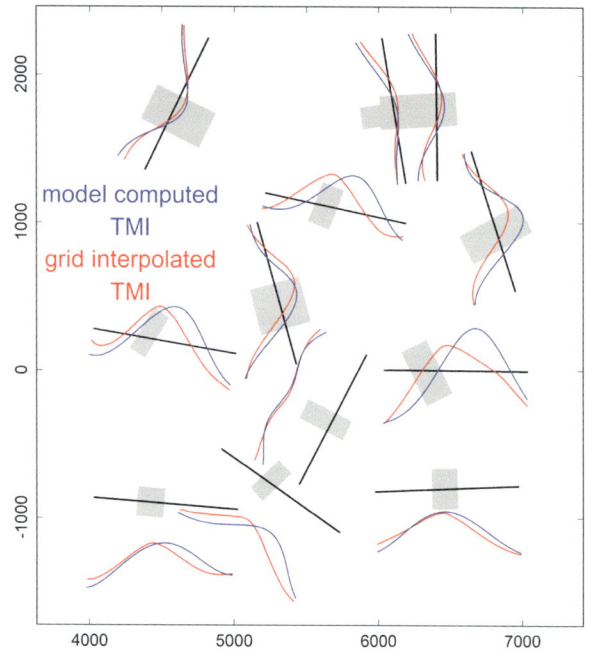

Fig. 3.16. Stacked profiles of the model field computed along the grid traverses (blue) and the field interpolated from gridding of the 400 m line-spaced data (red) with a plan view of the top of the inversion model bodies.

3.3.10 Conclusions of the synthetic data inversion study

This synthetic, noise-free study has established several key issues for estimation of depth-to-magnetisation:

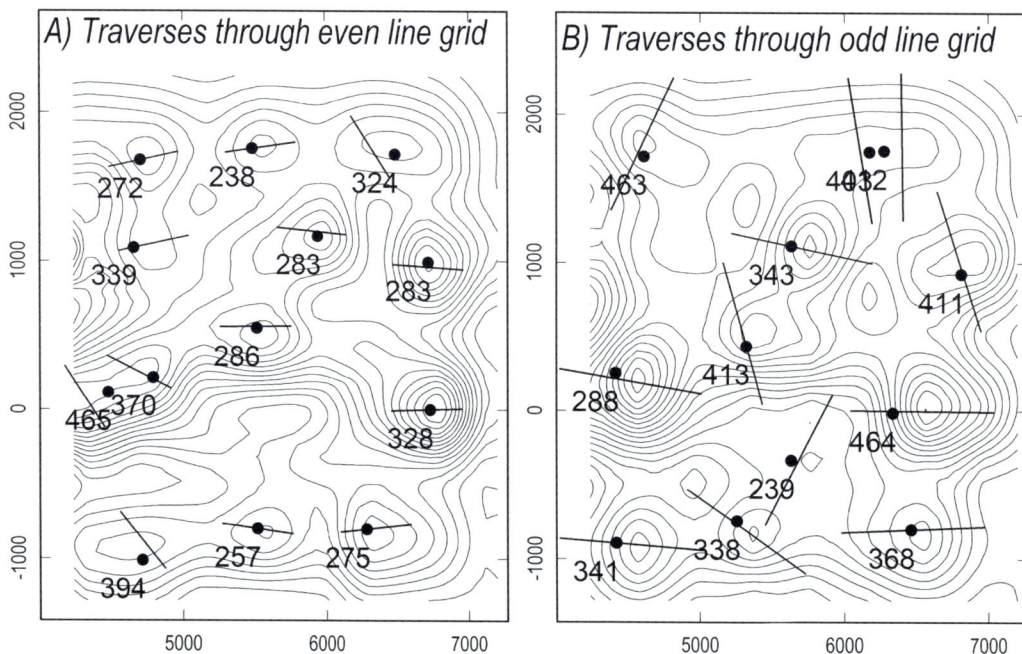

Fig. 3.17. Traverses and annotated depths (metres) from grid-traverse inversions of the two 400 m line-spaced grids.

1) Fundamental non-uniqueness resulting from unknown values of magnetic susceptibility, shape and size restricts confidence in depth-to-magnetisation estimates from isolated, well-defined anomalies to no better than 5%.

2) Anomaly isolation is a key source of additional uncertainty.

3) Individual focus on single anomalies provides superior depth estimates.

4) Field curvature is the most reliable indication of depth-to-magnetisation and is best estimated on well chosen individual profiles – ideally profiles of direct measurements.

5) Fidelity of field curvature is difficult to preserve in gridding – particularly of widely spaced data and this restricts reliability of depth estimates made from gridded data.

3.4 A DEPTH TO MAGNETISATION FIELD STUDY OF SEDAN, SOUTH AUSTRALIA

The Sedan detailed aeromagnetic survey (location shown in Fig. 3.18) was flown over an area with basement rocks belonging to the Delamerian Orogeny which include Cambrian-Ordovician mafic to felsic intrusives with some minor Cambrian metasediments. This basement is overlain by up to 100 m of weakly magnetic Tertiary cover. The shallowest strong magnetisation is expected to be at the top of fresh basement beneath a deep and variable weathering of the pre-Tertiary surface. The igneous rocks that source the highest amplitude

magnetic field variations are a suite of complex, zoned intrusions with high magnetisation in the more mafic zones and moderate to low magnetisation in the more granitic zones. In 1999 the Geological Survey of South Australia commissioned a regional aeromagnetic survey over the Sedan area. This survey (referred to here as the regional survey) was flown on east–west flightlines at a line spacing of 400 m and nominal terrain clearance of 80 m. At the same time the contractor flew a more detailed survey of a mineral tenement within the area. This survey (referred to here as the detailed survey) was flown on east–west flightlines at a line spacing of 80 m and nominal terrain clearance of 50 m. The two surveys were managed so that the regional survey flight and tie lines are horizontally coincident with detailed survey lines. Flightline maps of the two surveys are shown in Fig. 3.19.

There is a moderate variation in ground elevation of 160 m across the area and the range in TMI is ~1,200 nT. The two surveys provide a rare opportunity to compare data at different line spacings and elevations flown with the same aircraft, instruments and pilot and with the data processed identically. There are too few drill intersections of magnetisation to provide a meaningful ground-truthing of depths predicted from analysis and interpretation of the magnetic field data but nevertheless, these two surveys provide a test of the predictions of the previous synthetic data study. TMI images from gridding of each survey dataset are plotted in Fig. 3.20. Both TMI images show similar general features, with greater detail resolved by the closer line-spaced survey. Figure 3.21 shows very similar TMI profiles from the detailed and regional surveys along a horizontally coincident north–south tie-line (for location see Fig. 3.20). The short wavelength variations are of slightly higher amplitude in the lower elevation detailed survey than in the regional survey but this difference is almost removed by upward continuation of the detailed survey measurements by the average difference in reported elevation for the two lines of 24 m. Minor remaining long-wavelength differences between the two datasets may be due to differences in levelling of the data and/or along-profile variations of sensor elevation. Three sub-areas are outlined in Figs 3.19 and 3.20 for more detailed investigation.

Typical data complexity issues are illustrated in Fig. 3.22. The magnetometer elevation and horizontal position were monitored with global positioning system (GPS) instruments using real-time differential correction. There are variations in GPS elevation with

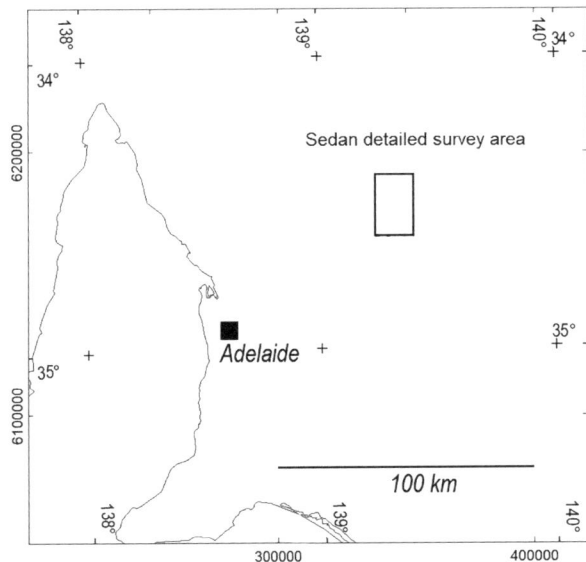

Fig. 3.18. Location of the Sedan detailed survey area.

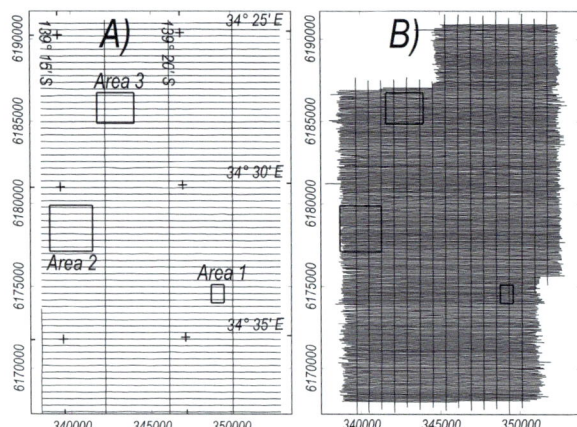

Fig. 3.19. Flight plans of A) the regional survey and B) the detailed survey.

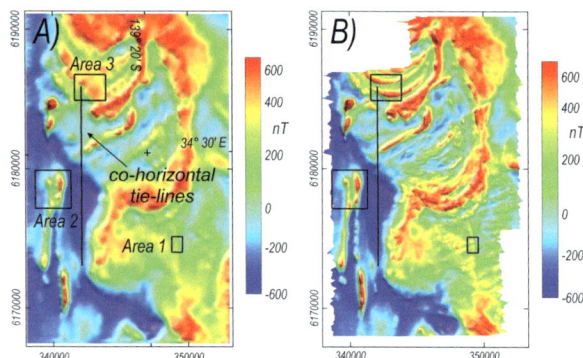

Fig. 3.20. TMI images from the regional survey (left) and detailed survey (right).

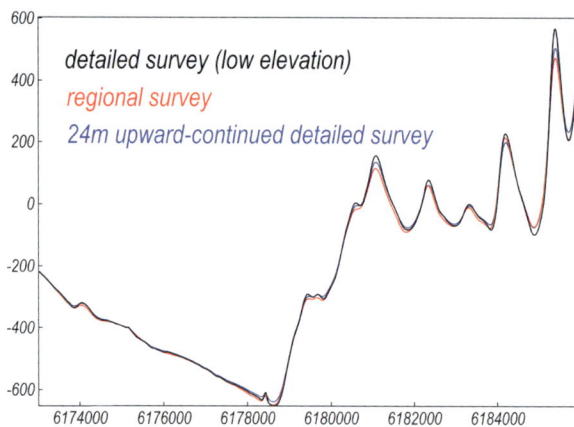

Fig. 3.21. Coincident regional and detailed survey N-S flightline TMI profiles (length 13 km -for location see Fig. 3.29).

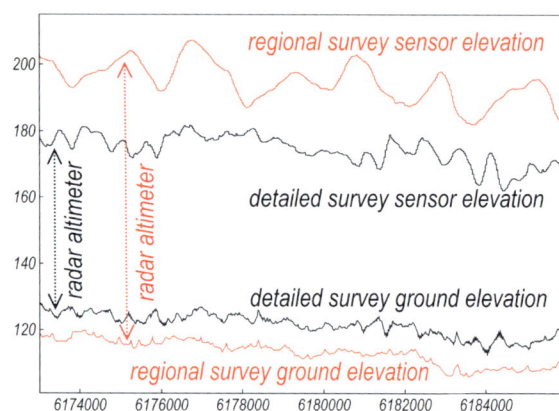

Fig. 3.22. Coincident regional and detailed survey N-S flightline GPS altimeter and DEM profiles (length 13 km – for location see Fig. 3.29).

wavelengths of up to 2 km and amplitudes up to 10 m in the detailed survey data and up to 15 m in the regional survey data that are uncorrelated between the surveys. Ideally use of the GPS channel provides compensation for these elevation differences in magnetic field computations, but exact linkage between the magnetic field and elevation is not preserved in the network adjustment (levelling) of the data. The surveys also measured elevation of the aircraft above the ground surface using a radar altimeter. Subtraction of the radar altimeter from the GPS altimeter should map the ground surface elevation. A digital elevation model (dem) of the ground surface is published as both a line data channel and a grid in the contractor-supplied survey data. An example of the line data is plotted in Fig. 3.22. The dem channels from the two surveys show common low-amplitude, long-wavelength variation and uncorrelated shorter wavelength variations of up to 5 m amplitude. However, the most prominent feature is the 8 m displacement between the curves which

should be coincident. This is most likely due to improper calibration of the radar altimeter, which is disappointing as the same instrument was used in the two surveys. Given that a primary objective of the detailed magnetic field survey is to map depth-to-magnetisation beneath the ground surface, the starting point of an inconsistency of 8 m between these survey datasets clearly adds to uncertainty in the subsequent estimate of depth to magnetisation below that surface.

3.4.1 Area 1

Figure 3.23 shows regional and detailed survey TMI images over Area 1 (for context see Fig. 3.20). The image colour scales are similar and the contour interval for both images is 20 nT. The centre of the detailed survey is dominated by a prominent anomaly of c. 100 nT peak

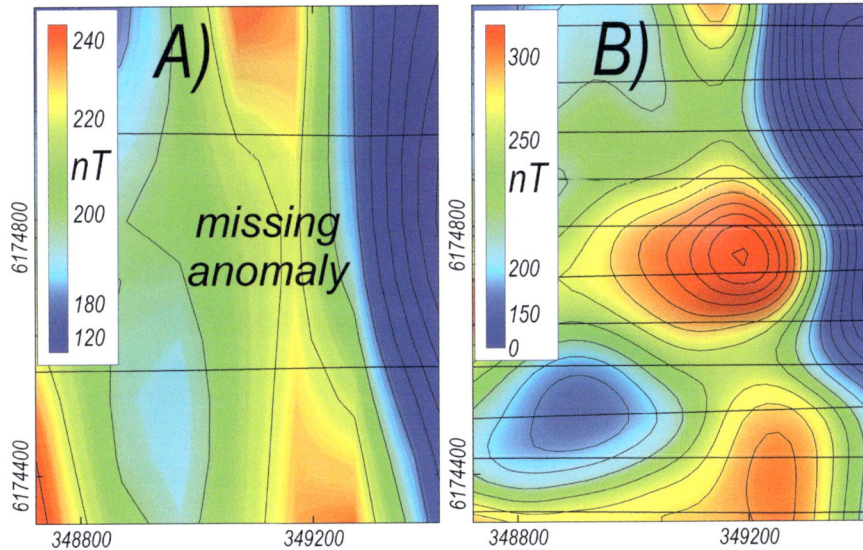

Fig. 3.23. TMI images of Area 1 (for location see Figs 3.19 and 3.20) from A) the regional survey and B) the detailed survey. For both images the contour interval is 20 nT.

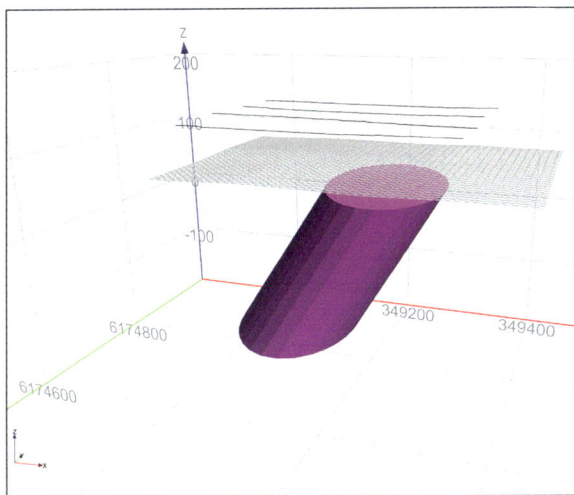

Fig. 3.24. Perspective of inversion model and flightlines.

amplitude. Two flightlines pass through the centre of this anomaly with the peak apparently halfway between them (Fig. 3.23B). The two adjacent lines to north and south define the flanks of the anomaly which has a north–south width of ~240 m. The anomaly is not evident in the regional survey because it is centred halfway between regional survey lines at 400 m spacing (Fig. 3.23A). A model generated from inversion of a segment of five detailed survey flightlines is shown in Fig. 3.24. The field computed from the model closely matches the measured field. The modelled top of magnetisation is 39 m below the ground surface and 89 m below the mean sensor elevation. The magnetisation appears to have

roughly circular cross-section with a diameter of just over 200 m, a depth extent of 180 m and a volume of over $5 \times 10^6 \text{ m}^3$.

The detailed survey anomaly is sufficiently defined to estimate a magnetisation direction. The best estimated magnetisation has intensity 0.9 A/m, declination 195° and inclination −69°. This direction is rotated from the local geomagnetic field direction by 44° which is a confident indication that remanent magnetisation is significant. Clearly none of this information can be recovered from the regional survey data.

This example indicates the inadequacy of wide line spacing for mapping shallow magnetisations of small extent. Anomalies of this size and shape might be of economic significance (e.g. diamondiferous kimberlite pipes) and if those were targets of interest the regional survey would be unreliable for finding them. If the regional survey had been flown with a 200 m north–south shift one of the flightlines would have passed over the anomaly and clearly detected it, but interpretation would still be difficult because it would be defined only on that single line. Regional 400 m line-spaced surveys have been flown over large areas in Australia. To investigate small but potentially significant anomalies detected by these surveys on only single flight lines there is a case for dedicated drone surveys that can detail the anomalies at low elevation and much closer line spacing. At lower elevation, shallow magnetisations produce significantly higher-amplitude anomalies allowing them to be mapped without the same precision required of aeromagnetic surveys to map low-amplitude field variations over large parts of their survey area. Another method to improve data coverage is to use cross-line, wing-tip gradiometry. Gradiometry would not have solved this example of extreme insufficiency in mapping magnetisation but it provides valuable additional information for less problematic cases of inadequate sampling. Unfortunately, increased cost of wing-tip gradiometry is commonly offset against flying surveys at wider line-spacing, in large part negating the potential advantage of the data.

3.4.2 Area 2 complete dataset inversions

Figure 3.25A shows TMI mapped from the regional survey data in Area 2 (see Fig. 3.20 for location). The anomalies are elongated, partially overlap and are aligned north–south perpendicular to the east–west flightlines. A corresponding image of TMI forward computed from an inversion model of the data is shown in Fig. 3.25B. The close match of measured and model computed fields suggests that this model is at least a candidate representation of the ground magnetisation, but in the previous synthetic data study similarly close data-fits were achieved by models with error in estimated depth of over 10%. This model of a geological magnetisation derived from measured data is likely to have more substantial failings. The values posted in Fig. 3.25B are estimated elevations of the top of magnetisation in metres below sea level. The datum is determined by referencing elevation to the GPS altimeter measurements. For each of the anomalies labelled in Fig. 3.25A the depth-to-magnetisation statistics are listed in Table 3.2 and summarised in Table 3.3. In the synthetic data study the top of magnetisation was at a constant elevation and the scatter of values was due completely to errors in the depth estimation. For this study the elevation of each magnetisation is unknown and is likely to vary between anomalies, so the range of estimated depth values is an unknown mixture of true depth variation between different magnetisations and errors in estimation of those values. For the five selected anomalies the mean elevation is 94 m BSL, the mean depth below sensor is 327 m and the mean depth below ground is 247 m, with a range of 148 m.

Figure 3.26A shows the TMI image over the same area generated from the detailed survey data, and Fig. 3.26B shows a matching TMI image forward computed from the model derived from inversion of that data. For the five selected anomalies the mean elevation is 89 m BSL, the mean depth below sensor is 283 m and the mean depth below ground is 233 m with a range of 71 m. Multi-flightline inversions of the two survey datasets provide consistent results. The difference between the mean elevation of the five anomaly source models from inversion of the two survey datasets is only 5 m, and the difference in their mean depths below ground is only 14 m (Table 3.3) that is 5% of the average depth below sensor. The average difference between corresponding pairs of estimates of depth to magnetisation below ground is 25 m (9% of the average depth below sensor).

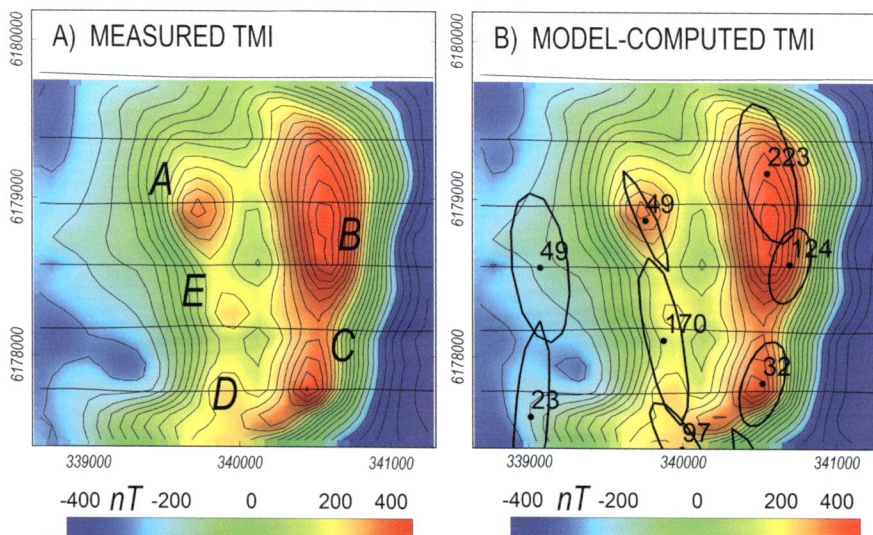

Fig. 3.25. Area 2 Regional survey A) measured TMI and B) TMI computed from the inversion model with body plan and annotated top of magnetisation elevation (metres BSL).

Table 3.2. Sedan Area 2 summary of depth to top of inversion models.

Survey	Data	Susceptibility (SI)	Top elevation (m ASL)	Sensor (m ASL)	Ground (m ASL)	Depth below ground (m)	Depth below sensor (m)
Anomaly_A							
regional	complete	0.336	–49	240	160	209	289
regional	traverse	.0872	68	229	149	81	161
regional	flightline	.137	55	234	146	91	179
detailed	complete	.342	–48	198	148	196	246
detailed	traverse	.535	14	208	158	144	194
detailed	flightline	.124	61	210	157	96	149
Anomaly_B							
regional	complete	0.621	–124	224	144	268	348
regional	traverse	.161	–75	213	133	208	288
regional	flightline	.211	–90	215	132	222	305
detailed	complete	.354	–110	181	131	241	291
detailed	traverse	.187	–16	191	141	157	207
detailed	flightline	.286	–37	191	142	179	228
Anomaly_C							
regional	complete	0.266	–32	227	147	179	259
regional	traverse	.572	99	225	145	46	126
regional	flightline	.189	9	207	141	132	198
detailed	complete	.509	–61	195	145	206	256
detailed	traverse	.291	33	195	145	112	162
detailed	flightline	.138	56	194	145	89	138
Anomaly_D							
regional	complete	0.356	–97	236	156	253	333
regional	traverse	.0443	46	229	149	103	183
regional	flightline	.156	15	228	149	134	213
detailed	complete	.377	–118	199	149	267	317
detailed	traverse	.195	14	202	152	138	188
detailed	flightline	.163	5	204	153	148	199
Anomaly_E							
regional	complete	0.256	–170	237	157	327	407
regional	traverse	.0942	68	228	148	80	160
regional	flightline	.0194	106	231	146	40	125
detailed	complete	.385	–107	197	147	254	304
detailed	traverse	.0848	10	205	155	145	195
detailed	flightline	.107	50	209	155	105	159

In the Sedan area the weak magnetisation contrast across the ground surface does not generate significant magnetic field variations in the aeromagnetic data (the ground is magnetically 'transparent') and performance of the magnetic source depth analysis is most appropriately referenced to depth below sensor. However, to users of the depth estimates the most relevant measure is depth below ground. Accumulation of uncertainty as a percentage of depth from the sensor to the ground surface is effectively a penalty for aeromagnetic data. For the Sedan surveys flown at nominal elevations of 80 and 50 m above the ground and with the top of magnetisation

Table 3.3. Summary of Area 2 depth-below-ground estimates.

Survey	Data	Mean depth (metres)	Samples	
regional	complete lines	247	5	
detailed	complete lines	233	5	
regional	grid-traverse	104	5	
detailed	grid-traverse	139	5	
regional	Single flightline	124	5	
detailed	Single flightline	123	5	
Differences between results				**Standard deviation (metres)**
regional-detailed	complete line set inversions	25	5	23
regional-detailed	grid-traverse inversions	56	5	13
regional-detailed	single flightline inversions	34	5	24
regional	grid-traverse and flightline inversions	36	5	30
detailed	grid-traverse and flightline inversions	29	5	15

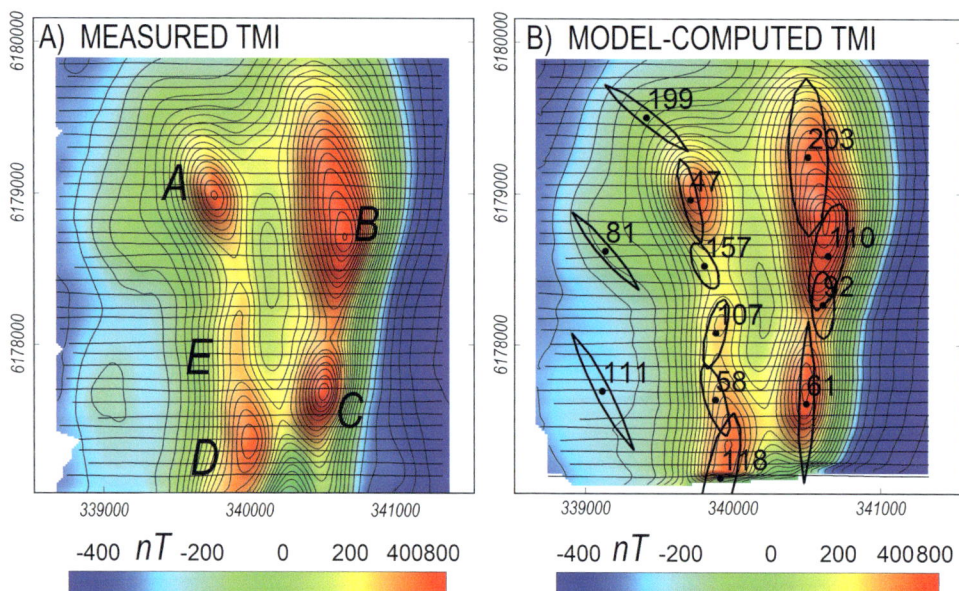

Fig. 3.26. Area 2 Detailed survey A) measured TMI and B) TMI computed from the inversion model with body plan and annotated top of magnetisation elevation (metres BSL).

in most places more than 100 m below ground this penalty is not prohibitive, but for substantially shallower magnetisations or greater flying heights the uncertainty accumulated between the sensor elevation and the ground surface can exceed the average depth of magnetisation below surface. In such cases depth-below-ground estimates generated from aeromagnetic data analysis have little meaning. For depths that must be known reliably (e.g. as a basis for making drilling decisions) a lower elevation or ground surface survey should be acquired to upgrade the aeromagnetic analysis results.

3.4.3 Area 2 single flightline and grid-traverse inversions

In the synthetic data study the most reliable depth-to-magnetisation estimates were derived from inversions of carefully selected segments of single flightlines. A corresponding example from the study of area 2 is shown in Fig. 3.27. In this example there is a subtle inflection on the western flank of the anomaly that is unlikely to be recognised as significant in a grid traverse (although the feature can just be seen as a minor bulge in the grid contours in Fig. 3.26A). This feature is confirmed more clearly in the upper vertical derivative track. This track is derived from FFT analysis of the closely spaced and directly measured flight line data. The transform is approximate because there are no estimates of the cross-line horizontal gradients, but those are relatively weak and the transform applied identically to the measured and model computed fields allows valid comparison in optimising the inversion with reduced concern for the regional field separation and the depth extent of the bodies. The main function of the weakly magnetised western body is to allow the more prominent easterly magnetisation to explain only the main anomaly. After inversion, the western body can optionally be retained or discarded. Individual profile inversion depths for the various anomalies derived from both the regional and detailed surveys are listed in Table 3.2.

Figure 3.28 shows an example south-east to north-west grid traverse through the detailed grid anomalies C and E. There is overlap between the flanks of the anomalies but this does not substantially change the sharpness of curvature over their centres and therefore has little impact on the depth-to-magnetisation estimates. The overlap does, however, introduce an opportunity to incorrectly assign overlapping field variations between the two magnetisations and change the asymmetry of each anomaly from which the apparent dip of the magnetised bodies is estimated. For both the flightline and grid-traverse inversions the field variation assigned to the regional might include contributions from deep extensions of the magnetisations, but this has little effect in estimation of depths to their tops. Figure 3.28 also shows the location and profile intersection of bodies derived from inversion of the adjacent Traverse C

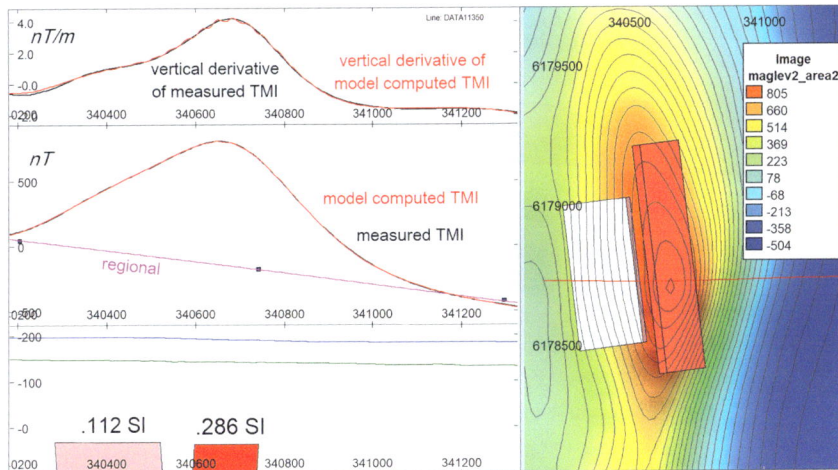

Fig. 3.27. Detailed survey flightline inversion through anomaly B in area 2 (for location see Fig. 3.26A).

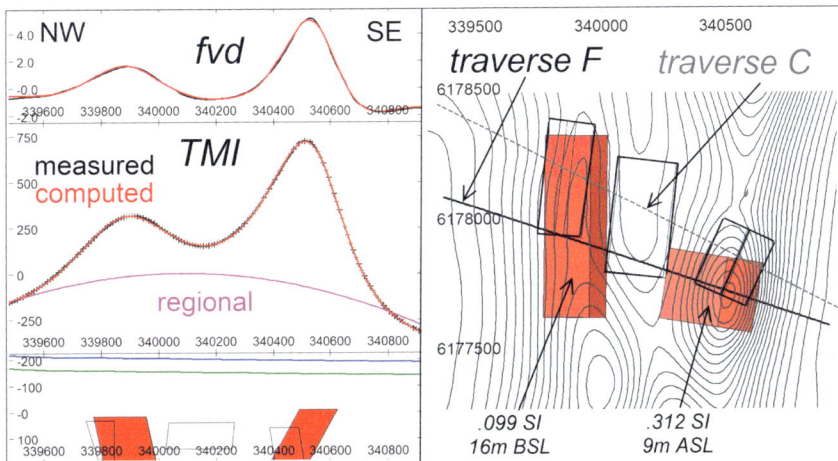

Fig. 3.28. Detailed survey flightline inversion through anomalies C and E in area 2 (for location see Fig. 3.26A).

through the regional survey grid. The two survey grids are representations of the same magnetic field (with a difference in elevation between them of 30 m) but as a result of the different sampling the two grids are best interrogated on differently located traverses. Tops of the detailed grid-traverse inversion bodies shown in red in Fig. 3.28 are both shallower than tops of the corresponding regional survey grid-traverse bodies (shown in outline).

As reported in Table 3.3, the mean grid-traverse depth below ground estimate from inversion of the regional survey data is shallower than the estimate from the complete-dataset regional survey inversion (104 m compared to 247 m) consistent with relationships established in the synthetic data study. The same relationship (but with a smaller discrepancy) is seen for the mean depths of the detailed survey inversions (139 m for the grid-traverses compared to 233 m for the complete dataset). From learnings of the synthetic data study the individual flightline segment inversions are considered the most reliable depths and these show little variation in mean values between the two surveys.

Relationships between depths derived by the various methods are also summarised in Fig. 3.29. Consistent with the results of the synthetic data study, the regional and detailed survey complete-dataset inversion depth-below-ground values (plotted in black in Fig. 3.29) are significantly deeper than the sweet-spot individual grid-traverse and single flightline inversion depths (plotted in red and blue). The detailed and regional survey single flightline inversion depths (red and blue crosses respectively) have almost identical mean values of 123 and 124 m but the mean grid-traverse depths for the two surveys differ by 35 m (139 and 104 m). Many of the cross-plotted

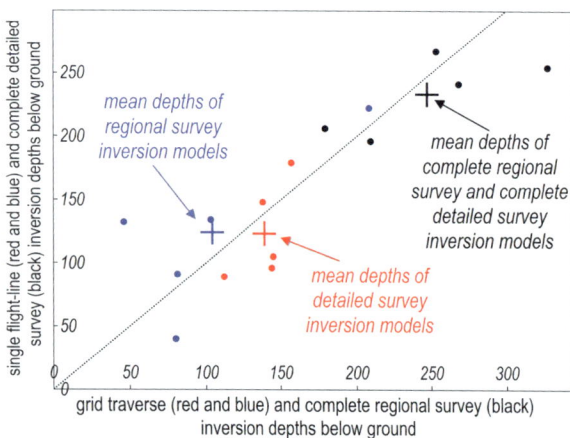

Fig. 3.29. Cross-plots of depth estimates in Area 2 derived from the two surveys and different data types.

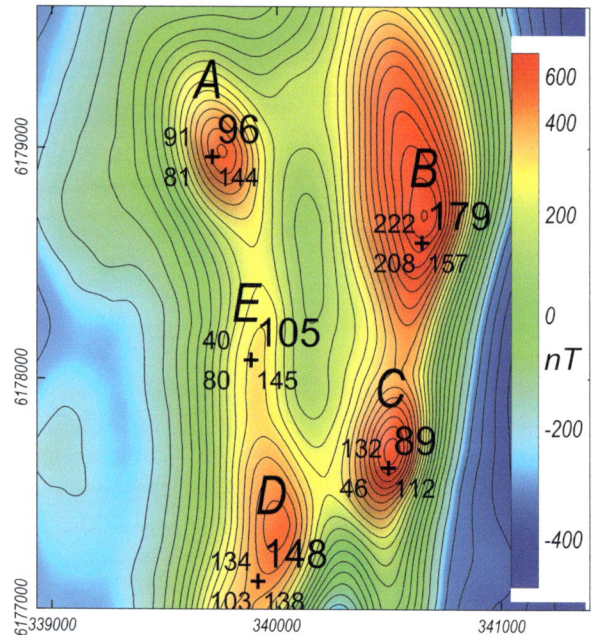

Fig. 3.30. Depth-below-ground estimates for each anomaly. The upper values are from the detailed survey and lower values from the regional survey. Left-hand values are from grid traverses and right-hand values are from flightlines. The bold upper right values from inversion of detailed survey flightline data are the preferred depths.

individual depth values are close to the 1:1 slope and hopefully the variations along that line are substantially due to true depth variation to the source of each anomaly. From lessons learnt in the synthetic data study the preferred depth values are those derived by single flightline inversions of the detailed survey data.

Individual grid-traverse and single flightline depth values for each anomaly are mapped in Fig. 3.30. The best estimate of mean depth below sensor is 175 m for the detailed survey and 204 m for the regional survey. For the flightline spacings of 80 m and 400 m this gives line spacing to source depth ratios of 1:2.2 and 1: 0.5 for the detailed and regional surveys respectively. This is a less satisfactory sampling of the magnetic field than was investigated in the synthetic data study and would be expected to give larger proportional depth errors. However, for this section of the Sedan survey the anomalies are elongate perpendicular to the flightline direction and this optimises fidelity with which depth estimates are recovered.

3.4.4 Area 3 regional and detailed survey TMI anomalies

Figure 3.31 shows TMI images from Area 3 (for location see Fig. 3.20). For Area 2 the TMI images of the regional

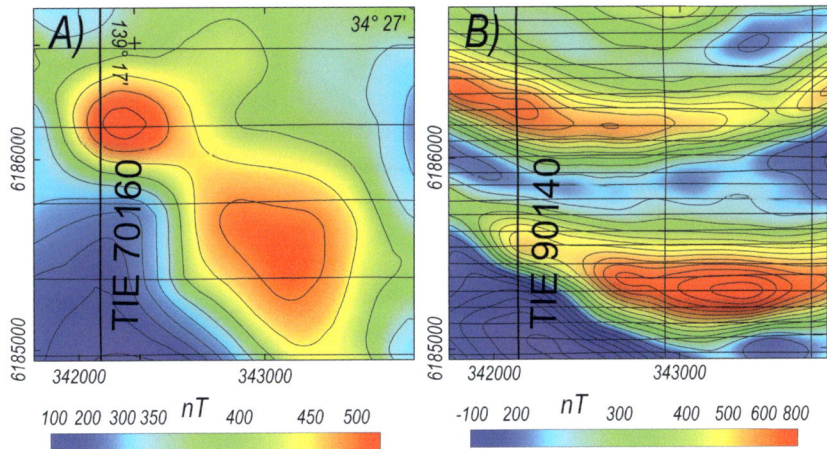

Fig. 3.31. Area 3 TMI and flight lines: A) regional survey and B) detailed survey.

and detailed surveys were broadly similar. For Area 3 it is difficult to recognise that the two TMI images are of the same area. Under-sampling at the 400 m line spacing of the regional survey completely distorts representation of the magnetic field. The anomalies in Area 3 are more elongate than those in Area 2 which should provide advantage in mapping the magnetic field. However, in Area 3 elongation of the anomalies is almost parallel to the flightline direction rather than perpendicular to it as in Area 2. Figure 3.32 shows a 2.3 km section of a horizontally co-located north–south tie-line with data from both surveys. The northern and southern anomalies in the detailed survey data have amplitudes of 450 and 410 nT respectively. Those peak values are smaller by 40 and 95 nT respectively for the regional survey measurements made at a higher elevation. A substantial part of this difference is removed by applying a 23.4 m upward continuation filter to the detailed survey data to compensate for the reported mean elevation difference between the profiles. The remaining differences between the profiles reduced to the same elevation are primarily long wavelength variations due to differences in elevation along the individual profiles or introduced in the two different survey network adjustments of the data. However, these longer wavelength differences do not significantly influence the local curvature of the field at the peaks of the anomalies and have little effect on the depth estimates.

Figure 3.33 also shows measured TMI along the profile, in this case together with interpolation of the regional and detailed TMI grids. The algorithm used in gridding both survey datasets ensures that the grid values along the flightlines approximately honour the profile data. However, tie-lines are discarded once they

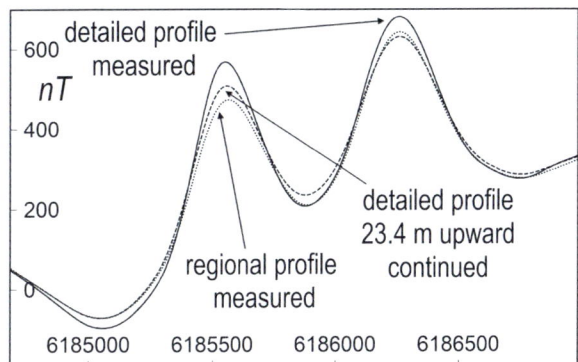

Fig. 3.32. TMI on a coincident regional and detailed survey flightline (located in Fig. 3.31).

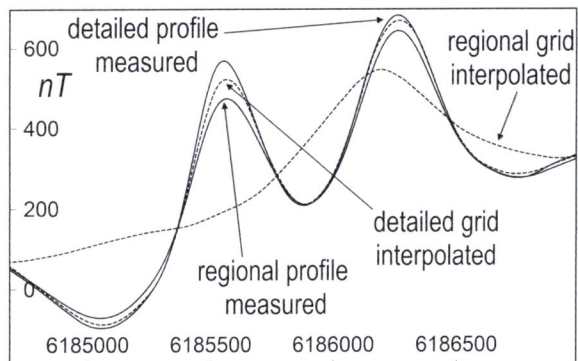

Fig. 3.33. TMI on a coincident regional and detailed survey tie-line (located in Fig. 3.31).

have been used for network adjustment and are not directly included in the gridding. The detailed survey grid interpolation reasonably matches the measured tie-line data from that survey with minor attenuation of the shorter wavelength anomalies (their peak to trough amplitudes are reduced slightly). However, as shown in

Fig. 3.33, for the regional survey the data interpolated from that grid onto the tie-line completely misrepresents the directly measured data. The segment of the regional survey tie-line in Fig. 3.33 includes only five survey flightline intersections to provide samples for generation of the grid. The regional survey TMI grid interpolation does not reveal the southern anomaly at all and the northern anomaly is much broader, of lower amplitude and horizontally displaced relative to the profile measurements. Clearly no depth can be estimated from the interpolated regional survey grid data for the missing southern anomaly and any depth estimate for the northern anomaly will be highly erroneous. In Fig. 3.31 the catastrophic under-sampling of the magnetic field by the regional survey grid is obvious by comparison with the detailed survey grid. The only evidence of that under-sampling from the regional survey data itself is the discrepancy between the measured and grid-interpolated data along the tie-line as shown in Fig. 3.33. Very few users of grid data interrogate the validity of a grid in this way and consequently significant deficiencies in grids go undetected. Any grid-based methods of depth estimation will only provide misleading results from this segment of the regional survey grid and second order derivatives of the gridded data as used in some depth estimation methods are completely fictitious.

3.4.5 Area 3 individual grid-traverse and single tie-line segment inversions

Inversion of single measured profiles in Area 3 from either survey is restricted to the tie-lines because the flight lines are too oblique to the field variations for individual analysis or inversion. Figure 3.34 shows inversion of a tie-line from the detailed survey. Also shown for comparison is the intersection of the model from inversion of a coincident (but higher) regional survey line. The models are in excellent agreement. In a total depth below sensor of over 200 m there is only a 13 m difference in depth between the tops of the southern magnetisation models (the detailed survey inversion gives the shallower result). As predicted from the synthetic data study, in compensation for the difference in depth the deeper body is narrower and of higher apparent susceptibility. Depth to magnetisation is well mapped at the tie-line intersections with the northern and southern anomalies but even for the detailed survey these tie-line intersections only occur at 800 m intervals. Any attempt to estimate depth to magnetisation from grid data in Area 3 is poorly justified. The quality of depth estimates can be improved for these linear anomalies by use of anisotropic gridding (Naprstek and Smith 2019; Davis 2022) as discussed in Chapter 2. However, inverting grid data away from the measurement lines

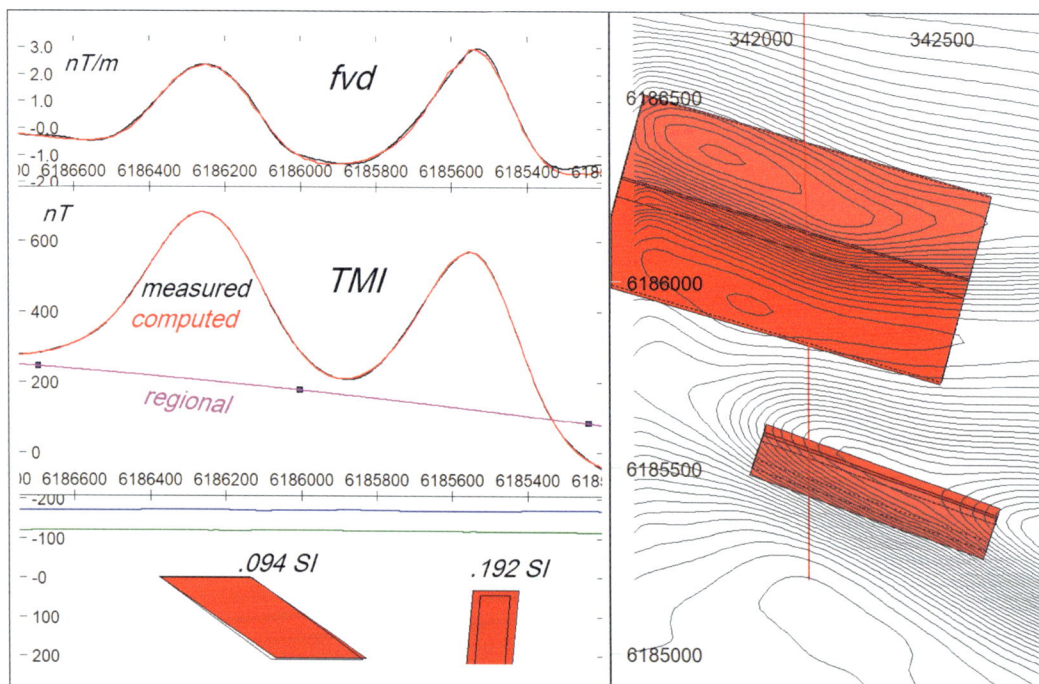

Fig. 3.34. Detailed survey tie-line 10940 inversion models (red) and intersections of models from inversion of the coincident regional survey tie-line.

tests only the fidelity of the gridding algorithm used and not the true magnetic field.

Even without knowledge of the true depths to magnetisation, the Sedan study has established limitations of mapping depth-to-magnetisation from regional 400 m line-spaced surveys over areas where magnetisation is at depths below sensor in the range of 200 m or less (100 m or less below surface). The surveys are able to reasonably determine depths to magnetisation at 50 m below ground for magnetisations that are elongate perpendicular to the flightline direction but are inadequate to determine magnetisation depth from small, equidimensional anomalies or from anomalies elongate oblique to the flightline direction. The detailed survey with closer line spacing and at lower elevation provides for some anomalies more consistent and reliable depth estimates than the regional survey and for other anomalies provides estimates that cannot be made from the regional survey.

3.5 CONCLUSIONS

The challenges and capabilities of magnetic source depth estimation are complex. The highly disruptive consequences of non-uniqueness are not just in not being able to know the correctness or otherwise of a solution but also therefore of not being able to assign a conventional statistical uncertainty to that solution. The capabilities of magnetic source depth estimation vary between projects dependant on the suitability of the geology and the quality and sufficiency of the magnetic field data. To better accumulate lessons of magnetic field source depth estimation requires development of a significant database of depth predictions that are subsequently tested by drilling. No such comprehensive database is presently available in published literature.

In this study I have focussed on the 'greenfields' challenge of estimating depth to an unknown magnetisation without access to independent information. The sweet-spot method I have illustrated can be applied to any reasonably defined discrete magnetic field anomaly, of which a typical aeromagnetic survey may contain tens to many thousands. Parametric inversion used in this study is well suited for any such studies. This chapter has investigated generation of estimates of depth to top of a magnetisation. In addition to this focus on the best expected depth, further studies can be designed to independently estimate the feasibility that the top of magnetisation is at some shallower depth or to estimate the greatest feasible depth.

REFERENCES

Almond R, Fitzgerald DJ (1998) Naudy based automodelling with trend enhancements. *Exploration Geophysics* **29**, 372–377. doi:10.1071/EG998372

Davis A (2022) Nested anisotropic geostatistical gridding of airborne geophysical data. *Geophysics* **87**, E1–E12. doi:10.1190/geo2021-0169.1

Hansen RO (2005) 3D multiple-source Werner deconvolution for magnetic data. *Geophysics* **70**, L45–L51. doi:10.1190/1.2073883

Ku CC, Sharp JA (1983) Werner deconvolution for automated magnetic interpretation and its refinement using Marquardt's inverse modelling. *Geophysics* **48**, 754–774. doi:10.1190/1.1441505

Marquardt DW (1970) Generalized inverses, ridge regression, biased linear estimation, and nonlinear estimation. *Technometrics* **12**, 591–612. doi:10.1080/00401706.1970.10488699

McGrath PH, Hood PJ (1970) The dipping dike case: a computer curve-matching method of magnetic interpretation. *Geophysics* **35**, 831–848. doi:10.1190/1.1440132

Nabighian MN (1972) The analytic signal of two-dimensional magnetic bodies with polygonal cross-section: its properties and use for automated anomaly interpretation. *Geophysics* **37**, 505–517. doi:10.1190/1.1440276

Naprstek T, Smith RS (2019) A new method for interpolating linear features in aeromagnetic data. *Geophysics* **84**, JM15–JM24. doi:10.1190/geo2018-0156.1

Naudy H (1971) Automatic determination of depth on aeromagnetic profiles. *Geophysics* **36**, 717–772. doi:10.1190/1.1440207

Pratt DA, White AS, Parfrey KL, McKenzie KB (2020) 'ModelVision User Guide Version 17.0.' Tensor Research (unpublished).

Reid AB (1980) Aeromagnetic survey design. *Geophysics* **45**, 482–516. doi:10.1190/1.1441102

Reid AB, Allsop JM, Granser H, Millett AJ, Somerton IW (1990) Magnetic interpretation in three dimensions using Euler deconvolution. *Geophysics* **55**, 80–91. doi:10.1190/1.1442774

Silva JBC, Barbosa VCF (2003) 3D Euler deconvolution: theoretical basis for automatically selecting good solutions. *Geophysics* **68**, 1962–1968. doi:10.1190/1.1635050

Thompson DT (1982) EULDPH: a new technique for making computer-assisted depth estimates from magnetic data. *Geophysics* **47**, 31–37. doi:10.1190/1.1441278

Vallée MA, Keating P, Smith RS, St-Hilaire C (2004) Estimating depth and model type using the continuous wavelet transform of magnetic data. *Geophysics* **69**, 191–199. doi:10.1190/1.1649387

Werner S (1953) 'Interpretation of magnetic anomalies at sheet-like bodies.' Sveriges Geologiska Undersok, Ser. C. Arsbok, 43, no. 6. Published by Sveriges geologiska undersöking, Stockholm.

4

Empirical measures of proximal and distal magnetic fields

C.A. Foss

ABSTRACT

At great distance from a magnetisation the information carried about it in the magnetic field is condensed to six parameters: the three coordinates of its centre, and the strength and orientation of its moment (the three-dimensional orientation of magnetic moment is specified by two angles). Further recession from the magnetisation does not cause loss of this information provided the field can still be detected and analysed. This is a distal field. Fields closer to the magnetisation that carry additional information about its distribution are to some degree proximal. There is a penalty that comes with the additional information in a proximal field: that it complicates and degrades recovery of information about its centre and direction of magnetisation that are (at least in theory) well conveyed by the distal field. This degradation is primarily the consequence of domination of the closest samples of the magnetic field by the closest zones of the magnetisation. It is important in magnetic field analysis and interpretation to understand the capabilities and limitations in recovering information from magnetic fields. The attempts I present in this chapter to quantify this gradual and continuous transition between distal and proximal fields are of limited success but hopefully provide insights to data-related capabilities and limitations in magnetic field analysis.

4.1 INTRODUCTION

A magnetic field is proximal to a magnetisation if it carries significant detailed information of its spatial distribution. If the only spatial information is the location of the centre of magnetisation then the magnetic field measurements are fully summarised by a dipole model and the field is completely distal. For distal magnetic fields the only two challenges in estimation of source magnetisation direction are correct isolation of the field due to that magnetisation and correct determination of the centre location of the magnetisation. As a proxy of transition between proximal and distal fields I compare misfits of dipole model fields with fields of simple alternative models rather than (as is usually done) with input measured fields. These misfits increase with proximity to the magnetisation and are dependent on the complexity of its distribution. For inversion of alternative models I assign complexity to the model bodies in three classes: (i) approximately equidimensional complexity, (ii) horizontal elongation and (iii) plunge.

To investigate inadequacy of dipole magnetisation models for a selected task I propose two analyses. The first is analysis of difference between the best-fit dipole model and the best-fit model of two independent vertically and horizontally polarised dipoles (I call this VHPD or 'vertically and horizontally polarised dipole' analysis).

The second is analysis of differences between the best-fit dipole and best-fit ellipsoid models (I call this DE or 'dipole-ellipsoid' analysis).

4.2 DIPOLES AND DISTAL AND PROXIMAL MAGNETIC FIELDS

There is continuous and gradual transition between proximal and distal fields and the degree to which a magnetic field is distal or proximal to its source depends not just on position of the source magnetisation relative to the field measurements but also on the spatial distribution of magnetisation and the spacing, precision and nature of the field measurements (e.g. total field, component, single gradient or tensor gradients). For highly proximal magnetic fields measured by low elevation or ground surface surveys over shallow subsurface magnetisations, discontinuities in the top surface of the magnetisation or near-top-of-magnetisation inhomogeneities cause sharp and high-amplitude field variations within the closest field measurements. Those local field variations are unlikely to be sufficiently sampled and their extreme amplitude variations disproportionately influence inversion, reducing confidence in the resulting models. At greater distances from the magnetisation these details of shape become much less significant and more reliable estimates of the total magnetisation can be obtained.

Descriptions of a magnetic field as proximal or distal (and the corresponding descriptions of magnetisation as 'compact' or 'distributed') specifically imply the level of information about the source magnetisation that can be recovered from a magnetic field. If a measured magnetic field can be broken into several discrete or only partially overlapping data regions within which the field is acceptably matched by dipole sources, then the field is purely distal and a list of the dipole statistics contains all of the information that can be recovered from it (note that for an ideal dipole the magnetic field becomes distal immediately external to it). A dipole magnetisation does not specify an intensity of magnetisation or a volume, only their product: the magnetic moment. Exactly the same dipole magnetic field variation due to a homogeneously magnetised sphere is generated by any co-centred sphere with the same magnetic moment and with purely radial variation in intensity and/or direction of magnetisation (i.e. consisting of shells of homogeneous magnetisation). Dipole source models leave some information unrecovered from the magnetic fields of irregular and complex (distributed) magnetisations.

The most pertinent question on first examining an image of a magnetic field is how distal or proximal various parts of the field are to their source magnetisations. This question is the starting point to evaluate how much information can be recovered from the magnetic field data and to develop strategies by which to achieve it. Note that meaningful evaluation of a magnetic field requires knowledge of the distribution of its individual measurement locations to appreciate how the level of detail in the field may be influenced by the density at which it is sampled. It is especially important to know the distribution of measurements before interpreting or inverting magnetic field grids because grid data is often separated from the primary measurements from which it was generated.

4.3 VERTICALLY AND HORIZONTALLY POLARISED DIPOLE (VHPD) AND DIPOLE-ELLIPSOID (DE) ANALYSES

We can attempt to match a magnetic anomaly that might be due to a dipole magnetisation with two dipoles of independent location and strength, one of a vertically directed magnetisation and the other with a magnetisation in the horizontal plane. If the anomaly is indeed due to a dipole source and has been correctly separated from other field variations then inversion should co-locate the two polarised dipoles, the vector sum of their magnetisations is the true dipole magnetisation, and the field is distal. If the dipoles are not co-centred or if there is a significant coherent residual misfit to the input field then the magnetisation is not completely dipolar and the field measurements are to some degree proximal to the magnetisation.

A second, separate dipole-ellipsoid (DE) test of inadequacy of a dipole model is the change in magnetisation parameters on replacing the best-fit dipole model with a more complex best-fit ellipsoid model. This test has the advantage that an ellipsoid is a physically meaningful model (which separated VHPD sources are not) and that ellipsoids supply what may prove significant additional parameter values and upgrades of dipole-based parameter estimates. VHPD and DE model analyses provide related but different measures of inadequacy of dipole models that may be able to contribute towards evaluation of a magnetic field as predominantly distal or predominantly proximal.

4.4 VHPD DECOMPOSITION OF A SYNTHETIC DATA DIPOLE ANOMALY

Figure 4.1 illustrates VHPD analysis of an input field from a dipole source. The input field (Fig. 4.1A) is due to a homogeneous sphere of radius 100 m, depth 200 m and magnetisation 2 Amp/metres with declination 135° and inclination +45°. The geomagnetic field setting of this model has a strength of 60,000 nT, declination 0° and inclination −60°. TMI is computed at 25 m intervals across a square array of 800 m side length centred on the source magnetisation. Figure 4.1B shows TMI computed from the pre-inversion polarised dipoles starting model. Both starting polarised dipoles have magnetisations of 1 A/m and the horizontal magnetisation has a declination of 80°. The bodies are displaced by 16 and 243 m from the source solution. The inversion model has nine free parameters – the X, Y and Z coordinates of the two bodies, the intensity of magnetisation of the bodies and the declination of magnetisation of the horizontally magnetised body. Despite the substantial misfit of the starting model this inversion successfully and rapidly converges to an almost perfect fit to the input field. The final body locations are coincident with the primary model to within 1 m both horizontally and vertically and the vector sum of their magnetisations agrees within 1° with the primary model. The individual fields of the final inversion polarised dipoles are shown in Fig. 4.1C and 4.1D. It is the summation of these fields that matches the input field shown in Fig. 4.1A. All other test inversions with input dipole model data have produced similarly successful results.

The purpose of VHPD analysis is to provide a model-based rather than a data-based measure of the insufficiency of a dipolar model and thereby indicate the degree of transition between a distal field and a proximal field. Several statistics can be derived from imperfection of a VHPD analysis as presented in the following synthetic

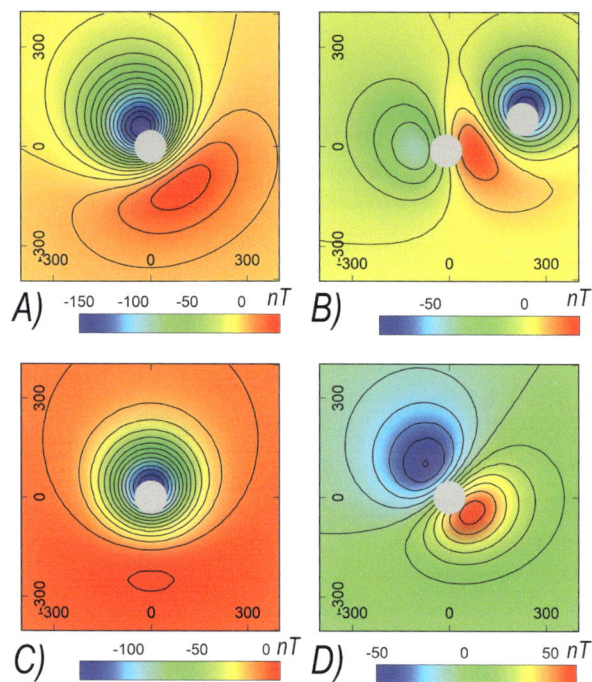

Fig. 4.1. A) TMI due to a dipole magnetisation of declination 135°, inclination +45°, B) inversion starting model TMI, C) TMI of the vertically polarised dipole and D) TMI of the horizontally polarised dipole.

data and measured case study inversions. The four statistics computed from VHPD analysis presented below each have a corresponding statistic in D-E analysis.

4.5 DIPOLE MODEL DEPARTURE STATISTICS

I present four statistics to indicate departure of a VHPD model and an ellipsoid model from that of a dipole model recovered from inversions of the same data. Parameters of the dipole, VHPD and ellipsoid models are listed in Table 4.1 followed by derivation of statistics quantifying differences between VHPD and dipole models and between ellipsoid and dipole models.

Table 4.1. Model parameter terminology.

Model	Magnetic moment	Intensity	Dec-lination	Inc.-lination	Centre E	Centre N	Centre Z	Radius	Volume
Best-fit dipole	$\mathbf{M}_d$	Int_d	Dec_d	$Inc._d$	X_d	Y_d	Z_d	r_d	vol_d
Vertically polarised dipole	$\mathbf{M}_{VPD}$	Int_{vp}	0	+/−90	X_{vp}	Y_{vp}	Z_{vp}	r_d	vol_d
Horizontally polarised dipole	$\mathbf{M}_{HPD}$	Int_{hp}	Dec_{VHPD}	0	X_{hp}	Y_{hp}	Z_{hp}	r_d	vol_d
Best-fit ellipsoid	$\mathbf{M}_e$	Int_e	Dec_e	$Inc._e$	X_e	Y_e	Z_e	$r_e{}^*$	vol_e
VHPD pair	$\mathbf{M}_{VHPD}$	Int_{VHPD}	Dec_{VHPD}	$Inc._{VHPD}$	X_{VHPD}	Y_{VHPD}	Z_{VHPD}	r_d	vol_d

* $r_e = (ra \times rb \times rc)^{1/3}$ where ra, rb, rc are the ellipsoid radii

The VHPD model of a pair of dipoles of identical volume, one with vertically and one with horizontally polarised magnetisation has the following effective magnetisation direction and centre of magnetisation for a field at a large distance compared to the dipole separation:

Intensity of magnetisation: $\text{Int}_{\text{VHPD}} = \sqrt{(\text{Int}_{\text{vp}}^2 + \text{Int}_{\text{hp}}^2)}$

Inclination of magnetisation: $\text{Inc}_{\text{VHPD}} = \text{asin}(\text{Mz}_{\text{VHPD}}, \text{Mh}_{\text{VHPD}})$

East centre of magnetisation: $X_{\text{VHPD}} = X_{\text{vp}} + (X_{\text{hp}} - X_{\text{vp}})^* \text{Int}_{\text{hp}}/(\text{Int}_{\text{vp}} + \text{Int}_{\text{hp}})$

North centre of magnetisation: $Y_{\text{VHPD}} = Y_{\text{vp}} + (Y_{\text{hp}} - Y_{\text{vp}})^* \text{Int}_{\text{hp}}/(\text{Int}_{\text{vp}} + \text{Int}_{\text{hp}})$

Depth to centre of magnetisation: $Z_{\text{VHPD}} = Z_{\text{vp}} + (Z_{\text{hp}} - Z_{\text{vp}})^* \text{Int}_{\text{hp}}/(\text{Int}_{\text{vp}} + \text{Int}_{\text{hp}})$

Direction of magnetisation: $(\text{Dec}_{\text{d}}, \text{Inc}_{\text{d}})$

The following analysis statistics describe differences between dipole and VHPD models and between dipole and ellipsoid models. They are designed to highlight differences between a dipole inversion model and VHPD or ellipsoid inversion models that arise from permitting additional degrees of freedom in the inversions.

4.5.1 Statistic 1 – angular separations between pairs of model magnetisation directions

S1_VHPD (measured in degrees) is the angular separation between the VHPD and dipole model magnetisation directions:

$$(\text{Dec}_{\text{VHPD}}, \text{Inc}_{\text{VHPD}}) - (\text{Dec}_{\text{d}}, \text{Inc}_{\text{d}})$$

S1_DE (measured in degrees) is the angular separation between the dipole and ellipsoid model magnetisation directions:

$$(\text{Dec}_{\text{e}}, \text{Inc}_{\text{e}}) - (\text{Dec}_{\text{d}}, \text{Inc}_{\text{d}})$$

4.5.2 Statistic 2 – magnitude of the magnetisation difference between pairs of models

VHPD_2: magnitude of the vector difference between the VHPD magnetisation and the dipole model

magnetisation normalised to the dipole model intensity and converted to a percentage:

$$\text{S2\_VHPD} = 100 * \left| M_{\text{VHPD}} - M_{\text{d}} \right| / \text{Int}_{\text{d}}$$

S2_DE: magnitude of the vector difference between the dipole and ellipsoid total magnetisations normalised to the dipole magnetisation and converted to a percentage:

$$\text{S2\_DE} = 100 * \left| M_{\text{d}} - \text{vol}_{\text{e}}/\text{vol}_{\text{d}} * M_{\text{e}} \right| / \text{Int}_{\text{d}}$$

4.5.3 Statistic 3 – horizontal separation between model magnetisation centres

S3_VHPD: summed horizontal separations of the VHPD magnetisations from the dipole centre weighted by their relative intensities, normalised to depth and converted to a percentage:

$$\text{S3\_VHPD} = 100 * ((\text{Int}_{\text{vp}} / \text{Int}_{\text{VHPD}}) * \sqrt{((X_{\text{vp}} - X_{\text{d}})^2 + (Y_{\text{vp}} - Y_{\text{d}})^2)} + (\text{Int}_{\text{hp}} / \text{Int}_{\text{VHPD}}) * \sqrt{((X_{\text{hp}} - X_{\text{d}})^2 + (Y_{\text{hp}} - Y_{\text{d}})^2)}) / Z_{\text{d}}$$

S3_DE: horizontal separation of the ellipsoid magnetisation from the dipole centre normalised to depth of the dipole and converted to a percentage:

$$\text{S3\_DE} = 100 * \sqrt{((X_{\text{e}} - X_{\text{d}})^2 + (Y_{\text{e}} - Y_{\text{d}})^2)} / Z_{\text{d}}$$

4.5.4 Statistic 4 – product of statistics 1 to 3

$$\text{S4\_VHPD} = \text{S1\_VHPD} * \text{S2\_VHPD} * \text{S3\_VHPD}$$

$$\text{S4\_DE} = \text{S1\_DE} * \text{S2\_DE} * \text{S3\_DE}$$

4.6 DIPOLE-ELLIPSOID (DE) ANALYSIS

A natural geometric extension of model complexity to improve the fit to a measured field from that achieved with a dipole (hopefully thereby achieving an improved representation of the true magnetisation) is to convert a spherical magnetisation model to an ellipsoid model with three axes of variable length. Analytic solutions to compute the magnetic field of an ellipsoidal magnetisation have been derived by Clark *et al.* (1986). An ellipsoid is a very versatile form. If the three radii have similar lengths the body is little different to a sphere and shares a weak sensitivity to individual values of volume and intensity of magnetisation. If one radius is much longer than the other two (the ellipsoid has a cigar shape) there is greater sensitivity to the length of the longest axis (that

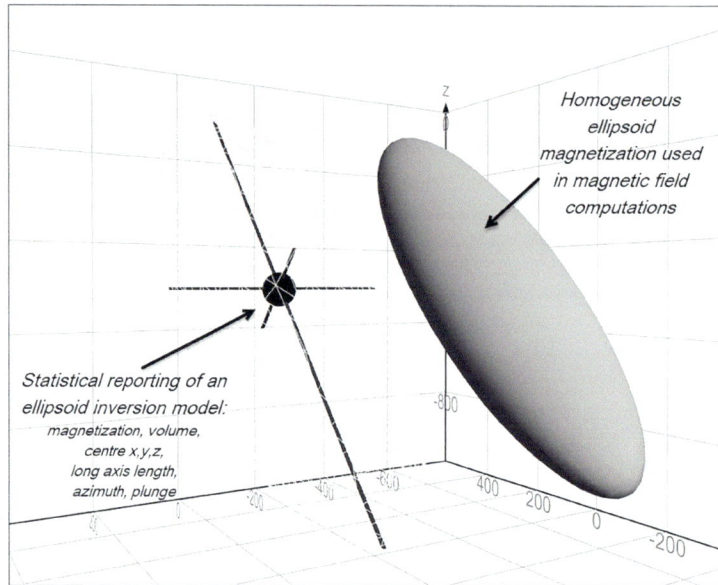

Fig. 4.2. Statistics summarising the distribution of magnetisation of an ellipsoid model.

can be oriented in any three-dimensional direction) than to the other two axes. If one radius (that again can have any three-dimensional orientation) is much shorter than the other two then the ellipsoid has a plate-like shape with reduced sensitivity to the length of the shortest radius. Note that as the shape of the ellipsoid progressively deviates from that of a sphere there are increasing issues of directional sensitivity to factors such as magnetisation direction and separation of the anomaly from the background field. Justification for increasing the complexity of the model from a sphere to an ellipsoid depends on the interpreted significance of the reduced misfit to the measured field which that achieves and the interpreted geological viability of the dipole or ellipsoid body shapes. Just as for a spherical model, an ellipsoid

model should be visualised as providing statistical parameters of the distribution of magnetisation rather than the directly representing it (see Fig. 4.2). Alternative extensions of model complexity can be achieved with a variety of shapes using faceted bodies.

4.7 VHPD AND DE ANALYSIS OF COMPLEX EQUIDIMENSIONAL MAGNETISATION

Figure 4.3 shows images of TMI at elevations 50 and 200 m above a complex equidimensional magnetisation of ~300 m side length. At the lower and upper elevations, separations between the top of magnetisation and the closest field measurements are respectively 17% and 67% of the horizontal extent of the magnetisation. It might

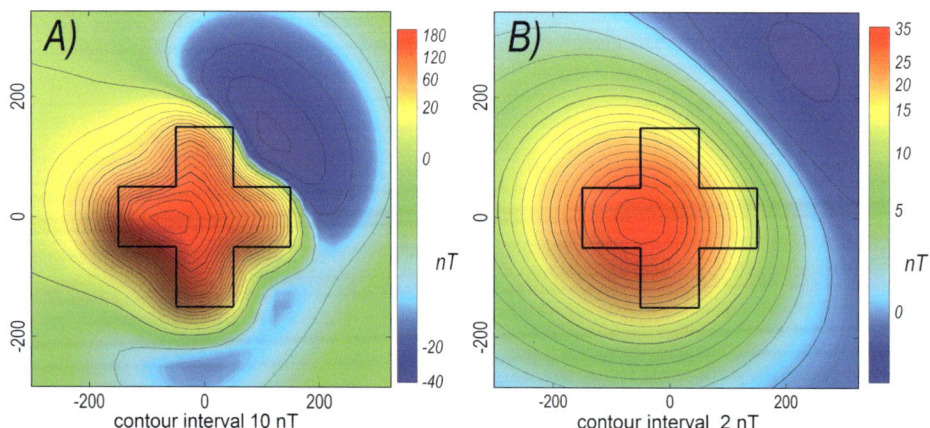

Fig. 4.3. TMI over a complex magnetisation of declination 225°, inclination −45° at elevations of A) 50 m and B) 200 m.

seem that such a directly pertinent statistic would be valuable for discriminating between distal and proximal fields, but this measure has no practical value in analysis of measured field data unless the true depth and horizontal extent of the source magnetisation are already known.

Figure 4.4 shows contours of the input model field and best-fit dipole inversion field at both elevations. At the higher elevation a dipole model clearly provides a closer fit to the field of the original model than it does at the lower elevation. At both elevations the horizontal centres of the dipoles are close to the horizontal centres of the input model and the magnetisation direction estimates provided by the dipole inversions of the fields are only different from the true magnetisation direction by 1.2° and 0.6° degrees at the lower and upper elevations respectively (see Table 4.2).

Differences in magnetisation direction and separations of the vertically and horizontally polarised

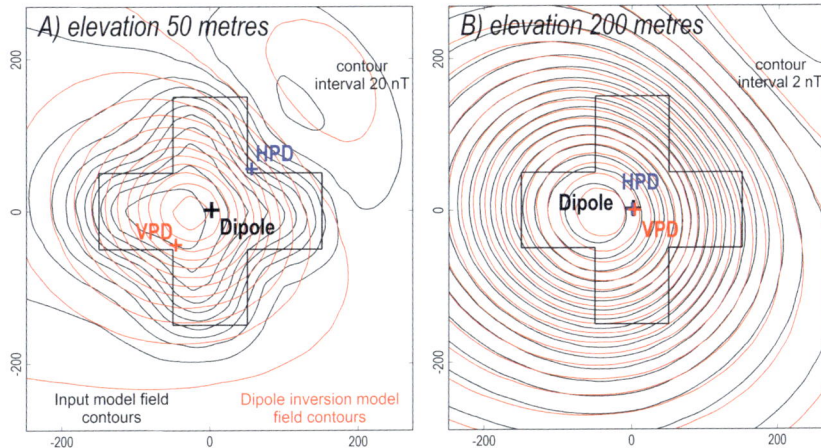

Fig. 4.4. Source (black) and dipole inversion model (red) TMI contours at A) 50 m and B) 200 m elevations above the top of magnetisation. Crosses mark the centres of the dipole and vertically and horizontally polarised dipoles.

Table 4.2. Equidimensional magnetisation study: model statistics from inversion of field data at 50 m and 200 m.

Model	Xc	Yc	Zc	J (A/m)	Vol (m³)	Dec	Inc.	Rotation	Data misfit %
Source_50	0	0	100	1	5,000,000	225	–45	–	–
Dipole_50	1.7	1.7	178	2.016	4,188,790	225	–43.8	1.2°	27
Dip_V_50	–46	–44	178	1.234	4,188,790	000	–90	–	–
Dip_H_50	56	55	178	0.928	4,188,790	224.6	0	–	–
VHPD_50	–2.2	–1.4	178	1.552	4,188,790	224.6	–53.3	8.3°	24
Ellipsoid_50	1.1	0.7	93.7	4.450	1,090,699	225.4	–44.3	0.8°	18
Source_200	0	0	250	1	5,000,000	225	–45	–	–
Dipole_200	1.5	0.6	290	1.394	4,188,790	225.5	–44.4	0.6°	5
Dip_V_200	3.8	0.2	290	0.977	4,188,790	000	–90	–	–
Dip_H_200	0.3	0.7	290	1.006	4,188,790	224.8	0	–	–
VHPD_200	2.0	0.5	290	1.403	4,188,790	224.8	–44.2	0.8°	5
Ellipsoid_200	0.5	0.4	247.8	1.347	3,694,336	225	–44.9	0.1°	1

Differences of the VHPD analysis and ellipsoid model from the dipole model

Model	Statistic1	Statistic2	Statistic3	Statistic4	% Data misfit
VHDP_50	9°	27	3	729	13
VHDP_200	0.6°	1	0.2	0.12	1
ellipsoid_50	0.6°	43	0.7	18	21
ellipsoid_200	0.6°	15	0.4	3.6	5

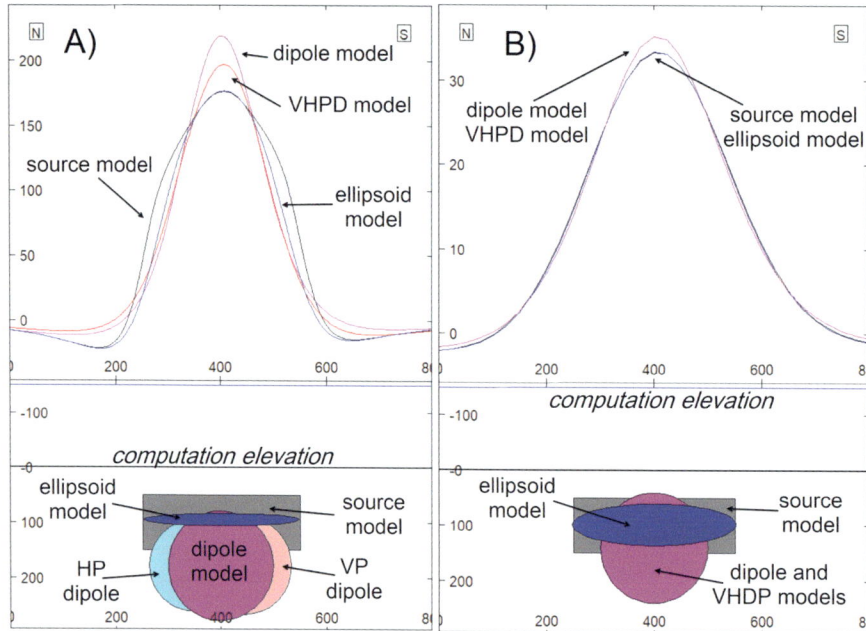

Fig. 4.5. Central north to south profiles at elevations of A) 50 m and B) 200 m above the top of the source model.

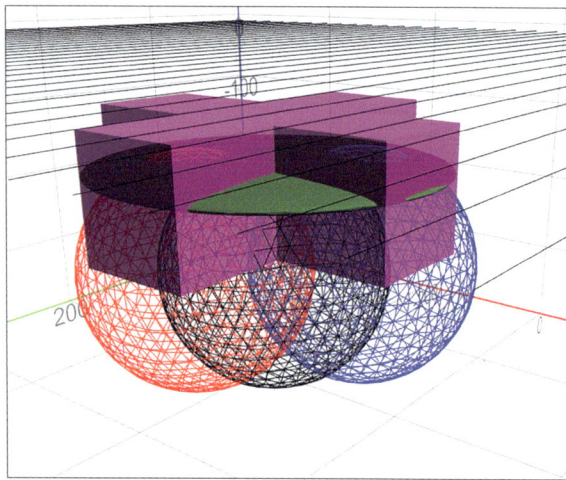

Fig. 4.6. Perspective of the source model (magenta), ellipsoid (green) and dipole (black mesh) models, and the vertically (red mesh) and horizontally (blue mesh) polarised VHPD bodies from inversions of the field at 50 m elevation.

dipoles for analyses at the lower and upper elevations show clear differences between more proximal and more distal fields. A central north–south section through the models and the computed fields at the two elevations is shown in Fig. 4.5. For inversions of the lower elevation data (Fig. 4.5A) the input model field is poorly matched by the single dipole and VHPD inversions. It is, however, more closely matched by the ellipsoid inversion model that is almost co-centred with

the input model. For inversions of the higher elevation data (Fig. 4.5B) the data misfits are all substantially reduced and the single dipole and the polarised dipoles are coincident and almost co-centred with the input model. A perspective view of the lower elevation inversion models is shown in Fig. 4.6. Because the input model has two axes of horizontal symmetry there is little shape bias to produce a significant horizontal displacement of the dipole or ellipsoid inversion models. In keeping with strong linkage between estimation of the horizontal location of a magnetisation and of its direction, lack of substantial horizontal displacement of these bodies is matched by small differences of their magnetisation directions from that of the input model.

4.8 VHPD AND DE ANALYSIS OF HORIZONTALLY ELONGATE MAGNETISATION

The previous study investigated analysis of fields due to a complex but equidimensional distribution of magnetisation. If a magnetisation is elongated along a horizontal axis there is reduced sensitivity in recovery of both its spatial and magnetisation properties. Figure 4.7 shows magnetic fields computed at an elevation 50 m above a homogeneous-magnetisation, elliptic-section pipe with horizontal top and base surfaces, a north-east–south-west long axis of 300 m length and a short axis of 100 m.

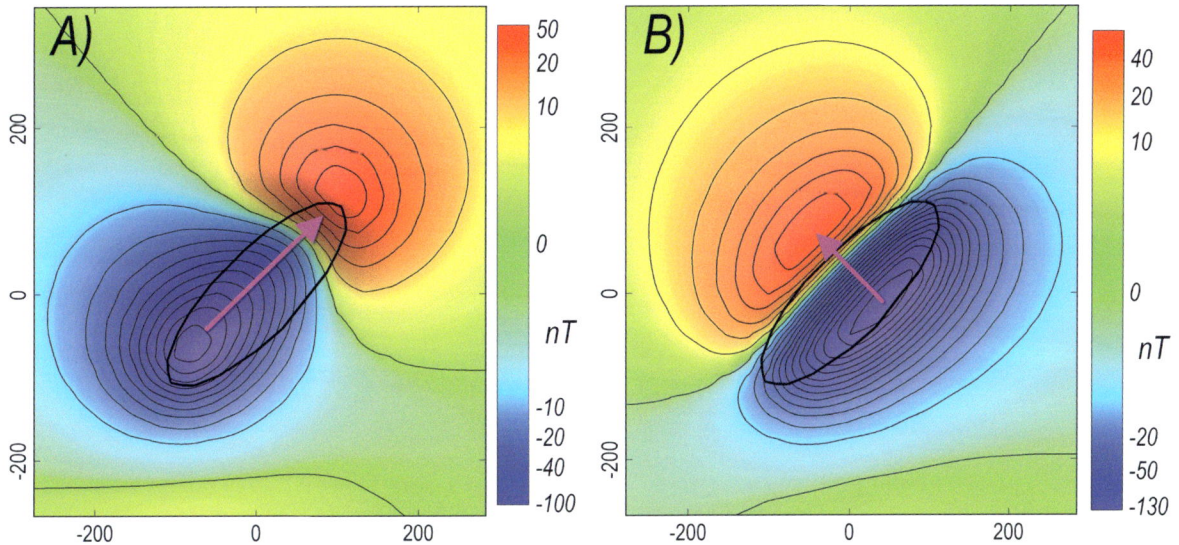

Fig. 4.7. TMI computed at an elevation of 50 m above a SW–NE trending elliptic pipe of axis lengths 150 and 50 m. Magnetisations: A) axial declination of 45° inclination +15°, B) transverse declination of 315° inclination +15°.

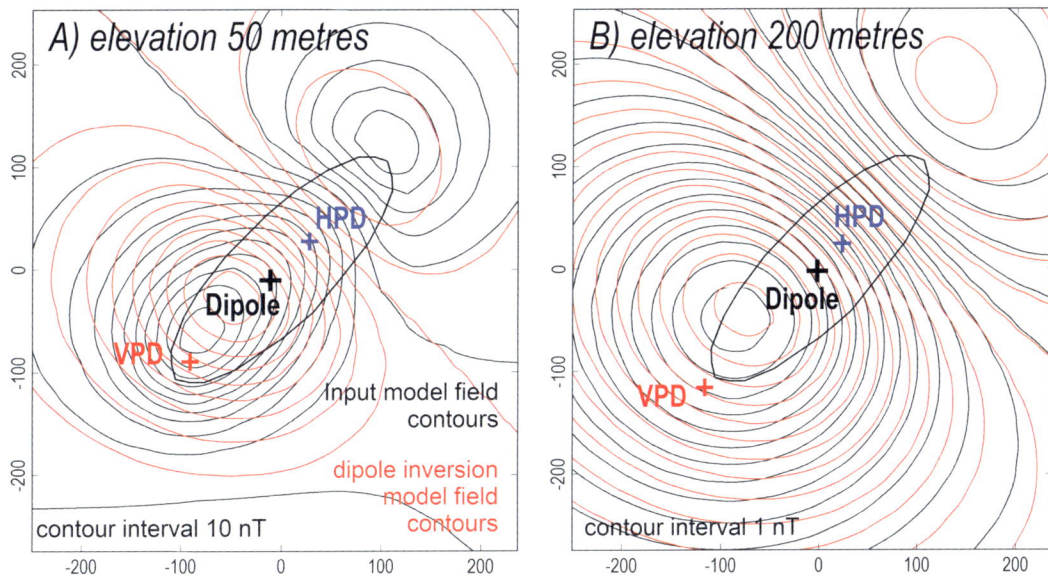

Fig. 4.8. TMI contours of the input model with a long-axis (045°) magnetisation (black) and dipole inversion model (red) at A) 50 m elevation and B) 200 m elevation.

In Fig. 4.7A the magnetisation is parallel to the long axis and in Fig. 4.7B it is parallel to the short axis. The ratios of the separation distance between the closest magnetisation and magnetic field computation at the two investigated elevations of 50 and 200 m to the long axis of magnetisation are 1:6 and 1:1.5 respectively. At the relatively steep geomagnetic inclination, regardless of body shape and orientation the TMI anomaly trough-to-peak azimuths approximately indicate the declination of magnetisation.

Figure 4.8 shows TMI contours (in black) for the input model field with magnetisation parallel to the long axis and (in red) for the dipole inversion model. The field at 50 m elevation is shown in Fig. 4.8A and the field at 200 m elevation is shown in Fig. 4.8B. The contour match is clearly closer at the higher elevation at which there is less influence of body shape. For the VHPPD analysis of this elongate source body there is significant separation of the two polarised dipoles at both elevations. Despite this, the error in estimated magnetisation direction of

Table 4.3. Elongate magnetisation study: model statistics from inversion of field data at 50 m and 200 m above the input model 'Source-50' with magnetisation parallel to elongation.

Model	Xc	Yc	Zc	J (A/m)	Vol (m³)	Dec	Inc.	Rotation	Misfit %
Source_50	0	0	100	1	2,344,250	045	+15	–	–
Dipole_50	–11	–10.2	167	0.848	4,188,790	044.1	23	8 °	34
Dip_V_50	–91	–90	167	0.323	4,188,790	000	90	–	–
Dip_H_50	28	28	167	0.584	4,188,790	044.3	0	–	–
VHPD_50	–14	–14	167	0.667	4,188,790	044.3	29	14°	27
Ellipsoid_50	0.3	0.2	82	0.792	2,634,354	045.0	14.6	0.4°	5
Source_200	0	0	250	1	2,344,250	045	+15	–	–
Dipole_200	–2.8	–2.3	285	0.642	4,188,790	044.8	16.2	1.2°	8
Dip_V_200	–116	–116	285	0.165	4,188,790	000	90	–	–
Dip_H_200	26	27	285	0.488	4,188,790	045	0	–	–
VHPD_200	–10	–9	285	0.515	4,188,790	045	18.7	4°	7
Ellipsoid_200	0.2	0.1	245	0.388	3,694,336	045	15.0	0°	0.4

Differences of the VHPD analysis and ellipsoid model from the dipole model

Model	Statistic1	Statistic2	Statistic3	Statistic4	% Data misfit
VHDP_50	6°	23	3	414	20
VHDP_200	2.5°	20	3	150	7
ellipsoid_50	8°	43	9	3,096	36
ellipsoid_200	1.2°	47	1.3	73	8

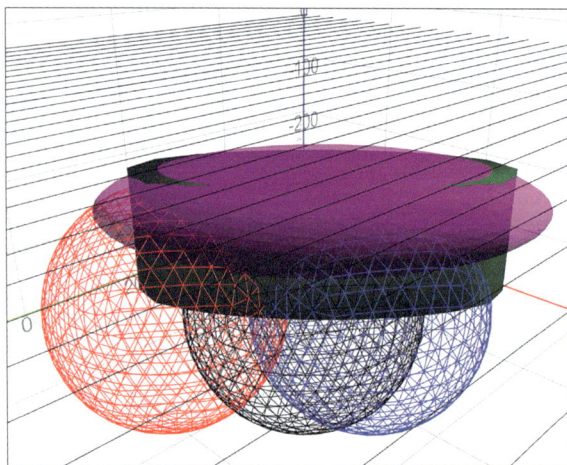

Fig. 4.9. Perspective of the source (green), ellipsoid (magenta), dipole (black mesh), vertical (red mesh) and horizontal (blue mesh) polarised VHPD bodies from inversions of the field computed at 200 above the source model.

the VHPD model is reduced from 14° in the lower elevation field to 4° in the higher elevation field as reported in Table 4.3. Errors in magnetisation direction of the dipole model at the two elevations are 8° and 1.2° respectively. Figure 4.9 shows a perspective view of all the models from inversion of the 200 m elevation data. Both the dipole and ellipsoid inversion models are horizontally co-centred with the input model but the additional shape complexity of the ellipsoid gives it the freedom to more closely match the field of the input model. The errors in magnetisation direction of the ellipsoid inversion model are only 0.4° at 50 m elevation and less than 0.1° at 200 m elevation (see Table 4.3).

Figure 4.10 shows (in black) TMI contours of the field computed from the input model with magnetisation parallel to the short axis, and (in red) TMI contours of the field computed from the dipole inversion. Figure 4.10A is at 50 m elevation and Fig. 4.10B is at 200 m elevation. Misfit between the two model fields reduces significantly from the lower elevation to the higher elevation. Model statistics are listed in Table 4.4. The statistics designed to indicate whether a field is more distal or more proximal show less variation with elevation for magnetisation parallel to the short axis of the body (Table 4.4) than for magnetisation parallel to body elongation (Table 4.3) revealing that alignment of magnetisation with the direction of elongation of the body accentuates the magnetic field expression of body shape.

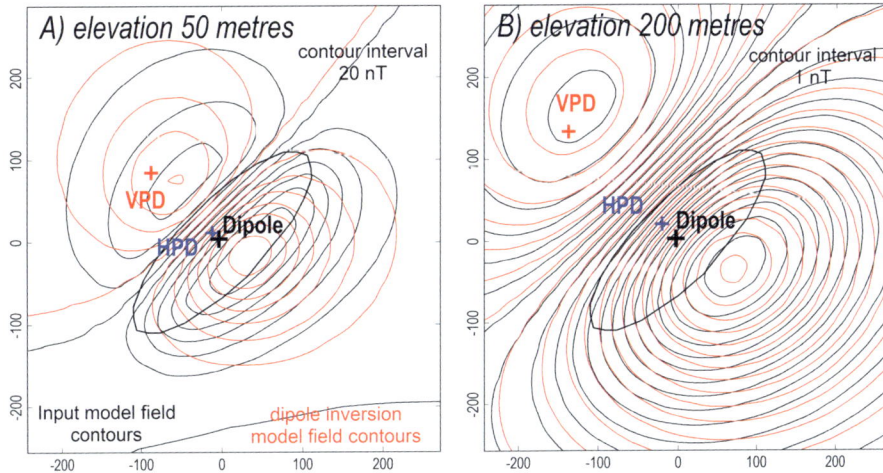

Fig. 4.10. TMI contours of the elliptic-pipe model with a short-axis (315°) magnetisation (black) and dipole inversion model (red) at A) 50 m elevation and B) 200 m elevation.

Table 4.4. Elongate magnetisation study: model statistics from inversion of field data at 50 m and 200 m above the input model 'Source-50' with magnetisation perpendicular to elongation.

Model	Xc	Yc	Zc	J (A/m)	Vol (m^3)	Dec	Inc.	Rotation	Misfit %
Source_50	0	0	100	1	2,344,250	315	+15	–	–
Dipole_50	–4	4	124	0.696	4,188,790	314.2	9.7	5°	30
Dip_V_50	–88	85	124	0.143	4,188,790	000	90	–	–
Dip_H_50	–12	12	124	0.825	4,188,790	314.3	0	–	–
VHPD_50	–23	23	124	.849	4,188,790	314.3	9.7	5°	28
Ellipsoid_50	0.2	–0.7	88	.602	3,554,152	315.1	15.4	0.4°	3
Source_200	0	0	250	1	2,344,250	315	+15	–	–
Dipole_200	–3	3	263	0.596	4,188,790	314.8	13.4	1.6°	7
Dip_V_200	–114	111	263	0.129	4,188,790	000	90	–	–
Dip_H_200	–17	17	263	0.653	4,188,790	314.7	0	–	–
VHPD_200	–33	32.5	263	0.666	4,188,790	314.7	11.2	5°	7
Ellipsoid_200	0.2	–0.2	248	1.424	1,634,620	315.0	15	0°	0.3

Differences of the VHPD analysis and ellipsoid model from the dipole model

Model	Statistic1	Statistic2	Statistic3	Statistic4	% Data misfit
VHDP_50	3.7°	20	22	1,628	31
VHDP_200	2.2°	12	16	422	11
ellipsoid_50	5.8°	28	5.1	828	12
ellipsoid_200	1.6°	7	1.7	19	8

4.9 SYNTHETIC-DATA VHPD AND DE ANALYSIS OF A PLUNGING MAGNETISATION

A key trade-off between spatial and magnetisation parameters in finding a magnetisation to explain a measured magnetic field variation is between the spatial plunge of the magnetisation and the inclination of its magnetisation. For semi-infinite thin sheets of magnetisation this trade-off is extreme with the two angles combining into a single term (Hood 1964). For more horizontally compact sources, such as pipes, the trade-off is less effective. However, even for a pipe of circular section an error in either angle can substantially compensate for a corresponding error in the other. This leads to collaborative reduction of sensitivity in

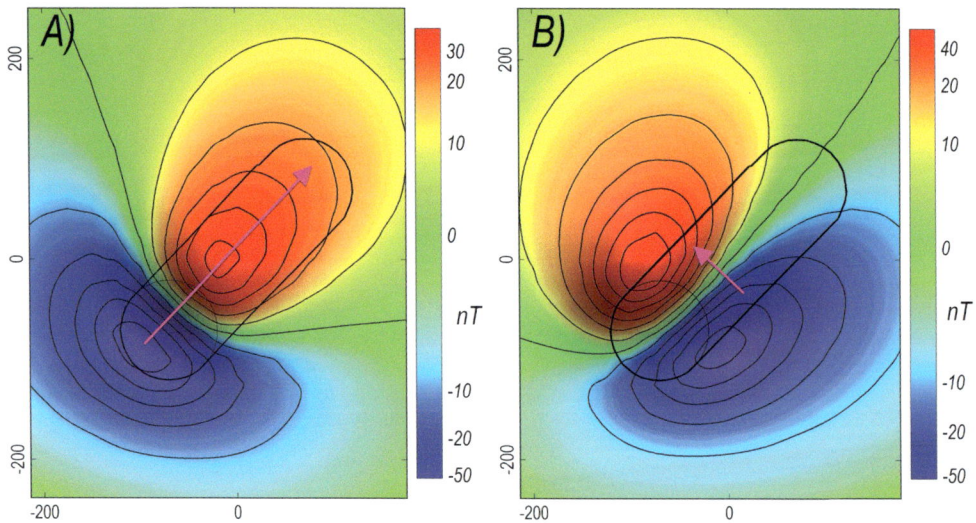

Fig. 4.11. TMI images at 50 m elevation over circular pipes plunging to the north-east with A) an axial declination of 45° and B) a transverse declination of 315°.

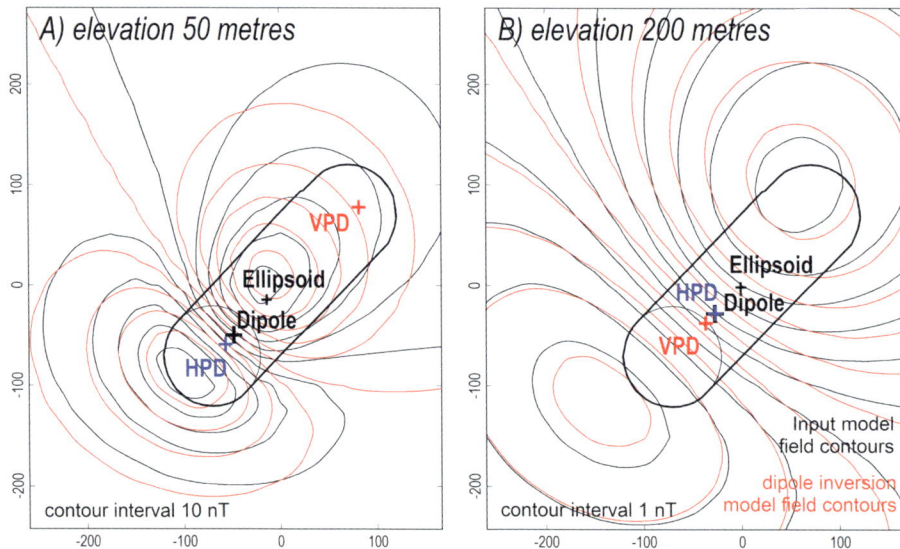

Fig. 4.12. TMI contours of the input model with a down-dip declination (45°) magnetisation (black) and dipole inversion model (red) at A) 50 m elevation and B) 200 m elevation.

estimation of both the inclination of magnetisation and spatial plunge. Figure 4.11 shows TMI images at an elevation of 50 m over a circular pipe plunging at 45° to the north-east with a declination of magnetisation in the down-plunge direction (Fig. 4.11A) and a declination of magnetisation in the north-west strike direction (Fig. 4.11B). In both cases the declination of magnetisation is reliably indicated by the trough to peak azimuth of the TMI anomalies. Figure 4.12 shows TMI contours for the input models (in black) and the single dipole inversion models (in red) for inversions of the lower, more proximal field data (Fig. 4.12A) and for inversion of the higher,

more distal field data (Fig. 4.12B). The single-dipole model from inversion of the more proximal data is displaced from the centre of the input model in the up-dip direction because of the higher weighting of that shallow section of magnetisation. The single-dipole model from inversion of the more distal field is closer to the centre of the input model because the differential weighting of the top of the magnetisation is reduced (for a truly distal field the shape influence attenuates completely and the apparent centre of magnetisation is at its true position). For the VHPD model there is a substantial horizontal displacement between the vertically and horizontally polarised

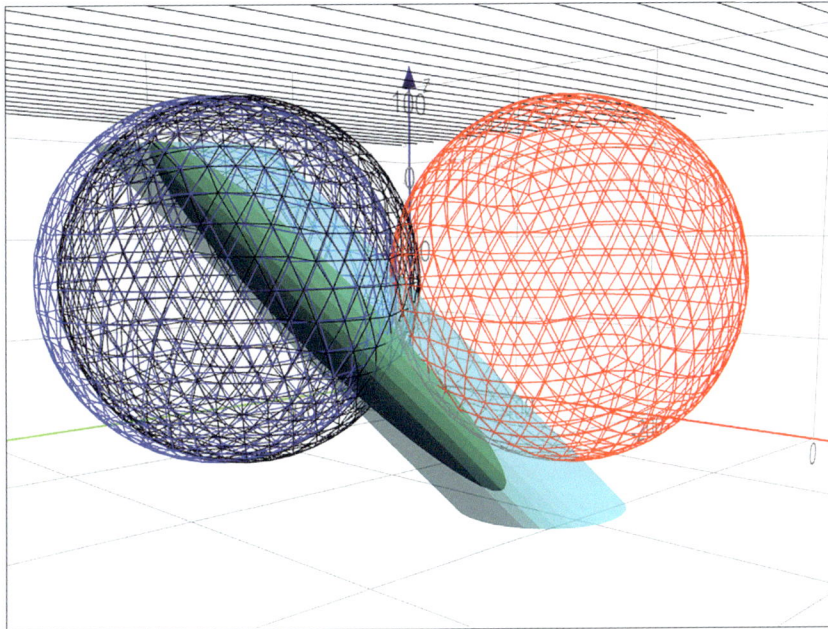

Fig. 4.13. Perspective of the plunging pipe source (solid light blue), ellipsoid (solid green) and dipole (black mesh) models, and the vertical (red mesh) and horizontal (blue mesh) polarised VHPD bodies from inversions of the field 50 m above the top of the source model.

dipoles in the lower, more proximal field (Fig. 4.12A) with reduced displacement in the more distal 200 m elevation field (Fig. 4.12B).

Figure 4.13 is a perspective view of the input and inversion models for the field at 50 m elevation. The ellipsoid inversion develops a cigar shape to match the input pipe model and the two bodies are approximately co-centred. The dipole model, however, cannot match the input model shape and is centred towards the top of the pipe model.

Figure 4.14 shows TMI contours for the input models (black) and the dipole inversion models (red) for magnetisations with declination in the along-strike direction at the lower 50 m elevation (Fig. 4.14A) and at the higher 200 m elevation (Fig. 4.14B). Just as for the horizontally elongate bodies, magnetisation in the transverse direction reduces the influence of distribution shape. This has a significant influence on characterisation of the field as being more proximal or more distal.

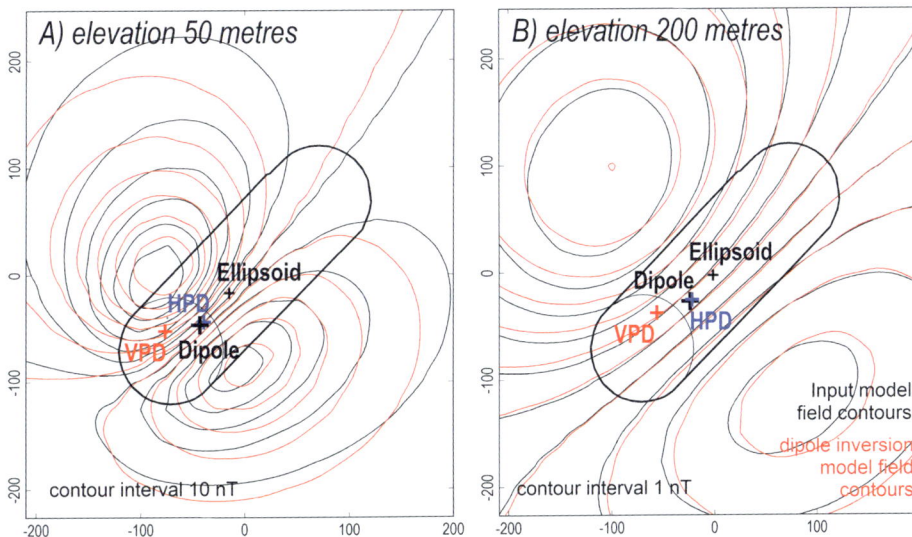

Fig. 4.14. TMI contours of the input model with an along-strike (declination 315°) magnetisation (black) and dipole inversion model (red) at A) 50 m elevation and B) 200 m elevation.

4.10 PERFORMANCE OF THE PROXIMAL/ DISTAL STATISTICS

As shown by these inversion studies, many factors influence the magnetic field expression of a magnetisation and therefore the transition between proximal and distal fields. The studies presented of complex but equidimensional magnetisations (Figs 4.3 to 4.6 and Table 4.2), magnetisations with elongation along a single horizontal axis (Figs 4.7 to 4.10 and Tables 4.3 and 4.4) and plunging magnetisations (Figs 4.11 to 4.14 and Tables 4.5 and 4.6) show that it is simplistic to consider that any single statistic can summarise the proximal/distal nature of a set of magnetic field measurements to a distribution of magnetisation. However, with the range of proposed statistics in Section 4.5 I attempt to approximately discriminate the degree to which a field is proximal or distal to its source magnetisation without prior knowledge of that magnetisation. Figure 4.15 plots values of statistics S1, S2 and S3 for VHPD and D-E analysis. The circle and cross symbols show the various statistics for more proximal and more distal fields respectively, as computed for the various classifications of magnetisation shape (columns A to E in Figs 4.15 and 4.16). The left side of Fig. 4.15 shows results from the VHPD analyses and the right side shows results of the DE analyses. Individual statistics behave differently between the classes of magnetisation distribution. This suggests that combination of statistics may provide better discrimination of distal and proximal fields than any individual statistic. Figure 4.16 similarly summarises the compound statistic 4 (given by the product of statistics 1 to 3) together with inversion data misfit statistics. What is required of the statistics plotted in Figs 4.15 and 4.16 to discriminate between distal and proximal fields is a reliable separation between their values in distal and proximal fields (the cross and circle symbols) and ideally a consistent separation for each class of magnetisation distribution. In Fig. 4.15 the model misfit statistics mostly (but not consistently) have higher values in the more proximal field. The data misfit statistics and compounded model misfit statistics (S4) plotted in Fig. 4.16 are more consistent indicators of proximal or distal fields but are still not reliable in all cases.

Table 4.5. Plunging magnetisation study: model statistics from inversion of field data at 50 m and 200 m above the input model 'Source-50' with magnetisation in the down-dip (45°) direction.

Model	Xc	Yc	Zc	J (A/m)	Vol (m^3)	Dec	Inc.	Rotation	Misfit %
Source_50	0	0	150	1	1,552,914	045	–15	–	–
Dipole_50	–49	–50	122	0.287	4,188,790	044.5	–9.0	6°	23
Dip_V_50	82	78	122	0.0353	4,188,790	000.0	–90	–	–
Dip_H_50	–57	–59	122	0.272	4,188,790	044.6	0	–	–
VHPD_50	–41	–43	122	.274	4,188,790	44.6	–7.3	8°	22
Ellipsoid_50	–14	–15	143	3.461	435,597	044.9	–14.6	0.4°	8
Source_200	0	0	300	1	1,552,914	045	–15	–	–
Dipole_200	–27	–28	292	0.361	4,188,790	044.7	–13.0	2°	8
Dip_V_200	–36	–38	292	.0814	4,188,790	000.0	–90	–	–
Dip_H_200	–27	–28	292	.360	4,188,790	044.8	0	–	–
VHPD_200	–29	–30	292	.369	4,188,790	044.8	–13	2°	8
Ellipsoid_200	–1	–2	302	13.84	116,831	045.0	–15.0	0°	0.9

Differences of the VHPD analysis and ellipsoid model from the dipole model

Model	Statistic1	Statistic2	Statistic3	Statistic4	% Data misfit
VHDP_50	1.6°	5	9	72	18
VHDP_200	2.0°	2	1	4	1
ellipsoid_50	5.6°	27	41	6,199	23
ellipsoid_200	2.0°	8	13	208	8

Table 4.6. Plunging magnetisation study: model statistics from inversion of field data 50 m and 200 m above the top of the input model with magnetisation in the strike (315°) direction.

Model	Xc	Yc	Zc	J (A/m)	Vol (m^3)	Dec	Inc.	Rotation	Misfit %
Source_50	0	0	150	1	1,552,914	315	–15	–	–
Dipole_50	–43	–47	115	.285	4,188,790	316.3	–12.6	3°	16
Dip_V_50	–77	–52	115	0.0685	4,188,790	0	–90	–	–
Dip_H_50	–39	–44	115	0.259	4,188,790	323.1	0	–	–
VHPD_50	–47	–46	115	0.268	4,188,790	323.1	–14.9	8°	15
Ellipsoid_50	–15	–17	129	0.197	6,982,113	315.0	–14.5	0.5°	6
Source_200	0	0	300	1	1,552,914	315	–15	–	–
Dipole_200	–24	–27	281	0.341	4,188,790	315.5	–14.2	0.9°	5
Dip_V_200	–57	–38	281	0.085	4,188,790	0	–90	–	–
Dip_H_200	–22	–25	281	0.322	4,188,790	319	0	–	–
VHPD_200	–29	–28	281	0.333	4,188,790	319	–14.8	4°	5
Ellipsoid_200	–2.1	–2.5	298	0.281	5,485,234	315.0	–15.0	0°	0.4

Differences of the VHPD analysis and ellipsoid model from the dipole model

Model	Statistic1	Statistic2	Statistic3	Statistic4	% Data misfit
VHDP_50	7°	13	4	364	4
VHDP_200	3.4°	6	2	41	3
ellipsoid_50	2.2°	16	36	1,267	16
ellipsoid_200	1°	8	12	96	5

The dipole and ellipsoid models (but not the virtual VHPD models) map apparent spatial distributions of magnetisation from which distances between magnetisation and the magnetic field measurements can be estimated. However, for an unknown intensity of magnetisation a dipole model only constrains the shallowest magnetisation to be no deeper than its centre, and ellipsoid models also provide uncertain estimates of the elevation of the top of magnetisation. None of the model-misfit statistics S1 to S4 specify elevation of the top of magnetisation. They therefore do not directly report the separation distances between magnetisation and magnetic field measurements and/ or computations; nevertheless, in combination with the statistics 1 to 4 as derived above they indirectly indicate the more general concept of proximity to magnetisation.

4.11 VHPD AND DE MAGNETIC FIELD ANALYSIS CASE STUDY NEAR KEMENDOK PARK, SOUTH-WEST NEW SOUTH WALES

Figure 4.17 shows the location of the study area close to the Kemendok National Park in south-west New South Wales, north of the Murray River and Victorian border. The local geomagnetic field has a declination of 10° and inclination −65°. The area is covered with a variable thickness of semi-consolidated, weakly magnetic Cenozoic Murray Basin sediments. The TMI image of the area shown in Fig. 4.18A is created from data acquired on the Murray Basin aeromagnetic survey flown in 2005 on east–west flightlines at a spacing of 400 m and nominal terrain clearance of 60 m. This data (survey P1105) is available for download from Geoscience Australia's Geophysical Archive Data Download (GADDS). The two anomalies analysed here can also be downloaded from The Australian Remanent Anomalies Database (ARAD numbers 355 and 356). Studies of their magnetisation directions are reported by Foss *et al.* (2024). Anomaly ARAD356 (Fig. 4.18A) has similar peak (+38 nT) and trough (−27 nT) amplitudes and a south-east directed trough to peak azimuth. In the steep southern hemisphere background field the similar peak and trough amplitudes suggest the resultant magnetisation has low inclination, and the trough to peak orientation suggests that the declination of magnetisation is to the south-east (c. 135°). Anomaly ARAD355 (Fig. 4.18A) in the south-east also has similar peak (+76 nT) and trough (−66 nT)

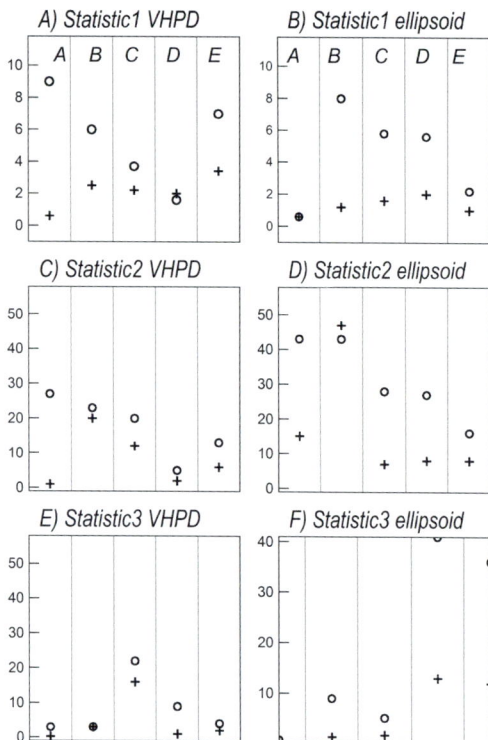

Fig. 4.15. Top to bottom S1 to S3 values, (left) VHPD analysis and (right) DE analysis. In each plot: column A is for equidimensional models, column B for models with magnetisation parallel to horizontal elongation, column C for models with magnetisation perpendicular to horizontal elongation, column D for models with magnetisation parallel to plunge azimuth and column E for models with magnetisation perpendicular to plunge azimuth. Circles are for 50 m elevation (more proximal fields) and crosses for 200 m elevation (more distal) fields.

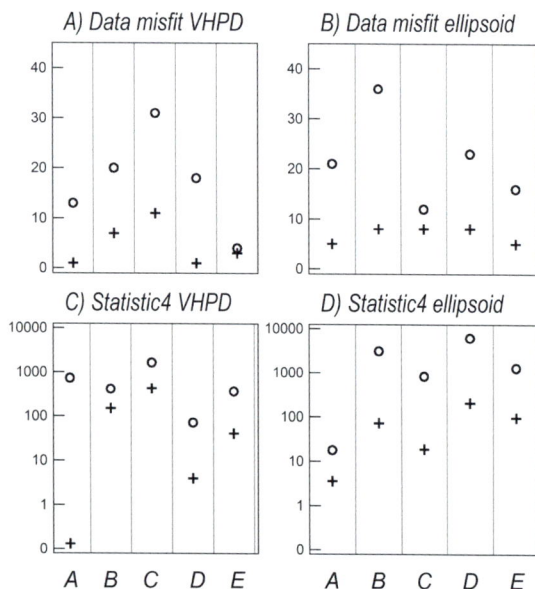

Fig. 4.16. Top row: data misfit and Bottom row: S4 values for (left) VHPD analysis and (right) DE analysis. Columns A to E and symbols are as in Fig. 4.15.

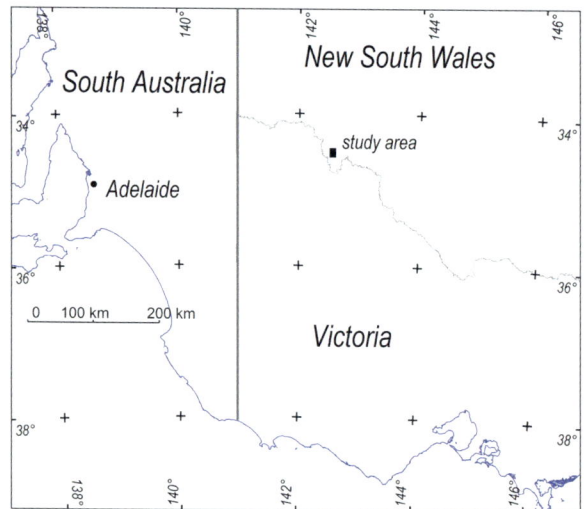

Fig. 4.17. Kemendok study area in south-west New South Wales.

amplitudes, in this case with south-south-west trough to peak azimuth, suggesting a low inclination south-south-west directed magnetisation. In the western half of the area are two less distinct, lower amplitude and almost completely negative anomalies (not investigated in this study) apparently due to steep positive inclination (reverse) magnetisations but these smaller, single-polarity anomalies are insufficiently sampled by the 400 m spaced flightlines for reliable magnetisation analysis.

The best estimates of source magnetisation for the two selected anomalies are derived from individual inversions to simultaneously find both the spatial distribution of magnetisation and its direction. Acceptable starting estimates of magnetisation direction are provided by the simple rules described above and the long-wavelength, low-amplitude regional field variations support reliable separation of the anomalous (residual) magnetic fields. Figures 4.19A and 4.19B show the close fits of measured and model-computed fields achieved using elliptic-section pipe models with horizontal top and base and with homogeneous magnetisation for the north-east and south-east anomalies respectively. Success of the models in matching the field does not prove them correct or that the magnetisations are homogeneous. The models should not be considered direct representations of the in-ground magnetisations but rather summary statistics of the magnetic moments and distributions of magnetisation. The horizontal-top faces of the models are interpreted as due to termination of the magnetisation at a sub-horizontal unconformity and the limited depth extents suggest that the bodies may

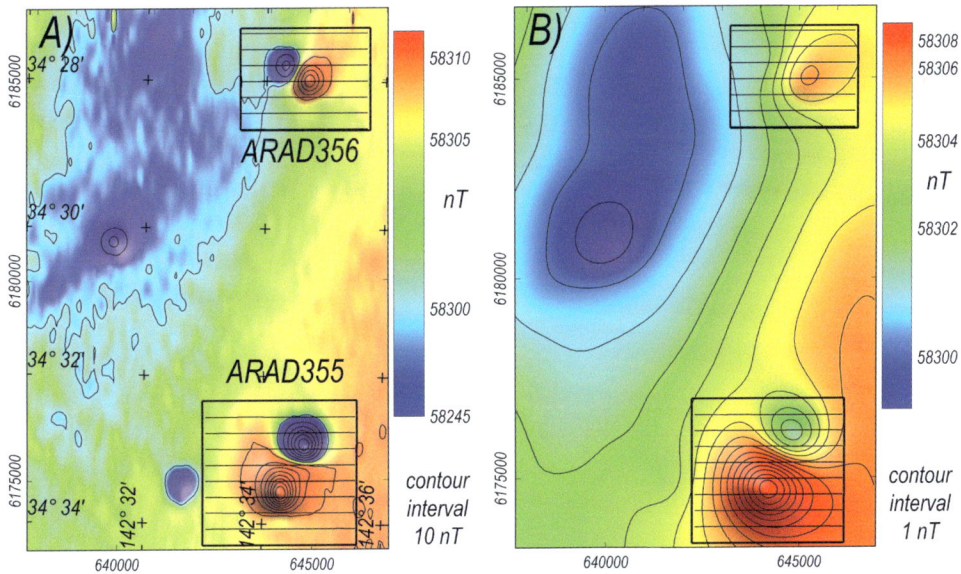

Fig. 4.18. TMI images of the Kemendok study area: A) measured TMI and B) upward continued by 800 m.

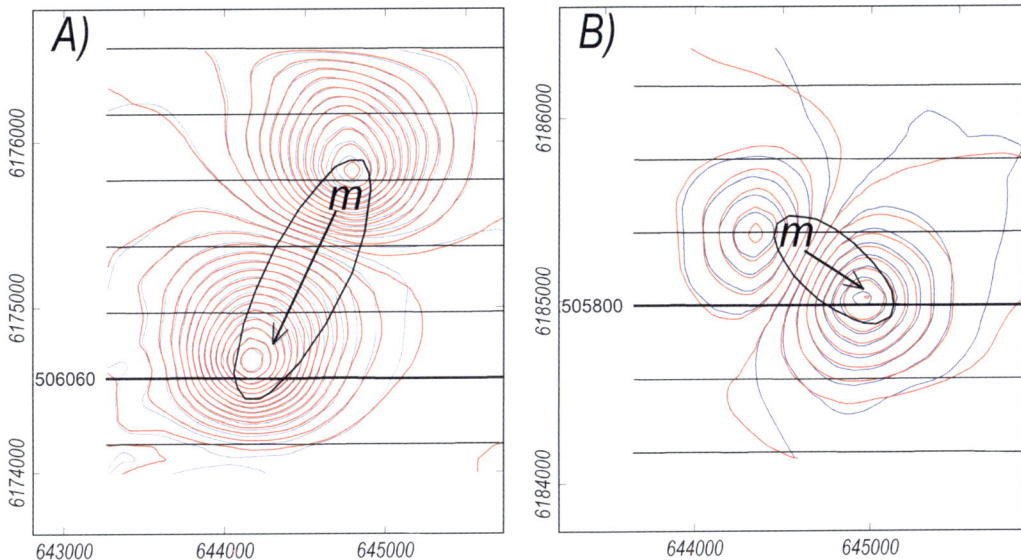

Fig. 4.19. TMI contours at 5 nT intervals measured (blue) and model-computed (red) of the best-fit elliptic-section pipes for anomaly ARAD355 (A) and ARAD356 (B) with orientation of the magnetisation vectors.

possibly be diatremes with only narrow undetected feeder zones rather than deep-going pipes. Declinations of resultant magnetisation derived by the inversions and plotted in Fig. 4.19 are parallel to elongation of the bodies. This most likely reveals a bias in estimating the shape of the anomalies arising from their magnetisation directions. Figure 4.20 shows contours of model-computed TMI fields from best-fit vertical circular pipe models. These models with fewer degrees of freedom do not match the measured anomalies as closely as the

elliptic pipe models; however, the key statistics of the computed field peak and trough amplitudes and their locations are almost identical between the elliptic and circular pipe models and the model magnetisation directions only differ by 1° for anomaly ARAD356 and by 3° for anomaly ARAD356. The inversion results with the different geometry bodies illustrate that for these moderately proximal anomalies the direction of magnetisation is far more influential than the distribution of magnetisation in determining anomaly characteristics.

The distribution of magnetisation influences anomaly details but inversion does not have the power to reliably recover source information from those details. The only feasible (but unlikely) major deviation from these models would be to explain the anomalies as due to similar net low-inclination magnetisations consisting of separated steep positive and steep negative inclination magnetisations beneath the anomaly troughs and peaks respectively.

Figures 4.21A and 4.21B show sections along key west–east flightlines through the two anomalies. The best estimated depths to the top of magnetisation below the elevation of the magnetic field sensors are 215 and 195 m for anomalies ARAD355 and ARAD356 respectively. The diameters of the circular pipe models for the

two anomalies are 1360 and 780 m, giving width to measurement separation ratios of 6 for anomaly ARAD355 and 4 for anomaly ARAD356, consistent with the anomalies being moderately proximal.

These proximal anomalies provide suitable tests of dipole inversions and VHPD and DE analyses. To contrast the analyses with those of a more distal field I upward continued the TMI grid by 800 m, interpolated the grid values onto the flight lines and inverted those channels, adding the 800 m upward continuation to the flying height as the elevation reference channel. The addition of 800 m to the estimated original sensor elevation of ~200 m above the top of magnetisation increased the virtual separation of the field measurements from the magnetisation by a factor of approximately 5,

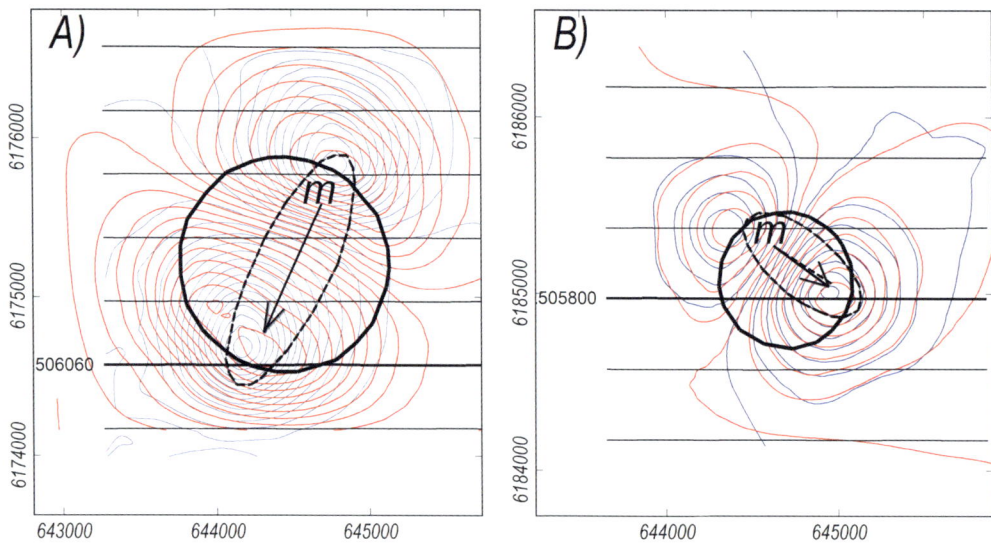

Fig. 4.20. TMI contours at 5 nT intervals measured (blue) and model-computed (red) of the best-fit vertical, circular-section pipes for anomaly ARAD355 (A) and ARAD356 (B) with orientation of magnetisation vectors. The elliptic pipe outlines are shown for comparison.

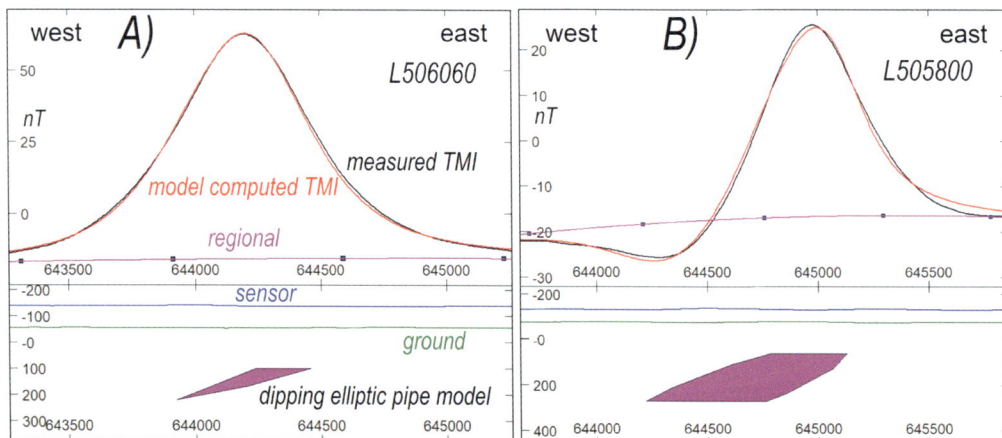

Fig. 4.21. Flightline model cross-section views for anomalies A) ARAD355 and B) ARAD356. The measured channel is in black, the background field in magneta and the model-computed field is in red.

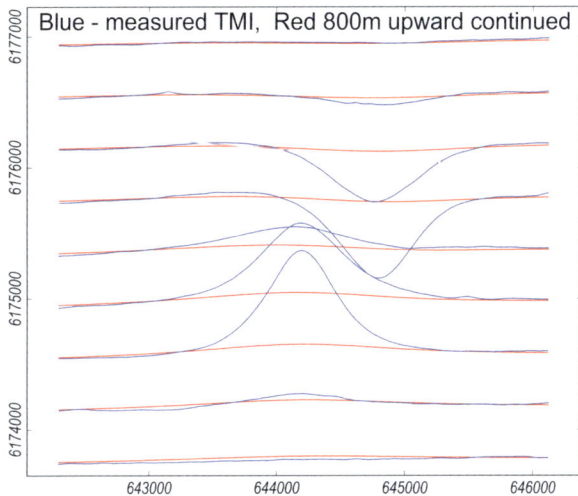

Fig. 4.22. TMI stacked profiles for Anomaly 355: measured (blue) and after 800 m upward continuation (red).

resulting in an anomaly width to magnetisation-field measurement separation ratio of ~1.2 for anomaly ARAD355 and ~0.8 for anomaly ARAD356. These values represent moderately distal fields with greatly reduced capacity to resolve detail of the distribution of magnetisation compared to the (already unreliable) resolution of spatial detail from the lower-level directly measured field. Attenuation of signal with the 800 m upward continuation is illustrated in Fig. 4.22 which shows stacked profiles of both the measured (blue) and upward continued (red) TMI channels. The attenuation is substantial

and reliable modelling and inversion of the upward-continued data requires valid anomaly separation from the background field. This separation could not be performed reliably on the upward continued data because of the reduced contrast in curvature of the anomalous and background fields, and so the upward continuation was applied to a residual separation of the fields performed at the lower elevation.

I inverted the two anomalies using dipole and ellipsoid models. Figure 4.23A shows stacked profiles of measured TMI for anomaly ARAD355, together with stacked profiles of computed TMI from the best-fit dipole and ellipsoid inversion models. The computed field from the dipole model (the red curves) is clearly too simple to acceptably explain the measured field (the black curves), but nevertheless the estimated magnetisation direction is within 1° of the estimate from much closer data-fits of the ellipsoid model. The computed field from the ellipsoid model (the blue curves) still reveals minor unexplained detail in departure from the measured curves. The equivalent curves after 800 m upward continuation of the residual measured field (the measured field with the regional field from the modelling subtracted) are shown in Fig. 4.23B. At this elevation the dipole model provides a much closer fit to the measured field, and the curves for the measured field and the field computed from the ellipsoid model are almost identical, showing the greatly reduced sensitivity in distal fields to the distribution pattern of source magnetisation.

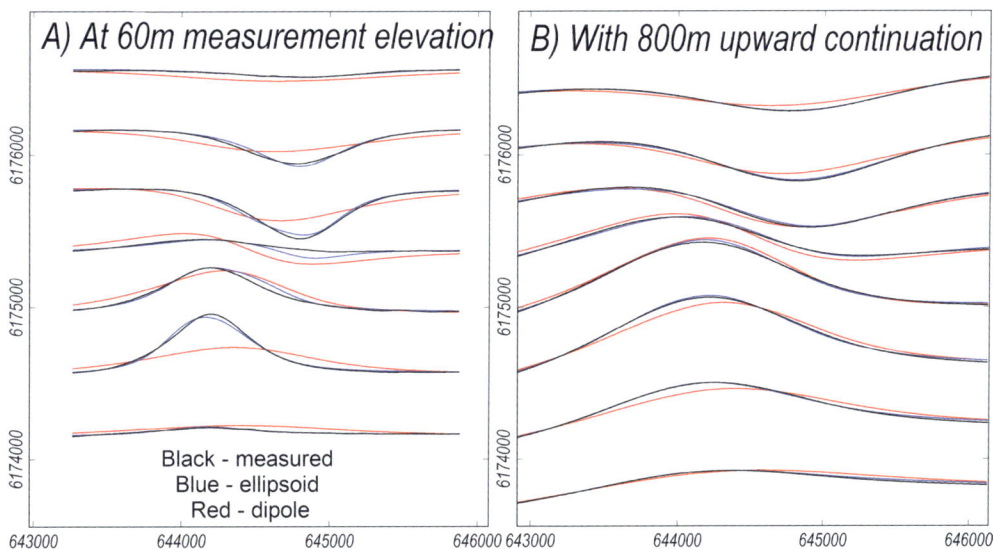

Fig. 4.23. TMI Stacked profiles for anomaly ARAD355: field data (black), computed from the elipsoid model (blue) and from the dipole model (red). A) at the initial measurement elevation of 60 m above ground and B) after an 800 m upward continuation.

Figure 4.24 shows in perspective view the best-fit elliptic pipe model for anomaly ARAD355 (in magenta), the best-fit dipole model (mesh in black), the best-fit ellipsoid model (dark green), and the vertical (red mesh) and horizontal (blue mesh) VHPD models. The dipole and VHPD model bodies are clearly poor representations of the distribution of magnetisation (that is best-estimated from the elliptic pipe model). The dipole model is centred close to the apparent centre of magnetisation but at significantly greater depth. The ellipsoid model (for which the computed field closely matches the measured field) is almost co-incident with the elliptic pipe model. Parameters of the various models and their analysis statistics are reported in Table 4.7 for field inversions of both the measured data and of the field after 800 m upward continuation. The percentage data misfits of the inversions at the two elevations do not change substantially. The estimated angular error of the VHPD inversion falls from 10° in the proximal field to 5° in the distal field, but the DE analysis increases from 0° to 4°. There are regular improvements in the VHPD and DE statistics from the proximal to distal inversion statistics, with significant reductions in the compound S4 statistic from 2640 to 45 for the VHPD analysis and from 226 to 35 for the DE analysis.

Figure 4.25 shows contours of measured field (black) and dipole model computed field (red) at the original survey height (Fig. 4.25A) and the 800 m upward continuation (Fig. 4.25B). The dipole model produces a reasonable fit to the measured field at the survey height and a closer fit in the more distal field at the upward continuation height. At the lower elevation the separation of the VHPD analysis dipoles of 900 m (Fig. 4.25A) is far less than the 1,300 m separation of the anomaly peak and trough. At the 800 m upward continuation height (Fig. 4.25B) the separation of the corresponding dipoles reduces to 88 m although the peak to trough separation increases to 1575 m. Figure 4.25 clearly shows the reduced influence of the distribution of magnetisation from the more proximal field (Fig. 4.25A) to the more distal field (Fig. 4.25B) and the influence of this in the location of the individual dipole virtual source analysis.

There are nearby anomalies from sources of similar size and at similar depth that appear to be due to steeply inclined magnetisations (although these anomalies might be influenced by poor sampling on the east–west flightlines). This introduces possible alternative models in which the preferred single body models are replaced with two bodies of opposite, steep magnetisation (much like the virtual VHPD models). Figure 4.26A and Fig. 4.26B show perspective models for anomalies ARAD355

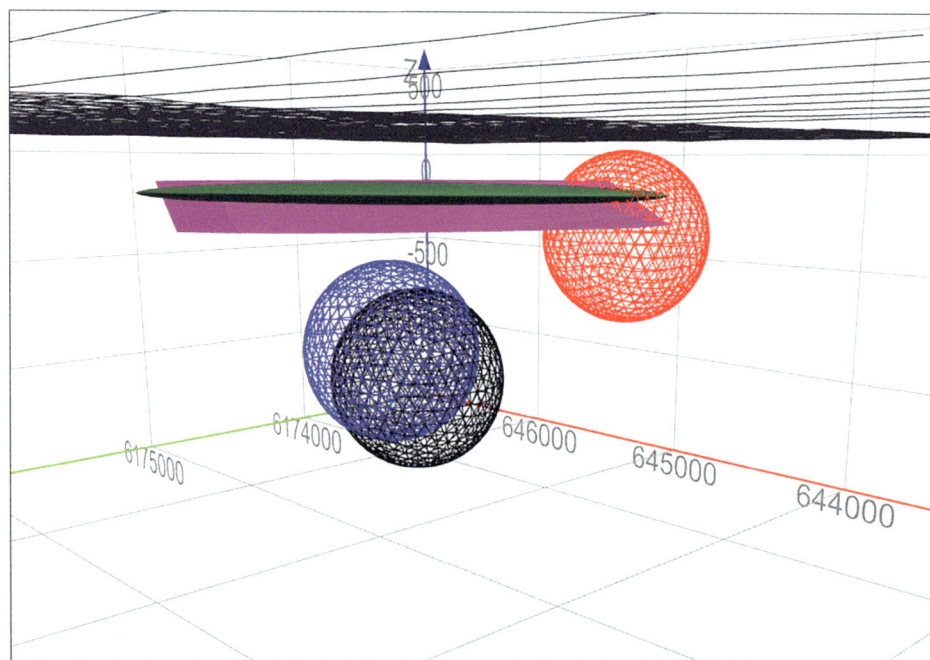

Fig. 4.24. Models for anomaly ARAD355: Elliptic-section pipe (magenta), ellipsoid (dark green), best dipole (black) and vertically polarised (blue) and horizontally polarised (red) combined sphere model.

Table 4.7. ARAD355 model and analyses statistics.
* rot_n is the angular rotation (in degrees) from the best-estimated magnetisation direction of the elliptic pipe inversion (undefined for the individual V and H dipoles).

Model	Top Xc	Top Yc	Z_{centre}	Depth to centre$_{below}$ sensor	J (A/m)	Vol (m^3)	Dec$_n$ degrees	Inc.$_n$ degrees	rot$_n$* degrees	Data misfit %
Best pipe (TMI)	644,636	6,175,170	–167	301	2.645	6.11x10^7	206	–2	–	7
Dipole	644,449	6,175,241	–747	884	3.516	11.31x10^7	206	–2	0°	10
Dip_V	644,129	6,174,519	–747	884	0.630	11.31x10^7	000	–90	–	–
Dip_H	644,466	6,175,274	–747	884	2.933	11.31x10^7	206	0	–	–
VHPD	644,410	6,175,136	–747	884	3.000	11.31x10^7	206	–12	10°	9
Ellipsoid	644,464	6,175,212	–139	886	3.252	4.767x10^7	206	–2	0°	4
Best pipe (UC800)	644,588	6,175,235	–73	1,010	0.579	23.7x10^7	207	–6	4°	6
Dipole	644,439	6,175,197	–403	1,340	1.606	11.31x10^7	207	–7	1°	12
Dip_V	644,528	6,175,194	–403	1,340	0.190	11.31x10^7	000	–90	–	–
Dip_H	644,440	6,175,196	–403	1,340	1.607	11.31x10^7	212	0	–	–
VHPD	644,449	6,175,196	–403	1,340	1.618	11.31x10^7	212	–7	5°	12
Ellipsoid	644,438	6,175,217	–100	1,037	0.771	18.05x10^7	207	–6	4°	5

Model	Statistic1	Statistic2	Statistic3	Statistic 4 (Product S1.S2.S3)	% Data misfit
VHDP 50	10°	22	12	2,640	9
VHDP 800	5°	9	1	45	12
Ellipsoid 50	1°	61	3.7	226	4
Ellipsoid 800	1°	23	1.5	35	5

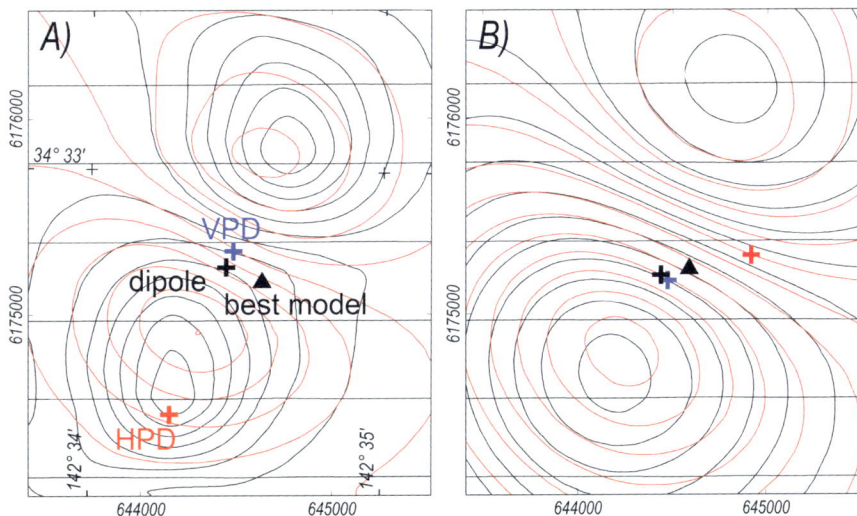

Fig. 4.25. TMI contours for anomaly ARAD355: measured field (black) and best-dipole computed field (red), A) at meaurement elevation of 60 m above ground and B) after 800 m upward continuation. Centres of the best pipe (black triangle), best dipole (black cross) and vertically polarised (red cross) and horizontally polarised (blue cross) dipole pair model are also plotted.

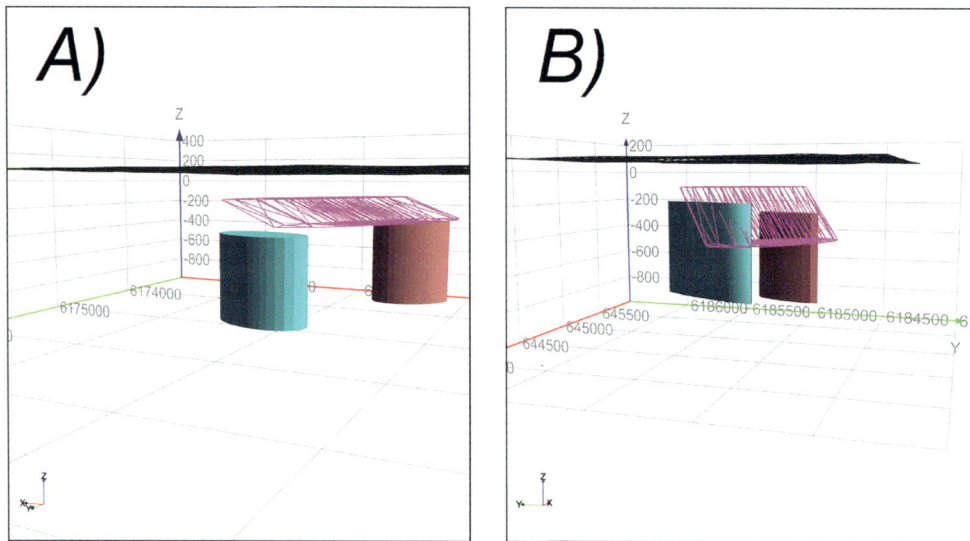

Fig. 4.26. Elliptic pipe (magenta) and dual-polarity vertical elliptic pipe (blue negative inclination and brown positive inclination) models for anomalies A) ARAD355 and B) ARAD356.

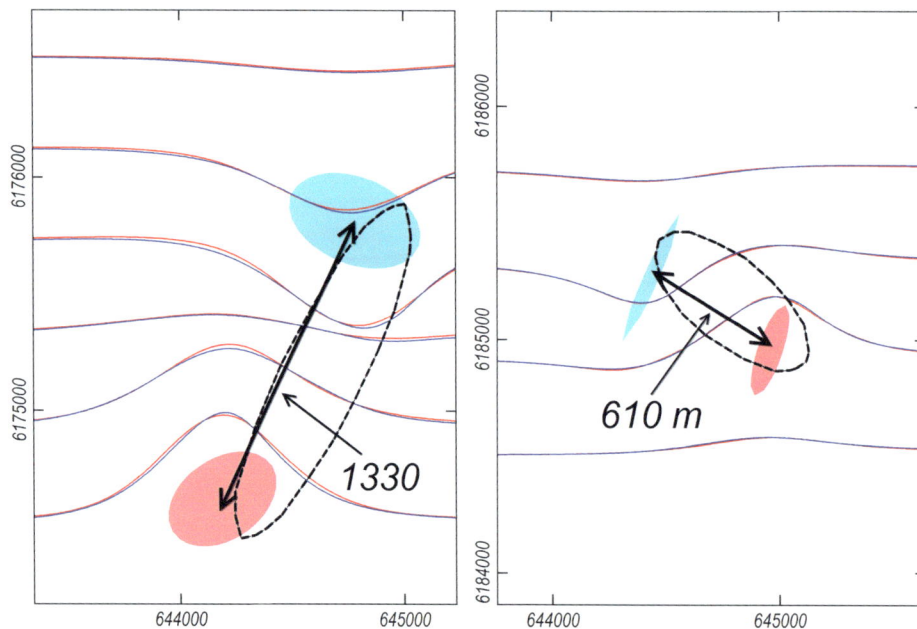

Fig. 4.27. Stacked profiles of fields computed from the best-fit elliptic pipe model (blue) and dual-polarity vertical elliptic pipe (blue negative inclination and brown positive inclination) models (red) for anomalies A) ARAD355 and B) ARAD3561.

and ARAD356 respectively, using two vertical pipes of steeply inclined opposite polarity magnetisation. These pipes are approximately centred beneath the anomaly peaks and troughs, and as shown in Fig. 4.27 are close to the ends of elliptic pipe models that are elongated (apparently artificially) in the magnetisation directions. Figure 4.27 shows stacked profiles of the field computed from the elliptic pipe models (that I believe to be the best representative models of the magnetisations) and of the

field computed from the dual vertical pipe bodies of opposite-polarity, steep-inclination magnetisation. The stacked profiles match closely, with slightly greater misfit for the more elongate anomaly ARAD355 (Fig. 4.27A) for which the measured field is effectively more proximal than for the more compact anomaly ARAD356 (Fig. 4.27B).

Fields computed after switching the positions of the bodies in the dual-pipe models (and thereby switching the peak and trough of that computed field) produces fields that can be closely matched using the original single elliptic pipe model with a rotation of 180° in its declination of magnetisation. This clearly demonstrates the role that the distribution of magnetisation plays in these models of extreme inhomogeneity of two magnetisations of very different direction. The ambiguity of possible multiple magnetisations extends to distal fields.

4.12 CONCLUSIONS

The concept of proximal and distal fields relates specifically to the information carried in a magnetic field about its source magnetisations and plays a critical role in evaluation of what information can be reliably recovered from magnetic field analysis. The ultimate distal field is of a dipole source that is completely specified by its centre coordinates and total magnetisation (including direction). Circular or elliptic anomalies in a magnetic field dataset can in the first instance be matched with a dipole model and if that is found insufficient then additional complexity can be progressively introduced as justified.

I propose vertically and horizontally polarised dipole (VHPD) analysis as a means to measure coherent departure of a measured magnetic field from that due to a dipole. The statistics of VHPD analysis can be incorporated with statistics derived from best-fit dipole and best-fit ellipsoid models as an aid to evaluate the degree of additional detail that is justified in a magnetisation model.

In this study I first used a synthetic model of a complex equidimensional magnetisation to show that a dipole model can reasonably recover both the horizontal centre of magnetisation and the direction of that magnetisation. The magnetic field expression of shape details reduces at higher elevations as the match with the best-fit dipolar model field improves. I then used dipoles, polarised dipoles and ellipsoids to invert magnetic fields of horizontally elongated magnetisations. It is difficult to estimate magnetisation directions of horizontally elongate bodies that can alternatively be represented as polarisations towards their ends, and additional challenges in recovering magnetisation direction are encountered in analysis of plunging magnetisations. At low elevations magnetisation at the shallowest end of a plunging elongate source dominates the magnetic field and may distort estimation of magnetisation direction. At higher elevations the magnetic field has a weaker bias to the influence of plunge and estimates of both the centre of magnetisation and of magnetisation direction are more dependable.

I have presented a case study of inversion and VHPD analysis of two magnetic anomalies measured in southwest New South Wales over what are believed to be volcanic bodies intruded above or into basement beneath a cover of younger sediments. The anomaly patterns reveal that the anomalies are predominantly due to remanent magnetisations differently directed to the local geomagnetic field. I inverted each anomaly independently, in both cases making an interpretive separation of the anomalous field due to the magnetisations from what I believe to be the background field. The best estimate of source magnetisation direction for the sources of these anomalies is derived from inversion using elliptic-section pipes with horizontal top and bottom surfaces. Alternative inversions using ellipsoid bodies produces models with similar spatial distribution and almost identical resultant magnetisation direction, but these bodies are less useful for estimation of depth to the top of magnetisation. Inversion using dipole models provided similar estimates of the centres and directions of magnetisation but coherent departures in matching the measured field reveal that the magnetisations are insufficiently represented by homogeneous spherical magnetisations and that the fields are therefore to some degree proximal. VHPD and DE analysis of these field anomalies applied to the original survey data and to more distal data derived from an 800 m upward continuation match the characteristics predicted from the synthetic modelling studies.

REFERENCES

Clark DA, Saul SJ, Emerson DW (1986) 'Magnetic and gravity anomalies of a triaxial ellipsoid'. *Exploration Geophysics* 17, 189–200. doi:10.1071/EG986189

Foss CA, Hope JA, Patabendigerada S (2024) 'Regional magnetic depth estimates for the Eastern Resources Corridor (ERC), Officer Basin, Tasmania, and Northeast Queensland'. eCat No. 149239.

Hood PJ (1964) 'The Königsberger ratio and the dipping dyke equation'. *Geophysical Prospecting* 12, 440–456. doi:10.1111/j.1365-2478.1964.tb01916.x

SECTION 2:
ESTIMATION OF MAGNETISATION DIRECTION FROM MAGNETIC FIELD ANALYSIS AND INVERSION

5

Expression of remanent magnetisation in magnetic field data: recognition, analysis and inversion

C.A. Foss

ABSTRACT

The first question to address in analysis or inversion of magnetic field data is whether that magnetisation is oriented parallel to the geomagnetic field or in a different direction. For a well-isolated magnetisation it may be possible to make this determination from the pattern of its magnetic field variation but in more complicated cases the field of the magnetisation must first be separated from overlapping fields of adjacent magnetisations. This is an interpretive process and if the separation is incorrect then the resulting magnetisation direction estimates will also be incorrect. Challenges in separation of magnetic fields increase with complexity in the distribution of magnetisation and with complexity of the fields with which they overlap. However, for correctly separated magnetic fields there is considerable tolerance in estimation of the mean magnetisation direction, and reliable estimates of magnetisation direction should be obtained from the wide range of available analysis and inversion methodologies. Expectation should only be to recover a single representative magnetisation direction for each discrete, compact feature in the magnetic field.

5.1 IS ESTIMATION OF MAGNETISATION DIRECTION FROM MAGNETIC FIELD DATA JUSTIFIED?

Until the start of the new millennium there was widespread scepticism that magnetisation direction could be reliably recovered from magnetic field data. This doubt was the legacy from times when magnetic field analysis was largely confined to profile data using master curves computed for highly elongated 'two-dimensional' magnetisations. For thin planar sheets of magnetisation the dip of the sheet combines with the inclination of magnetisation into a single angular term from which neither can be resolved without knowledge of the other (Hood 1964). It has long been known that reverse polarity magnetisations generate magnetic field anomalies of opposite sign to those expected from normal polarity magnetisation, but these anomalies were considered mostly in terms of their polarity (as an apparent negative magnetic susceptibility) rather than more generally in terms of direction of magnetisation. Relationships between magnetisation direction and the pattern of magnetic field variation were reported in a three-dimensional modelling study by Zietz and Andreasen (1967). In the steep northern geomagnetic fields they studied, they noted that declination of magnetisation is revealed by the azimuth of the line from the anomaly peak to trough and inclination is revealed by the peak and trough amplitude ratio.

The only analytic proof that magnetisation direction can be recovered from magnetic field data was developed by Helbig (1963). The two conditions of Helbig's analytic proof are that the magnetisation is a dipole and that its horizontal location is known. An analysis of magnetisation direction based on this proof is illustrated in Fig. 5.1. The three orthogonal magnetic field components can be

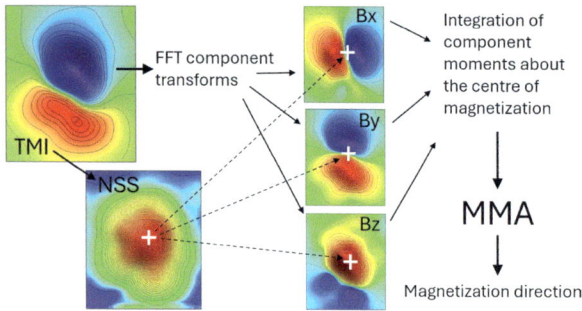

Fig. 5.1. Schematic of Helbig magnetic moment analysis (MMA) using integrals of components obtained by FFT phase transform of TMI data about the centre of magnetisation derived from the NSS transform.

mapped by appropriate phase transforms of the TMI grid (Lourenço and Morrison 1973; Blakely 1995). The moments of these components are then determined by numerical integration about a supplied horizontal centre of magnetisation, and those integrals are substituted into Helbig's equations. Estimation of magnetisation is over-prescribed from the three integrals, providing consistency measures. The integrals should be taken to infinity but experimentation shows that magnetisation direction can be correctly estimated from small field samples provided they are correctly centred. Schmidt and Clark (1998) and Clark (2014) provide correction factors for the magnitude of the magnetic moment as a function of the ratio of grid side length to depth to the centre of magnetisation. Helbig's analysis was not widely applied until Schmidt and Clark (1997) proposed the computerised form of the analysis illustrated in Fig. 5.1. Foss and McKenzie (2011) investigated Helbig analysis in a study of the Black Hill Norite anomaly and were able to recover magnetisation directions similar to those determined from inversion of the same magnetic field data and consistent with nearby direct palaeomagnetic measurements (Rajagopalan *et al.* 1993; Schmidt *et al.* 1993). Using the same magnetic field data Phillips (2005) published an automated version of Helbig analysis that attempted to isolate magnetisation directions by searching for stable directions in analysis of arrays of overlapping windows of different sizes. However, because of the high sensitivity to the centre of the analysis, this search method was only partially successful.

The two-dimensional total gradient transform (also known as the two-dimensional analytic signal) peaks over a compact magnetisation irrespective of its magnetisation direction (Nabighian 1972). The three-dimensional total gradient is not perfectly independent of magnetisation direction but has low sensitivity to it,

particularly for compact magnetic field anomalies. The normalised source strength (NSS) (Beiki *et al.* 2012) has proven independence to magnetisation direction for a dipole anomaly. Therefore, if an NSS transform is applied to a TMI grid of a dipole magnetisation to locate its centre, Helbig analysis can be performed using that estimated centre.

Since the turn of the millennium there has been a sharp increase in the number of determinations of magnetisation using parametric or voxel inversions. This marks an abrupt change to widespread confidence that magnetisation direction can be automatically recovered from any sample of magnetic field data. I propose that meaningful magnetisation direction estimates can only be made from suitable magnetic field measurements above suitable concentrations of magnetisation (magnetisation direction 'sweet-spots'). Empirical tests have established that modest departure of the distribution of magnetisation from a dipole (the proven case for Helbig analysis) causes only modest reduction in resolution of magnetisation direction. If, for instance, the spherical source of a dipole is relaced by an ellipsoid, the reliability in estimating magnetisation direction by Helbig analysis decreases gradually with increasing difference in ratio of the axes (particularly the horizontal axes). Extremely large axial ratios, as noted for thin sheets, are least suitable for estimation of magnetisation direction from magnetic field data. Location of the centre of magnetisation from total gradient or NSS transform of magnetic field data similarly loses resolution gradually as the distribution of magnetisation progressively departs from that of a dipole.

Figure 5.2 is a schematic representation of the population of magnetic anomalies in an area as a function of

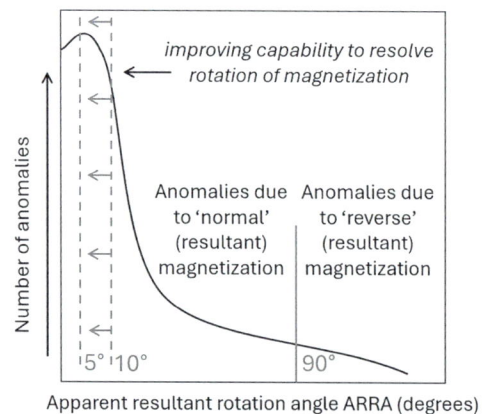

Fig. 5.2. Schematic of the population of anomalies within an area with magnetisation directions rotated away from the geomagnetic field.

difference in their magnetisation direction from the local geomagnetic field (the apparent resultant rotation angle, 'ARRA'). The shape of a curve varies with the geological ages of the magnetisations in an area, with the spatial distributions of those magnetisations and with the resolution with which the magnetic field is mapped. Many anomalies are due to magnetisations dominated by induced and/or 'soft' viscous remanent magnetisation of the same direction and have resultant magnetisation directions close to the geomagnetic field direction. For angular separations up to at least 90° the number of anomalies falls with increasing separation from the geomagnetic field. In areas of recent volcanic or subvolcanic rocks there may be a secondary peak in the population marking reverse remanent magnetisation. As indicated in Fig. 5.2, the significance of apparent deviation of a magnetisation direction from the geomagnetic field tests an increasing number of anomalies if analyses or inversions can be performed with high sensitivity. At present the resolution at which it is feasible to assign meaning to a magnetisation direction or compare magnetisation directions derived from magnetic field data is of the order of 5° in optimum cases but is more typically between 10° and 15°. An extensive review of the wide range of methods for determination of magnetisation direction is given by Clark (2014) and the highest resolution is achieved by dedicated user-guided inversion of isolated anomalies.

The widespread concept that magnetisation direction can be mapped continuously is a fallacy. Models with continuous variation in magnetisation direction may match the measured field well and may even closely represent the true magnetisation of the ground but they cannot be known to be valid. From stepwise demagnetisation of palaeomagnetic samples we know that the 1 inch diameter, 1 inch high cylinders can carry multiple over-printed remanent magnetisations of quite different direction that from their external magnetic field appear to be a single homogeneous magnetisation, and the same vector summation applies to closely adjacent magnetisations at much larger scales. However, without independent information there is no justification to propose multiple magnetisations to explain a magnetic field feature that can be acceptably explained by a single magnetisation. In the case that two magnetisations of different direction are present, local magnetisations due to their combination do not generate a cloud of directions, but directions smeared along the segment of a great circle between the two directions.

5.2 KEY POINTS RELATING MAGNETISATION DIRECTION AND MAGNETIC FIELDS

There are several key points in analysis of magnetisation direction from magnetic field data:

- Spatial variation in magnetic fields is due to spatial variation in either intensity or direction of magnetisation

- Many magnetisations causing prominent, discrete magnetic field variations are sufficiently strong that magnetic field analysis provides reasonable estimates of their absolute value rather than just their contrast with surrounding magnetisation

- The external magnetic field of a magnetisation is an expression of the vector sum of its induced and remanent magnetisations. In most cases, induced magnetisation is parallel to the local geomagnetic field but it can be deflected by anisotropy of magnetic susceptibility (AMS) or self-demagnetisation effects. The direction of remanent magnetisation depends on its acquisition age and any subsequent rotations by continental drift or tectonic forces

- Values of the Koenigsberger ratio of remanent to induced magnetisation are highly variable but many measurements of magnetic susceptibility and remanent magnetisation return values between 0.3 and 3.0. In magnetic field inversion it is unjustified to ignore remanent magnetisation because it is unknown or is perceived to be problematic

- As a scalar, the Koenigsberger ratio of the strength of remanent to induced magnetisation is insufficient to describe the vector relationships between remanent, induced and resultant magnetisations and this statistic should be supplemented with the Apparent Resultant Rotation Angle (ARRA) that is a measure of the rotation of the resultant magnetisation away from the local geomagnetic field direction

- ARRA is the statistic most critical in determining the ease of recognition of remanent magnetisation in magnetic field data. A large rotation angle requires a moderate to large Koenigsberger ratio, but a large Koenigsberger ratio does not necessarily produce a large rotation angle

- Coarse-grained, multi-domain magnetite generally carries a viscous remanent magnetisation in the present field direction. This remanent magnetisation cannot be discriminated from induced magnetisation in (static) magnetic field data and is undetected by magnetic susceptibility measurements. Inversions imposing magnetisation based only on magnetic susceptibility measurements constrains those inversions to give incorrect answers even if the remanent magnetisation does not rotate the resultant magnetisation direction

- There are only selected locations in a magnetic field ('sweet-spots') where the magnetic field data support reliable detection of magnetisation direction. Each sweet-spot supports estimation of a single magnetisation direction (the mean direction of that magnetisation)

- In steep northern geomagnetic inclinations and for equidimensional anomalies the declination of resultant magnetisation is indicated by the anomaly peak-to-trough azimuth, and in steep southern geomagnetic inclinations it is indicated by the anomaly trough-to-peak azimuth

- In the same steep-inclination fields, a high peak-to-trough amplitude ratio indicates a steep-inclination magnetisation with the same polarity as the field and a high trough-to-peak amplitude ratio indicates a steep-inclination magnetisation with opposite polarity to the field. A peak-to-trough amplitude ratio close to parity indicates a low-inclination magnetisation

- These relationships support estimation of magnetisation direction from visual inspection of the magnetic field in high geomagnetic inclination fields. At low geomagnetic inclination it is also possible to predict magnetisation from the pattern of well-defined anomalies but the relationships are more complex

- If you cannot recognise the presence of remanent magnetisation from visual inspection of well-imaged magnetic field data (especially with the benefit of subdued or extinguished sun-shading and a contour overlay) you should not expect inversion to find it for you

- It is a fallacy that voxel inversion can recover magnetisation direction of each voxel in a model. The fields computed from those models may match the measured magnetic field but their apparent detail cannot be justified from magnetic field data alone

- The greatest challenge in estimation of magnetisation direction from magnetic field data is to effectively separate the field variation due to that magnetisation from other field variations. Field separations are invariably interpretive, even if analytical methods are applied to effect their separation

- A major trade-off against magnetisation direction in the inversion of magnetic field data is its horizontal position. If a magnetic field is inverted with an incorrect magnetisation direction the estimated horizontal location of the magnetisation will be incorrect, and conversely if horizontal location of the magnetisation is incorrect then inversion will give an incorrect magnetisation direction

- Incorrect estimation of the structural dip or plunge of a magnetisation also trades off against estimation of its magnetisation direction. This is because change of the dip or plunge of a magnetisation causes change in its horizontal position with depth. The problem is more extreme for a thin sheet of magnetisation than for a narrow pipe

- Vertical location and depth extent of magnetisation do not strongly trade-off against estimated direction. However, if error in estimated depth or depth extent gives rise to substantial misfit between measured and model-computed fields those misfits can accommodate error in estimating magnetisation direction even though they do not cause it

- The influence of shape of a distribution of magnetisation complicates estimation of its magnetisation direction. This problem is more acute in 'proximal' fields close to the magnetisation and eases with distance in 'distal' fields. In the ultimate distal field an anomaly (if still detectable and sufficiently defined for reliable analysis) is identical to that of a co-centred dipole magnetisation of identical magnetic moment

- If we can improve capability to determine source magnetisation direction from analysis or inversion of magnetic field data then correlation of magnetisation direction estimates will become a valuable method to map the regional extent of igneous, metamorphic, thermal and mineralising events

- The Australian remanent anomalies database is a resource to report magnetisation direction estimates recovered from magnetic field analysis or inversion. Knowledge of the direction of a magnetisation is more valuable if there are other magnetisations in the area with which it can be compared

5.3 RESULTANT (REMANENT AND INDUCED) MAGNETISATIONS

Two alternative descriptions of three-dimensional vectors are provided by the amplitudes of three orthogonal components (e.g. north, east and down) or by amplitude with declination and inclination angles. Figure 5.3 shows identical descriptions of the magnetic field and

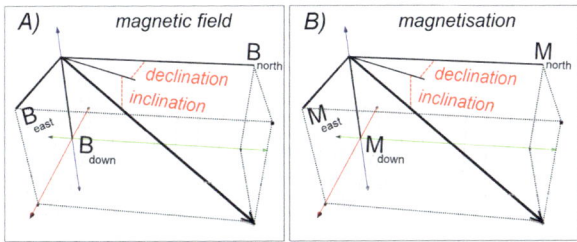

Fig. 5.3. Identical vector component and magnetisation direction (declination, inclination) descriptions of A) the geomagnetic field and B) magnetisation.

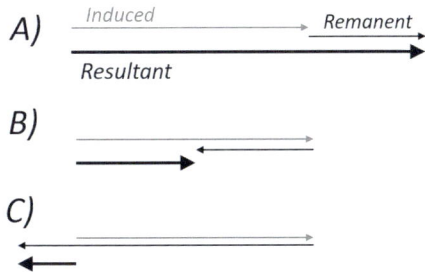

Fig. 5.4. Induced and remanent magnetisation vectors: A) parallel, B) anti-parallel (induced > remanent), C) anti-parallel (remanent > induced).

magnetisation in these two formats. If a magnetisation is purely induced with no anisotropy or self-demagnetisation effects then the declination and inclination angles of the magnetic field and magnetisation are identical.

As described in Chapter 1, magnetic field variations are expressions of contrast between magnetisations that are vector sums of induced and remanent components. Figure 5.4 shows the vector relationships for three of the four conditions in which induced and remanent magnetisation are either parallel or anti-parallel to each other. For parallel magnetisations (Fig. 5.4A) the amplitudes sum, with no rotation of direction. In static magnetic field surveys these magnetisations are indistinguishable from purely induced magnetisations of the same total strength.

Magnetisations that are closely anti-parallel exist in recent volcanic rocks that cooled during a period of reverse geomagnetic polarity. Surface and near-surface volcanic units that commonly carry these magnetisations have significant expression in magnetic field data (even if they are of small volume). Figure 5.4B represents the case that remanent magnetisation is weaker than the induced magnetisation (the Koenigsberger ratio is less than 1) and for which case resultant magnetisation is anti-parallel to the induced magnetisation. If the reverse remanent

magnetisation is stronger than the induced magnetisation there is a 180° rotation of the resultant magnetisation direction as represented in Fig. 5.4C. If the strength of the remanent magnetisation only just exceeds the induced magnetisation the resultant intensity is low, but for Koenigsberger ratios of more than 2 it exceeds the strength of the induced magnetisation. Some volcanic episodes span both normal and reverse periods of the geomagnetic field with one or more reversals of the field, and these rock units may contain magnetic field expressions of mixed normal and reverse polarity. The fourth case not shown in Fig. 5.4 is that normal and reverse components have identical strengths and their resultant is zero.

Older rocks carry remanent magnetisations different in direction from the present geomagnetic field because of rotation by either post-intrusion tectonics or translation with continental drift. This leads to a much wider range of magnetisation directions. Induced, remanent and resultant magnetisations are co-planar and lie on a great circle in a stereographic projection. Figure 5.5A shows the relationship between these vectors for a Koenigsberger ratio less than 1 and Fig. 5.5B for a Koenigsberger ratio greater than 1. The Koenigsberger ratio is a scalar measure and does not completely describe the vector relationship between induced and remanent magnetisation. It cannot be determined directly from magnetic field data and therefore is of limited application in magnetic field analysis. For analysis of magnetisation direction recovered from magnetic field data the Koenigsberger ratio should be augmented with the apparent rotation angle (ARRA) between induced and resultant magnetisations that can be recovered directly from analysis or inversion of magnetic field

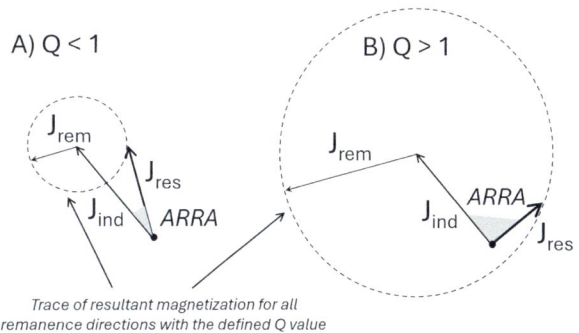

Fig. 5.5. Coplanar induced (Jind), remanent (Jrem) and resultant (Jres) magnetisations for A) Koenigsberger ratio (Q) > 1 and B) Q > 1. The apparent resultant rotation angle (ARRA) is between induced and resultant magnetisation vectors.

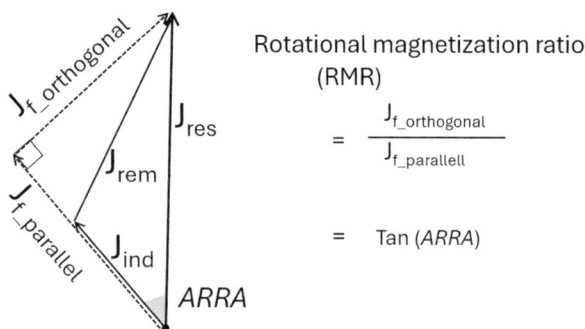

Fig. 5.6. Classification of magnetisation from magnetic field analysis as field-parallel (Jf_parallel) and field-perpendicular (Jf_orthogonal) components rather than laboratory classification as induced (Jind) and remanent (Jrem) magnetisations.

data. In Fig. 5.5 the trace of the resultant magnetisation vector for different directions of remanent magnetisation is mapped by the dashed circles. If the Koenigsberger ratio is less than 1 (Fig. 5.5A) there is a limited range for the resultant direction. If the Koenigsberger ratio is greater than 1 (Fig. 5.5B) the resultant can lie anywhere along the great circle that includes the induced and resultant magnetisation directions. Only if magnetic susceptibility is known or the Koenigsberger ratio is assumed can the remanent magnetisation strength and direction be found by completing the vector triangle.

Splitting a magnetisation into induced and remanent contributions is consistent with its genesis and provides a direction for the remanent magnetisation that should have geological meaning, but this cannot be done from magnetic field data alone. For magnetic field analysis, as shown in Fig. 5.6, a more practical separation of the magnetisation is in terms of the component parallel to the field that includes induced magnetisation with addition or subtraction of the parallel remanent contribution, and the remanent component perpendicular to the field (that can only be due to remanent magnetisation). The ratio of these components is given by the tangent of the ARRA angle, from which a maximum bound can be placed on the Koenigsberger ratio.

5.4 RECOGNITION OF MAGNETISATION DIRECTION IN MAGNETIC FIELD IMAGERY

Visual detection of differently directed magnetisations in magnetic field imagery provides a starting point for their analysis. If the field variations all appear to be consistent with magnetisations parallel to the geomagnetic field then a standard analysis can proceed. For instance, reduction-to-pole (RTP) transforms can be applied with limited concern of distortion by inappropriate magnetisation direction. If any of the magnetic field variations appear to be inconsistent with field-parallel magnetisation then two different approaches to their analysis are to minimise the influence of magnetisation direction by applying a total gradient or NSS transform to the data, or to highlight influence of magnetisation direction by submitting those features to a dedicated analysis or inversion to determine their magnetisation direction.

Figure 5.7 shows alternative images of a TMI grid measured in a geomagnetic inclination −60°, declination 0°. Sun-shading is a popular magnetic field display option designed to accentuate subtle, low-amplitude features. Figure 5.7.A is an image with subdued sun-shading and Fig. 5.7.B is a 'flat' image without sun-shading and with a contour overlay to indicate the true amplitude variations. Both images have a common histogram-equalised colour stretch (the colour scale is adjusted so that each colour is assigned to an equal area). This reveals maximum detail across the image (commonly applied to adress the highly dynamic range of magnetic field data that includes values substantially higher and substantially lower than mot of the data) but even by reference to a colour scale bar it is difficult to evaluate the absolute amplitude range of individual features. The discrete anomalies scattered across the map area have three different magnetisation directions (labelled 'a' to 'c' in Fig. 5.7.B). The different anomaly patterns caused by these three magnetisations are quite subtle in Fig. 5.7.A but with the help of the contour overlay are more easily discriminated in Fig. 5.7.B. Recognition of these differences in magnetisation may be critical to the magnetic field interpretation if, for instance, known mineralisation is associated with one direction whereas the other directions are association with barren bodies, or if the different magnetisations indicate different ages of the source bodies. Note that for the smooth background field of Fig. 5.7 in which the discrete anomalies produce the sharpest field gradients, the anomaly patterns can also be effectively highlighted by application of a vertical derivative filter that preferentially expresses the local anomalies and supresses longer wavelength gradients in the background field. A sound approach to magnetic field interpretation is to start with piecewise inspection of each discrete feature to best determine the information it carries, and then to proceed to relate those features in a geological synthesis.

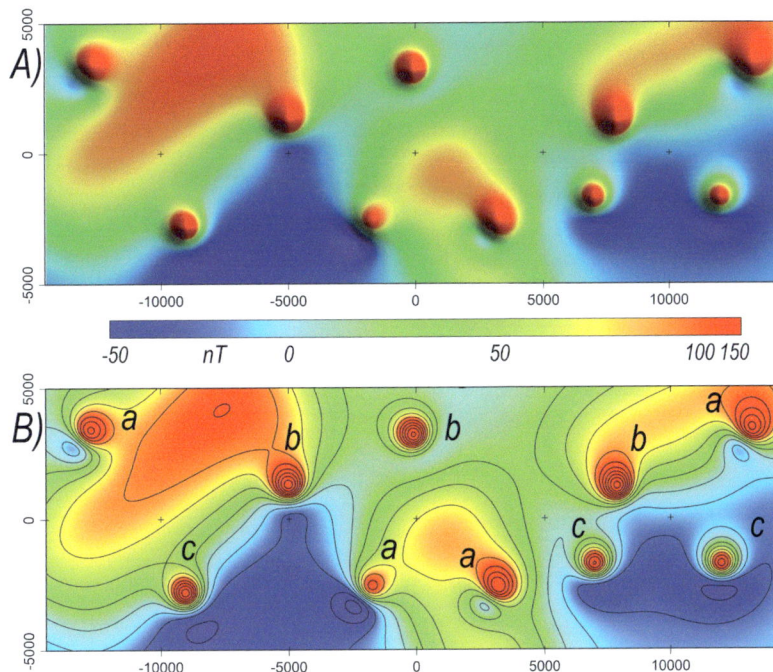

Fig. 5.7. TMI A) with north-east sun-shading, B) with 20 nT contours. Magnetisations: a) inclination −20°, declination 40°, b) inclination −80°, declination 300°, a) inclination −60°, declination 0°.

5.5 EXPRESSION OF MAGNETISATION IN PROXIMAL AND DISTAL FIELDS

Chapter 6 illustrates the influence of position, shape and plunge of a magnetisation on its magnetic field expression and the trade-offs arising in compensation by these factors for any errors in estimation of magnetisation direction. The influence of shape and plunge are most pronounced in a proximal magnetic field immediately surrounding a magnetisation and these influences attenuate in more distal fields. Figure 5.8 shows central north–south profiles over horizontally co-centred spherical and ellipsoidal magnetisations with inclination −90° in a northerly directed magnetic field of inclination −60°. In Fig. 5.8A TMI is computed at an elevation 100 m above the tops of ellipsoidal magnetisations and 340 m above the top of a spherical magnetisation of identical volume. All bodies have identical vertical magnetisations of intensity 1 A/m. The ellipsoids have a long axis of 400 m radius and the other two axes of 100 m radius, with dips vertical and at 45° to north and south. All four curves have different shape. The horizontal locations of the anomaly peaks for the sphere and vertical ellipsoid are co-located close to the centre of magnetisation, but over the plunging ellipsoids there is considerable horizontal offset of the anomaly peaks in the up-plunge direction. Figure 5.8B shows the curves

computed 400 m higher. This is an increase in height by a factor of 5 from the shallowest magnetisations of the three ellipsoids. The increase in elevation causes substantial reduction in amplitude of the computed fields as shown by the solid-line profiles in Fig. 5.8B. To more easily compare these curves the magnetisation intensities were increased to normalise all curves to a peak of 100 nT as shown by the dashed-line profiles. The horizontal displacements between the peaks of the various curves are reduced substantially with the increase in elevation of the computed field, and the curves have more consistent pattern (the anomaly pattern is less revealing of plunge). The curves of the three ellipsoids could be brought to a closer match by applying slight horizontal displacements to counteract the bias introduced by differences in horizontal location of their shallowest magnetisations. If the fields were computed at even higher elevations, the amplitudes of the curves would weaken further and the shape and position of the curves would converge more closely. At extreme elevations all four bodies would produce near-identical (very low amplitude) field variations, and in each case inversion of those anomalies should use a spherical (dipole) model because there would be no justification to represent any apparent axial elongation or plunge that had little or no influence on the computed magnetic field.

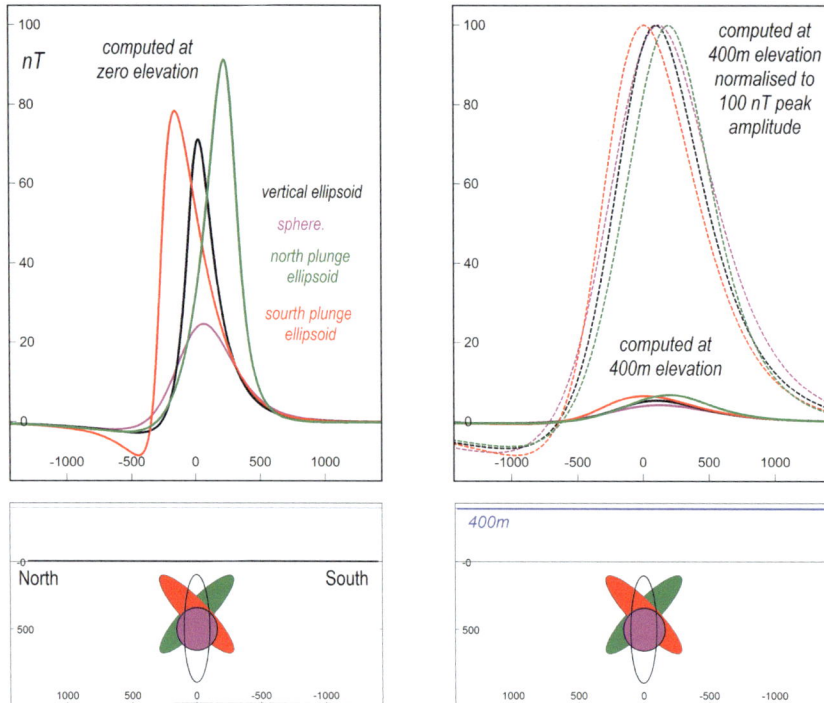

Fig. 5.8. North–south central profiles over horizontally co-centred sphere and ellipsoid magnetisations computed at A) 100 m above tops of the ellipsoid bodies and B) 500 m above the tops of the ellipsoid bodies.

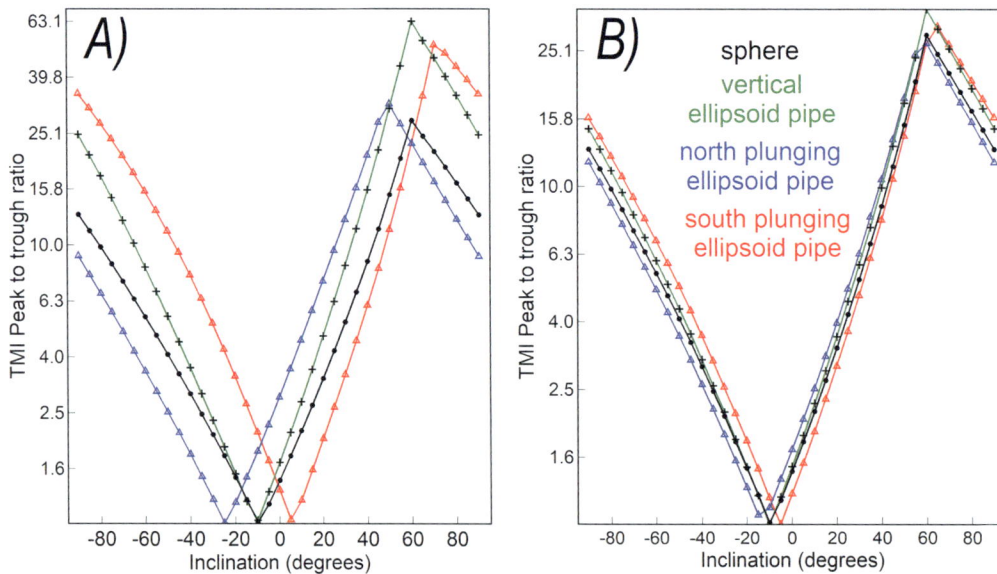

Fig. 5.9. Peak to trough amplitude ratios for the bodies shown in Fig. 5.6 and computed at the same lower elevation A) and higher elevation B) for north–south magnetisations with inclinations −90° to +90°.

Figure 5.9 plots peak-to-trough ratios for the four bodies shown in Fig. 5.8 at the two different field elevations and computed for the complete range of inclinations of magnetisation from −90° to +90°. On a logarithmic amplitude scale the curves can be split into approximately straight-line segments convenient for prediction of inclination of magnetisation from measurement of the anomaly peak and trough amplitudes. For the lower elevation measurements (Fig. 5.9A) curves for the different models form a wide band and recovery of the angle of inclination of magnetisation would require knowledge or independent determination of its plunge. At the higher elevation (Fig. 5.9B) the influence of plunge of the magnetisation is greatly reduced and would supply only a fine-tuning in estimation of the inclination angle by up to ~5°. As explained in Chapter 14, an improved estimate of inclination of magnetisation from the peak to trough amplitude ratio of a magnetic field anomaly can be achieved after application of a transform to provide the vertical component of the field B_z or its vertical derivative $B_{z,z}$.

5.6 RTP TRANSFORM OF FIELDS DUE TO REMANENT MAGNETISATION

The reduction to pole (RTP) transform is widely used in magnetic field interpretation to compensate for inclination effects of the geomagnetic field and approximately centre magnetic field variations over their source magnetisations (Baranov and Naudy 1964; Blakely 1995). The transform has limited influence in steep geomagnetic inclinations and has instability issues when applied to low inclination fields. The RTP applies phase transforms to bring the geomagnetic field and the source magnetisation to vertical. A standard RTP transform assumes that the initial magnetisation is parallel to the local geomagnetic field but this can be amended for magnetisations of different known direction. RTP with a different input magnetisation direction is generally restricted to single anomalies because it is unusual to have multiple magnetic field variations directed consistently other than parallel to the geomagnetic field. Before applying an RTP transform it is important to carefully inspect the TMI image using the cues described in section 5.4 to recognise magnetic field expression of magnetisation direction. It is more difficult to later recognise the expression of magnetisation direction in the output of an RTP that has already been applied. Trial application of RTP with different magnetisation directions has been shown capable of recovering magnetisation direction by cross-correlation with a transform such as the total gradient that approximately centres field variation over a magnetisation similarly to a successful RTP (Dannemiller and Li 2006). There have also been associated attempts to estimate magnetisation direction by correlation of gravity and transformed magnetic fields; however, geological complexity and limited correlation of density and magnetisation contrasts generally restricts this analysis (see Chapter 1).

Figure 5.10A shows the TMI Black Hill Norite northwest anomaly and Fig. 5.10B shows a standard RTP transform using the assumption of magnetisation parallel to the geomagnetic field. The background field has a relatively steep inclination of −67° and the standard RTP creates only a minor adjustment and southerly displacement of the anomaly. Figure 5.11 shows parametric inversion of the anomaly with a polygonal sheet model using Modellvision software (Foss and McKenzie 2011). The free parameters are the location of the model, its size and shape (defined by vertex coordinates), depth to the top, depth extent, plunge, resultant magnetisation intensity and direction. The best-estimated magnetisation has intensity 39 A/m, declination 232° and inclination +8°. Figure 5.12 shows the output of an RTP transform using the inversion-estimated magnetisation direction. This produces a compact, almost positive-only anomaly consistent with expectation of an anomaly in a vertical geomagnetic field over a compact vertical

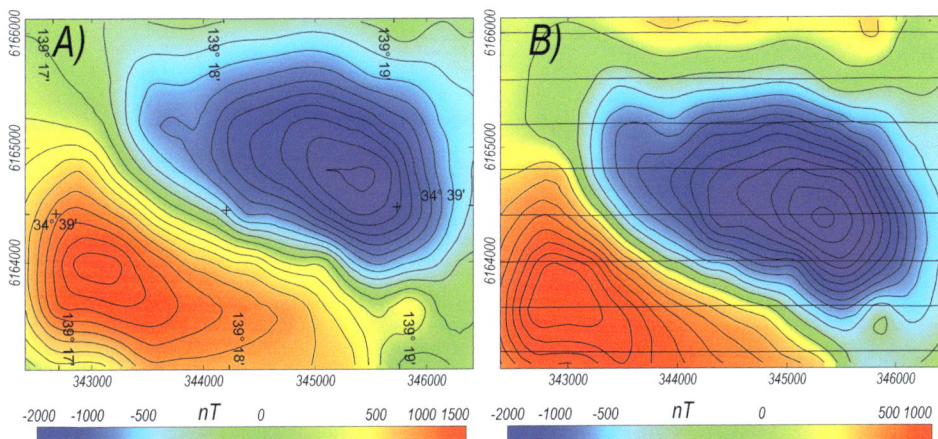

Fig. 5.10. Black Hill Norite north-west anomaly A) TMI and B) standard (induced magnetisation) RTP.

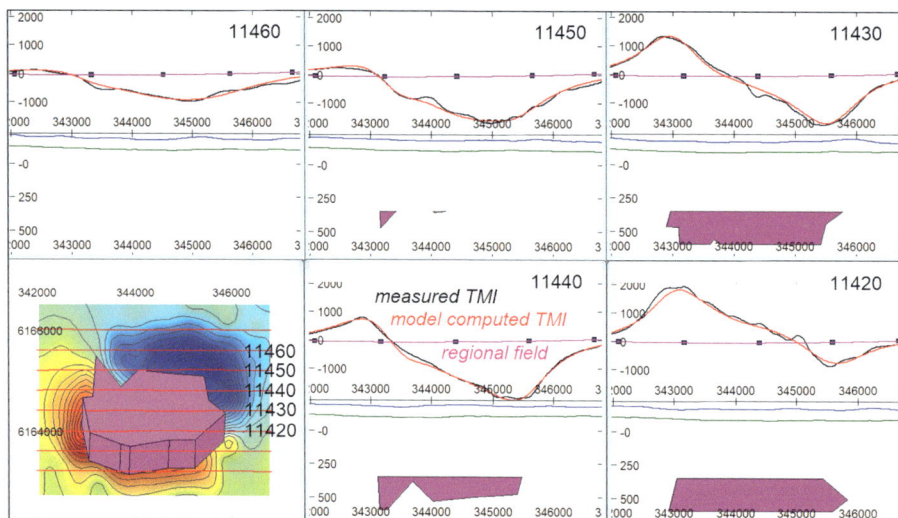

Fig. 5.11. Black Hill Norite north-west anomaly inversion model sections. Model magnetisation intensity 39 A/m, declination 232°, inclination +8°.

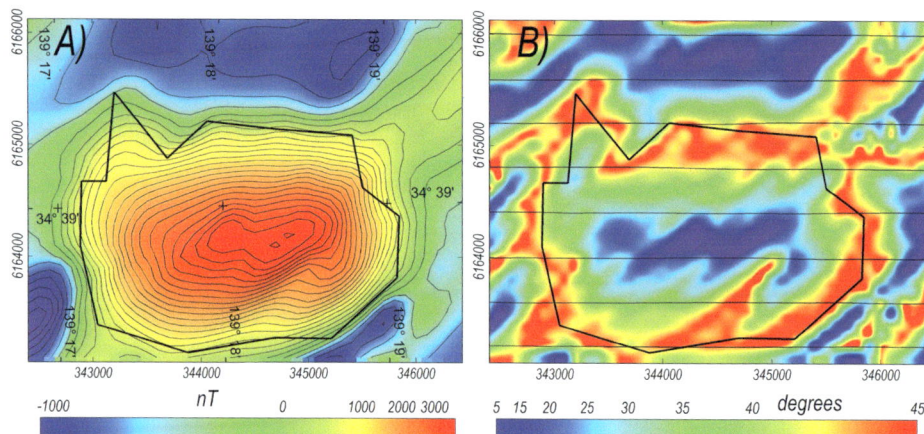

Fig. 5.12. Black Hill Norite north-west anomaly A) RTP using the inversion resultant magnetisation direction, with an overlay of the edge of the inversion model and B) edge inversion of the RTP anomaly.

magnetisation. The RTP anomaly location and shape is also consistent with the inversion model as shown by the trace of the model outline on the RTP image in Fig. 5.12A, and on an edge filter transform of the anomaly in Fig. 5.12B (from Foss and McKenzie 2011). This correspondence does not prove the transform correct because it is derived with input from the inversion model, but the close agreement is strong support for both the estimated magnetisation direction and the spatial model. While the standard RTP transform using an induced direction magnetisation (a rotation of 23° to the vertical) generated only a modest change to TMI (Fig. 5.10B), RTP with the estimated resultant magnetisation direction (a rotation of 82° to the vertical) produced the radically different RTP output (Fig. 5.12A). In this case the substantial rotation of the sub-horizontal magnetisation direction

to vertical is clearly more significant in the RTP process than the slight change in magnetic field direction.

Foss *et al.* (2021) performed an analogue model test of the RTP of the Black Hill Norite north-west anomaly. A palaeomagnetic rock core of the Black Hill Norite itself (a cylinder with height 2.5 cm and radius 2.5 cm) was given a saturation magnetisation along its axis and placed in a Rubens cage (Fig. 5.13) oriented to replicate the low-inclination south-west directed resultant magnetisation of the Black Hill Norite, with the applied field set to the direction at the Black Hill Norite site. The magnetic field was measured with a three-component magnetometer drawn along 'flightlines' with the sample sequentially moved perpendicular to the flightlines to generate the survey. Individual flightline data were levelled to replicate what would normally be achieved with

Fig. 5.13. Rubens cage in which the magnetometer was run along the yellow track above the rock cylinder sample carrying a magnetisation in the Black Hill Norite direction. The background field was set to be identical to the geomagnetic field at the Black Hill Norite location. The survey was performed by moving the sample perpendicular to the track after recording each flightline.

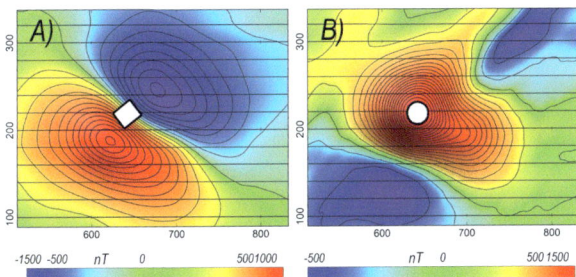

Fig. 5.14. TMI maps from surveys in the Rubens cage emulating A) the Black Hill Norite north-west anomaly and B) the RTP anomaly created by rotating the magnetisation to a vertical direction and setting the background field to vertical.

tie-line levelling and then residual misfits were attenuated with micro-levelling. The resulting TMI image (Fig. 5.14A) closely matches the measured Black Hill Norite north-west anomaly (Fig. 5.10A) revealing that for this anomaly, magnetisation direction is much more influential in determining anomaly shape than the distribution of magnetisation. An analogue RTP (Fig. 5.14B) was performed by repeating the Rubens cage survey with the core rotated to have a vertical magnetisation and the background field set to vertical. The measured field of this 'polar' survey is an almost positive-only anomaly consistent with that expected from a successful RTP transform. RTP of both the measured Black Hill Norite TMI anomaly and the synthetic anomaly generated in the Rubens cage using the same input magnetisation direction were successful, but both show striations parallel to the magnetisation direction. These striations are attributed to distortion caused by the low inclination of

magnetisation (Foss *et al.* 2021). The analogue experiment reinforces the substantial influence of magnetisation direction and the validity of including it in the RTP transform.

Well-known instability of the RTP transform for low-inclination magnetic fields (Blakely 1995) arises from extreme amplification of what should be near-zero terms. The synthetic Black Hill Norite Anomaly experiment demonstrates that this instability also applies to RTP transform of low-inclination magnetisation in high-inclination fields. At any given geomagnetic inclination, RTP transform of reliable TMI data should be stable but instabilities are introduced for noisy data, data with errors such as levelling busts or incorrect representation by gridding because the flightlines are too widely spaced.

The RTP transform does not substantially change the power spectrum of the data. Long wavelength components of the field (generally due to deeper magnetisations) have the greatest displacement and should move towards the pole. RTP transform of TMI grids with substantial north–south variation in the background field can be problematic. In these cases it is especially important that the data are well padded to a considerable distance beyond the grid border (preferably by including the true field data in that region). Note that for small survey areas and deep magnetisations, the measured magnetic field may not have any influence from the magnetisation directly beneath the survey area.

5.7 INVERSION OF HORIZONTALLY COMPACT, STEEPLY PLUNGING ('PIPE') MAGNETISATIONS

The key trade-off against error in assigned magnetisation direction is the horizontal location assigned to a magnetisation. However, for a magnetic field measured closely above a magnetisation, horizontal position is effectively locked in place. Dominance of the influence of the shallowest section of a magnetisation reduces the confidence with which deeper sections can be investigated by magnetic field analysis or inversion. In consequence of these relationships, an effective compensation for error in magnetisation direction is horizontal misplacement of magnetisation at depth in the form of an incorrect apparent plunge. Even if a magnetisation does not have significant extension to depth, inversion can create apparent extension. If a body is elongate in one horizontal direction (a sheet-like body) the compensation by apparent plunge is quite effective and neither plunge nor magnetisation direction can be reliably estimated. For a body that is horizontally compact (for instance a pipe) compensation for error in magnetisation direction by an erroneous apparent plunge still applies but is less effective.

We know that inversion solutions are non-unique and that close fits to measured data do not necessarily qualify an inversion model as a reliable representation of the true magnetisation. Even allowing for this uncertainty, the only practical measure that we have for acceptance of an inversion model (excluding highly interpretive geological appraisal) is the goodness of fit between its computed magnetic field and the measured magnetic field, and the most reliable result of inversion is exclusion of models that produce unacceptable data misfits. In practice, discomfort in acceptance of an inversion model increases gradually with increasing data misfit but there is no diagnostic test of whether a model should be accepted or rejected. Data misfit is often either the only factor or the major factor by which a model is evaluated and is commonly the key or sole driving factor in its inversion. To better understand inversion we can investigate success in recovering known models from inversion of the fields forward computed from them, starting with an inappropriate model and tracking its (hopefully progressive) return to the input model as inversion reduces the data misfit.

Figure 5.15 is a schematic of the inversion process. A starting model is shown in the top left of the figure where it has both a large data misfit and a poor representation of the true magnetisation. In this study I measure success of an inversion by the spatial overlap of the model with the known magnetisation. With the objective of

reducing data misfit, inversion can only proceed in a downward direction in Fig. 5.15. If the model change is directly downwards then the inversion does not improve the model but only reduces the discrimination for any subsequent inversion that will start with a smaller data misfit. The ideal of inversion is that the reduced data misfit is achieved by true improvements to the model, in which case the inversion proceeds diagonally downwards and to the right towards a model that is a closer representation of the ground and has a low data misfit as a consequence of this.

Figure 5.16 shows north–south and east–west profiles through the centre of an equidimensional anomaly due to a vertical pipe with induced magnetisation in a north–south geomagnetic field of inclination −60°. The complete anomaly from which these profiles have been extracted was inverted with models of an induced magnetisation sphere and ellipsoid. The data misfit of these models is because they have inappropriate geometries (different to the input pipe model). The spherical model is simpler than the ellipsoid model, has a larger data misfit and a smaller overlap with the input model. The data misfit of the spherical model might be recognised to betray a significant misrepresentation of the true magnetisation, and that might be the incentive to attempt inversion with the more complex, ellipsoidal model. It is unlikely for measured field data and a geological source body that meaning would be assigned to the residual data misfit as small as that of the ellipsoidal model, but the input model is already reasonably represented (other than for the detail of depth to its top). The

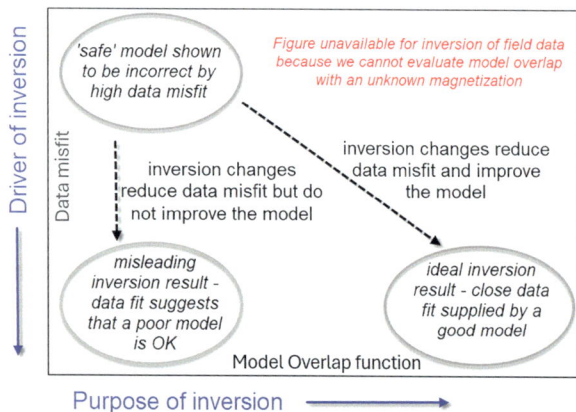

Fig. 5.15. Schematic of the relationship between the objective function of an inversion to reduce data misfit and the purpose of inversion to better represent the source magnetisation.

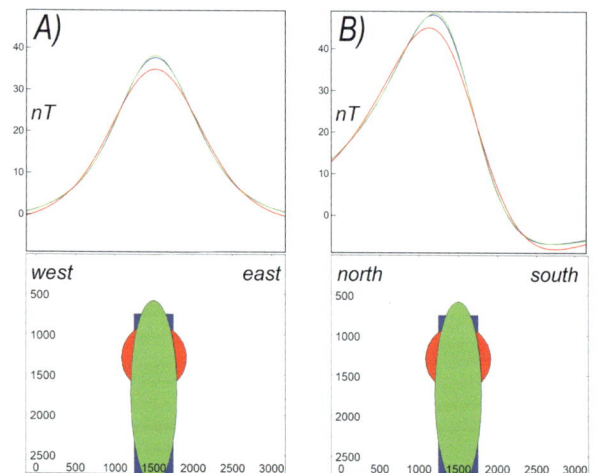

Fig. 5.16. Best-fit TMI profiles over a vertical cylinder (blue), spherical inversion model (red) and ellipsoid inversion model (green). A) east–west and B) north–south. All bodies have magnetic susceptibility 0.05 SI.

progression of an inversion from the spherical to the ellipsoidal model represents a productive inversion as a move downwards and towards the right in Fig. 5.15.

Following the simple case of Fig. 5.16 we can proceed to the more complex situation that the input vertical pipe model has a magnetisation differently directed to the magnetic field but that inversion is forced to proceed on the assumption that magnetisation is induced. Figure 5.17 shows TMI computed from three vertical pipe models with magnetisations divergent by 30° from the geomagnetic field direction of inclination −60°. The pipes have radii of 100 m, depth extents of 2000 m and the field is computed 150 m above their tops. The westernmost pipe has a steeper magnetisation of inclination −90°, the central pipe has an eastward rotation of magnetisation of declination 62.35° and the easternmost pipe has a shallower magnetisation of inclination −30°. From visual inspection of this ideal computed data it should be evident that the magnetisations are likely to be differently directed to the magnetic field and inversion should allow for that. However, with irregularities of measured survey data and widespread reluctance to introduce complexities of remanent magnetisation, many anomalies such as these are inverted on the assumption of induced-only magnetisation.

Figure 5.18 shows a perspective view of models produced by different inversions of the anomalies imaged in Fig. 5.17. The input vertical circular-section pipe models with rotated magnetisations are shown in magenta. Inversions using spherical models with free magnetisation direction generated the bodies shown in red in Figs 5.18 and 5.19. The spheres are horizontally co-centred with the source magnetisations and recover their

magnetisation directions faithfully. The relationship between these spheres and the input pipe magnetisations is identical to that between the induced-magnetisation sphere and pipe in Fig. 5.16. Inversions for magnetisations with known vertical plunge have no penalty from allowing a free magnetisation direction. However, if inversion forces an induced magnetisation (a direction 30° different to the true magnetisation direction for all three anomalies in Fig. 5.17) spherical models are horizontally displaced from the input magnetisations as shown by the light blue spheres in Figs 5.18 and 5.19.

Inversions using models assumed to have the same induced magnetisation but as vertically extended pipes produce the bodies shown in green in Figs 5.18 and 5.19 that are horizontally co-located with the induced magnetisation spherical models. This demonstrates that magnetisation direction is more effective than details of shape in estimation of the horizontal position of a magnetisation. The induced magnetisation spherical and vertical pipe models have large data misfits that should reveal that they are inappropriate models.

Inversions can also allow the magnetisation to plunge (adding two free inversion parameters of plunge angle and azimuth). In Figs 5.18 and 5.19 plunging induced magnetisation pipe models are shown in dark blue. To compensate for the 30° error in magnetisation direction that is too steep, horizontally rotated or too shallow, the models develop plunges from the vertical of 31°, 30° and 27° respectively with azimuths that vary systematically with input magnetisation direction as shown in Figs 5.18 and 5.19. In compensation for the input magnetisation that is too steep the plunge azimuth is towards the pole (to the south for this southern hemisphere example); for

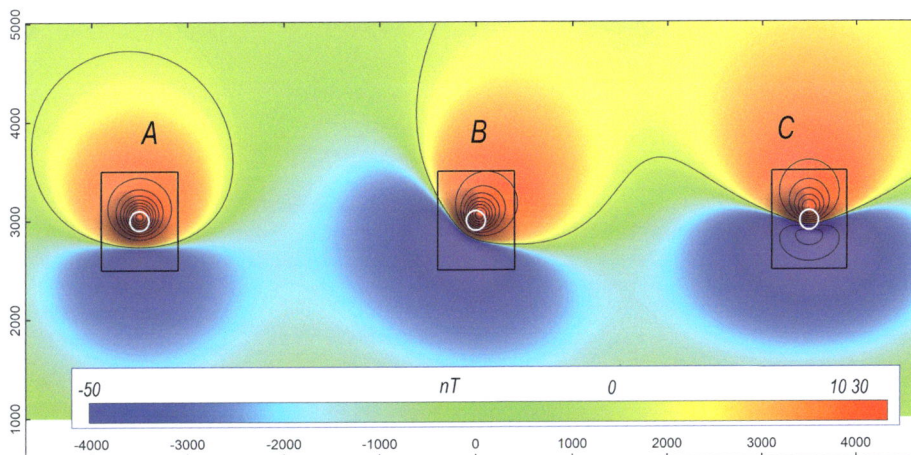

Fig. 5.17. TMI anomalies over vertical cylinders: A) inclination −90°, B) inclination −60°, declination 62°, C) inclination −30°, declination 0°. The background field has declination 0°, inclination −60°.

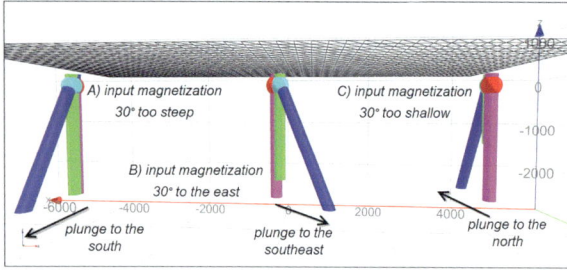

Fig. 5.18. Input remanently magnetised vertical pipes (magenta) and inversion models: free magnetisation spheres (red), induced spheres (light blue), induced magnetisation vertical pipes (green) and induced magnetisation plunging pipes (dark blue).

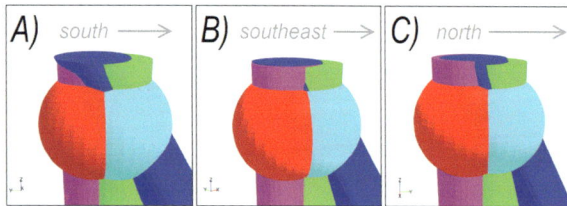

Fig. 5.19. Detail of the tops of the models shown in Fig. 5.18 that highlight the systematic influence of error in magnetisation direction and model geometry. Each view is perpendicular to the plunging induced magnetisation models (shown in dark blue).

the input easterly directed magnetisation the plunge is to the south-east; and for the input magnetisation that is too shallow the plunge is away from the pole (in this case to the north). Figure 5.19 shows the input and inversion models in perspective views perpendicular to the plunge of the induced pipes and illustrates the consistent effect of the magnetisation errors, with offset of the induced spheres and vertical pipes in the down-plunge direction. At this shallow measurement elevation the tops of the plunging pipes are approximately locked in place and the erroneous plunge takes the deeper sections of the inversion models progressively further from the true magnetisation. This reveals the weak concept of a centre-of-magnetisation for deep-going bodies. The base of the bodies contributes very little to the magnetic field variation and the magnetisation has a shallower effective centre-of-magnetisation than its true depth. The difference between effective and true centre of magnetisation depends on the elevation and nature of the measurements (whether they are of the field or its gradients). Shallower measurements and gradient measurements focus magnetic field analysis preferentially on the shallower sections of magnetisation.

Figure 5.20 shows contours of the magnetic field computed for the input vertical pipes with 30° rotated magnetisations (blue) and the inversion-derived

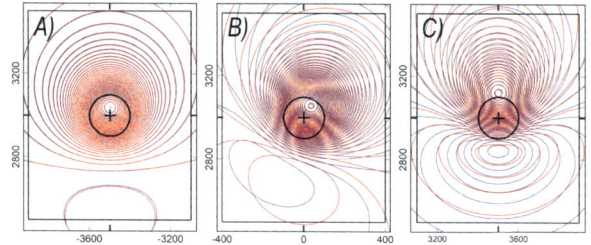

Fig. 5.20. Details of TMI in the areas outlined in Fig. 5.17: TMI contours (blue) of vertical pipes with magnetisations: A) declination 0° inclination −90°, B) declination 62.35° inclination −60° and C) declination 0° inclination −30° and (red) of best-fit plunging magnetisation pipes of declination 0°, inclination −60°. Circles show the edges of the vertical input models.

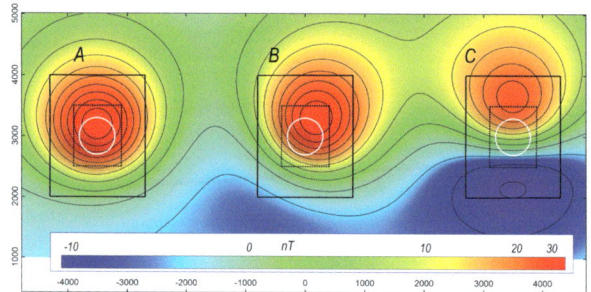

Fig. 5.21. The same magnetisations as in Fig. 5.17 (but in bodies three times wider) computed at an elevation of 1,000 m above the top of magnetisation. White circles show outlines of the input magnetisations. Dashed rectangles are extents of the previous shallow magnetisation studies and larger rectangles are the areas shown in Fig. 5.22.

plunging induced magnetisation pipes (red). The misfit between these fields is small and with even the best survey data it would not be feasible to attribute meaning to the residual data misfits, nor would those residual differences reliably drive further inversion changes to the model as they are smaller than would be anticipated from geological irregularities and data imperfections. The data misfit is least for the steeper magnetisation (Fig. 5.20A), meaning that model error is more readily accommodated for magnetisation directions that are too steep, and this provides a bias to inversion results.

Figure 5.21 shows the lower amplitude TMI field computed over deeper magnetisations with tops 1,000 m below the computation surface. The magnetisations are identical to the shallower bodies previously studied. Depth extents are the same but the pipes are wider (300 m radius) and volumes are nine times larger. Small and deep bodies are rarely detected reliably in magnetic field surveys unless they have extremely high magnetisation intensities. From images of this distal field it is more challenging to predict the horizontal location of the tops of the magnetisation (plotted as white circles in Fig. 5.21)

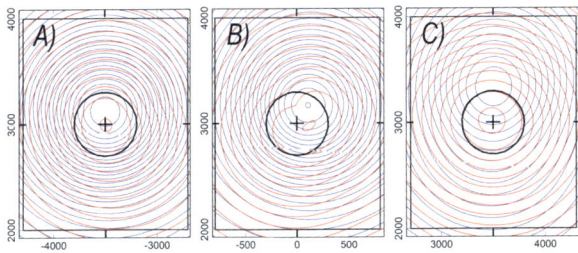

Fig. 5.22. Details of TG (blue) and NSS (red) contours computed from the TMI grid of Fig. 5.17. The black circles are the magnetisation outlines. In all cases the NSS is superior to the TG in locating the magnetisation.

because the proportional influence of the shallowest magnetisation reduces with increased source depth. Enhancement transforms can be used to assist in locating source bodies irrespective of their magnetisation direction. Figure 5.22 shows contours of TG (blue) and NSS (red) with the centres and margins of the magnetisations marked by the black crosses and circles respectively. For all three magnetisation directions the NSS peak closely marks the centre of magnetisation (because the magnetisations have vertical plunge) but the TG transform performs less well. For the horizontally rotated magnetisation the TG peak is close to the edge of the magnetisation and for the shallow magnetisation it plots outside the magnetisation (these details vary with the geometry of the source magnetisation, its depth and magnetisation direction).

I performed the same series of inversions for these deep magnetisations as for the shallower magnetisations and evaluated performance of both sets of inversions using criteria as plotted schematically in Fig. 5.15. The data misfit for each inversion is measured as the standard deviation of the data misfit normalised to the standard deviation of the input data and converted to a percentage. The success of the inversion is measured as the product of overlap between the input and inversion models normalised to the volume of the input model (a measure of what proportion of the input model is correctly located) and the same overlap normalised to the inversion model volume (a measure of what proportion of the inversion model correctly locates magnetisation). This measure is also converted to a percentage. The second term (the proportion of the inversion model that correctly locates magnetisation) is required to discriminate against large volume inversion models that include the input magnetisation but have little value in localising it. Overlap volumes are computed only across the depth range of the top 10% of the input model. No significance should be ascribed to models beneath that depth.

Deep-going magnetisations such as those investigated here are almost invariably tested by drilling only in their shallow sections. It is unlikely that bodies of depth extent 2,000 m such as these would be drilled to depths of more than 200 m below their tops, and even if they were to be, it is evident that the inversion models do not provide reliable guidance to those depths because the shallower sections of magnetisation completely dominate the measured magnetic field variations. As noted above, the induced magnetisation models with the lowest data misfit achieve that by a compensating plunge that takes their deeper sections further from the input magnetisation, with no overlap of those deeper sections.

Some of the key conclusions from the inversions are illustrated in Fig. 5.23 in cross-plots consistent with the inversion process schematic of Fig. 5.15. Figure 5.23A shows the results of the spherical model inversions. The spheres were restricted to have intensity of magnetisation identical to the input magnetisation because there is no sensitivity to that parameter, only to its product with volume to give the magnetic moment. The spherical inversion model geometry is very different to the input magnetisation and this limits the possible overlap to no more than ~50% as plotted in Fig. 5.23A. In Figs 5.23 and 5.24 induced magnetisation inversion results are plotted with closed symbols and free magnetisation direction inversion results are plotted with open symbols. Shallow model results are plotted in blue and deep model results are plotted in red. At both depths the free magnetisation models produce the lower data misfits and have the larger overlap with the input magnetisation, qualifying inclusion of the free variable of magnetisation direction as a successful inversion strategy. As already noted, these output models are centred on the axis of the input magnetisation, whereas the induced magnetisation inversion models are off-centred.

Figure 5.23B shows the influence of allowing free plunge in the induced-magnetisation-only inversions. The vertical inversion models (the triangular symbols in Fig. 5.23B) are compatible with the vertical input magnetisation but are offset because of their incorrect magnetisation direction. At both depths, freedom of plunge allows the inversion to reduce data misfit (as shown by the square symbols). For the shallow models there is significant reduction of data misfit but almost no increase in overlap resulting from inversion with free plunge. These are ineffective inversions (although that would not be known for inversions with unknown

Remanent and induced
spherical inversion models

Induced vertical and plunging
pipe inversion models

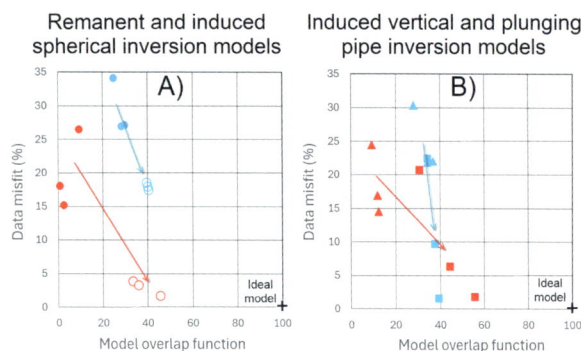

Fig. 5.23. Data misfit against model overlap for inversions with A) induced and remanent spherical models and B) induced vertical and plunging pipe models. Symbols are explained in the text.

Remanent and induced
spherical inversion models

Induced vertical and plunging
pipe inversion models

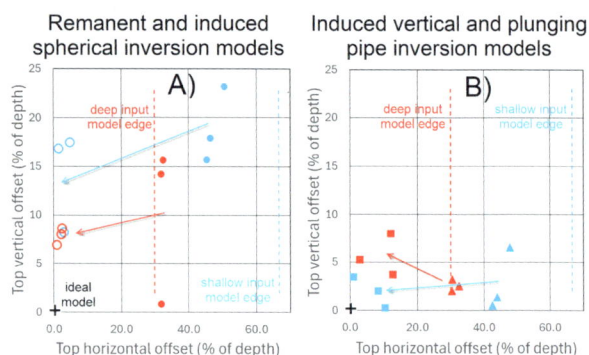

Fig. 5.24. Vertical and horizontal offsets for A) induced and free-magnetisation spherical inversions and B) vertical and plunging pipe induced magnetisation inversions. Symbols are as in Fig. 5.23.

sources). For the deep models the reduction of data misfit on allowing the magnetisation to plunge gives rise to an increase in overlap with the top 10% of the input magnetisation, but only to about twice that of the induced magnetisation spherical models and similar to the free-magnetisation spherical models. Note that in Fig. 5.23B inversion of this ideal synthetic data with a free-magnetisation pipe model would recover a model with a very high overlap.

Figure 5.24 shows horizontal misfit between input magnetisations and inversion models cross-plotted against vertical misfit. Horizontal misfit is measured between input magnetisation axes with spherical model centres in Fig. 5.24A and with top-face centres of pipes in Fig. 5.24B. Vertical misfit is measured between the top of input magnetisation and the top of the inversion models. Both the horizontal and vertical offsets are measured as percentages of depth to the top of the input magnetisation. Note that in Fig. 5.24 the ideal model (zero horizontal and vertical misfit) is plotted in the bottom left corner rather than the bottom right corner in Fig. 5.23.

The most significant differences between offsets of the spherical inversion models in Fig. 5.24 are between the small horizontal offsets of the free-magnetisation inversion models and the much larger horizontal offsets of the induced magnetisation inversion models. The free-magnetisation inversion bodies all have horizontal offsets of less than 5% but the centres of the deep induced-magnetisation inversion bodies lie just outside the extent of the source magnetisations. The vertical misfits for the spherical inversion models of Fig. 5.24A are similar for induced and free-magnetisation inversions, and both sets of inversions show a reduction in depth-weighted vertical misfit by ~50% between the shallow and deep models.

Figure 5.24B shows the results of induced magnetisation inversions with vertical pipes (triangular symbols) and plunging pipes (square symbols). All inversions have significantly less vertical misfit than for the spherical models in Fig. 5.24A. This emphasises the key importance of the shape of magnetisation in magnetic source depth estimation. Centres of the deep vertical pipe inversion models lie slightly beyond the margin of the input magnetisation (close to the centres of the spherical induced magnetisation models plotted in Fig. 5.24A). Including freedom of plunge in the inversions (the square symbols in Fig. 5.24B) reduces horizontal misfit of the top faces of 50% and 30% for the shallow and deep models respectively to only 10%. The spatial goodness of fit statistics of c. 10% horizontal offset and c. 6% vertical offset for the centre of the top face of the plunging induced magnetisation models in Fig. 5.24B present a more favourable report of the inversion results than the overlap measures for the same bodies of 30% to 60% over the shallowest 10% of the magnetisation as plotted in Fig. 5.23B. Note again that with pipe models, inversion of the noise-free synthetic data allowing freedom of magnetisation direction and an appropriate model type should achieve almost perfect representation of the input magnetisation.

5.8 A SENSITIVITY-EVALUATION CASE STUDY IN RECOVERY OF MAGNETISATION DIRECTION

Inversion applied to investigate unknown magnetisations are generally reviewed only according to data misfit. Verification of a model and evaluation of inversion errors comes only with any subsequent drilling. This case study reviews the sensitivity with which resultant magnetisation direction and depth to magnetisation were recovered from inversion of a compact negative anomaly and the prediction of depth to magnetisation made in design of a drillhole. Figure 5.25A shows a negative TMI anomaly in south-west South Australia measured by the 2015 Coompana Survey (P1270) on seven east–west profiles at a ground clearance of 80 m. The anomaly has an approximate range of 820 nT, from 780 nT below the background level to 40 above it, although the anomaly separation is not well defined. Figure 5.25B shows the model-computed field resulting from a free-magnetisation inversion. Flightline model sections are shown in Fig. 5.26. The background field was given a near-linear west to east slope. Two independent inversions were performed, one using a horizontal-topped elliptic-section pipe (shown in red in Fig. 5.26) and one with a more versatile polygonal-section pipe (shown in blue in Fig. 5.26) for which the increased degrees of freedom achieve a marginally improved data-fit (only visually obvious on profile 30180 in Fig. 5.26). The two inversion models mostly overlap but have a difference in their top elevations of 40 m (13% of mean depth below sensor and 18% of mean depth below ground). Both bodies have similar widths and dips towards the east, with resultant magnetisation intensities and directions of 34 A/m, declination 1°, inclination +23° and 21 A/m, declination 2°, inclination +24° for the elliptic and polygonal section pipes respectively. For simplicity only the top of the ellipsoid model is plotted in the map views in Fig. 5.25. An RTP performed using the inversion-derived magnetisation direction produced the field imaged in Fig. 5.25C, that is almost exclusively a positive anomaly

as expected of a successful RTP, with a peak coincident with the inversion model. The NSS transform of TMI is imaged in Fig. 5.25D. This image is independent of the inversion modelling but provides a very similar mapping of the horizontal distribution of magnetisation. Paine et al. (2001) used combined vertical integration and TG transforms (in both sequences) to invert TMI data with reduced sensitivity to magnetisation direction. Foss (2006) also used a transform image in staged inversion to approximately locate a source body, invert it for magnetisation direction, and then adjust both the spatial and magnetisation parameters in a final combined inversion (but found this approach unnecessary). Multiple synthetic and case studies have since established that provided the anomaly separation is appropriate and a reasonable starting model is used (guided if required by a suitable data transform) well-constrained inversion simultaneously resolves spatial and magnetisation parameters to give stable and reproducible solutions.

The estimated width of the top face of the polygonal section pipe inversion model is 190 m with an approximate length of 1100 m. Elongation of the body reduces confidence in the magnetisation direction estimate and the limited width of the body raises concerns that it might be missed with drilling. If the top of the body is narrowly missed in the up-dip direction the body will not be intersected, whereas if the top of the body is narrowly missed in the down-dip direction the borehole should still intersect it at a greater depth. Borehole CDP007 (Dutch et al. 2017a) into the source of this anomaly was designed with a westerly plunge based on the magnetic field inversion model results. This borehole

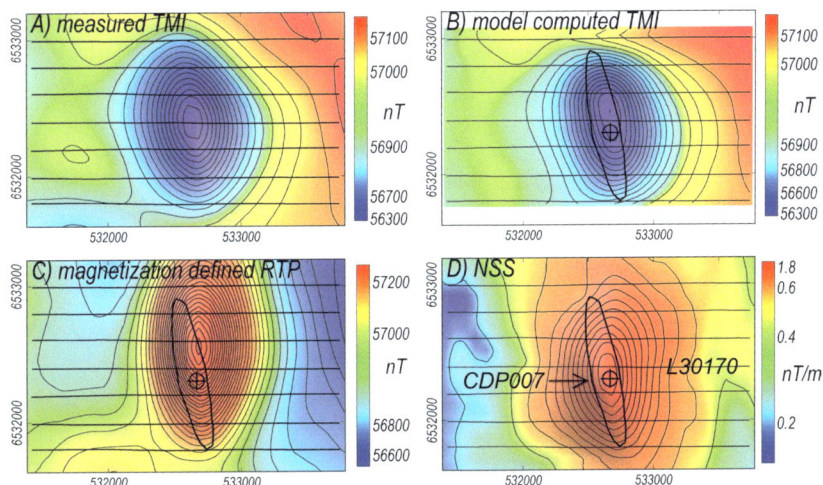

Fig. 5.25. Coompana ARAD anomaly 275: A) measured TMI, B) model-computed TMI, C) RTP derived with the supplied resultant magnetisation direction, D) NSS of TMI.

intersected the body within 10% of the depth below sensor of both inversion model predictions and at the horizontal location shown in the maps in Fig. 5.25. The borehole is also shown projected onto the nearest flightline section (line 30170) in Fig. 5.26.

Design of boreholes based on inversion studies benefits from understanding the confidence of those predictions. Non-uniqueness prevents a conventional measure of uncertainty in any inversion prediction but we can at least quantify the sensitivity of model-dependent estimations. We do this by first finding the model of a specified type that most closely fits the data (in a least-squares sense). This model is by definition the best combination of parameter values (including parameters of the background field) for that body type. Discovery of the parameter set is a multi-dimensional optimisation task, with each free parameter introducing an additional dimension. Sensitivity of the model to individual parameters is investigated by offsetting that parameter value and determining how effectively the resulting increase in data misfit can be compensated by revision of the other parameter values. In this study I investigated sensitivity to declination and inclination of magnetisation as well as to depth to the top of magnetisation. Figure 5.27 maps

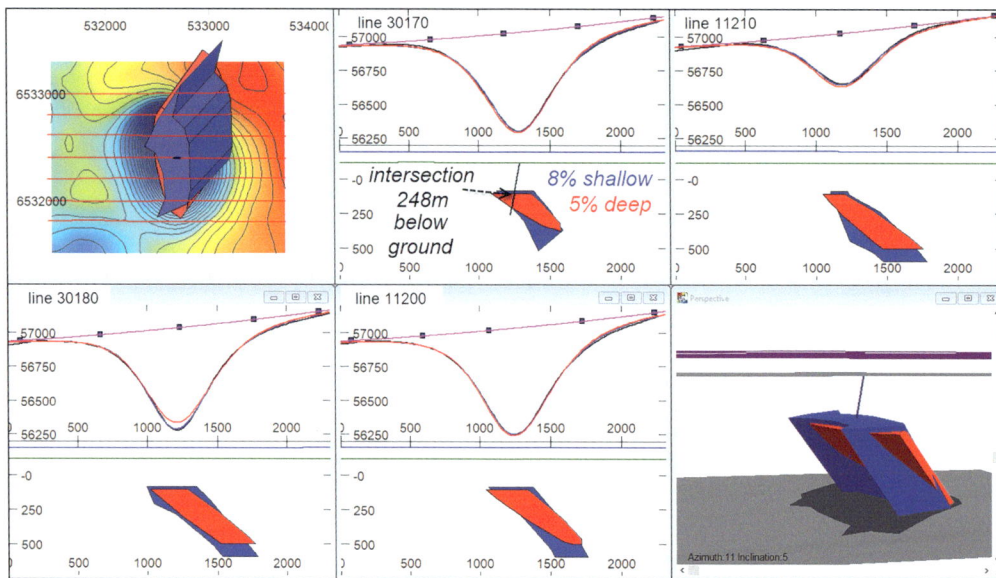

Fig. 5.26. Coompana ARAD anomaly 275 flightline model cross-sections with the elliptic pipe (red) and polygonal pipe (blue) models and their computed fields in the same colours. The magenta curve is the assigned background field.

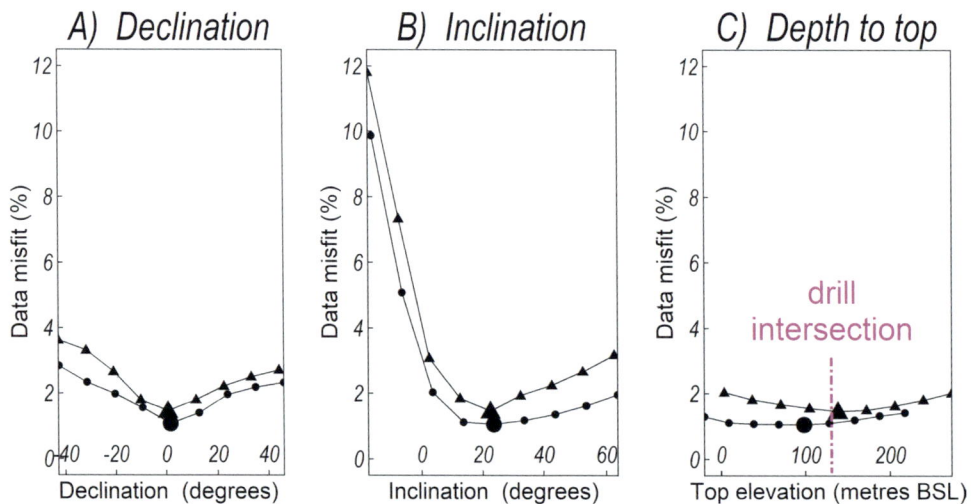

Fig. 5.27. Sensitivity curves for A) declination of magnetisation, B) inclination of magnetisation and C) depth to the top of magnetisation.

the increase in data misfit as those parameters progressively diverge from their values in the optimum parameter assemblage. The solutions determined by inversion to have minimum data misfits for the elliptic and polygonal section pipes (hopefully the global minimum solutions for those bodies) have misfit values (percentage standard deviation of the misfit normalised by standard deviation of the input data) of 1.1% and 1.5% respectively. These values represent field variations too complex or for some other reason unsuitable for explanation by a combination of background and model fields in the selection searched by the inversion. The absolute value of a residual data misfit does not have specific interpretational meaning, although large values raise questions about the suitability of proposed models to explain those field variations.

In Figs 5.27A and 5.27B misfit values are plotted for declination and inclination of magnetisation across a range of angles in steps of 10° about the optimum values. For the more complex polygonal-section pipes the values are plotted with circular symbols and for the elliptic-section pipes the values are plotted with triangular symbols. The difference in best estimated declination and inclination values between the two bodies are in both cases only 1°, although the error misfit curves in Fig. 5.27 suggests that these angles should not be interpreted to resolutions less than 5° to 10°. The section of the curves illustrating underestimation of inclination indicates a much greater sensitivity to this error than overestimation of inclination or error in declination. This is consistent with the synthetic data analysis in section 5.7. I also performed a sensitivity analysis of depth to the top of the magnetisation models in steps of 30 m. Angular sensitivity of magnetisation direction in degrees cannot be exactly related to spatial sensitivity of depth in metres, but nevertheless the shallower curvature of the depth sensitivity curves suggests that depth to the top of magnetisation is less reliably recovered from the inversions than the direction of magnetisation. This is consistent with the difference in depth to the top of the two models of 40 m compared to the difference in their magnetisation directions of ~1°. Horizontal offset of magnetisation is linked to angular error in estimated magnetisation direction but there is no equivalent linkage for vertical offset of magnetisation. It is not feasible to apply significance to the sensitivity curves in Fig. 5.27, but the clear data misfit minima for the magnetisation direction curves provided confidence that these narrow models were sufficiently well qualified to confidently design the drillhole.

5.9 RESOLUTION OF COMPLEXITY OF MAGNETISATION

Voxel inversion models are being created that have both incremental and large stepwise variation in magnetisation direction between immediately adjacent elements. An example is an inversion that reports three joined magnetisations spanning a range of direction of 133° as the source of the Black Hill Norite north-west anomaly (MacLeod and Ellis 2016, their fig. 7). This is the same anomaly investigated in section 5.6 that was already shown to be well explained by a single magnetisation model compatible with nearby rock magnetisation measurements. The three-magnetisation model may well match the anomaly, but its unnecessary complexity cannot be justified and should not be promoted on the grounds that it was generated by a sophisticated computer algorithm. For a well-defined anomaly, a compact single-magnetisation model that successfully matches the anomaly has the only feasible mean magnetisation direction (typically within uncertainties of less than 15°).

Complex distributions of magnetisation can be mapped from magnetic field data if there is at least partial separation between those magnetisations to support their separation. These complex magnetisations are not mapped as reliably as a single magnetisation would be and their analysis and inversion requires greater attention and strategy than for a single magnetisation. Figure 5.28A shows an 11 km diameter complex magnetic field anomaly in the Coompana area of south-west South Australia, measured by the same survey as the anomaly investigated in the previous section. The first issue on inspection of the TMI image is to design a starting model. The anomaly is primarily negative and is due to a mostly reverse but diverse magnetisation. I represent the magnetisation of the central zone of the anomaly (marked 'a' in Fig. 5.28B) with a single horizontal sheet of magnetisation with its margin digitised to the step change in the field (and represented by body 'a' in the post-inversion model in Fig. 5.29C). There is an outer second step of similar amplitude but mostly sharper gradients that seems to approximately map the extent of a wider body of reverse magnetisation (marked 'b' in Figs 5.28B and 5.29C) and I digitised the margin of a second sheet to that magnetic field step. It is not certain from inspection of the TMI anomaly whether the two sheets are vertically or horizontally zoned. In the starting model the two sheets were given similar depths of 500 m and depth extents of 2,000 m, so that inversion

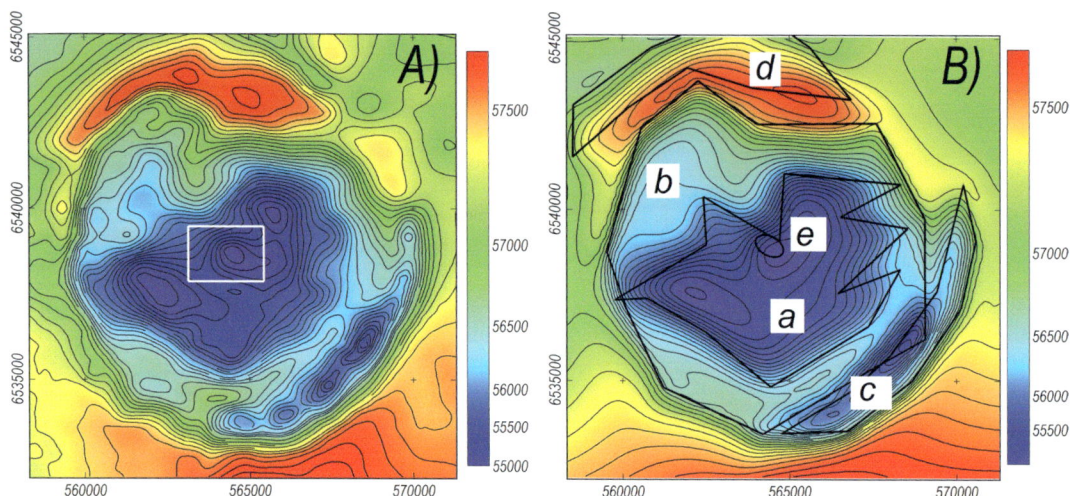

Fig. 5.28. A) TMI anomaly of the south-east complex anomaly in Coompana, south-west South Australia and B) model-computed TMI with outlines of the top extent of the model bodies.

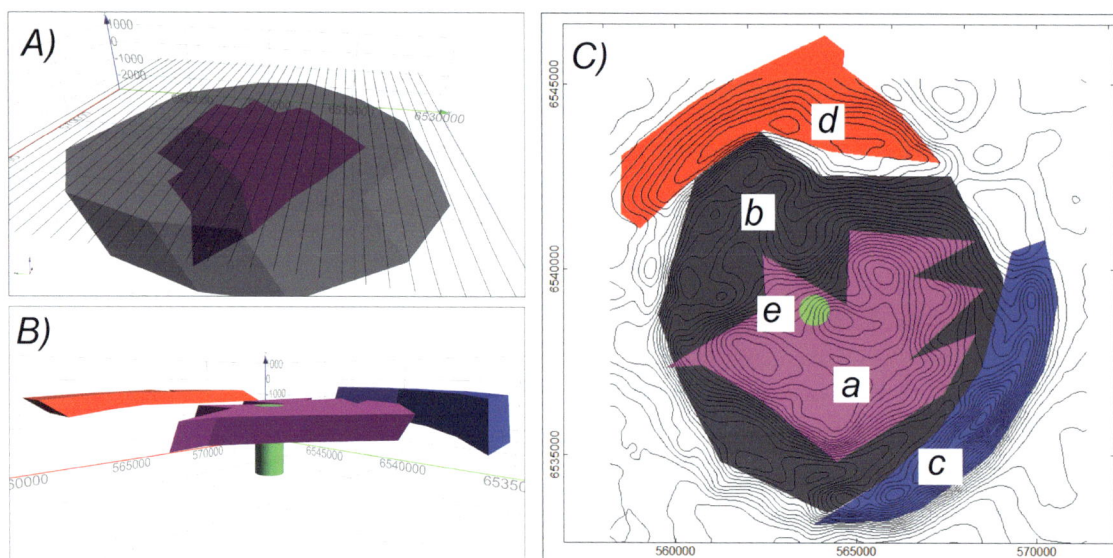

Fig. 5.29. Model components A), B) in perspective view and C) in map view (with TMI contours). Overlap of the inner and outer sheets is shown in A) and in B) the outer sheet is excluded to more clearly show the other components.

could adjust depth to the top and depth extent of each as favoured by the magnetic field data. The major remaining features are a sharp magnetic low to the south-east and a positive to the north, marked 'c' and 'd' respectively in Fig. 5.28B and 5.29C. I then managed the inversion (performed with the ModelVision software) in a series of stages, applying manual forward-modelling adjustments as advantageous to encourage conversion to a meaningful model. I call this process of prior design of a model and its manipulation throughout the inversion process 'user-guided' inversion. Those who trust in full automation of inversion see human intervention as a contamination of the inversion process and an unwanted introduction of human bias. My argument is that inversion of this degree of complexity benefits from guidance of a higher authority, and that inversion code is only a tool to be applied with understanding and skill to generate a geologically acceptable explanation of the measured magnetic field. Documentation of the inversion and decision history (such as provided here) is important for evaluation of the models that are produced.

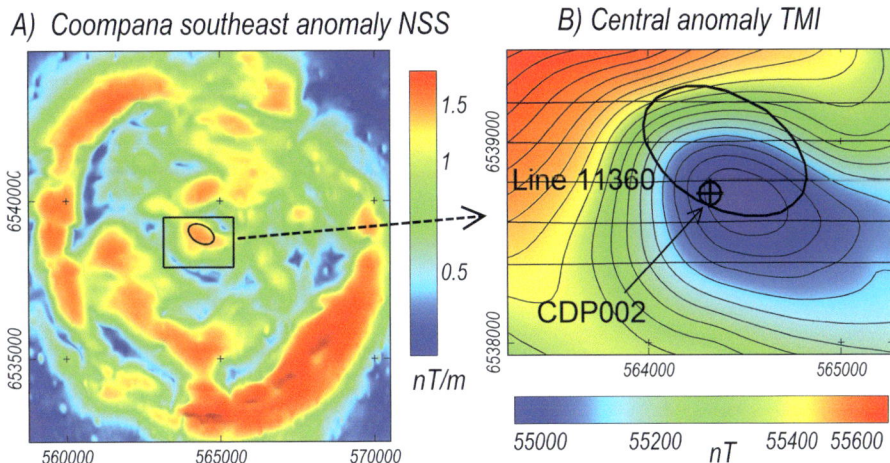

Fig. 5.30. A) NSS of TMI of the south-east complex anomaly in Coompana and B) TMI of the inset box in A with east–west flightlines, outline of the top of the elliptic-section pipe model and location of drillhole CDP002.

I progressed the inversion in a series of steps, starting with addressing the major features and ending with addition of details as justified and required. The first step allowed adjustment of the depth, depth extent, magnetisation strength, magnetisation direction and background field parameters. This produced a surprisingly close match to the measured field, confirming that the basic model concept was feasible. The inversion assigned a greater depth and smaller depth extent to the central sheet than to the outer one, so that the two sheets appear to have similar base depths (although these depths are poorly constrained). The central sheet overlaps the outer sheet in all locations, and in consequence there is no need to introduce additional complexity of a contact between them. Instead the central sheet has the summed magnetisations of both overlapping components. After minor adjustments, in the next step I allowed the bodies to move horizontally and to plunge, and subsequently I allowed the bodies to reshape by adjusting the horizontal coordinates of their vertices. Each step of inversion included multiple iterations. Bounds were set on each parameter to encourage a steady convergence by multiple small steps. In any one iteration, if a reduction in data misfit was achieved the inversion adjusted to search a reduced model space in more detail. If no improvement was achieved the inversion investigated a larger model space to evaluate if it was stalled in a local minimum. After each change all model views in map, cross-section and/or perspective windows are updated, together with all displays of the computed field. In this way I was able to track live progress of the inversion,

interrupt it to stop unwanted changes or to plan the next set of changes.

In the last stages of the inversion, the local minima in the centre of the anomaly began to emerge as significant residual data misfits and I introduced an additional vertical circular pipe to represent a possible feeder zone (at location 'e' in Fig. 5.29B and shown as body 'e' in the inversion model in Fig. 5.29C). Even at this late stage of the inversion the influence of this body is so diluted within the complete model that it is not reliably constrained. This local feature is one of several that have distinct expressions in the NSS transform (Fig. 5.30A) and was chosen as the location for a drillhole to test the structure. The model of the complete anomaly, covering an area of more than 100 km^2 has the objective of resolving major geological structure and is not suited to define detail at any specific location. To locate and design the borehole a data clip was extracted (shown by the box in Figs 5.28A and 5.30A) that was inverted separately. The flightline cross-sections through this detailed model are shown in Fig. 5.31. The best horizontal-topped plunging elliptic-section pipe returned by the inversion is very different to the corresponding (poorly constrained) body in the full anomaly inversion (depth 900 m) with a depth of 400 m. From the results of the detailed inversion the drillhole (CDP002, Dutch *et al.* 2017b) was located. This drillhole intersected the top of magnetisation (a gabbro) at 366 m below surface, 9% shallower than predicted by the detailed inversion. This is encouraging as the anomaly is not confidently separated from the complex background field and is poorly suited for depth estimation.

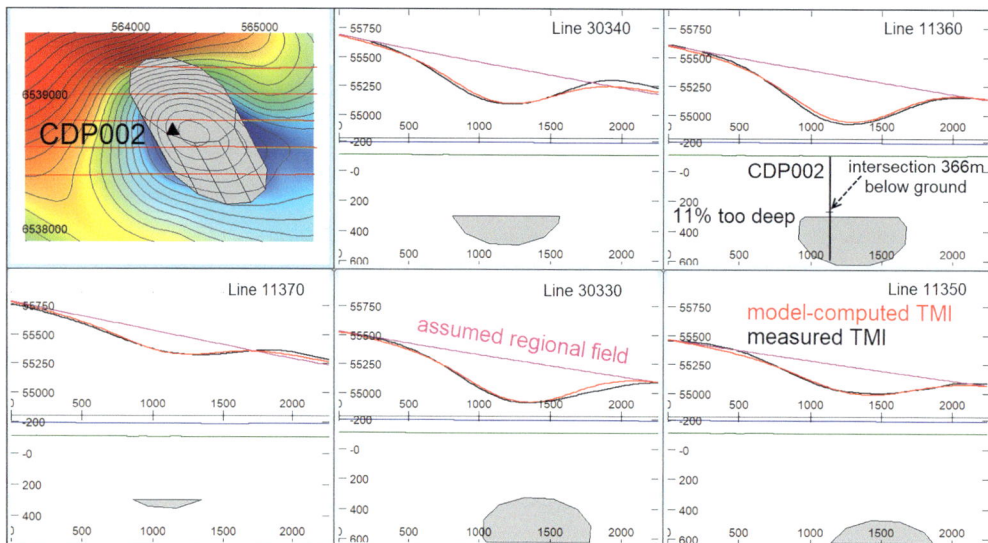

Fig. 5.31. Flightline sections of the central detailed inversion model to design borehole CDP002.

5.10 CONCLUSIONS

The key finding of the main inversion model is to confirm that the magnetic anomaly is consistent with a model that can be geologically interpreted as a (probably differentiated) basic, sub-volcanic intrusive complex, including both sub-horizontal sheets (model bodies 'a' and 'b') and arcuate fracture infillings (bodies 'c' and 'd'). This model can be incorporated into mineral exploration programs, and components of the model such as possible feeder zones and contacts between internal units or with surrounding basement can be targeted in concept-driven models of possible areas of mineralisation. The detailed model of the central section of the anomaly successfully predicted a depth for design of a borehole to test that feature.

For this complex model, inversion should not be considered complete but to have reached a stage at which further development is not warranted until answers to any specific questions are required and a targeted upgrade of the model can be performed. There are still imperfections in terms of unwanted overlap of model components at a few locations and residual data misfits. Magnetisations of the different model components are listed in Table 5.1 (including adjustment of magnetisation values for body overlaps where appropriate).

An argument against inversions such as these and favouring greater automation and less supervision is that automated inversions are performed much more quickly and require less geological understanding of the operator. However, the time required for a user-guided inversion such as this example should not just be compared against the computation time of an automated inversion but in context of the time and expense to plan, contract and acquire the survey, process and

Table 5.1. Volume, top depth (metres below surface) and resultant magnetisation of the model bodies.

Model	Description	Volume (km³)	Depth to top (m)	Intensity A/m	Declination	Inclination
a	Inner sheet	17.2	750	12.9	30	+45
a*	Corrected for b overlap			16.1	18	+52
b	Outer sheet	79.8	170	3.5	322	+46
c	Negative southern	7.4	360	8.5	15	+40
d	Positive northern	5.0	490	6.0	186	−14
e	Central pipe	0.94	910	26.5	266	+4
e*	Corrected for b overlap			28.3	270	+11
e'	Dedicated local inversion	0.123	405	7.8	33	+51
e'*	Corrected for b overlap			11.2	12	+58

analyse the data, make the necessary interpretation decisions based on the results and make the financial commitments for any drilling. Saving a few days of inversion time at the cost of less informed decisions is a false economy given that even minor details of inversion results might be of critical importance to the project objectives and determine between success and failure.

REFERENCES

Baranov V, Naudy H (1964) Numerical calculation of the formula of reduction to the magnetic pole. *Geophysics* **29**, 67–79. doi:10.1190/1.1439334

Beiki M, Clark DA, Austin JR, Foss CA (2012) Estimating source location using normalized magnetic source strength calculated from magnetic gradient tensor data. *Geophysics* **77**, J23–J37. doi:10.1190/geo2011-0437.1

Blakely RJ (1995) 'Potential Theory in Gravity and Magnetic Applications'. Cambridge University Press. pp. 441.

Clark DA (2014) Methods for determining remanent and total magnetisations of magnetic sources – a review. *Exploration Geophysics* **45**, 271–304. doi:10.1071/EG14013

Dannemiller N, Li Y (2006) A new method for determination of magnetisation direction. *Geophysics* **71**, L69–L73. doi:10.1190/1.2356116

Dutch RA, Jagodzinski EA, Pawley MJ, Wise TW, Tylkowski L, Lockheed A, McAlpine SRB, Heath P (2017a) 'Drillhole CDP007 preliminary field-data report'. PACE Copper Coompana Drilling Project: Report Book 2017/00044. Department of the Premier and Cabinet, South Australia, Adelaide.

Dutch RA, Jagodzinski EA, Wise TW, Pawley MJ, Tylkowski L, Lockheed A, McAlpine SRB, Heath P (2017b) 'Drillhole CDP002 preliminary field-data report'. PACE Copper Coompana Drilling Project: Report Book 2017/00038. Department of the Premier and Cabinet, South Australia, Adelaide.

Foss CA (2006) 'Evaluation of strategies to manage remanent magnetisation effects in magnetic field inversion'. 76th Annual International Meeting. *SEG Expanded Abstracts*, 938–942.

Foss CA, McKenzie KB (2011) Inversion of anomalies due to remanent magnetisation: An example from the Black Hill Norite of South Australia. *Australian Journal of Earth Sciences* **58**, 391–405. doi:10.1080/08120099.2011.581310

Foss CA, Leslie K, Schmidt PW (2021) 'If you can't go to the anomaly then let the anomaly come to you'. Extended abstract, AEGC Brisbane, pp. 1–4.

Helbig K (1963) Some integrals of magnetic anomalies and their relation to the parameters of the disturbing body. *Zeitschrift für Geophysik* **29**, 83–96.

Hood PJ (1964) The Königsberger ratio and the dipping dyke equation. *Geophysical Prospecting* **12**, 440–456. doi:10.1111/j.1365-2478.1964.tb01916.x

Lourenço JS, Morrison HF (1973) Vector magnetic anomalies derived from measurements of a single component of the field. *Geophysics* **38**, 359–368. doi:10.1190/1.1440346

MacLeod I, Ellis R (2016) 'Quantitative Magnetisation Vector Inversion'. ASEG Extended Abstracts 2016, 1–6. doi:10.1071/ASEG2016ab115

Nabighian MN (1972) The analytic signal of two-dimensional magnetic bodies with polygonal cross-section: its properties and use for automated anomaly interpretation. *Geophysics* **37**, 505–517. doi:10.1190/1.1440276

Paine P, Haederle M, Flis M (2001) Using transformed TMI data to invert for remanently magnetised bodies. *Exploration Geophysics* **32**, 238–242. doi:10.1071/EG01238

Phillips JD (2005) Can we estimate total magnetisation directions from aeromagnetic data using Helbig's integrals?. *Earth, Planets, and Space* **57**, 681–689. doi:10.1186/BF03351848

Rajagopalan S, Schmidt PW, Clark DA (1993) Rock magnetism and geophysical interpretation of the Black Hill Norite, South Australia. *Exploration Geophysics* **24**, 209–212. doi:10.1071/EG993209

Schmidt PW, Clark DA (1997) Directions of magnetisation and vector anomalies derived from total field surveys. *Preview* **70**, 30–32.

Schmidt PW, Clark DA (1998) The calculation of magnetic components and moments from TMI: a case study from the Tuckers igneous complex, Queensland. *Exploration Geophysics* **29**, 609–614. doi:10.1071/EG998609

Schmidt PW, Clark DA, Rajagopalan S (1993) An historical perspective of the Early Palaeozoic APWP of Gondwana: new results from the Early Ordovician Black Hill Norite of South Australia. *Exploration Geophysics* **24**, 257–262. doi:10.1071/EG993257

Zietz I, Andreasen GE (1967) Remanent magnetisation and aeromagnetic interpretation. *Mining Geophysics* **2**, 569–590.

6

The influence of source location, shape and plunge in recovery of magnetisation direction from magnetic field data

C.A. Foss

ABSTRACT

Parametric magnetic field inversion requires simultaneous solution of multiple source values. The estimated values for magnetisation (strength and direction) are subject to error in compensation for corresponding errors in the spatial parameters and in this chapter I investigate systematic relationships between error in magnetisation and spatial parameters. Vertical location and extent of a magnetisation does not substantially influence its estimated direction. There is, however, considerable trade-off between horizontal position and magnetisation direction. For well-defined magnetic fields at moderate to large distances from the magnetisation we should expect errors in magnetisation direction associated with mis-positioning of the magnetisation to be of the order of 5° or less. Controls on the horizontal distribution of magnetisation by plunge are the most problematic single-parameter trade-off for magnetisation direction, giving rise to uncertainties in magnetisation direction of up to 7°. Errors in estimated magnetisation direction increase with proximity to the magnetisation where shape factors become more prominent, with increasing complexity of the distribution of magnetisation, and with increasing challenges of isolating the magnetic field of the magnetisation.

6.1 INTRODUCTION

Many magnetic field analyses and inversions ignore the complexity arising from magnetisation direction. Those that seek to address it (Roest and Pilkington 1993; Fedi *et al.* 1994; Schmidt and Clark 1998; Paine *et al.* 2001; Dransfield *et al.* 2003; Phillips 2005; Dannemiller and Li 2006; Foss 2006; Lelièvre and Oldenburg 2009; Foss and McKenzie 2011; Clark 2014; Pratt *et al.* 2014; Fullagar and Pears 2015) either act to include magnetisation direction directly in analysis or inversion or to partially mitigate its effect. However, there has been no systematic evaluation of the capabilities and limitations with which analysis or inversion of magnetic field data can succeed without prior knowledge of its direction. The only analytic proof that magnetisation direction can be reliably recovered from magnetic field data is by Helbig (1963) derived specifically for dipole analysis of a magnetisation at a pre-defined horizontal location. For inversion of distal field data of a compact magnetisation the key challenge is to correctly separate the field to be analysed from any other superimposed field variations. Closer to the magnetisation, field variations include expressions of the shape and plunge of the magnetisation that disrupt reliable estimation of its direction.

A magnetic field variation generated by a discrete subsurface magnetisation depends on its horizontal position, depth, size, shape, orientation, magnetisation intensity and direction. Many magnetic field measurements have substantial further complication that they include contributions from multiple sources that cannot be uniquely separated. I investigate the fortunate but still challenging case that an isolated magnetisation produces a field variation sufficiently well defined that it can be confidently isolated from the background field. The analyses are restricted to compact magnetisations that have maximum extent no greater than twice the closest approach of the magnetic field measurements and computations. Bodies with a single large vertical axis (pipes) or with at least two large axes (sheets) require separate analysis. Parametric inversion is well suited to analysis of the sensitivities of the relationships between magnetisations and their external magnetic fields. It allows investigation of the influence of individual model parameters while keeping others fixed, and also supports investigation of how two or more parameters interact. These investigations illustrate fundamental relationships between a magnetisation and its magnetic field that are relevant to parametric and voxel inversion and any related analysis. The findings of this synthetic data study are well confirmed in a study of aeromagnetic data acquired near Rylstone, New South Wales.

6.2 INFLUENCE OF SOURCE LOCATION ALONE

In magnetic field inversion there is strong trade-off between estimation of horizontal source position and magnetisation direction. To investigate this relationship, I first attempt to recover magnetisation direction for a displaced equidimensional source with no complication of shape or orientation. Figure 6.1 shows a series of dipole

(homogeneous spherical magnetisation) centre-points in north–south and east–west traverses over images of TMI (Fig. 6.1A) and total gradient of TMI (Fig. 6.1B) computed for an induced dipole magnetisation in a geomagnetic field of inclination −60°. It is difficult to accurately select the horizontal centre of magnetisation from the TMI image. A reduced to pole (RTP) transform can be used to centre the magnetic field anomaly over the magnetisation but only if the magnetisation direction is already known. Alternatively, the total gradient transform (also known as the analytic signal or modulus of the analytic signal) can be used to reduce the influence of magnetisation direction (Keating and Sailhac 2004; Li 2006; Cooper 2014). However, as can be seen in Fig. 6.1 the total gradient peak is still displaced from the centre of magnetisation by ~12% of source depth. As a further option, the normalised source strength (Beiki *et al.* 2012) provides superior mapping of the horizontal centre of magnetisation that peaks directly over the centre of a dipole magnetisation. Figure 6.2 shows a north–south traverse for sources with horizontal to the north, and vertical-up and vertical-down displacements of 25% of the centre depth. The magnetic field profiles of these sources are inverted with free magnetisation direction in attempted compensation for the enforced spatial offsets (note that for this and all other model profiles shown in this paper inversions are best-fitted not just to data on the displayed individual profile but to the complete anomaly). The enforced horizontal source displacements shown in

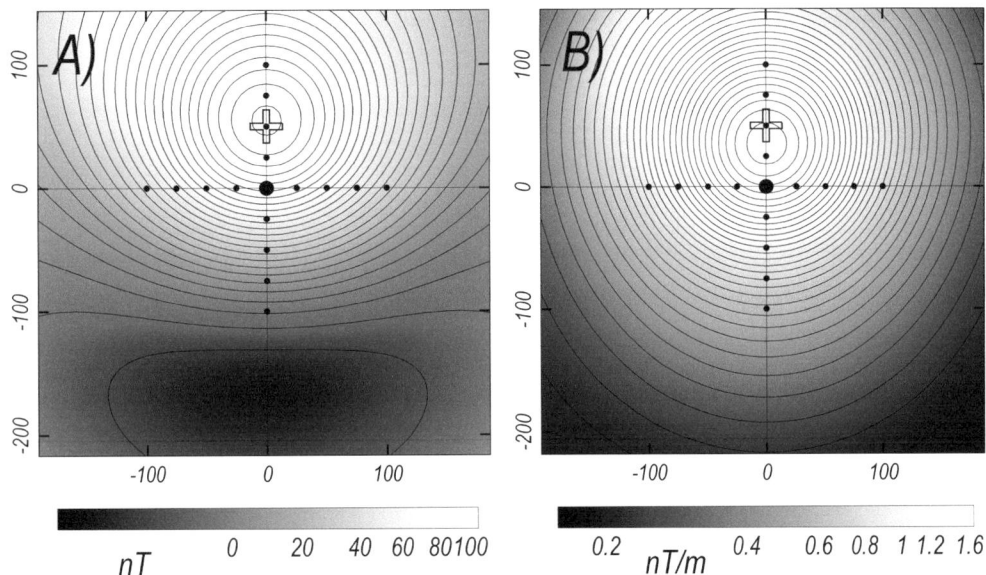

Fig. 6.1. Test dipole centre locations plotted over A) TMI, and B) the total gradient of TMI, for a magnetic dipole (magnetisation 0°, −60°) with centre (0,0, depth 200 m) in a geomagnetic field of (0°, −60°). The cross marks the location of the horizontal-offset dipole in Fig. 6.2.

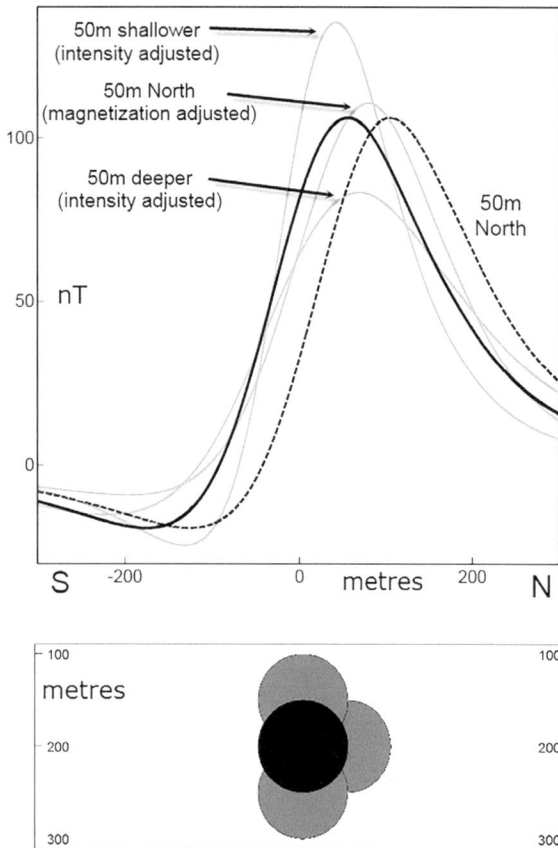

Fig. 6.2. South to north section and TMI profiles through dipole anomalies: reference (black solid line), 50 m north offset (dashed line) and post-inversion compensation by magnetisation changes (grey lines).

Fig. 6.2 are partially compensated by adjustment of magnetisation direction but not to the extent that the magnetic field of the correctly positioned and spatially offset sources are closely matched. The change in magnetisation direction to best compensate for the offset returns the location of the peak of the anomaly by only about one-half of its initial displacement and there is substantial residual data misfit. Vertical offsets of the dipole produce changes predominantly in amplitude and sharpness of the anomaly. The influence of a vertical displacement is partially compensated by adjustment of the intensity but not direction of magnetisation.

Plots of post-inversion data misfit against spatial offset of the source are shown for a range of horizontal and vertical displacements in Fig. 6.3. In most cases there is an almost linear increase of data misfit with the offset distance. Inversion of magnetisation intensity reduces data misfit by ~50%. At a conservative detection level of a normalised data misfit of 0.1, horizontal source offsets are less than 4% of source depth. This reveals that change of magnetisation direction can only compensate effectively for small errors in horizontal location of a magnetisation. Figure 6.4 shows that the magnetisation directions from inversion of horizontally offset sources are systematically rotated according to the direction of spatial offset. Horizontal offsets of 50% of source depth to the north, north-west or north-east cause rotation of the estimated magnetisation by over

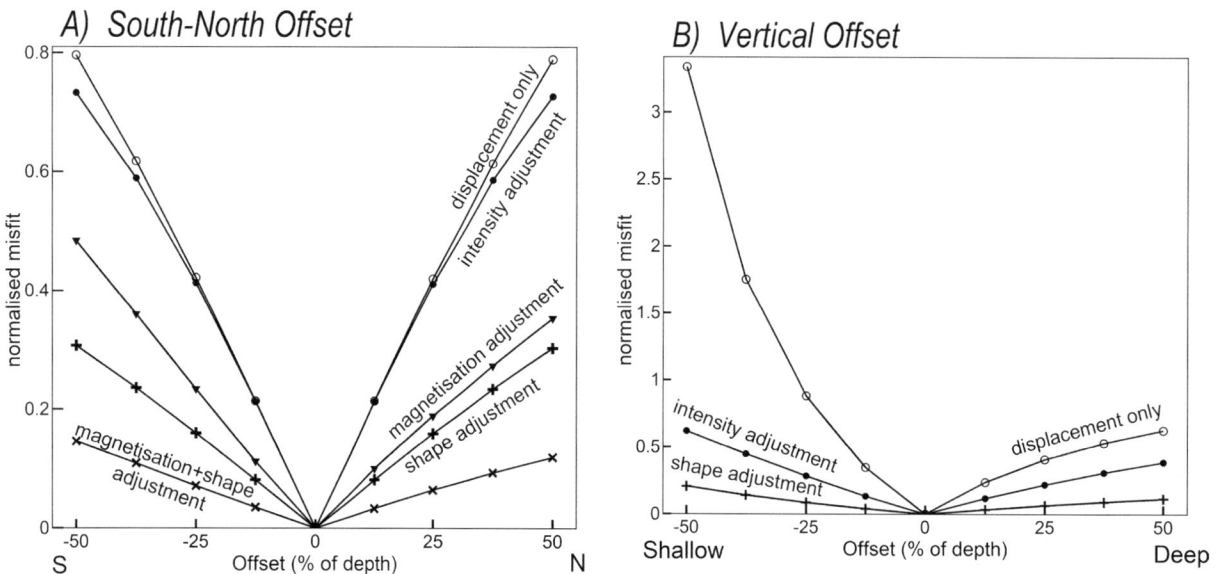

Fig. 6.3. Normalised misfit between the TMI fields of a reference dipole and north–south horizontal and vertical offset dipoles following inversions of intensity of magnetisation and total magnetisation.

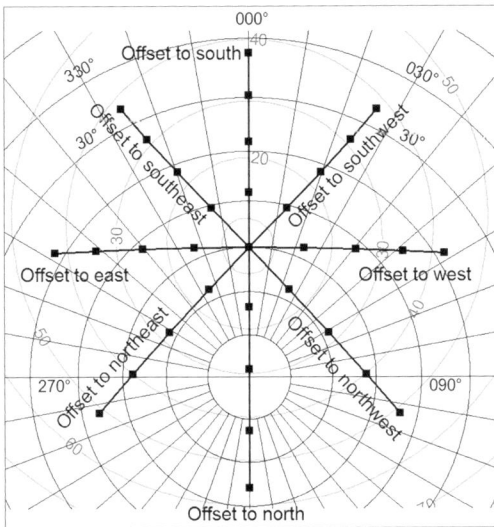

Fig. 6.4. Magnetisation directions recovered from inversion of TMI fields of horizontally displaced dipoles, with contours of angular separation from the reference (0°, −60°) magnetisation direction.

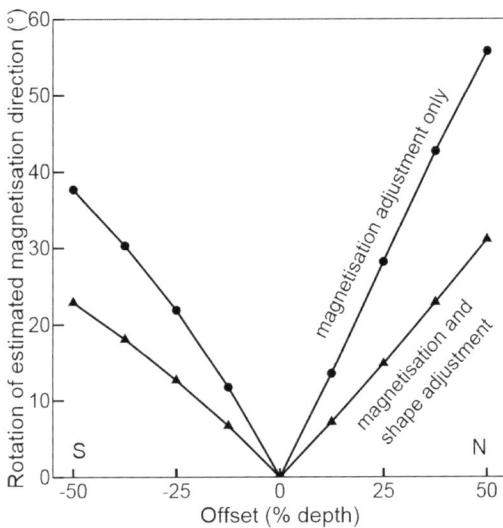

Fig. 6.5. Angular rotation of estimated magnetisation direction from inversions of magnetisation only, and magnetisation and shape for south-north horizontal dipole displacements.

50° but with substantial residual data misfits that clearly reveal inadequacy in representation of the magnetisation. Figure 6.5 plots the angle of rotation of magnetisation direction to best compensate for a horizontal source displacement along a south to north traverse. The rotation is of the order of one degree for a horizontal offset of 1% of source depth. Together with the maximum expected error in horizontal position of 4% of depth, this suggests a corresponding maximum unrecognised error in magnetisation direction of ~4° associated with

source displacement only. Change of magnetisation direction alone is not an effective compensation for mis-location of a source, and conversely source mis-location by itself should not cause substantial undetected error in estimation of magnetisation direction.

6.3 INFLUENCE OF SOURCE SHAPE ALONE

For an unknown magnetisation we need to consider to what extent its shape distribution might influence estimation of magnetisation direction. I classify shape as an elongation or shortening of the body along horizontal or vertical axes. I classify elongation or shortening along plunging axes separately as plunge. Figure 6.6 shows tabular bodies and ellipsoids with horizontal and vertical axes of 3:1, longest axis to depth-to-centre ratios of 1.5:1, and longest axis to depth-to-top of ≥ 2:1. The axial ratios of these ellipsoids and tabular bodies are identical but the tabular body is a more substantial departure from a dipole because it retains constant thickness along the extent of each axis whereas ellipsoids are centrally weighted and taper along their axes.

Figure 6.7 shows TMI maps for a dipole and the tabular bodies shown in Fig. 6.6 with magnetisation parallel to the geomagnetic field vector of (0°, −60°). The tabular body with vertical elongation (Fig. 6.7D) produces an anomaly with similar appearance to that of the dipole. The TMI map images of the bodies with the 3:1 ratio of horizontal axes (Figs 6.7B and 6.7C) clearly indicate the orientation of those bodies. Figure 6.8 shows a central section through a dipole and co-centred bodies of identical magnetic moment with both vertical and horizontal north–south elongation. The vertically

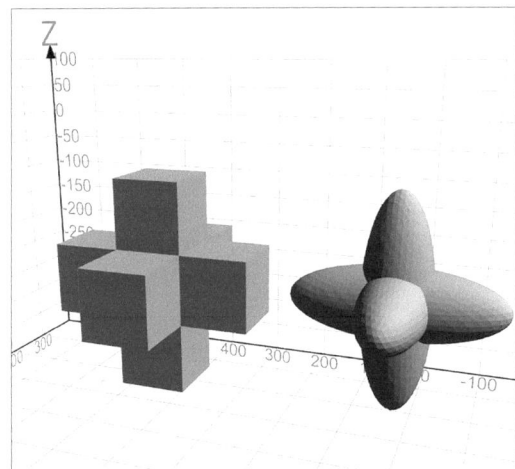

Fig. 6.6. Perspective view of 300:100:100 m axis ellipsoids and tabular bodies.

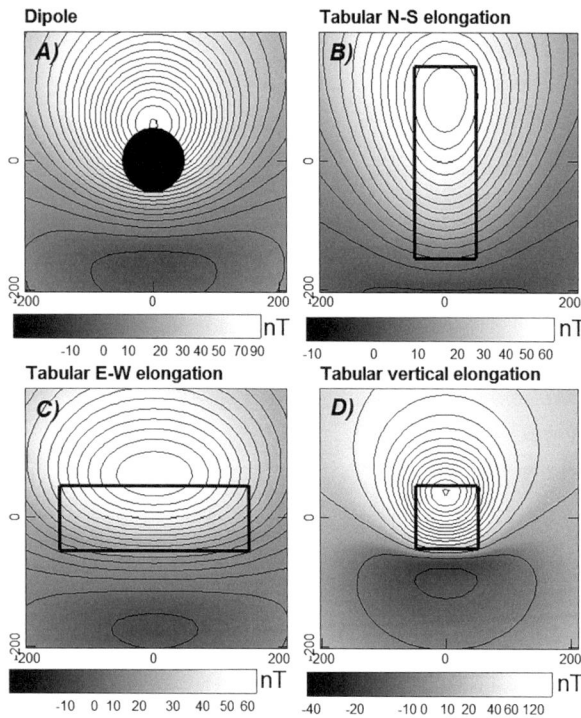

Fig. 6.7. TMI maps (in a geomagnetic field 0°, −60°) for A) dipole, and 3:1:1 elongated tabular bodies: B) horizontal north–south, C) horizontal east–west, and D) vertical.

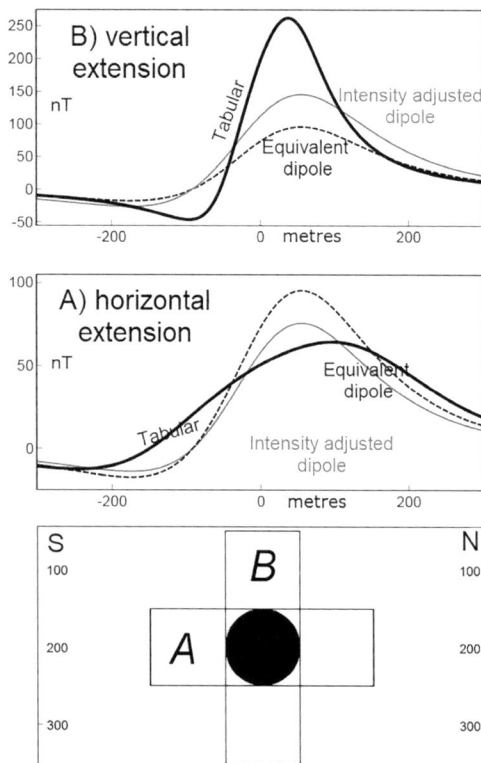

Fig. 6.8. South to north section and TMI profiles for co-centred dipole and pre- and post-inversion elongate tabular bodies and co-centred dipole. TMI profiles of the tabular bodies (inversions of the full anomalies not just the data in this section).

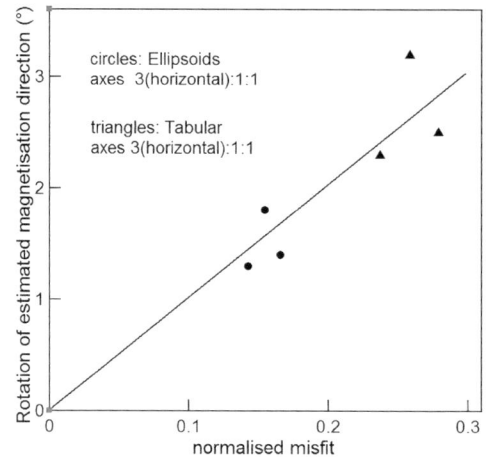

Fig. 6.9. Rotation of estimated magnetisation direction and post-inversion normalised misfit of computed TMI for co-centred N-S, E-W and SW-NE elongated ellipsoids and tabular bodies.

elongated magnetisation produces a field variation with substantially higher amplitude than the dipole due to its extension to shallower depth, but this difference is partially compensated by adjusting the intensity of magnetisation without change of magnetisation direction. For the horizontal elongation the best match to a co-centred dipole is also primarily by adjustment of the intensity of magnetisation, but in this case a slight additional improvement of fit is achieved with minor rotation of the estimated magnetisation direction of between 1.3° and 1.8° for the ellipsoids and between 2.3° and 3.2° for the tabular bodies. As shown in Fig. 6.9, the associated data misfits all have normalised values greater than 0.1 that should be recognised as significant. Extrapolation of the results in Fig. 6.9 suggests that normalised data misfit values of < 0.1 are all associated with rotations of magnetisation direction < 1°. Therefore, within the definition bounds of a compact source, uncertainty of the shape is not by itself a concern in estimating magnetisation direction.

6.4 INFLUENCE OF SOURCE PLUNGE ALONE

3D analysis of compact magnetisation does not suffer from the same exact trade-off between magnetisation direction and plunge as two-dimensional analysis but there is nevertheless quite effective cross-compensation between these parameters. Figure 6.10 shows ellipsoids and tabular bodies with axis ratios 2.5:1:1, long axis to depth-to-centre ratios of 1.25:1, and long axis to shallowest depth ratios of 3.2:1 (outside the conservative definition of a compact source). These bodies plunge 45° to north, south, east and west. The TMI maps for the tabular bodies plotted in Fig. 6.11 show clear departures from a dipole anomaly pattern, but only subtle characteristics by which that departure can be recognised as due to plunge rather than horizontal shape. To investigate rotation of estimated magnetisation direction due to plunge, I best-matched anomalies of the plunging ellipsoids and tabular bodies using dipole, vertical ellipsoid and vertical tabular sources. Figure 6.12 shows the central profile through a

Fig. 6.10. Perspective view of 250:100:100 m axis ellipsoids and tabular bodies with 45° plunge to North, South, East and West.

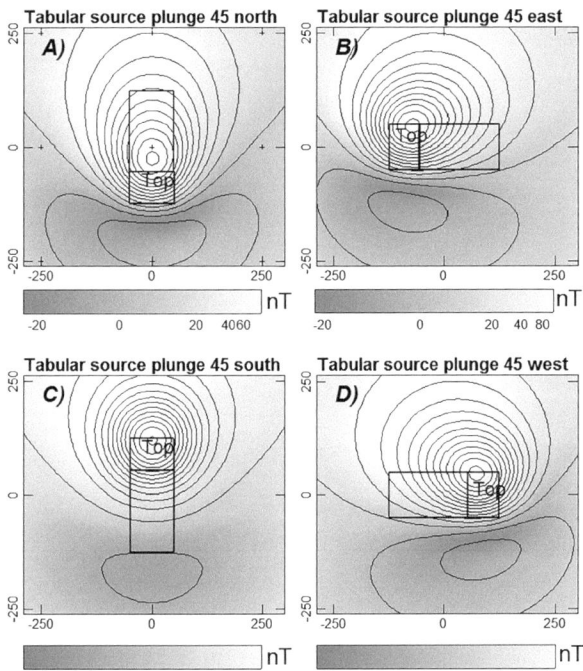

Fig. 6.11. TMI maps for 45° plunging tabular bodies with magnetisation directions (0°, −60°) in a geomagnetic field (0°, −60°). Plunges: A) North, B) East, C) South, D) West.

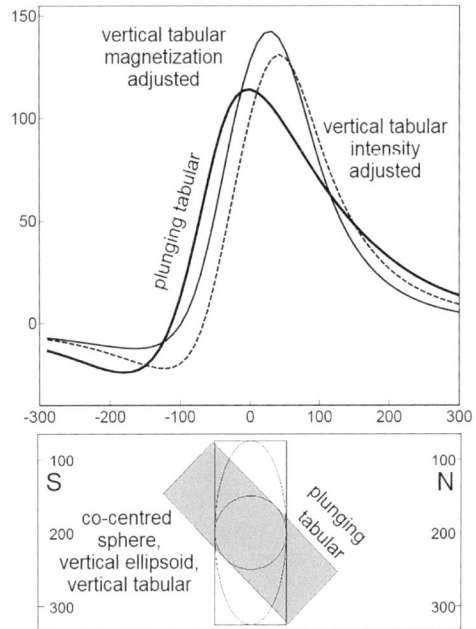

Fig. 6.12. South to north section through a 45° north plunging body and co-centred dipole, vertical ellipsoid and vertical tabular body. TMI profiles are shown for the plunging body, and for the intensity inverted and total-magnetisation inverted vertical tabular bodies.

tabular body plunging to the north and the best-fit co-centred dipole, vertical ellipsoid and vertical tabular bodies. The rotation of magnetisation in attempted compensation for the incorrect plunge of the body is 15° but the normalised data misfit is 0.23 and the models are clearly unacceptable.

Figure 6.13 is a cross-plot of the pre- and post-inversion normalised misfits between computed fields of plunging ellipsoid and tabular bodies with fields of co-centred dipole and vertical ellipsoid and tabular bodies. Figure 6.15A shows results from inversion of intensity of magnetisation alone, and Fig. 6.15B shows results from inversion of both intensity and direction of magnetisation. For both the ellipsoid and tabular bodies, inversion of the intensity of magnetisation reduces the initial misfit by ~20% and inversion of intensity and direction of magnetisation reduces it by a further 20%. The error in estimated magnetisation direction increases with data misfit as shown in Fig. 6.14 with a trend that reveals that normalised misfits of < 0.1 are restricted to errors in magnetisation direction of < 7°.

Figure 6.15A plots the apparent magnetisation direction for dipole and vertical tabular and ellipsoid bodies best-fitted to co-centred ellipsoid or tabular bodies with 45° plunge and magnetisation direction of

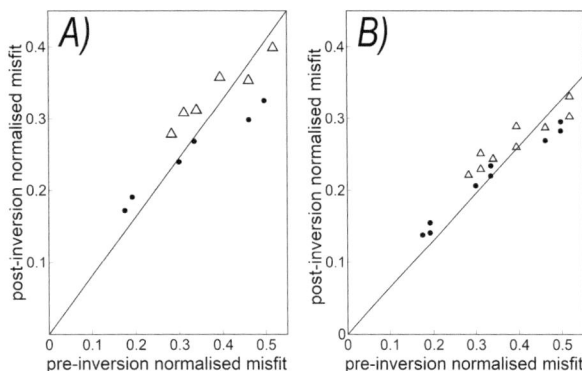

Fig. 6.13. Normalised data misfit between TMI fields computed from 45° plunging ellipsoid and tabular bodies and best-fit dipole and vertical bodies: A) after inversion of intensity of magnetisation, B) after inversion of total magnetisation.

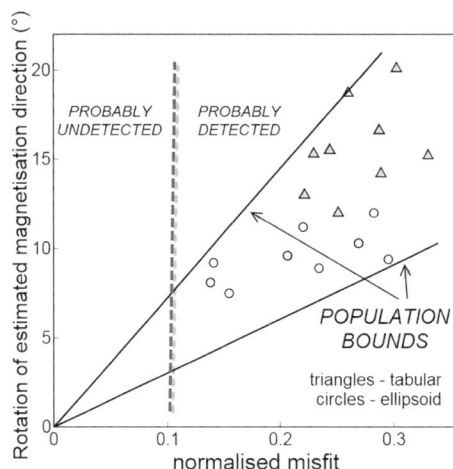

Fig. 6.14. Rotation of magnetisation against normalised misfit after inversion of total magnetisation of dipoles and vertical bodies to match the fields of 45° plunging ellipsoid and tabular bodies. The extrapolated maximum rotation of magnetisation at a normalised misfit of 0.1 is ~7°.

declination 0°, inclination −60° (the triangular symbol in Figs 6.15A and 6.15B). Rotation of apparent magnetisation direction away from the true direction to compensate for enforcement of erroneous plunge is of variable magnitude for different combinations of input and inversion model body geometries, but is in all cases anti-parallel to the direction of plunge. The relationships shown in Fig. 6.15B are discussed in a later section.

Figure 6.16 shows the relationship between error in plunge and corresponding error in estimated magnetisation direction for a range of plunge angles from vertical to horizontal in a north-east direction. Rotation of magnetisation direction is a maximum for a 45° plunge

as already investigated, and reduces almost symmetrically to zero for near-vertical and near-horizontal plunge. The results of this study into the ability of error in estimated plunge to compensate for error in estimated magnetisation direction indicate that error in plunge is by itself unlikely to cause unrecognised error in magnetisation direction any greater than 7°, with a maximum for a plunge angle of 45°. Magnitudes of error in estimated magnetisation direction accommodated by variation of other single source parameters as investigated above are listed in Table 6.1.

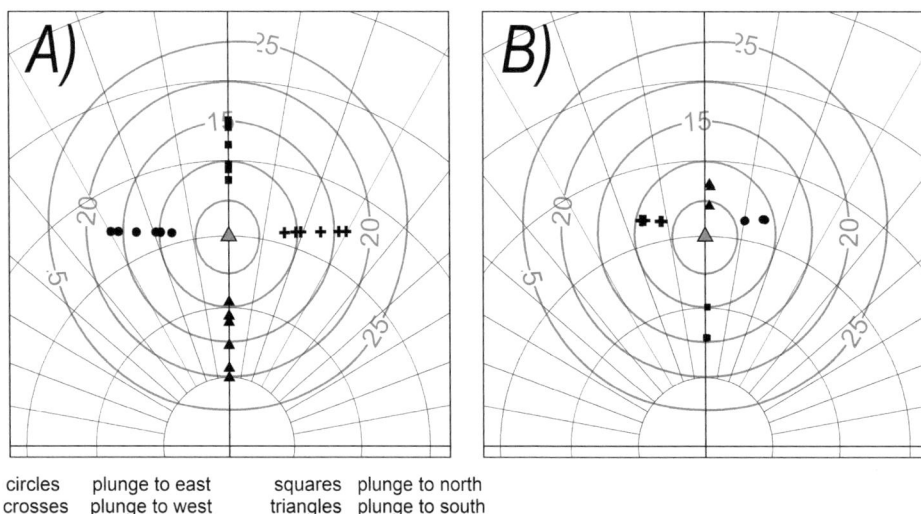

circles — plunge to east   squares — plunge to north
crosses — plunge to west   triangles — plunge to south

Fig. 6.15. Magnetisation direction plots of dipoles and vertical ellipsoid and tabular bodies best-fitted to 45° plunging ellipsoids and tabular bodies of true magnetisation direction 0°, −60° (marked by the triangle) with contours of angular rotation from that direction for inversion of A) magnetisation only, B) position and magnetisation.

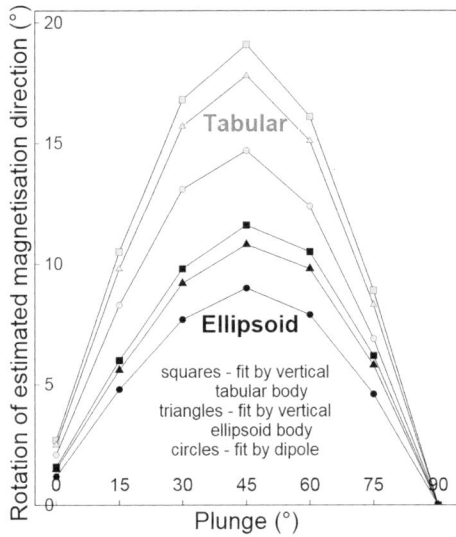

Fig. 6.16. Rotation of magnetisation direction of dipole, vertical ellipsoid and vertical tabular body inversions to match the fields of ellipsoids (black symbols) and tabular bodies (grey symbols) plunging to the north-east.

Table 6.1. Individual contributions of other single parameters to estimation of magnetisation direction for a compact source generating a well defined anomaly.

Model parameter	Uncertainty contribution
Horizontal position	4°
Vertical position	0°
Shape – horizontal extent	1°
Shape – vertical extent	0°
Shape – plunge	7°

6.5 COMBINED INFLUENCE OF SOURCE POSITION AND SHAPE

Inversion of magnetic field data over buried sources requires simultaneous solution of all source parameters rather than isolated solution of individual parameters as previously considered and summarised in Table 6.1. Figures 6.17 and 6.18 show inversion for a spatially offset dipole (as illustrated in Fig. 6.1) for the case that shape is also unknown. For this I replace the spherical model of a dipole with an ellipsoid model that can then be reshaped by the inversion. Results for horizontal offsets are shown in Fig. 6.17. The TMI misfit introduced by the horizontal offset of the magnetisation is substantially reduced by inversion of shape introducing a resulting plunge such that the magnetisation is shallower in the direction from which the body has been offset. Inversion of magnetisation intensity and direction is less effective in reducing data misfit than is inversion of shape. A slight further

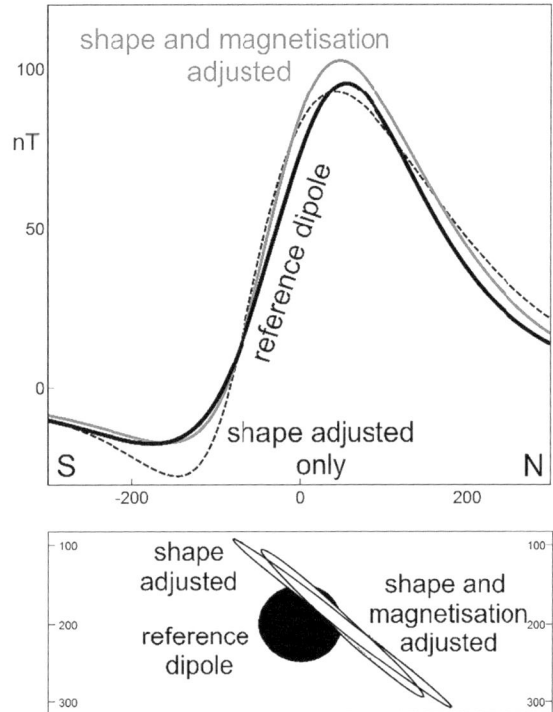

Fig. 6.17. South to north section and TMI profiles for a reference dipole and horizontally (northward) displaced ellipsoids of inversion-adjusted shape and magnetisation.

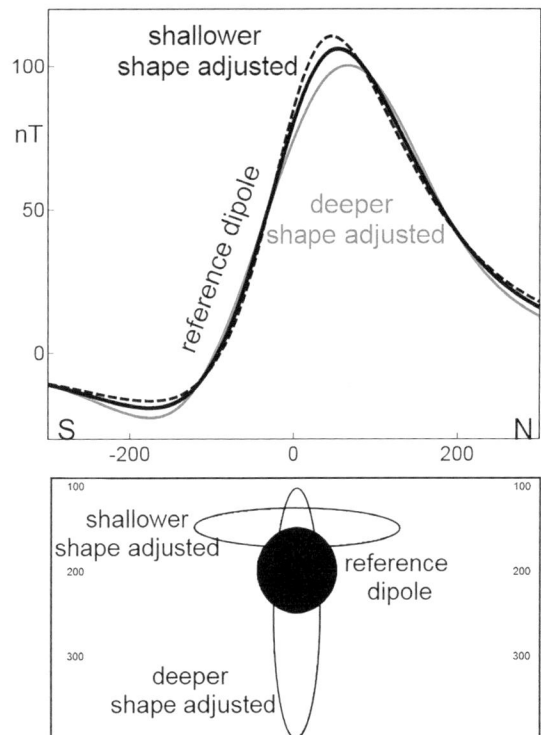

Fig. 6.18. South to north section and TMI profiles for a reference dipole and vertically displaced ellipsoids of inversion-adjusted shape.

reduction of data misfit is, however, achieved by inverting simultaneously for both shape and magnetisation direction. Figure 6.18 shows that in compensation for vertical displacement, inversion of shape is also more effective than inversion of magnetisation. For a vertical displacement, inversion of magnetisation introduces change exclusively in its intensity with no change in its direction.

Figures 6.19 and 6.20 plot the magnitude of post-inversion misfit as a function of offset distance for shape-and-magnetisation inversions. For vertical source offsets (Fig. 6.19) the results clearly show the more effective change of shape rather than magnetisation in compensation for vertical offset. Change of magnetisation direction does not reduce misfit and is not plotted. For horizontal source offsets (Fig. 6.20) change of shape is also more

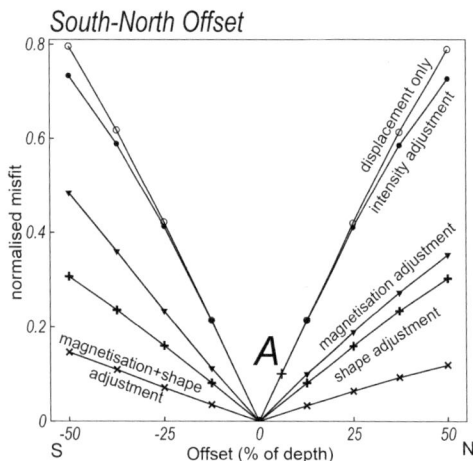

Fig. 6.21. Rotation of magnetisation direction in compensation for horizontal offset of a magnetisation without shape adjustment.

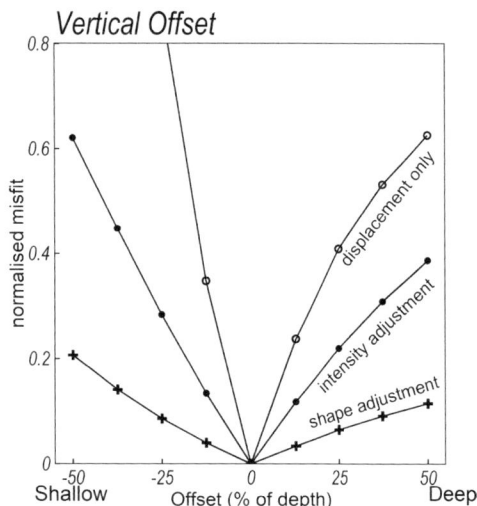

Fig. 6.19. Normalised misfit of inversions to compensate for vertical source offset.

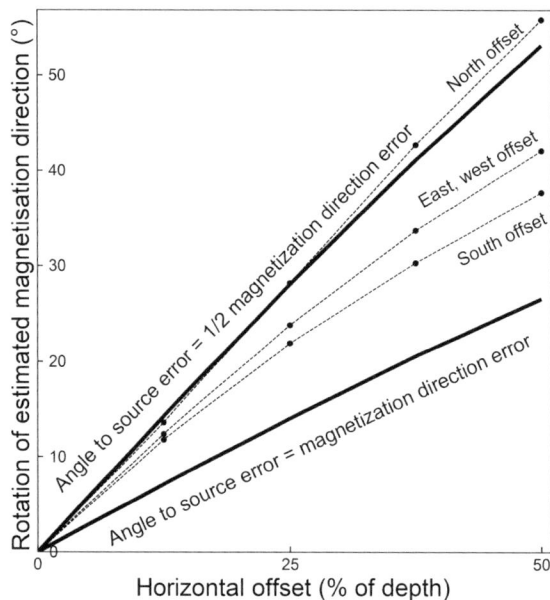

Fig. 6.22. Rotation of magnetisation direction for horizontally offset magnetisations including shape adjustment.

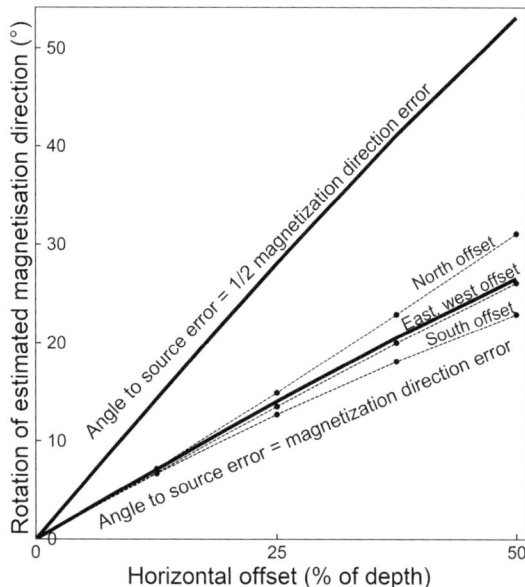

Fig. 6.20. Normalised misfit for inversions to compensate for horizontal source offset.

effective than change of magnetisation in compensating for horizontal displacement of magnetisation. In the examples studied, combined shape and magnetisation changes, halve the residual data misfit due to change of shape or change of magnetisation individually.

The magnitude of rotation of magnetisation direction to compensate for horizontal displacement as plotted in Fig. 6.21 varies with direction of horizontal displacement. For the magnetisation direction investigated of inclination −60°, inclination 0° a northerly displacement is compensated by rotation of magnetisation direction at a rate corresponding to approximately twice the angle to the centre of spatially displaced magnetisation. From the northly displacement-only curve in Fig. 6.20 a normalised misfit of 0.1 corresponds to a horizontal offsets of ~7% (point 'A') and therefore (from Fig. 6.21) a rotation of magnetisation of ~8°. The same data misfits for displacements to east, west and south correspond to smaller magnetisation rotations of 5° to 6°.

Rotation of magnetisation to compensate for horizontal displacement while also allowing simultaneous adjustment of shape is plotted in Fig. 6.22. The additional freedom in this inversion reduces data misfit by 50%, making displacements more difficult to detect. However, the corresponding rotations of magnetisation are also reduced by a factor of 2 and therefore sensitivity to magnetisation direction does not change significantly. The results in Fig. 6.22 fall close to the line of equivalence between angle of offset to source and angle of rotation of estimated magnetisation, suggesting that it is a reasonable approximation that (correcting if necessary for flying height of aeromagnetic data) angular uncertainty in targeting the centre of a compact source giving rise to a well-defined magnetic anomaly is similar to the angular uncertainty of its magnetisation direction. The angles themselves depend on the quality and sufficiency of the data which defines the anomaly, and in favourable circumstances may be of the order of 5° to 8°.

6.6 COMBINED INFLUENCE OF SOURCE POSITION, SHAPE AND PLUNGE

The previous section considers combined variation of magnetisation, position, and shape but from the earlier investigations the most influential individual spatial parameter is plunge. Detection of error in magnetisation direction associated with simultaneous inversion of source shape, plunge and magnetisation is addressed in Fig. 6.23. Inverting for source shape and plunge together with magnetisation intensity and direction for displaced sources (the closed symbols in Fig. 6.23) reduces both the data misfit and the rotation of estimated magnetisation direction from a magnetisation-only inversion (the open symbols). Unfortunately, because inversion of both parameters together reduces the data misfit at a greater rate than it reduces the compensating change in magnetisation direction, there is a resulting net reduction in sensitivity to magnetisation direction. Most of the

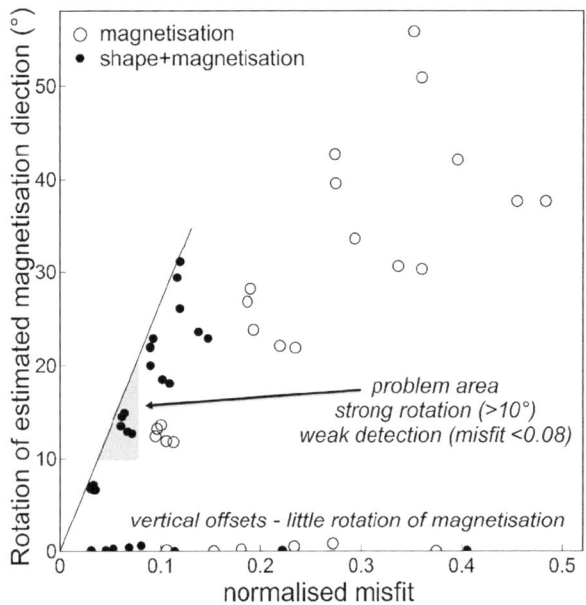

Fig. 6.23. Rotation of estimated magnetisation direction against data misfit for magnetisation-only inversions (open circles), and combined magnetisation and spatial inversions (closed circles). Note that there is almost no rotation of estimated magnetisation direction for vertical source displacements.

combined inversion results have either small rotations in estimated magnetisation direction or high data misfits, but there are a few results with rotations > 10° and misfits < 0.08 (the shaded region in Fig. 6.23) that can pose problems in magnetic field interpretation. These solutions are for horizontal offsets of approximately one-quarter of the depth to their centres, for which (as shown in Fig. 6.17) the optimum source solutions develop a false plunge.

6.7 SUMMARY OF THE SENSITIVITY STUDY

Table 6.2 reviews what I believe to be the key features of this complex analysis. Points are mostly listed in the sequence in which they are addressed in the text. I have tried to make this a comprehensive investigation of the magnetic field expressions of inter-relationships between different features of a magnetisation and its spatial distribution, but it samples only a small part of the full multi-parameter model space. Specific details of the relationships vary with geomagnetic inclination and direction of magnetisation. However, the general conclusions regarding influence of source position, shape and plunge are applicable across all geomagnetic inclinations and for most source magnetisation directions.

Table 6.2. Summarised results of the parametric studies.

1	Estimation of magnetisation direction is fundamentally a three-dimensional problem.
2	Vertical displacement of a dipole source does not in itself introduce significant undetectable error in estimated magnetisation direction.
3	Horizontal displacement introduces rotation of estimated magnetisation direction at the approximate rate of 1° for horizontal offsets of 1% of depth.
4	For a well-defined anomaly due to a compact magnetisation, errors due to incorrect positioning and associated incorrect magnetisation direction should become apparent at horizontal offsets of ~4% of depth and rotations of 4° respectively.
5	For a compact source, the effect of erroneous vertical elongation does not rotate the estimated magnetisation direction. For erroneous horizontal elongation there is only slight shape effect on estimated magnetisation with rotations of < 3° at normalised misfits < 0.1.
6	Erroneous plunge of a body causes greater uncertainty in estimated magnetisation direction than erroneous horizontal elongation, but for compact sources uncertainty in estimated magnetisation direction due to plunge alone is in most cases < 7° at normalised misfits < 0.1.
7	The uncertainty in estimated magnetisation direction most relevant to magnetic field interpretation is simultaneous solution of source location, shape, plunge and magnetisation. Most compact source inversions that acceptably fit the synthetic field data have magnetisation direction consistently recoverable to within 10° (many within 5°) but there are some cases of difficult-to-detect errors which may reach 15° at normalised misfits < 0.1.
8	Estimation of magnetisation direction for equidimensional bodies is relatively robust because it is a bulk characteristic rather than a detail (such as depth to top, or shape), and also because it does not have a particularly close pairing with another parameter or parameter set (as for instance do intensity of magnetisation and volume, which are so closely paired that there is generally little sensitivity to either value, only to their product). For bodies of larger depth extent, however, there is an approximate trade-off with dip or plunge that reduces confidence with which magnetisation direction can be recovered.

Note that since detail of source shape is shown to have little effect in estimation of magnetisation direction for compact sources (distal field data), that similar magnetisation direction estimates should be recovered from both parametric and voxel inversions provided they use similar anomaly separations.

The analytic capabilities of this study are provided by parametric inversions that support isolation and study of single components of complex interactive problems. This study suggests that for isolated, well-defined anomalies due to compact sources of homogeneous magnetisation, uncertainty in source position, shape and plunge should not preclude recovering an estimate of that magnetisation direction from inversion of magnetic field data to within a range of 5° to 10°. For less favourable cases, however, errors in magnetisation direction of up to 15° may occur as a result of combined errors in estimation of source location and plunge. Expected errors in estimated magnetisation direction also increase with decreasing data quality or sufficiency and with increasing complexity or diffuseness of the magnetisation and problems of anomaly separation

Relationships established in this study are more cryptic in complex voxel models, but the relationships are valid for a compact magnetisation and its magnetic field regardless of the model description used. If magnetisations are significantly more distributed or with significant internal variation of magnetisation, then the relationships between magnetisation and the magnetic field degrade and inversion results lose significance. Discrete features in the magnetisation that result in discrete features in the magnetic field are essential for inversion to resolve distribution of magnetisation and its direction.

6.8 THE RYLSTONE CASE STUDY

Rylstone lies close to the margin of the Western Coalfield of the Sydney Basin in New South Wales (Fig. 6.24). There are multiple igneous events in the region across a range of ages, including stocks, sills and diatremes of Jurassic to Tertiary age that intrude the Triassic coal measures (Yoo *et al.* 2001) and give rise to prominent magnetic anomalies strongly influenced by both normal and reverse remanent magnetisation. In 1997 the Geological Survey of New South Wales commissioned a 100 m spaced, nominal 50 m terrain clearance heli-mag survey of this rugged area to map the igneous bodies and assess their influence on the coal resource. The survey was flown by Tesla Airborne Geoscience and interpreted by Encom Technology (Encom Technology 1997). Variation in the geometry and remanent magnetisation of the igneous bodies provides a range of anomalies suitable for testing magnetic field analytic and inversion

Fig. 6.26. TMI image with flightlines and elliptic pipe model outline.

Google Earth™ aerial image. The anomaly coincides with a clearing that is possibly indicative of volcanic-derived soil but there is no obvious expression of outcrop. The TMI anomaly, that is imaged in Fig. 6.26 has a prominent expression on seven flightlines, a peak to trough range of 1330 nT, and a width of almost 600 m.

6.8.1 Source shape influence in estimation of magnetisation direction

The previous synthetic modelling establishes that estimates of magnetisation direction for a compact source show little sensitivity to the detail of shape. Figure 6.27 shows stacked profiles of TMI computed from the best-fit inversion model of a vertical circular cylinder, together with three of the central flightline sections through the model. All spatial parameters were free to vary in the inversion, as was intensity and direction of magnetisation. The resulting fit to the measured anomaly has a normalised data misfit (standard deviation of the difference between computed and measured data divided by the standard deviation of the measured data) of 0.18. The body has a diameter of 331 m, depth extent of 143 m and minimum depth below measurement sensor of 79 m. The magnetisation direction is declination 347°, inclination −76°. The maximum dimension of the body of 360 m gives a maximum-extent to measurement-proximity ratio of 4.6:1. Also shown in Fig. 6.27 are computed TMI stacked profiles and model sections for the best-fit dipole model. A dipole is a source either so compact that its shape requires no specification, or a

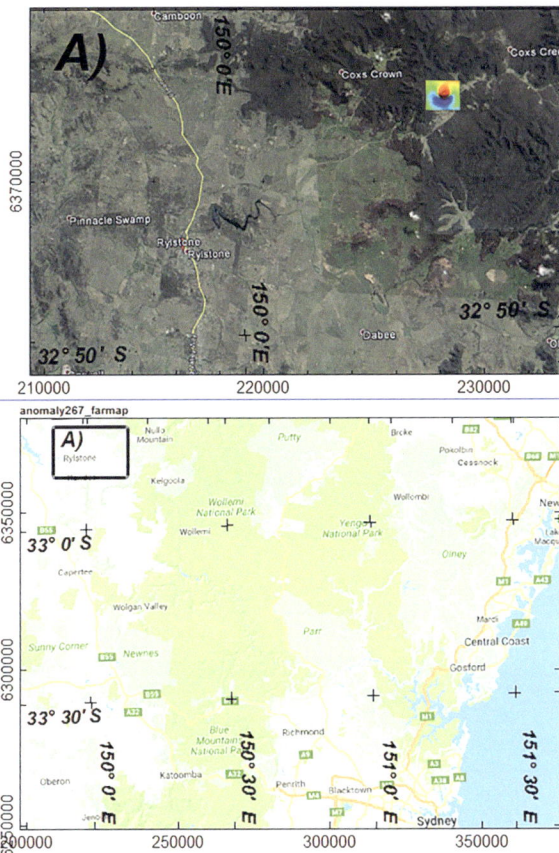

Fig. 6.24. Rylstone Anomaly 267 TMI image over local Google Earth imagery, and B) Regional location of Rylstone area.

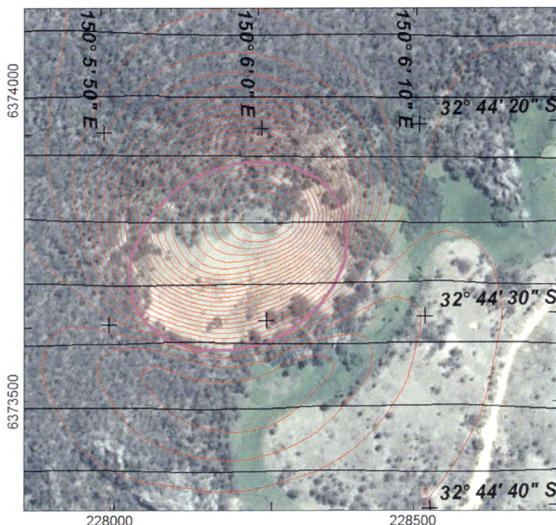

Fig. 6.25. Survey flightlines, TMI contours and elliptic pipe model outline over Google Earth image.

algorithms. In this study I focus on a single anomaly, anomaly 267 in the Australian Remanent Anomalies Database. Figure 6.25 shows the anomaly contours, flightlines and an outline of the modelled source over a

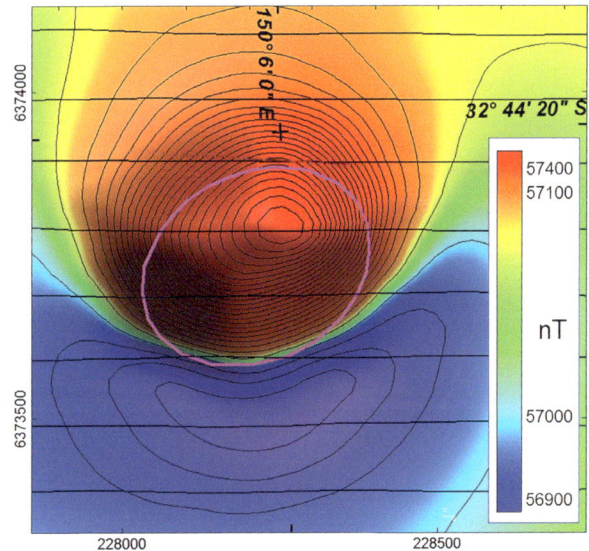

143

homogeneous sphere of any radius less than the distance from its centre to the ground surface. To constrain the size of the spherical source body (which by itself has no influence on the inversion results) I gave the spherical source the same intensity of magnetisation as the best-fit vertical circular-section cylinder model. The dipole model has a higher inversion misfit statistic of 0.32, and the estimation of magnetisation direction is 10° from that of the vertical circular-section cylinder model. The limitation of this model in matching the data is best illustrated on the highest amplitude flightline 100660 (the topmost model section in Fig. 6.27) which reveals that the dipole model is too centrally weighted to match the data well, producing a sharper anomaly than is measured. Evaluation of the inversion model is subjective, but the improved fit from the vertical circular-section cylinder model suggests that the magnetisation is likely to have a more horizontal distribution (for instance a horizontal top and base) than is represented by a sphere.

To be considered successful, a magnetisation model must explain the complete anomaly. Specific details of the misfit may assist interpretation as illustrated in Fig. 6.27. Figure 6.28 shows model sections and computed fields along the most critical flightline 100660 for best-fit tabular and ellipsoid models. Although those body types are quite different, there is no significant difference between the matches of their computed fields to the measured anomaly and no strong

grounds from the inversion results to prefer one model over the other as a representation of the sub-surface magnetisation. Figure 6.29 is a downloaded form from the Australian Remanent Anomalies Database with the details of four alternative source models, each of different geometry. Difficulty in discriminating between these various source model geometries highlights the insensitivity to details of shape in inversion of a compact magnetisation. For an identical anomaly separation a voxel inversion should recover the same magnetisation direction.

The Rylstone area has both sills which can be reasonably represented as horizontal sheets and diatremes that are saucer or funnel-shaped layers of breccia or lava above a thin neck (Yoo *et al.* 2001). The inversion source models are broadly consistent with either of these geological concepts. As shown in Fig. 6.30, each of the source geometries occupies almost the same space. All are near-surface with horizontal extents ~300 m and vertical extents of 50–100 m. The exact geometry of these models should not be interpreted literally but as approximate bounds on the expected distribution of magnetisation. Furthermore, the only constraint on internal inhomogeneity of the magnetisation is that the total magnetisation and its centre location should approximately match those of the homogeneous model. These restrictions arise because the magnetic field measurements are not close enough to the magnetisation to resolve shape details or minor variations of the

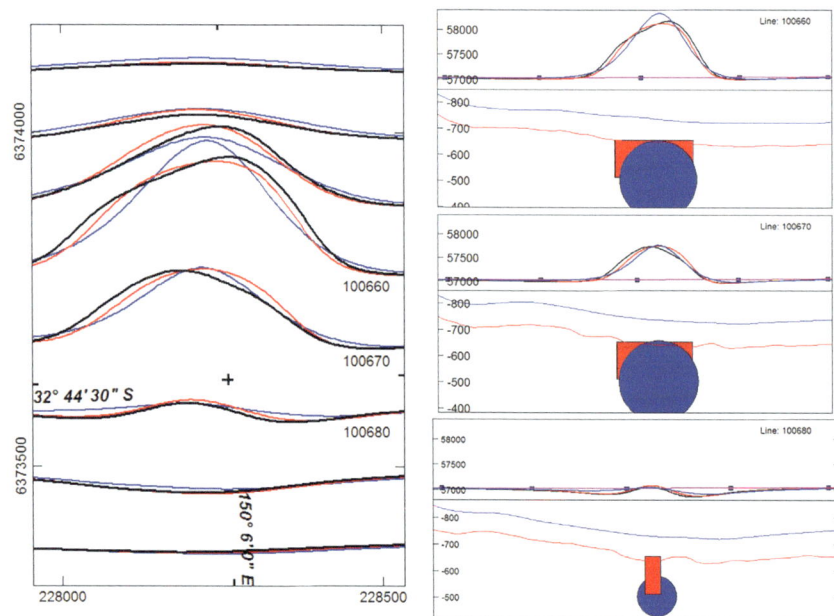

Fig. 6.27. Stacked profile and central model sections for the sphere model (blue) and the vertical circular pipe model (red).

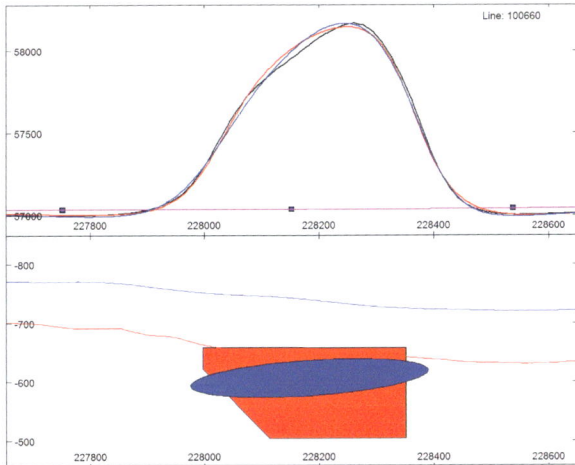

Fig. 6.28. Section along flightline 100660 through plunging tabular (red) and ellipsoid (blue) models.

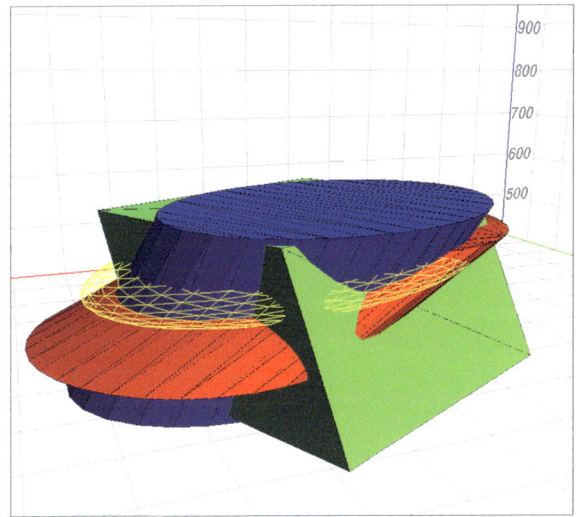

Fig. 6.30. Three-dimensional view of alternative source models.

Australian Remanent Anomalies Database
Anomaly 267

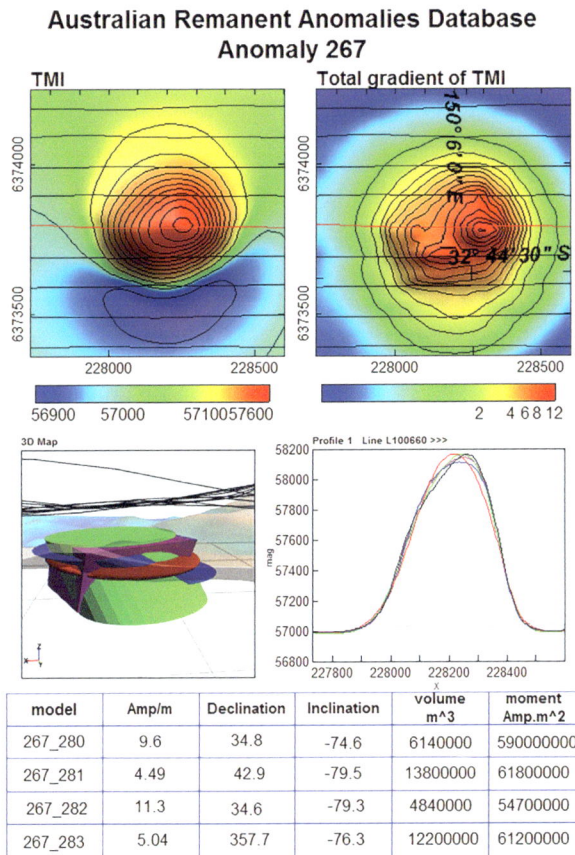

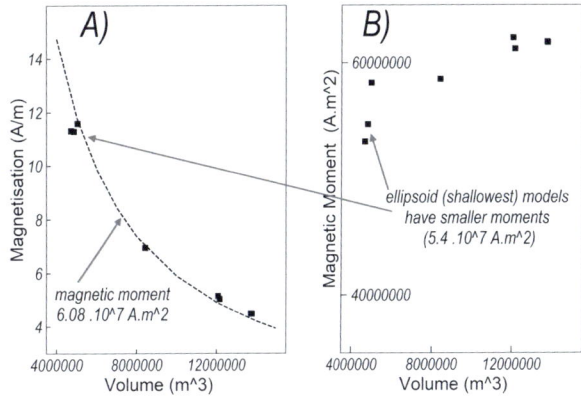

model	Amp/m	Declination	Inclination	volume m^3	moment Amp.m^2
267_280	9.6	34.8	-74.6	6140000	590000000
267_281	4.49	42.9	-79.5	13800000	61800000
267_282	11.3	34.6	-79.3	4840000	54700000
267_283	5.04	357.7	-76.3	12200000	61200000

Fig. 6.29. Anomaly 267 source inversion summary from the Australian Remanent Anomalies Database (model 280 – green, model 281 – blue, model 282 – red, model 283 – purple).

Fig. 6.31. Magnetic moment of best-fit inversion models.

internal distribution of magnetisation. Figure 6.31 shows plots of magnetisation and magnetic moment against volume for these source models. There is a variation of more than a factor of 2.5 in both volume and magnetisation intensity, but the magnetic moments calculated as the products of these two factors vary by less than 10%. The magnetisation directions of the various model bodies are plotted in Fig. 6.32 (the directions are listed in Fig. 6.29). The directions vary by less than 10° and all are less than 7° from the mean direction of declination 26.6° and inclination −78.0°. The 15° separation between the mean direction and the IGRF geomagnetic field for the site suggests that the magnetisation direction has a slight but significant rotation away from the geomagnetic field, most probably due to a remanent magnetisation component. The magnetisation models are not particularly compact with respect to the proximity of the magnetic field measurements. The extent to proximity ratios for the various models are 4.6:1 for the best ellipsoid, 6.0:1 for the best circular-section pipe, 6.4:1 for the best tabular body and 6.8:1 for the best

145

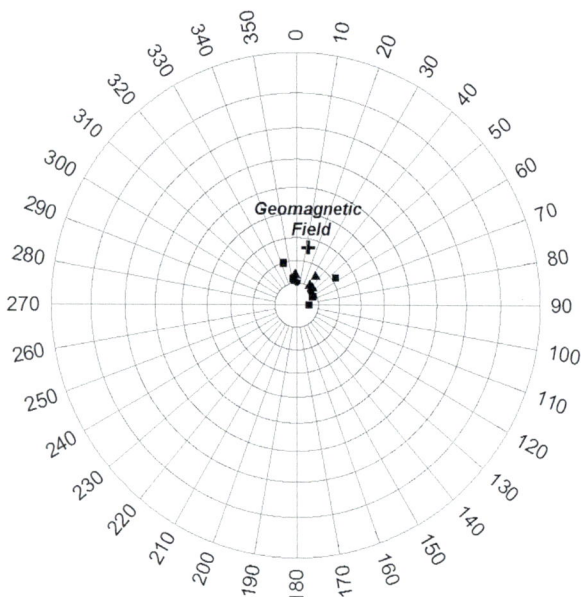

Fig. 6.32. Model (ellipsoid, tabular, circular, elliptic pipes) magnetisation directions from inversion of all lines (triangles), odd-number lines (circles), even-number lines (squares).

elliptic-section pipe. However, even at these values the inversion results are still in good agreement with the conclusion that magnetisation direction can be consistently recovered for compact sources with little influence of body shape (Foss 2017). The corollary of this finding is that for these sources shape is a detail that cannot be reliably recovered from the magnetic field inversions. Unfortunately, because I do not have direct measurement of the Anomaly 267 magnetisation I can only establish consistency and not correctness of the estimated magnetisation direction.

6.8.2 Source position influence in estimation of magnetisation direction

Each of the model inversions presented in the previous section, while primarily testing influence of source shape, was also free to independently position those sources. The consistency of the model results therefore also testifies to the stability in simultaneously estimating source position and magnetisation direction. Figure 6.33 shows the centre points of the various source models together with the outline of the elliptic-section cylinder model over images of measured TMI, computed TMI, total gradient of TMI and the 1:100,000 scale geological map (Yoo 1998). The centres of the source models all plot in an area mapped as undifferentiated Jurassic or Tertiary volcanics which provides a compelling explanation for the anomaly. There is approximately a 50 m spread in the various model centres, which

is only 50% of the survey line spacing and less than 15% of the estimated horizontal extent of the source. Surprisingly, the maximum dispersion is along the flightline direction rather than perpendicular to it. There is slight disagreement between the edges of the body as mapped geologically and by the magnetics but this apparent difference may be due to limited (or no) outcrop, inexact location in the pre-GPS geological mapping, or because the geological mapping is intended to note the presence of the volcanics rather than to map their exact extents.

Figure 6.34 shows in more detail the model centres plotted on the total gradient of TMI. There is a broad high in total gradient values at 7 to 8 nT/m, the edge of which coincides approximately with mapping of the extent of magnetisation by the elliptic-section pipe model. The local peak of the total gradient is sharp and irregular, possibly mapping local concentrations of near-surface magnetisation but clearly also strongly influenced by the survey flightline locations.

Figure 6.35 shows the centre points of the various models in 3D space, which reveals that the estimate of vertical centre of magnetisation is also consistently (although not necessarily correctly) recovered. This consistency of position and magnetisation direction supports the previous synthetic data study which established that inversion can reasonably resolve both the centre location and direction of magnetisation of compact sources.

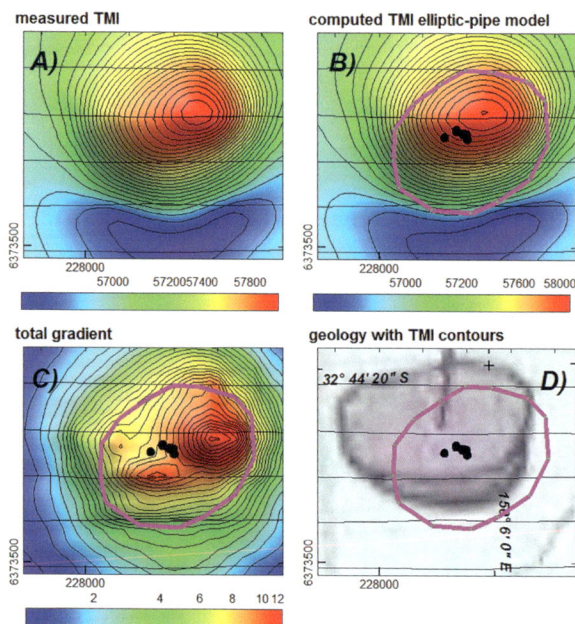

Fig. 6.33. A) measured TMI, B) TMI computed from the elliptic-pipe model, C) total gradient of TMI, ams D) pipe outline over the geological map.

Total Gradient of TMI

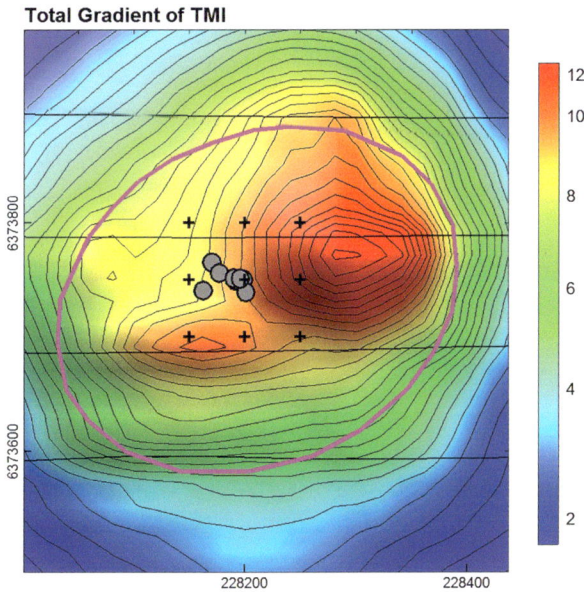

Fig. 6.34. Centres of inverted source models (circular points) and a grid of test body centres (crosses) over a contoured image of total gradient, with the outline of the best elliptic pipe model (magenta).

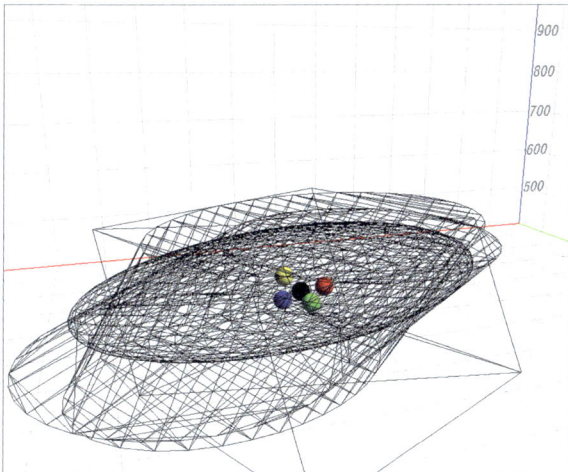

Fig. 6.35. Wireframe perspective of the four source models with their centre points (colour coded as in Fig. 6.30). The black symbol is the mean of the four estimates.

I investigated the influence of source position by imposing displacements on the source models, as in the previous synthetic data study. A three-by-three 50 m grid centred on the best-estimated horizontal centre of magnetisation is plotted in Fig. 6.36. Ellipsoid, circular-section pipe, elliptic-section pipe and tabular models were each centred at the various grid points and inverted to determine the best estimate of magnetisation direction for sources at those locations (the inversions retained the bodies as vertical to prevent migration of

the magnetisation centres with plunge). The resulting magnetisation directions from the inversions are plotted in Fig. 6.36. These directions are tightly grouped according to source displacement, with only minor variation between the various body types at any one location. This confirms the synthetic data study finding that for compact sources estimated location is much more strongly linked to magnetisation direction than is shape. The mean elevation difference between the measurements and the centre of magnetisation is 140 m, giving a rotation of estimated magnetisation direction of ~80% of the angle of spatial displacement, consistent with estimates from the synthetic data studies.

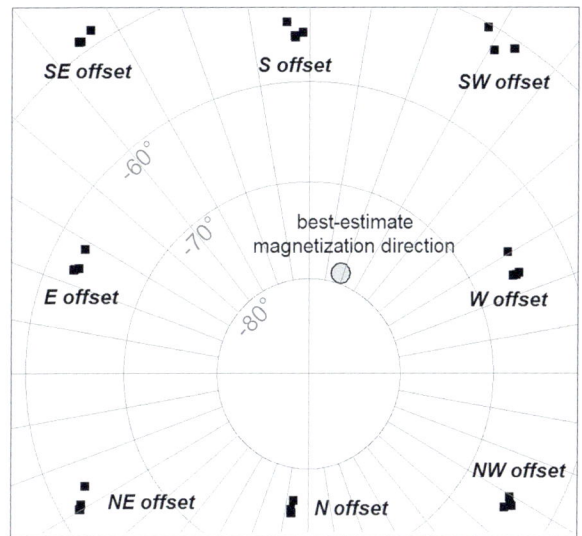

Fig. 6.36. Post-inversion magnetisation direction of bodies with imposed horizontal offsets.

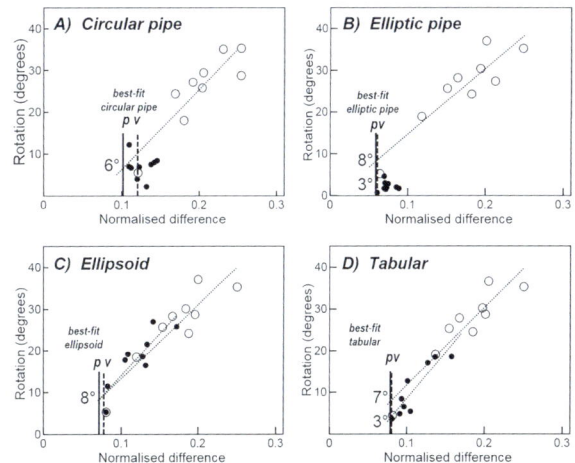

Fig. 6.37. Cross-plots of rotation of magnetisation direction against data misfit for horizontal offset sources: open circles are vertical sources, points are plunging sources.

In Fig. 6.37 rotation angle and data misfit are cross-plotted from the inversions of each of the four body types (both plunging and vertical) offset in the eight directions at 45° intervals. These straight-line relationships are extrapolated to the data misfit values of the best-fit inversions of each body type to estimate the angular error that can be accommodated by that data misfit. The angles range only between 6° and 8° for the different body types and are likely to be overestimates of error in magnetisation direction because much of the data misfit arises from local and uncorrelated irregularities in the data.

6.8.3 Source plunge influence on estimation of magnetisation direction

The previous synthetic data study found that variation in plunge of a body can be traded against variation in its magnetisation direction. Initial inversions of the horizontally offset sources retained their plunge as vertical. I subsequently re-ran the inversions allowing those bodies to plunge. Figure 6.38 shows a map of the axes of the post-inversion circular-section pipe bodies over the shallowest 10 m of their depth extent (this restriction is to reduce confusion where the axes cross each other). The trend of the body axes converge towards the centre of magnetisation, revealing that the inversions utilise their freedom of plunge to compensate for their imposed horizontal offsets. The cluster of the centroids about the best-estimated

centre of magnetisation confirm this behaviour (individual centroids overshoot the centre-point to compensate for mis-location of the more influential shallowest sections of the bodies). The same relationships are illustrated in perspective view in Fig. 6.39.

The post-inversion magnetisation directions for each body type are plotted in Fig. 6.40. The various source geometries behave differently to compensate for the enforced horizontal offsets in inversion of magnetisation and plunge. The elliptic-section pipes (Fig. 6.40B) have an average rotation of magnetisation from the best estimated direction of 26° for inversions in which they are not allowed to plunge, but that is reduced substantially to

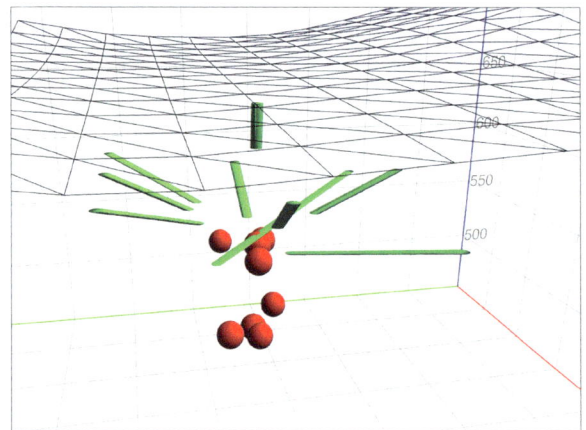

Fig. 6.39. Axes of horizontally offset bodies after inversion allowing plunge (green – shown to only 10 m depth) and centres of magnetisation of those bodies (red).

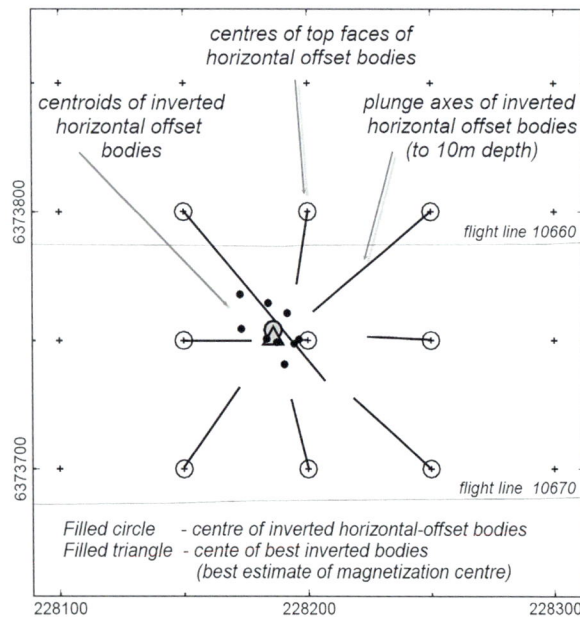

Fig. 6.38. Map of the top face centres (open circles), centroids (points), and plunge axes of the inverted horizontally offset.

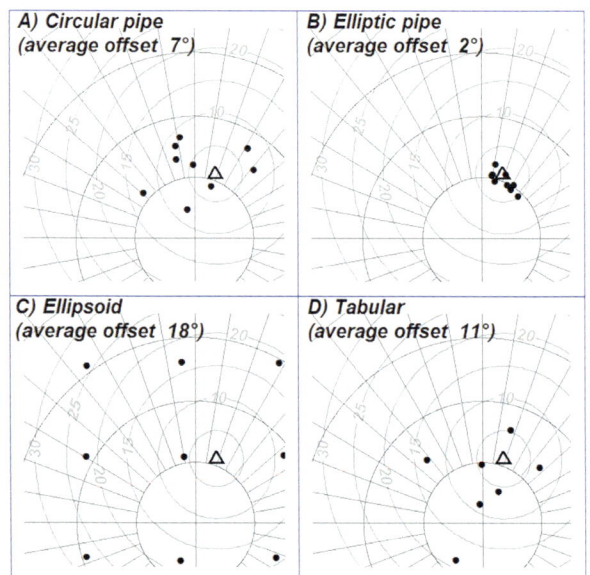

Fig. 6.40. Directions of magnetisation of horizontally displaced sources after inversion of magnetisation direction and plunge.

less than 3° if the bodies are allowed to plunge as well as to change magnetisation direction. The circular-section pipe model inversions (Fig. 6.40A) have fewer degrees of freedom and perform less well, but the average rotation of magnetisation is still reduced from 26° to 8°. Following inversion with free plunge and magnetisation, tabular bodies (Fig. 6.40D) have greater scatter of magnetisation direction and a marginally higher average rotation of 11°. Ellipsoid bodies (Fig. 6.40C) have the largest average rotation of 19° and retain the systematic rotation of magnetisation determined by the direction of imposed horizontal offset for each body. This ineffective compensation of plunge for horizontal offset ellipsoids is because the locked reference point for these bodies is their centre, which does not allow plunge to migrate the centre point (for the other body types the locked reference point is the centre of the top face and plunge migrates the centre of magnetisation). This difference in behaviour disappears for inversions that simultaneously allow position to change as well as plunge.

6.8.4 Sensitivity tests of estimated magnetisation direction

In previous sections I have analysed the influence of source location, shape and plunge on recovery of magnetisation direction to provide a case study test of the synthetic data inversions. An alternative approach to evaluate recovery of magnetisation direction is to perturb magnetisation direction and investigate how effectively other source parameters can compensate for that imposed deflection. Figure 6.41 shows test

magnetisation directions offset from the best-estimated direction 10° and 20° towards lower inclination (A), steeper inclination (B), and in orthogonal trends (C and D). The post-inversion data misfits for magnetisations in these offset directions are plotted in Fig. 6.42. The previous synthetic data studies established that post-inversion normalised misfit increases almost linearly in all directions away from the true magnetisation direction (for which the misfit value was zero). For the real data of this case study the minimum normalised data misfit is between 0.06 and 0.11 for different body geometries. The average increase of misfit for each curve over the 20° excursions is 54% of the minimum value, giving curves which still monotonically increase from the minimum but are more of a 'U' rather than 'V' section. With only a single exception, the curves for the different body types share common minima at the 10° sampling interval. These results suggest that by careful inversion, the best magnetisation direction with which the source of Anomaly 267 is represented as a compact homogeneous magnetisation has been resolved to between 5° and 10°. Non-uniqueness does not allow that success of the proposed model necessarily proves that this is a true representation of the subsurface magnetisation. Indeed, most geological bodies are expected to have moderate to extreme internal inhomogeneity, although that may not be detectable in analysis or inversion of even moderately distal magnetic field data.

The lowest data misfit values of three of the models are close to .05. The previous noise-free synthetic data study suggests that at this level of misfit magnetisation direction of compact sources is constrained to better than 5° to 10° (with some rare exceptions). The irreducible data

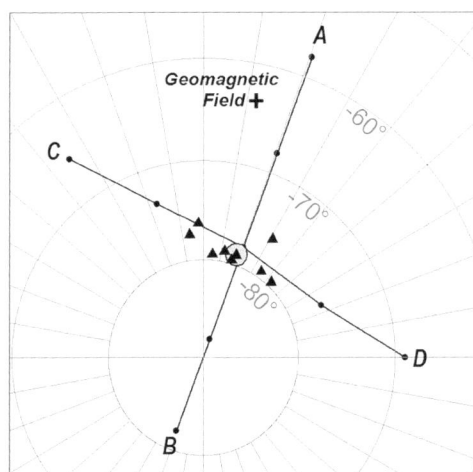

Fig. 6.41. Post-inversion data misfit for rotations of magnetisation: (crosses) circular-section pipe, (squares) tabular body, (triangles) ellipsoid, and (circles) elliptic-section pipe models.

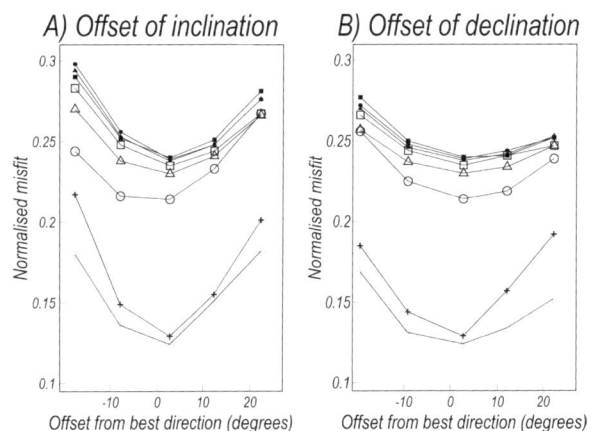

Fig. 6.42. Post-inversion model magnetisations (triangles), mean direction (filled circle) and sensitivity test traverses A-B and C-D with offsets of 10° and 20°.

misfits for the simple homogeneous models in this case study are most probably due to irregularities in the local distribution of magnetisation and/or short-wavelength errors in the magnetic field data acquisition and processing. The maximum extent to data-proximity ratio of between 4.6:1 to 6.8:1 is well outside the bounding value of 2 assumed for a compact source in the synthetic data study, suggesting that magnetisation direction estimates would be even more robust for truly compact sources.

Figure 6.43 shows stacked profiles and a grid image of residual misfit for the best elliptic pipe model. These reveal that post-inversion residual data misfits are local short-wavelength features that are either data artefacts or imperfections of the model at a shallow level. Figure 6.44 shows grid images of difference in the fields computed from the best-fit model and from inversion models with magnetisations 5° steeper and 5° shallower. These images show similar short-wavelength features as in Fig. 6.43, but also low-amplitude, longer wavelength features that extend over at least six flightlines. These features are of opposite polarity for the shallower and steeper magnetisation and are interpreted to be expressions of the misfit caused by enforcement of those inappropriate magnetisation directions. The absence of these features in the residual misfit image of the best model shown in Fig. 6.43 further suggests that the best magnetisation direction has been located to within 5°. Note that these broad, low amplitude features also emphasise the importance of anomaly separation and the sensitivity of the model inversion results to that separation.

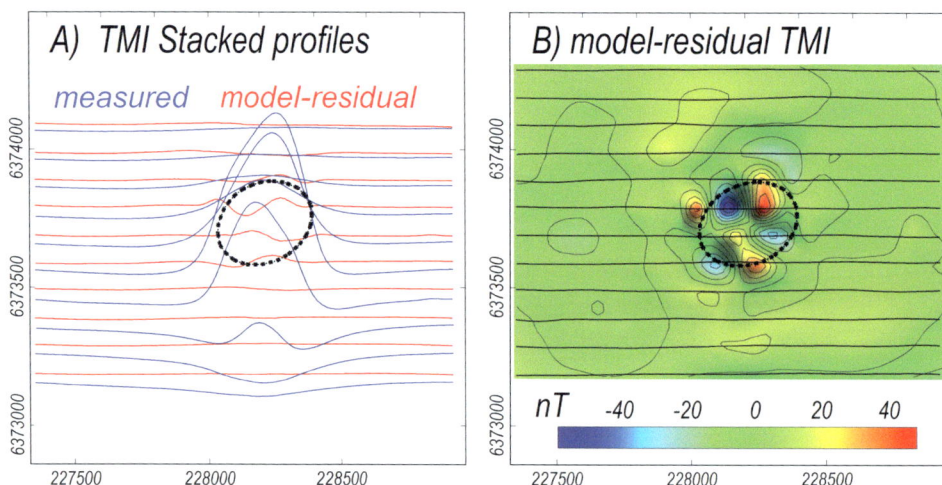

Fig. 6.43. (top) Stacked profiles of measured TMI (blue) and residual misfit (red) for the elliptic pipe model; (bottom) residual misfit from the elliptic pipe model. The dashed line outlines the top of the model.

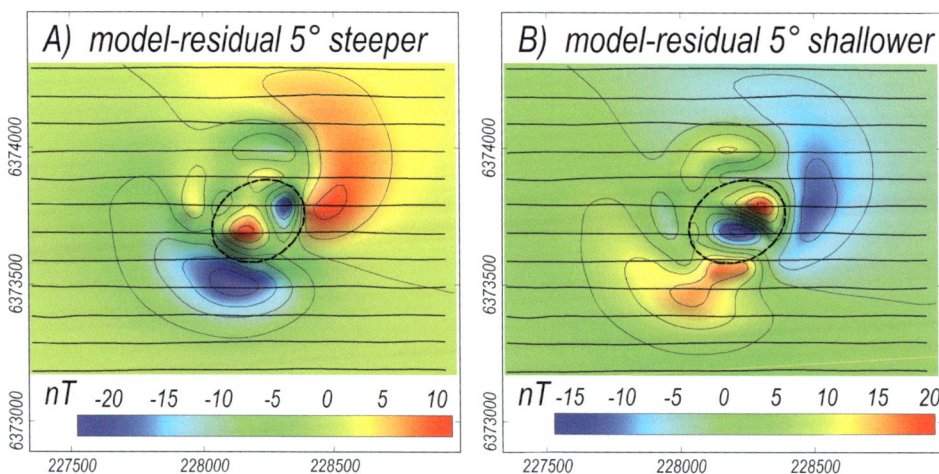

Fig. 6.44. Differences between fields computed from the best elliptic pipe model and best matching elliptic pipes with magnetisation 5° steeper (top) and 5°shallower (bottom).

6.8.5 Data sufficiency tests of estimated magnetisation direction

As discussed above, for this case study it seems feasible with multiple inversions to consistently recover the best-fit homogeneous magnetisation direction to within 5° to 10°. One of the major concerns for small anomalies such as this is that while it may be possible to find the best magnetisation direction to explain the available data, the data itself may be an insufficient or misleading sampling of the complete anomaly. As shown in Fig. 6.26, Anomaly 267 is defined on eight lines, with a significant expression on five of those. North–south lines would have provided a superior sampling of the anomaly because the positive and negative lobes are both sampled on each line. However, with even a coverage of five east–west lines at 100 m line spacing there is enough data redundancy to investigate sufficiency of a 200 m line-spaced survey by taking even and odd line subsets from the current survey. It has already been noted that the most diagnostic match of measured and computed data is on one specific flightline across the centre of the anomaly (line 100660), that is included in only one of the alternate-line subsets.

Figure 6.45 shows TMI images generated by gridding the two separate 200 m spaced line sets, together with stacked profiles of the TMI data and of TMI forward computed from elliptic-section pipe models inverted to fit the individual datasets. The two TMI images appear quite similar, but with an apparent offset, because in either case gridding of the data does not displace the central peak of the anomaly far from an included flightline. For both datasets the stacked profiles of measured TMI and TMI forward-computed from the elliptic-section pipe model inversions show closer agreement than the corresponding match for the complete 100 m spaced set of flightlines. This is because the models have a less demanding task in fitting these sparser datasets. The model centre-points plotted in Fig. 6.45 are well clustered for inversion of different model geometries for each individual line dataset and are consistent between the two datasets. Figure 6.46 shows the model centre-point distribution in more detail. The points are all confined within a 100 m span of one flightline spacing (for the individual inversions this is equivalent to one-half of a 200 m flightline spacing) and are mostly clustered within a radius of 25 m about the best-estimate body centre. There is significantly larger scatter for the even line-number inversion models than for the odd line-number inversion models. The even line-number set includes the central flightline but has only that one line through the anomaly peak, whereas the odd line number subset has two (more marginal) lines through the anomaly peak.

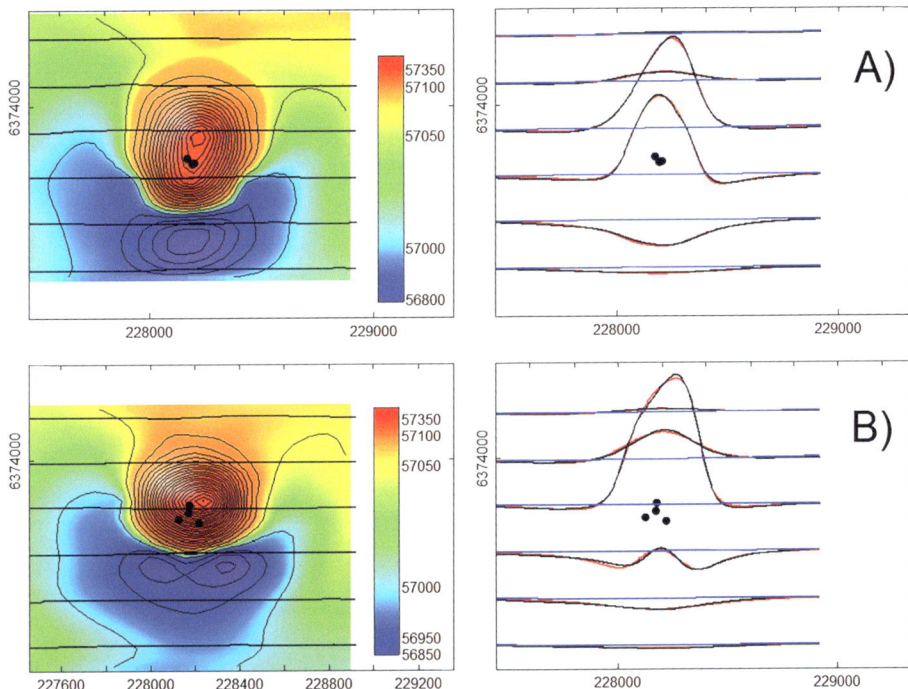

Fig. 6.45. TMI images and stacked profiles (black – measured, blue – interpreted regional, red – computed from elliptic pipe model on A) odd and B) even lines. Points mark centres of four inversion models (ellipsoid, tabular, circular and elliptic pipes).

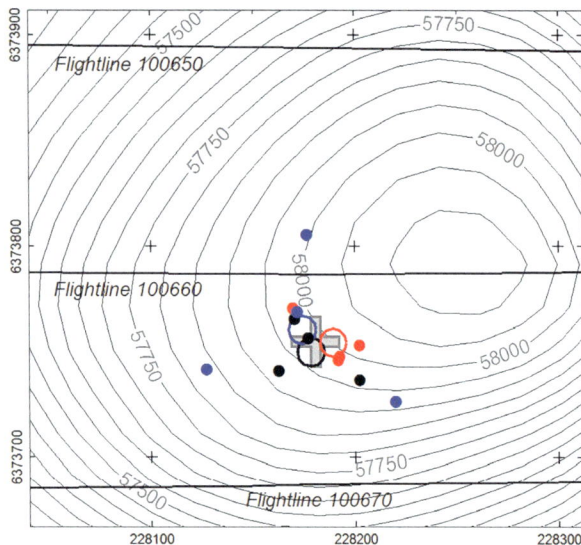

Fig. 6.46. Best inversion model centres from inversion of all lines (black points), odd-number lines (red points), even-number lines (blue points), with respective mean centres (open circles), global mean (grey cross), flightlines and TMI contours.

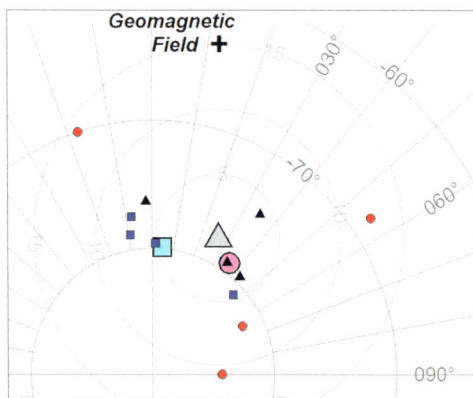

Fig. 6.47. Stereonet plot of model magnetisation directions from inversion of all lines (triangles), odd-number lines (red circles), and even-number lines (blue squares) with contours of rotation from the best direction (degrees).

Magnetisation directions recovered from the different model geometry inversions of the individual line datasets are plotted in Fig. 6.47. The different geometry inversions of the complete (100 m line-spaced) dataset cluster within 4° of the population best direction. The mean directions from different geometry model inversions of the individual line datasets are also within 4° of the best direction. Magnetisation directions of the different geometry inversions of the odd line number data are grouped within 5° of their mean direction, whereas directions for the even line number data are more scattered within 10° of their mean direction, consistent with the larger scatter of the model centre-points. Had the

Rylstone survey been flown at 200 m line-spacing instead of 100 m, the best-estimated (multiple inversion) magnetisation direction should have been recovered within 5° of the best-estimated direction that was obtained from the 100 m line-spaced survey.

This study raises the question of what ground investigation would be required to validate the results. Many magnetic field anomalies in Australia are from sources buried beneath cover and no petrophysical or palaeomagnetic measurements are available unless there is oriented borehole core. The magnetic anomaly most regularly related to a measured magnetisation is the Black Hill Norite (Rajagopalan *et al.* 1993, 1995; Schmidt *et al.* 1993; Foss and McKenzie 2011) but those measurements were made on only a few samples from a single quarry outcrop which is not part of the source of the studied anomaly. The only case study I am aware of in which a model based on petrophysical measurements of susceptibility and remanent magnetisation has been quantitatively matched to its magnetic field anomaly is the Rover 3 anomaly studied by Austin and Foss (2014). That study used samples from only a single borehole, but the body is horizontally compact and the measured magnetisations assigned to horizontal units defined by abrupt and significant magnetisation contrasts produce an acceptable explanation of the measured anomaly. The volcanic unit causing the Rylstone anomaly is more likely to be irregularly inhomogeneous, with magnetisation intensities possibly varying by orders of magnitude over a range of scales, and with irregular and unpredictable distribution. The volume estimates for the magnetisation vary between 5×10^6 m^3 and 1.4×10^7 m^3, with horizontal widths between 350 and 450 m and depth extent from 70 to 160 m. Ground-truthing with direct palaeomagnetic measurements would depend on finding outcrop free from pervasive weathering and/or extensive lightning strike re-magnetisation. Even extensive suitable outcrop over the top of the body would provide a very limited sampling of the complete body. That would require oriented core from several boreholes to 100 or 200 m depth. The present inversion model volume and magnetisation estimates (Fig. 6.29) vary by more than a factor of 2, and models can be found to well match the anomaly with at least another variation factor of 2 to 5. To compare measured and inverted total magnetisations requires boreholes to intersect the base of magnetisation as well as effective constrains on the margins of the body. The concept of meaningfully constraining magnetisation models with petrophysics measurements, as widely promoted in published literature, rarely take into account the variability of rock magnetisation and massively under-represents the challenge. Magnetisation is not a true bulk property of a rock – it depends completely on minerals that typically constitute only a few per cent of the rock, and the magnetisation of those minerals varies considerably with oxidation, iron/titanium ratios and grain size and shape. If measurements on core were to establish that there is high variability of magnetisation, then the homogeneous magnetisation intensity provided by an inversion model is only a bulk statistic for that rock unit and may not represent the true magnetisation at any point within it. The principal value of petrophysics measurements would be to relate magnetisation to mineralogy and geological processes in the history of the rock. Direct magnetisation measurements are also required to resolve the resultant magnetisation investigated in magnetic field analysis into its remanent and induced components.

6.9 CONCLUSIONS

In this chapter I provide empirical justification for recovery of magnetisation estimates from inversion of anomalies due to compact magnetisations, for which the maximum extension in any one direction is less than twice the closest point of measurement and/or computation of the magnetic field. The analytic capabilities of this study are provided by parametric inversion that permits isolation and study of single components of complex interactive problems. This study suggests that for isolated, well-defined anomalies due to compact sources of homogeneous magnetisation, uncertainty in source position, shape and plunge should not preclude recovering an estimate of that magnetisation direction from inversion of magnetic field data to within a range of 5° to 10°. For less favourable cases, however, errors in magnetisation direction of up to 15° may occur as a result of combined errors in estimation of source location and plunge.

The complexity of voxel inversion models does not clearly present the relationships established in this study, but the fundamental relationships between magnetisations and their magnetic fields apply to all inversion results provided magnetisations are compact and their fields are well isolated. If magnetisations are more distributed or more inhomogeneous than the discrete magnetisations represented by the parametric models of this study then the relationships degrade and inversion

results lose significance. Existence of discrete features (sweet-spots) is essential for inversion to resolve distribution of magnetisation and its direction.

A comprehensive inversion study of Anomaly 276 in Rylstone, New South Wales has established the capability to recover estimates of magnetisation direction consistent to within 5° to 10°. The magnetisation direction of this source is not independently known so the magnetisation directions, while consistent are not presently proven correct. However, the sensitivity of estimated magnetisation direction to the distribution of magnetisation, its location and plunge are all shown to be in close agreement with relationships established from the synthetic data study. The best estimate of the maximum-extent to measurement-proximity ratio for the Rylstone anomaly is between 4.6:1 and 6.8:1. This is significantly larger than the value of 2 suggested as a conservative bound on compact sources. This promisingly suggests that compact source behaviour, favourable for estimation of source magnetisation direction, extends to a wide range of geological magnetisations and survey data. The magnetisation direction estimate is also shown to be stable through sub-sampling of the anomaly from the 100 m line spacing to 200 m.

ACKNOWLEDGEMENTS

I would like to thank the Geological Survey of New South Wales for supply of the data. All modelling and inversion studies were conducted with Tensor Research's ModelVision^{TM} software package.

REFERENCES

Austin JR, Foss CA (2014) The paradox of scale: reconciling magnetic anomalies with rock magnetic properties for cost-effective mineral exploration. *Journal of Applied Geophysics* **104**, 121–133. doi:10.1016/j.jappgeo.2014.02.018

Beiki M, Clark DA, Austin JR, Foss CA (2012) Estimating source location using normalized magnetic source strength calculated from magnetic gradient tensor data. *Geophysics* **77**, J23–J37. doi:10.1190/geo2011-0437.1

Clark DA (2014) Methods for determining remanent and total magnetisations of magnetic sources – a review. *Exploration Geophysics* **45**, 271–304. doi:10.1071/EG14013

Cooper GRJ (2014) Reducing the dependence of the analytic signal amplitude of aeromagnetic data on the source vector direction. *Geophysics* **79**, J55–J60. doi:10.1190/geo2013-0319.1

Dannemiller N, Li Y (2006) A new method for determination of magnetisation direction. *Geophysics* **71**, L69–L73. doi:10.1190/1.2356116

Dransfield M, Christensen A, Liu G (2003) Airborne vector magnetics mapping of remanently magnetized banded iron formations at Rocklea, Western Australia. *Exploration Geophysics* **34**, 93–96. doi:10.1071/EG03093

Encom Technology (1997) 'Interpretation of airborne geophysical data over the Rylstone Area.' The Coal and petroleum Geology Branch, The Geological Survey of New South Wales (unpublished), 23 pages.

Fedi M, Florio G, Rapolla A (1994) A method to estimate the total magnetisation direction from a distortion analysis of magnetic anomalies. *Geophysical Prospecting* **42**, 261–274. doi:10.1111/j.1365-2478.1994.tb00209.x

Foss CA (2006) 'Evaluation of strategies to manage remanent magnetisation effects in magnetic field inversion.' 76th Annual International Meeting, SEG, Expanded Abstracts, 938–942.

Foss CA (2017) Resultant-magnetisation based magnetic field interpretation. In *Proceedings of Exploration 17: Sixth Decennial International Conference on Mineral Exploration.* (Eds V Tschirhart, MD Thomas) pp. 637–648.

Foss CA, McKenzie KB (2011) Inversion of anomalies due to remanent magnetisation: an example from the Black Hill Norite of South Australia. *Australian Journal of Earth Sciences* **58**, 391–405. doi:10.1080/08120099.2011.581310

Fullagar PK, Pears GA (2015) 'Remanent magnetisation inversion.' 24th ASEG Conference, Extended Abstracts. doi:10.1071/ASEG2015ab188

Helbig K (1963) Some integrals of magnetic anomalies and their relation to the parameters of the disturbing body. *Zeitschrift für Geophysik* **29**, 83–96.

Keating P, Sailhac P (2004) Use of the analytic signal to identify magnetic anomalies due to kimberlite pipes. *Geophysics* **69**, 180–190. doi:10.1190/1.1649386

Lelièvre PG, Oldenburg DW (2009) A 3D total magnetisation inversion applicable when significant, complicated remanence is present. *Geophysics* **74**, L21–L30. doi:10.1190/1.3103249

Li X (2006) 'Understanding 3D analytical signal amplitude. *Geophysics* **71**, L13–L16. doi:10.1190/1.2184367

Paine J, Haederle M, Flis M (2001) Using transformed TMI data to invert for remanently magnetised bodies. *Exploration Geophysics* **32**, 238–242. doi:10.1071/EG01238

Phillips JD (2005) Can we estimate total magnetisation directions from aeromagnetic data using Helbig's integrals? *Earth, Planets, and Space* **57**, 681–689. doi:10.1186/BF03351848

Pratt DA, McKenzie KB, White TS (2014) Remote remanence determination (RRE). *Exploration Geophysics* **45**, 314–323. doi:10.1071/EG14031

Rajagopalan S, Schmidt PW, Clark DA (1993) Rock magnetism and geophysical interpretation of the Black Hill Norite, South Australia. *Exploration Geophysics* **24**, 209–212. doi:10.1071/EG993209

Rajagopalan S, Clark DA, Schmidt PW (1995) Magnetic mineralogy of the Black Hill Norite and its aeromagnetic and palaeomagnetic implications. *Exploration Geophysics* **26**, 215–220. doi:10.1071/EG995215

Roest WR, Pilkington M (1993) Identifying remanent magnetization effects in magnetic data. *Geophysics* **58**, 653–659. doi:10.1190/1.1443449

Schmidt PW, Clark DA (1998) The calculation of magnetic components and moments from TMI: a case study from the Tuckers igneous complex, Queensland. *Exploration Geophysics* **29**, 609–614.

Schmidt PW, Clark DA, Rajagopalan S (1993) An historical perspective of the Early Palaeozoic APWP of Gondwana: new results from the Early Ordovician Black Hill Norite of South Australia. *Exploration Geophysics* **24**, 257–262. doi:10.1071/EG993257

Yoo EK (1998) 'Western Coalfield Regional Geology (northern part) 1:100 000, 1st edition.' Geological Survey of New South Wales, Sydney.

Yoo EK, Tadros NZ, Bayly KW (2001) 'A compilation of the geology of the Western Coalfield.' Geological Survey of New South Wales, Report GS2001/204 (unpublished).

7

Estimation of magnetisation direction from axial magnetic field component and gradient tensor element ratios of a dipole source

C.A. Foss, K.B. McKenzie and D.A. Clark

ABSTRACT

Anomalies in the measured magnetic field mark the location of subsurface magnetisations. We show that using a dipole model the direction of the contrast of that magnetisation against the surrounding material can be estimated by analysis of the components or tensor gradients of the magnetic field directly above the magnetisation. We propose the use of the normalised source strength (NSS) as an appropriate method to locate the centre of magnetisation and show that the NSS is superior to the total gradient (TG) that has traditionally been used for this application. Imperfections in applying the analysis arise from imprecision in locating the centre of magnetisation and inaccuracies in assigning a dipole model to represent non-dipole magnetisations. We investigate the effect of non-dipole characteristics according to the shape of the distribution of magnetisation, any plunge of that distribution, and the distance at which the magnetic field is measured. Finally, we show that for a magnetic field anomaly measured in Queensland, Australia the NSS peak coincides with the centre of magnetisation estimated by parametric inversion, and that magnetisation direction estimates made independently by tensor analysis and inversion agree to within 5°. For this anomaly the TG peak is significantly displaced from the centre of the inversion model and the direction of magnetisation estimated by tensor analysis at the TG peak differs from the inversion direction by 38°.

7.1 INTRODUCTION

The external magnetic field of a source body is a function of its resultant (induced plus remanent) magnetisations. Helbig (1963) provides a proof that over an extensive horizontal surface the direction of magnetisation of a compact source can be recovered from integral moments of the magnetic field components about the point directly above its centre. Helbig analysis itself provides a practical methodology to recover source magnetisation estimates from measured magnetic fields (Schmidt and Clark 1998; Phillips 2005; Phillips *et al.* 2007; Caratori Tontini and Pedersen 2008; Clark 2014) but the magnetic component moment integrals are highly sensitive to imperfection in removing background fields. The analysis requires isolation of the field due to the source and provision of the horizontal centre of magnetisation. Magnetic field components are derived by FFT phase transforms from a TMI grid (Lourenço and Morrison 1973; Purucker 1990; Blakely 1995; Schmidt and Clark 1998; Clark 2013). Phillips *et al.* (2007) provided an approximate relationship to use derivatives of the field components and reduce dependence on isolation from background fields. Many other methods have been established to estimate magnetisation direction from magnetic field data (Fedi *et al.* 1994; Dannemiller and Li 2006; McKenzie *et al.* 2012) including parametric inversion (Foss 2006; Foss and McKenzie 2011; Pratt *et al.* 2014) and voxel inversion (Paine *et al.* 2001; Lelièvre and Oldenburg 2009; Fullagar and Pears 2015; Li 2012).

Here we propose an approximate method to estimate the magnetisation direction of moderately compact sources from the ratios of their magnetic field components or tensor gradient elements directly above their centres. We first apply the NSS transform to a TMI grid and select suitable near-circular NSS anomalies. At the centre points of those anomalies we sample the Cartesian components of the field and/or its tensor gradient elements derived from FFT of the measured TMI field. Component analysis of measured fields requires that anomalies are separated from the background field (a regional-residual separation) but this is generally not required for tensor analysis because the anomalous field contributions tend to be the dominant local field gradients. Conversely, short wavelength variations due to measurement imperfections or near-surface magnetisations can be disruptive to gradient analysis and may require pre-processing such as by a mild upward continuation. The NSS derivation is complex and computationally intense but can be applied either to complete multi-anomaly field sets or to individual anomaly separations. We also include comparison of analyses performed at locations where the total gradient is a maximum.

7.2 ANALYTICAL BASIS OF THE FIELD COMPONENT RATIO METHOD

The expression for the magnetic field vector $\mathbf{B}(\mathbf{r})$ at an observation point $\mathbf{r} = (x, y, z)^T$ due to a point dipole source possessing a magnetic moment $\mathbf{m}$ and located at the origin is given by Blakely (1995, p. 75) as:

$$\mathbf{B}(\mathbf{r}) = \frac{C_m}{r^3} \{3(\mathbf{m} \cdot \hat{\mathbf{u}}_r) \hat{\mathbf{u}}_r - \mathbf{m}\}, \text{ (Eqn 7.1)}$$

where $\hat{\mathbf{u}}_r$ is the unit vector in the direction of the line joining the dipole to the measurement point $\mathbf{r}$

$$\hat{\mathbf{u}}_r = \left(u_{rx}, u_{ry}, u_{rz}\right)^T = \left(\frac{x}{r}, \frac{y}{r}, \frac{z}{r}\right)^T; r = |\mathbf{r}| = \sqrt{x^2 + y^2 + z^2}.$$

and where C_m is a constant which depends on the system of electromagnetic units used (see Blakely 1995, pp. 67–68). In the International Standard (SI) system of units,

C_m = 100 nH/m or 100 nTm/A for magnetic fields expressed in nanotesla (nT) and magnetisations expressed in ampere per metre (A/m). The coordinate system adopted here follows the International Geomagnetic Reference Field convention, i.e. *x* is North, y is East and +z is vertically down. The Cartesian components of the magnetic field vector are

$$B_x(\mathbf{r}) = \frac{C_m}{r^3} \{3(\mathbf{m} \cdot \hat{\mathbf{u}}_r)u_{rx} - m_x\}$$
$$= \frac{C_m}{r^3} \left\{ 3\left(\frac{m_x x^2}{r^2} + \frac{m_y xy}{r^2} + \frac{m_z xz}{r^2} \right) - m_x \right\}, \text{ (Eqn 7.2)}$$

$$B_y(\mathbf{r}) = \frac{C_m}{r^3} \{3(\mathbf{m} \cdot \hat{\mathbf{u}}_r)u_{ry} - m_y\}$$
$$= \frac{C_m}{r^3} \left\{ 3\left(\frac{m_x xy}{r^2} + \frac{m_y y^2}{r^2} + \frac{m_z yz}{r^2} \right) - m_y \right\}, \text{ (Eqn 7.3)}$$

$$B_z(\mathbf{r}) = \frac{C_m}{r^3} \{3(\mathbf{m} \cdot \hat{\mathbf{u}}_r)u_{rz} - m_z\}$$
$$= \frac{C_m}{r^3} \left\{ 3\left(\frac{m_x xz}{r^2} + \frac{m_y yz}{r^2} + \frac{m_z z^2}{r^2} \right) - m_z \right\}. \text{ (Eqn 7.4)}$$

The expressions in Eqns 7.2–7.4 for the magnetic field components due to a point dipole or uniformly magnetised sphere are completely general and apply to any external observation point. However, for an observation station located at a height *z* directly above the dipole or above the centre of a magnetised sphere, the expressions for the magnetic field components become greatly simplified since $\hat{\mathbf{u}}_r = (0, 0, -1)^T$. Thus

$$B_x(0,,0,-z) = -\frac{C_m m_x}{r^3}.; B_y(0,,0,-z) = -\frac{C_m m_y}{r^3}.; \text{ and}$$

$$B_z(0,,0,-z) = \frac{C_m m_z}{r^3}\left[\frac{3z^2}{r^2} - 1 \right] = \frac{2C_m m_z}{r^3}.$$

The corresponding expressions for the magnetic field components produced by a magnetised sphere of radius *a*, volume *v*, magnetisation $\mathbf{M}$ and magnetic moment $\mathbf{m} = \mathbf{M}v$ are

$$B_x(0,,0,-z) = -\frac{C_m m_x}{r^3} = -\frac{C_m M_x v}{r^3} = -\left(\frac{4\pi}{3} \right)\frac{C_m M_x a^3}{r^3}.$$
$$\text{(Eqn 7.5)}$$

$$B_y(0,0,-z) = -\frac{C_m m_y}{r^3} = -\frac{C_m M_y v}{r^3} = -\left(\frac{4\pi}{3} \right)\frac{C_m M_y a^3}{r^3}.$$
$$\text{(Eqn 7.6)}$$

$$B_z(0,0,-z) = -\frac{2C_m m_z}{r^3} = \frac{2C_m M_z v}{r^3} = \left(\frac{8\pi}{3} \right)\frac{C_m M_z a^3}{r^3}.$$
$$\text{(Eqn 7.7)}$$

By inspection of Eqns 7.5 and 7.6, an expression for the declination of magnetisation D_M (or declination of magnetic moment) at any point $\mathbf{r} = (0,0,-z)^T$ above the point dipole or centre of a magnetised sphere is obtained from the ratio of the horizontal field components, namely,

$$D_M = \arctan\left(\frac{M_y}{M_x}\right) = \arctan\left(\frac{-B_y}{-B_x}\right) \quad \text{for } 0 \leq D_M \leq 2\pi.$$

$$\text{(Eqn 7.8)}$$

Similarly, by inspection of Eqns 7.5–7.7, the inclination of magnetisation I_M is obtained from the ratio of the horizontal field component to half the vertical field component, namely,

$$I_M = \arctan\left(\frac{M_z}{M_h}\right) = \arctan\left(\frac{B_z}{2\sqrt{B_x^2 + B_y^2}}\right) \quad \text{for } -\frac{\pi}{2} \leq I_M \leq \frac{\pi}{2}.$$

$$\text{(Eqn 7.9)}$$

The expression for the magnetic gradient tensor at a measurement point $\mathbf{r} = (x, y, z)^T$ due to a point dipole with magnetic moment $\mathbf{m}$ centred at the origin is given by Wynn *et al.* (1975) and Wilson (1985), namely,

$$B_{ij}(\mathbf{r}) = u_{ri}\mu_j + u_{rj}\mu_i + (\boldsymbol{\mu} \cdot \hat{\mathbf{u}}_{\mathbf{r}})\delta_{ij} - 5(\boldsymbol{\mu} \cdot \hat{\mathbf{u}}_{\mathbf{r}})u_{ri}u_{rj} \quad \text{(Eqn 7.10)}$$

for $i,j = 1,2,3$ or x,y,z and where $\boldsymbol{\mu} = (\mu_x, \mu_y, \mu_z)^T$ is the scaled magnetic moment and δ_{ij} is Kronecker delta.

For a uniformly magnetised sphere of radius a, volume v and magnetisation $\mathbf{M}$, the scaled magnetic moment $\boldsymbol{\mu}(\mathbf{r})$ is

$$\boldsymbol{\mu}(\mathbf{r}) = \frac{3C_m\mathbf{m}}{r^4} = \frac{3C_m\mathbf{M}v}{r^4} = \frac{4\pi C_m a^3 \mathbf{M}}{r^4}, \quad \text{(Eqn 7.11)}$$

The expressions in Eqns 7.10 and 7.11 for the gradient tensor due to a dipole or uniformly magnetised sphere are completely general and apply to any external observation point. However, for an observation station P(0, 0, −z) located at a height |z| directly above a point dipole or magnetised sphere centred at the origin, the expressions for its gradient tensor become greatly simplified since $\hat{\mathbf{u}}_{\mathbf{r}} = (u_{rx}, u_{ry}, u_{rz})^T = (0,0,-1)^T$. Hence all terms in Eqn 7.10 not involving u_{rz} are identically zero. Therefore, at any observation station $\mathbf{r} = (0, 0, -z)^T$, the elements of the gradient tensor $B_{ij}(\mathbf{r})$ are

$$B_{xx} = B_{11} = \left(\boldsymbol{\mu} \cdot \hat{\mathbf{u}}_{\mathbf{r}}\right)\delta_{11} = -\mu_z = -\frac{4\pi C_m a^3}{r^4}M_z. \quad \text{(Eqn 7.12)}$$

$$B_{xy} = B_{12} = 0. \quad \text{(Eqn 7.13)}$$

$$B_{xz} = B_{13} = u_{r3}\mu_1 = -\mu_x = -\frac{4\pi C_m a^3}{r^4}M_x. \quad \text{(Eqn 7.14)}$$

$$B_{yx} = B_{21} = 0 = B_{xy}. \quad \text{(Eqn 7.15)}$$

$$B_{yy} = B_{22} = (\boldsymbol{\mu} \cdot \hat{\mathbf{u}}_{\mathbf{r}})\delta_{22} = -\mu_z = -\frac{4\pi C_m a^3}{r^4}M_z. \quad \text{(Eqn 7.16)}$$

$$B_{yz} = B_{23} = u_{r3}\mu_2 = -\mu_y = -\frac{4\pi C_m a^3}{r^4}M_y. \quad \text{(Eqn 7.17)}$$

$$B_{zx} = B_{31} = u_{r3}\mu_1 = -\mu_x = -\frac{4\pi C_m a^3}{r^4}M_x = B_{xz}.$$

$$\text{(Eqn 7.18)}$$

$$B_{zy} = B_{32} = u_{r3}\mu_2 = -\mu_y = -\frac{4\pi C_m a^3}{r^4}M_y = B_{yz}.$$

$$\text{(Eqn 7.19)}$$

$$\begin{aligned}B_{zz} = B_{33} &= 2u_{r3}\mu_3 + (\boldsymbol{\mu} \cdot \hat{\mathbf{u}}_{\mathbf{r}})\delta_{33} - 5(\boldsymbol{\mu} \cdot \hat{\mathbf{u}}_{\mathbf{r}})u_{r3}^2 \\ &= -2\mu_3 + (-\mu_3)\delta_{33} - 5(-\mu_3)u_{r3}^2 \\ &= -3\mu_3 + 5\mu_3 = 2\mu_3 = 2\mu_z = \frac{8\pi C_m a^3}{r^4}M_z. \quad \text{(Eqn 7.20)}\end{aligned}$$

Hence the magnetic gradient tensor $B_{ij}(0, 0, -z)$ is

$$\begin{aligned}\mathbf{B}(0,0,z) &= \begin{bmatrix} B_{xx} & B_{xy} & B_{xx} \\ B_{xy} & B_{yy} & B_{yz} \\ B_{zx} & B_{zy} & B_{zz} \end{bmatrix} \\ &= \frac{4\pi C_m a^3}{r^4}\begin{bmatrix} -M_z & 0 & -M_x \\ 0 & -M_z & -M_y \\ -M_x & -M_y & 2M_z \end{bmatrix}. \end{aligned} \quad \text{(Eqn 7.21)}$$

By inspection of Eqns 7.14 and 7.17, the declination of magnetisation D_M at any point above the point dipole or centre of a magnetised sphere (for $|z| > a$) is obtained from the ratio of the B_{xz} and B_{yz} tensor elements as follows:

$$D_M = \arctan\left(\frac{M_y}{M_x}\right) = \arctan\left(\frac{-B_{yz}}{-B_{xz}}\right) \quad \text{for } 0 \leq D_M \leq 2\pi.$$

$$\text{(Eqn 7.22)}$$

Similarly, by inspection of Eqns 7.14, 7.17 and 7.20, the inclination of magnetisation I_M at any point above the point dipole or centre of the magnetised sphere is derived from the B_{xz}, B_{yz} and B_{zz} tensor elements as follows:

$$I_M = \arctan\left(\frac{M_z}{M_h}\right) = \arctan\left(\frac{B_{zz}}{2\sqrt{B_{xz}^2 + B_{yz}^2}}\right) \quad \text{for } -\frac{\pi}{2} \leq I_M \leq \frac{\pi}{2}.$$

$$\text{(Eqn 7.23)}$$

In this study we use peaks in both the normalised source strength (NSS) and total gradient (TG) to estimate the horizontal centre of magnetisation. As proposed by Beiki *et al.* (2012) and Clark (2012) the NSS is a generalisation of the scaled or normalised magnetic moment of a dipole source (Wynn *et al.* 1975; Wilson 1985). Denoted by μ, the NSS is defined in terms of the eigenvalues of its magnetic gradient tensor, namely,

$$\mu = \sqrt{-\lambda_2^2 - \lambda_1 \lambda_3} \qquad \text{for} \quad \lambda_1 \geq \lambda_2 \geq \lambda_3, \quad \text{(Eqn 7.24)}$$

where $\lambda_1 > 0$ is the first eigenvalue which is always positive, λ_2 is the second or intermediate eigenvalue which has the smallest absolute value and $\lambda_3 < 0$ is always negative. The traceless property of the tensor, together with the ordering of its eigenvalues, ensures that the NSS is always real and positive definite. Importantly, Clark (2012) and Beiki *et al.* (2012) showed that the NSS peaks directly over the source for several useful elementary models and that it is completely independent of magnetisation direction for arbitrary two-dimensional sources, as well as for spheres and axially magnetised narrow plunging pipes. This conclusion approximately holds for any compact source with a reasonably coherent magnetisation direction for which the external field is dominated by the dipole moment.

The total gradient $G(r)$ of the total magnetic field intensity $T(r)$ at an observation point $r = (x, y, z)$ is defined as follows:

$$G(r) = \sqrt{\left(\frac{\partial T(r)}{\partial x}\right)^2 + \left(\frac{\partial T(r)}{\partial y}\right)^2 + \left(\frac{\partial T(r)}{\partial z}\right)^2}. \quad \text{(Eqn 7.25)}$$

TG peaks towards the horizontal centre of a compact magnetisation and towards the margins of wider magnetisations, and Roest *et al.* (1992) demonstrated that it can be used to map the location of magnetisation contrasts.

7.3 SOURCES OF ERROR IN THE ANALYSIS

7.3.1 Imprecise location of the NSS peak

Our analysis depends on location of the horizontal centre of magnetisation. On any horizontal plane above a dipole the divergence of component and gradient ratios from the values estimated about its centre increases consistently with offset of the analysis from that centre. Figure 7.1 shows NSS contours over TMI images of low- and high-inclination magnetisations in a southern hemisphere geomagnetic field of inclination −60°. The TMI anomalies of these two magnetisations are quite different but the NSS anomalies are identical. Figure 7.2 plots the error in estimation of magnetisation direction from analysis of the field component ratios (Eqns 7.8 and 7.9). This error is zero directly above the dipole. For the low-inclination magnetisation the error increases most rapidly for azimuths parallel and anti-parallel to the declination of magnetisation (Fig. 7.2A). For the steep-inclination magnetisation there is less dependence on direction of the offset of the analysis (Fig. 7.2B) but in all directions the error increases more rapidly than for the low-inclination magnetisation. For the low-inclination magnetisation the error in estimating magnetisation direction is < 5° for horizontal errors of centre offset up to 10% of source depth, but for the steep magnetisation there is a threefold increase in this error. Gradient anomalies are sharper than their corresponding field anomalies and in consequence equivalent imprecision in locating the horizontal centre of an anomaly gives

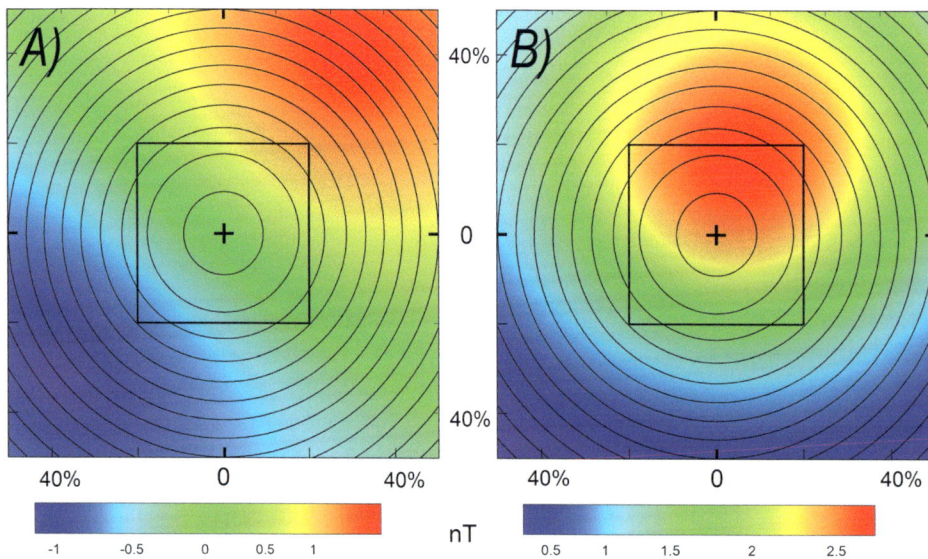

Fig. 7.1. NSS contours over TMI images for dipole magnetisations with declination 45° and inclination a) −15° and b) −75°. The horizontal scale is percentage of depth to the centre of magnetisation. The squares outline the area imaged in Fig. 7.2.

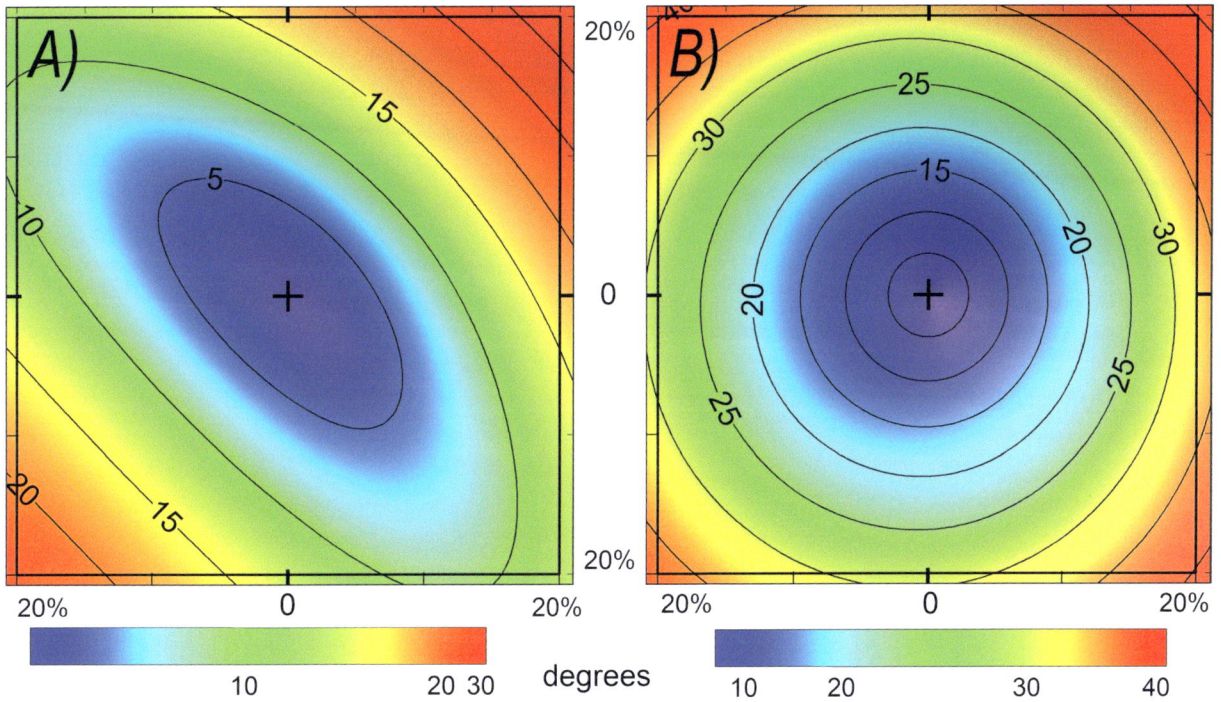

Fig. 7.2. Contoured error in magnetisation direction of the field component analysis for off-centred locations: a) low-inclination magnetisation (−15°) and b) steep-inclination magnetisation (−75°).

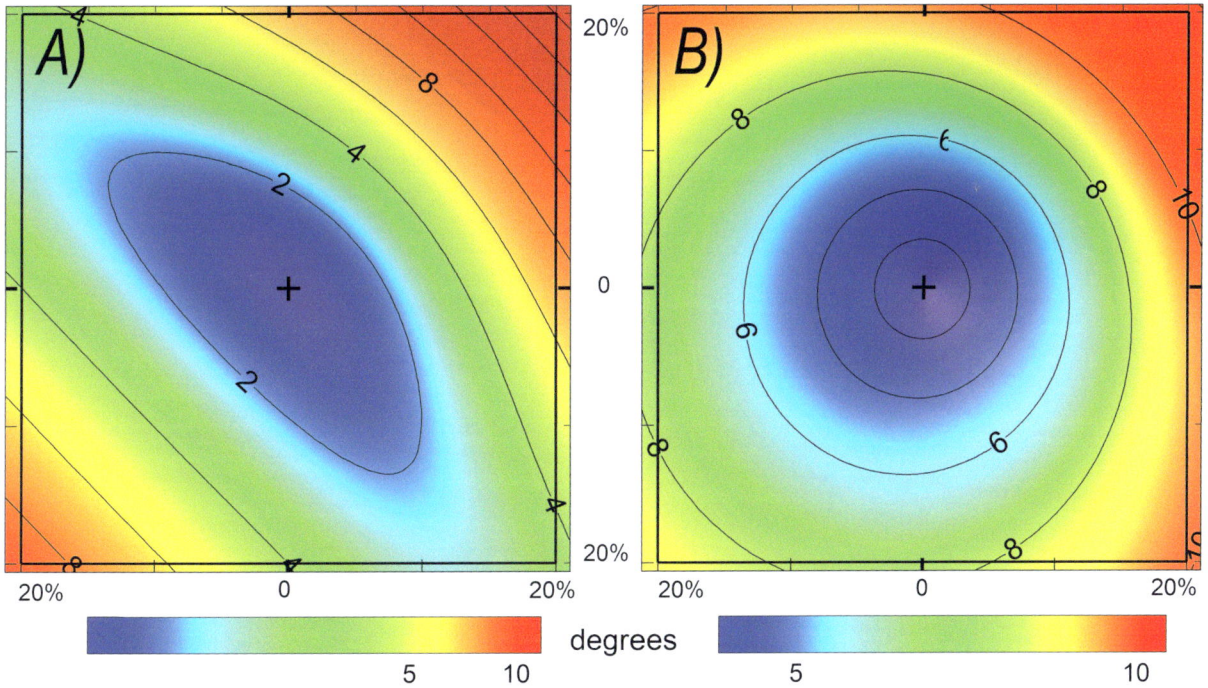

Fig. 7.3. Increase in error of magnetisation direction from the field component to the tensor gradient analysis for: a) low inclination (−15°) and b) steep inclination (−75°) magnetisations. Note that these errors are additional to those for the component analysis plotted in Fig. 7.2 and represent a penalty of using gradient rather than component analysis.

rise to larger errors in magnetisation direction from the gradient tensor analysis (Eqns 7.22 and 7.23) than from the component analysis (Eqns 7.8 and 7.9). Figure 7.3 plots the increase in error in changing from using component analysis to using gradient analysis. For location errors < 10% of source depth (the zone within which we hope the analysis will be performed) this increase is < 2° for the low-inclination magnetisation and < 5° for the steep-inclination magnetisation. We anticipate that in many cases this c. 30% loss of resolution from the use of gradients will be more than compensated by the advantage of lower sensitivity to the separation of the anomaly from its background field.

7.3.2 The influence of shape

Any departure from an ideal dipole compromises the model assumptions of relationships between magnetisation and the field or its gradients. For computations of non-dipole sources we have used expressions for the magnetic gradient tensor due to a uniformly magnetised general triaxial ellipsoid given in McKenzie (2020) and equations for the magnetic field components due to a uniformly magnetised general triaxial ellipsoid formulated in Clark *et al.* (1986). Figure 7.4 shows perspective views of several ellipsoids of axial ratios 1:1:0.5 and

1:0.5:0.5 and of centre depth to major axis length ratio 1:1. The departures of these bodies from dipoles distort the analysis in two ways: first the assumed dipole relationships are only approximate for these shapes, and second the NSS peak for these bodies may no longer mark their centre of magnetisation. We can investigate both effects by computing the dipole relationships over the known centre of magnetisation and at the NSS peak. We also include analysis at the alternative estimate of the centre of magnetisation provided by the total gradient transform. Results of these studies vary according to distribution shape and orientation of the magnetisation and its magnetisation direction.

Figure 7.5 is a map of error in a dipole tensor analysis. The figure shows an image of the angular difference between the input source magnetisation direction and the direction estimated by dipole tensor analysis (Eqns 7.22 and 7.23) from the magnetic field of an ellipsoid with horizontal north–south major axis of length 200 m, minor axes of 100 m, magnetisation inclination −75° and declination 45° and depth to centre 120 m. Note that for this non-dipole distribution of magnetisation there is a displacement between the centre of magnetisation (0,0) and the point at which by cancellation of errors the dipole tensor analysis provides the correct estimation of

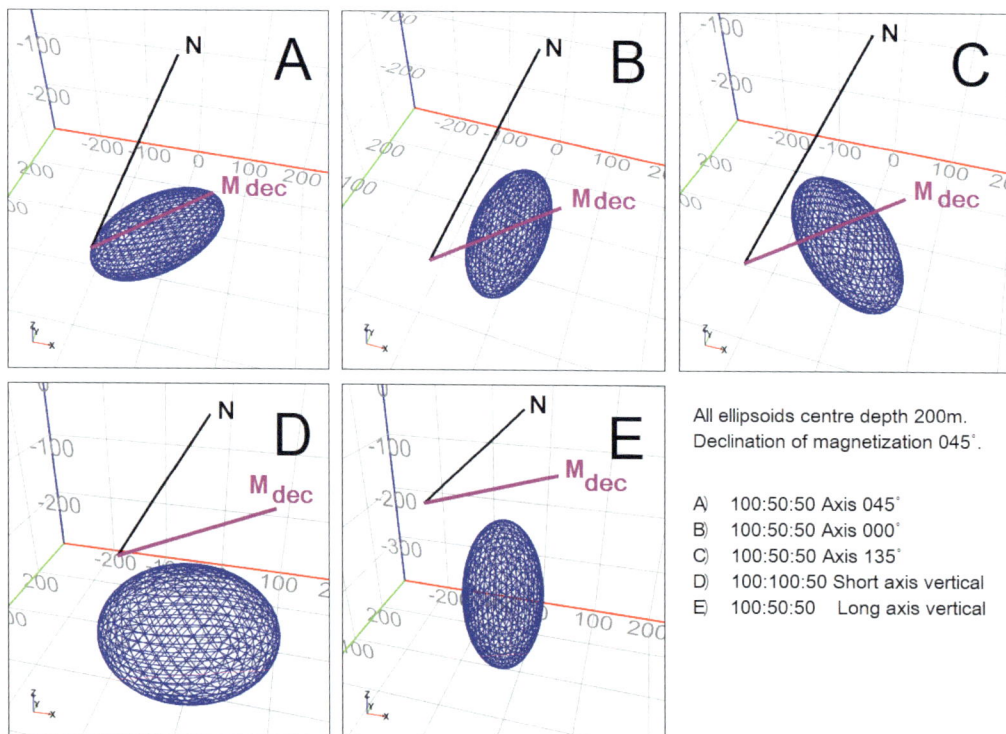

Fig. 7.4. Ellipsoids of different shape and orientation (but with horizontal and vertical axes) used to investigate the influence of shape in locating the centre of magnetisation and estimating magnetisation direction at those apparent centres.

magnetisation direction. Figure 7.5a shows the TG peak at which the tensor analysis has an error of 27°. Figure 7.5b shows a smaller displacement from the centre of magnetisation for the NSS peak at which the tensor analysis has an error of only 8° (Fig. 7.5b).

Figure 7.6 is a cross-plot of dipole component and tensor analysis errors for the various ellipsoids shown in Fig. 7.4. The symbols are shape coded to represent the various ellipsoid source models and colour coded to represent

inclination of magnetisation. Tensor analysis at the NSS peaks has slightly larger error than component analysis. For the vertical ellipsoid (shown in Fig. 7.4E), flattened horizontal ellipsoid (shown in Fig. 7.4D) and ellipsoid with long axis perpendicular to the magnetisation direction (shown in Fig. 7.4C) both analyses have errors of less than 1°. Larger errors of up to 13° are found for various combinations of moderate- or steep-inclination magnetisations and ellipsoids with long axis perpendicular to the

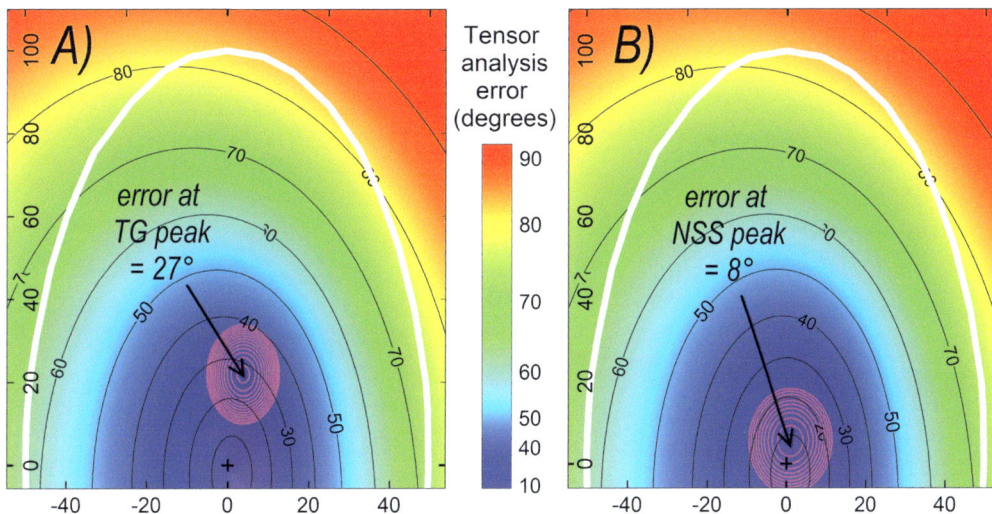

Fig. 7.5. Image of error in the tensor analysis of magnetisation direction overlaid by A) contours of the TG anomaly peak and B) contours of the NSS anomaly peak. The ellipsoid outline is shown in magenta.

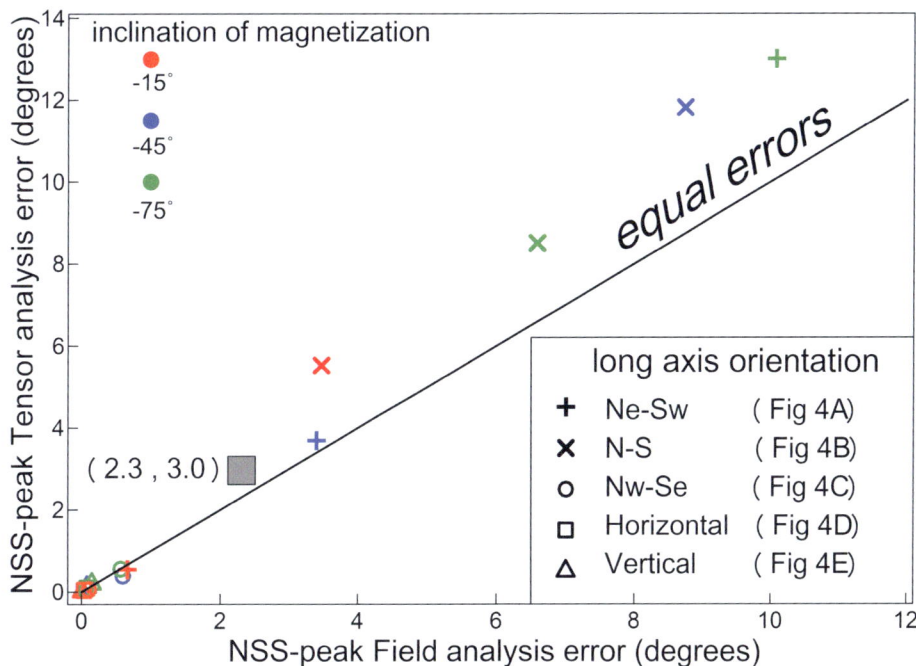

Fig. 7.6. Error in magnetisation direction at the NSS peak from tensor gradient and field component analyses for the ellipsoids show in Fig. 7.4. Note that many cases have low errors and plot in a cluster near the origin.

declination of magnetisation (shown in Fig. 7.4A) or at 45° to it (shown in Fig. 7.4B). Component and tensor analysis estimates of magnetisation direction have average errors of 2.3° and 3.0° respectively.

Figure 7.7 shows strong correlation between the displacement of the NSS peaks from the centre of magnetisation and the magnitude of error in magnetisation direction at those peaks estimated from the dipole tensor analysis. This strong correlation suggests that imperfect location of the centre of magnetisation is the dominant contribution to error in the analysis.

Figure 7.8 clearly shows the superiority of NSS over TG in estimating the centres of the various distributions of magnetisation. In all cases displacement of the TG

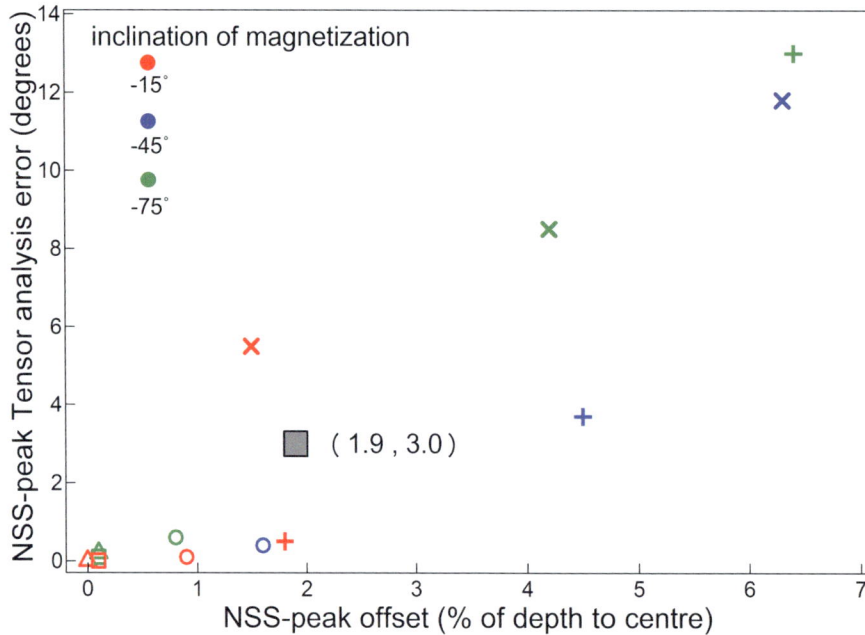

Fig. 7.7. Correlation of the offset of the NSS anomaly peak from the centre of magnetisation and the error in estimation of magnetisation direction by tensor gradient analysis at that peak. Symbols as in Fig. 7.6.

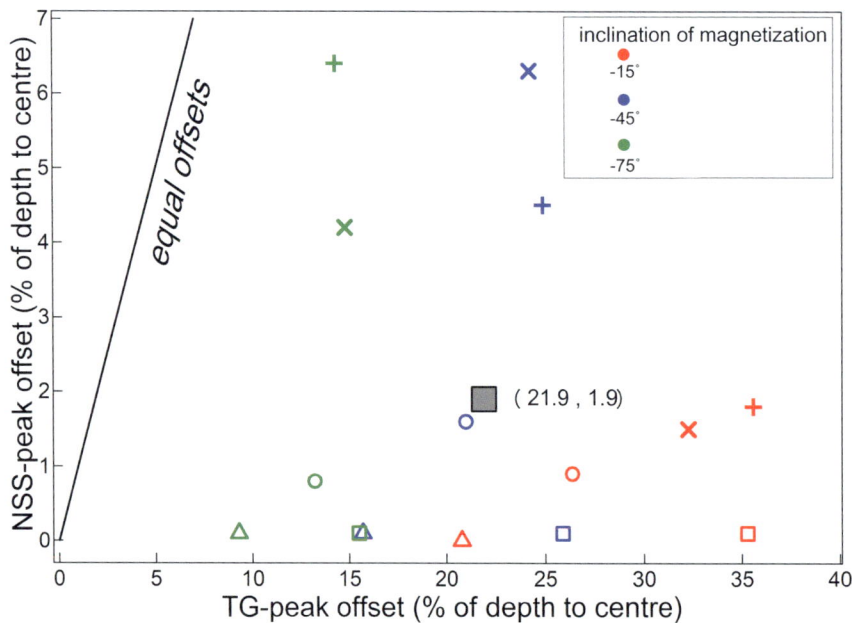

Fig. 7.8. Horizontal offset of the NSS and TG anomaly peaks from the centre of magnetisation. Symbols as in Fig. 7.6.

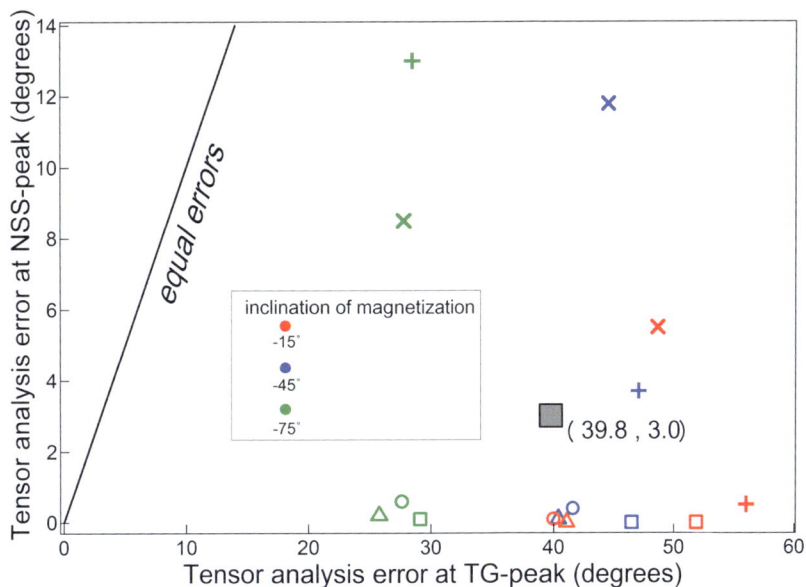

Fig. 7.9. Error in estimation of magnetisation direction by tensor gradient analysis at the NSS and TG anomaly peaks. Symbols as in Fig. 7.6.

peak is at least twice the displacement of the NSS peak. The average horizontal displacement of the NSS peak from the centre of magnetisation is 2% of depth to centre, and for the TG peak the average displacement is 20% of depth to centre. Consistent with our conclusion that error in locating the centre of magnetisation is the major source of error in the magnetisation analysis, corresponding errors in dipole tensor analysis at the NSS and TG peaks plotted in Fig. 7.9 follow a very similar pattern to offset of those peaks plotted in Fig. 7.8. The average error of the magnetisation analyses is 3° at the NSS peaks and 40° at the TG peaks.

7.3.3 The influence of plunge

For elongate thin sheet ('two-dimensional') magnetisations identical anomalies are produced by a wide range of interchangeable dip and inclination of magnetisation angles (Hood 1964). Ellipsoids best approximating segments of planar thin sheets ('plate' shapes) have a single short axis with plunge determined by the deviation of that short axis from the vertical. For ellipsoids with a single elongate axis ('cigar' shapes) plunge is determined by the deviation of the long axis from the horizontal. For a plunging source the shallower, up-plunge segment of the magnetisation produces higher amplitude, shorter wavelength contributions to the above-surface magnetic field than the deeper, down-plunge segment. This effect is most pronounced for plunging magnetisations of large

depth extent. Plunge of a body is an intrinsic parameter, but if we consider it as secondary to other shape and position parameters we can recognise that in some circumstances the error in magnetisation direction due to plunge increases error due to other parameters and in other circumstances it reduces it. Plunge is therefore a complicating parameter. Figure 7.10 shows images of dipole tensor analysis error mapped for the magnetic fields of the same north–south elongated ellipsoid investigated in Fig. 7.5 but with a 15° plunge down to the north (Fig. 7.10a) and 15° down to the south (Fig. 7.10b). In each case the point at which dipole tensor analysis provides the correct magnetisation direction is shifted from the point above the centre of the ellipsoid towards the up-plunge direction. For this geometry, plunge and magnetisation direction, the TG peaks are further from the centre of magnetisation than the NSS peaks and in consequence the dipole tensor analysis error is larger at the TG peaks than at the NSS peaks. At the NSS peaks the dipole tensor analysis error (which was 8° for the horizontal body) is 5° (a change of −3°) for the northward plunge and 12° (a change of +4°) for the southward plunge. At the TG peak the error (which was 27° for the horizontal body) is 20° (a change of −7°) for the northward plunge and 34° (a change of +7°) for the southward plunge. These values vary with body shape, orientation, plunge, magnetisation direction and elevation at which the magnetic field is measured.

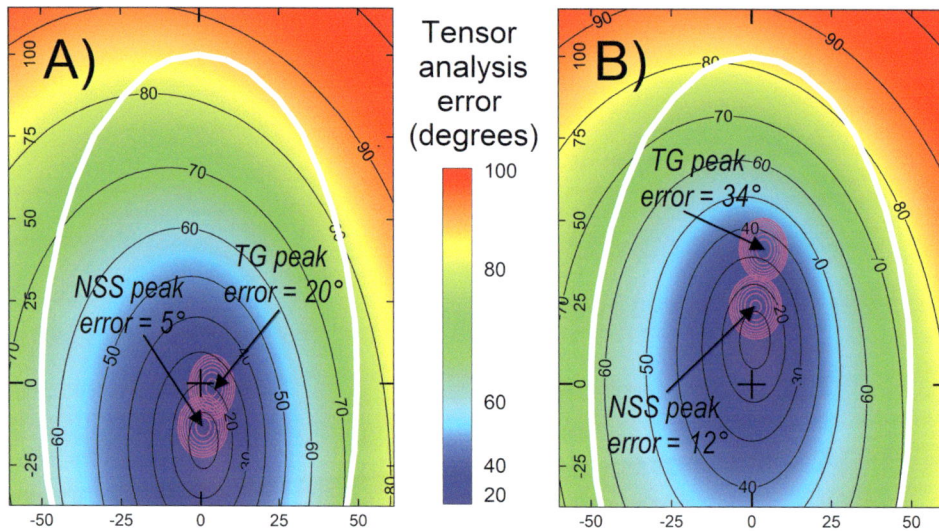

Fig. 7.10. NSS and TG peak contours over images of error in dipole tensor gradient analysis for the ellipsoid magnetisation of Fig. 7.5 with 15°plunge down a) to the north and b) to the south. The ellipsoid outline is shown in magenta.

7.3.4 The influence of elevation

The magnetic field expression of body shape and plunge varies with distance from the body, requiring the concept of 'compactness' (Clark 2014; Foss 2017; McKenzie 2020) to summarise the influence of these factors. One measure of compactness is the ratio between the longest axis of the body and distance to the closest measurement location, but no single statistic is sufficient to summarise the multi-factor variables describing the distribution of magnetisation and magnetic field measurements (see Chapter 4). At progressively greater distance the magnetic field of a body approximates closer to that of a dipole. Outside a spherical surface that entirely encloses an arbitrary magnetic source, the magnetic field can be described in terms of a multipole expansion (Jackson 2007, p. 145). Unless the dipole moment vanishes identically, the far-field is always dominated by the dipole term. For finite uniformly magnetised bodies of orthorhombic, or higher, symmetry (e.g. ellipsoids, rectangular prisms, right circular or elliptic cylinders) the quadrupole moment is zero and the lowest non-dipole term is usually the octupole contribution. As an extreme case, consider an axially magnetised long thin pipe, which behaves like a bar magnet with equal and opposite point poles at the ends. At a distance from the centre, r, that is greater than three times the length of the pipe, the non-dipole field in all directions contributes only ~5% of the dipole field, and this falls off as ~$1/r^2$ at greater distances. For sources that are more equidimensional, the dipole approximation is acceptable at shorter

ranges than for an elongated source. For highly symmetric sources (cube, octahedron, cylinder with height/diameter ~0.9, etc.) the octupole term also vanishes and the source is well represented by a point dipole at relatively short ranges. For complex shapes such as pipes with horizontal top faces it may be convenient to separately consider a 'near-field' close to the source and a 'far-field' at greater distances, but for the ellipsoids which we have used in this study the influence of distance from source is more smoothly continuous. To investigate the influence of elevation at which the magnetic field is measured we computed the fields of a dipole and the same ellipsoid used in the study of plunge. NSS peaks directly above the dipole at all elevations and (as already proven) dipole tensor analysis at that peak exactly recovers magnetisation direction. For the dipole the TG peak has a consistent horizontal displacement from the magnetisation centre of 12% of its depth (Fig. 7.11a) and error of the dipole tensor analysis at that peak is consistently 27° (Fig. 7.11b). These values vary for different magnetisation directions and different geomagnetic field inclinations. For the north and south plunging ellipsoids NSS and TG peaks have almost converged to those of the otherwise identical horizontal ellipsoid at a depth of 400 m (twice the length of the longest axis). At this depth the dipole tensor analysis error for all three ellipsoids at their NSS peak is 8° and this reduces to 4° at a depth of 800 m. The dipole tensor analysis error for all three ellipsoids at their TG peak is ~27° across this depth range (the same as for the dipole).

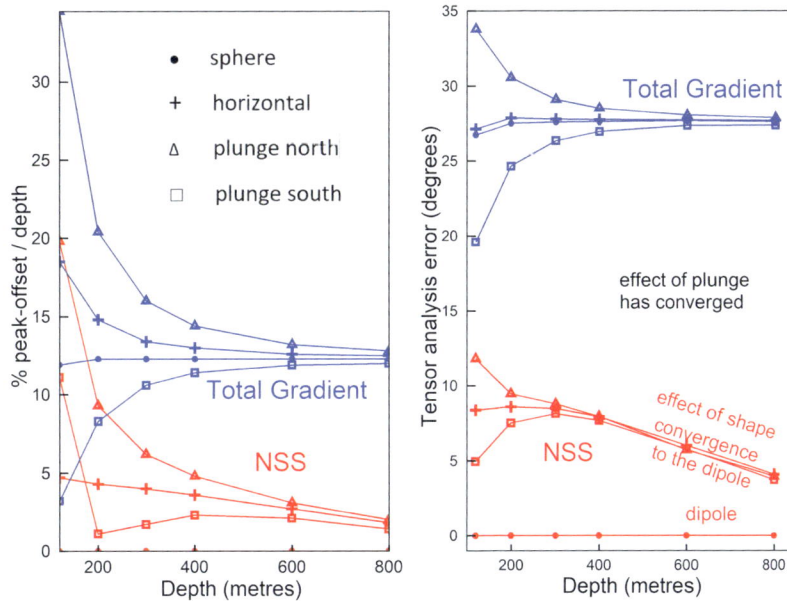

Fig. 7.11. a) NSS and TG peak offsets from the centre of magnetisation and b) error in dipole tensor magnetisation analysis at the NSS and TG peaks for a dipole and the ellipsoids of Figs 7.5 and 7.10.

An upward continuation of 100 to 200 m of the fields due to magnetisations at 120 m depth would reduce displacements of the NSS peak for the plunging ellipsoids with corresponding improvement of magnetisation direction estimates at those peaks. However, more substantial upward continuation is likely to achieve only slight further improvements due to substantial reduction in amplitude of the fields and gradients for these compact sources.

7.4 CASE STUDY

Figure 7.12a shows a measured TMI anomaly at Ethabuka, central Queensland, Australia from a survey with 400 m spaced east–west flightlines flown at a ground clearance of 80 m. There are no overlapping adjacent anomalies but there is some ambiguity in separating the anomaly from the background field. The east-south-east trough to peak orientation of this southern hemisphere anomaly indicates the declination of magnetisation, and the similar amplitudes of the peak and trough in this moderately steep geomagnetic field (inclination −65°) indicates a low inclination of magnetisation. Figure 7.12b shows an image of model-computed TMI for a parametric inversion of a steeply plunging elliptic cylinder (outline shown in Fig. 7.12b and 7.12c) with a top 740 m below sensor. The inversion simultaneously resolves the spatial and magnetisation properties of the model. The best estimated

magnetisation direction is declination 100.8°, inclination +18.2°, consistent with expectations from visual analysis of the anomaly. The close match between the computed field in Fig. 7.12b and the measured field in Fig. 7.12a does not prove the inversion model correct but does justify it as at least an equivalent source suitable for use in data transforms. One such transform is an RTP using the recovered magnetisation direction, the output of which is shown in Fig. 7.12c. The compact positive-only RTP anomaly strongly supports the estimated magnetisation direction. The box in Fig. 7.12c is the area shown in greater detail in Fig. 7.13.

Figures 7.13a and 7.13b show TG and NSS peak contours over an image of the difference between magnetisation direction estimated by inversion and that computed at each grid cell using the tensor analysis algorithm. Tensor analysis at the NSS peak recovers an estimated magnetisation direction (completely independent of the inversion) only 5° different to the inversion estimate, and the NSS contours are closely co-centred with the RTP contours of Fig. 7.12c. The difference in estimated magnetisation direction between tensor analysis at the TG peak and the inversion result is 38°. Estimation of the location of magnetisation from the inversion model is also more compatible with the NSS peak than the TG peak. These relationships are further supported by almost identical results obtained by applying TG and NSS transforms to the model

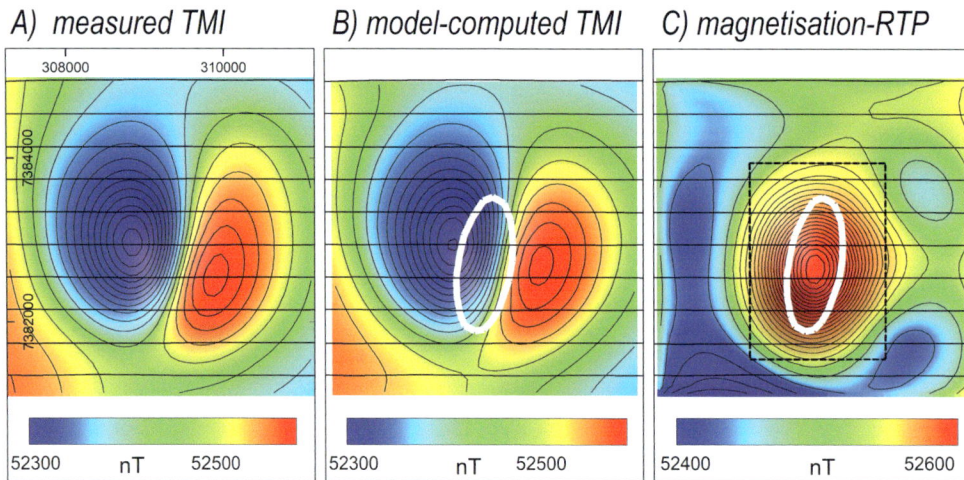

Fig. 7.12. The Ethabuka anomaly: a) measured TMI, b) model computed TMI with outline of the top of the model, and c) reduced to pole (RTP) TMI using the inversion model magnetisation direction.

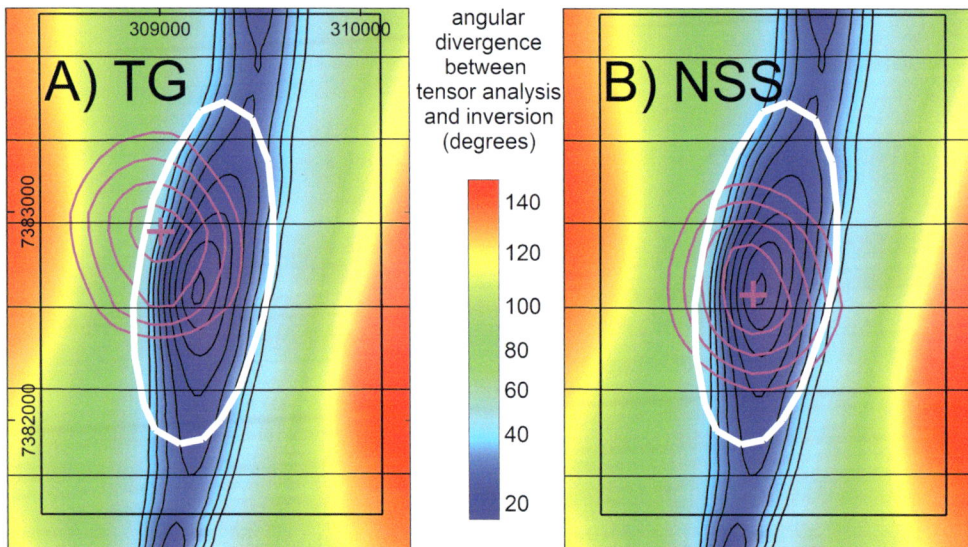

Fig. 7.13. Image and contour map (black with 5° contour interval) of the difference in magnetisation direction from inversion and tensor analysis with overlays (black lines) of a) TG peak contours and b) NSS peak contours. The outline of the inversion model is shown.

computed field of Fig. 7.12b for which the (virtual) magnetisation is of known direction and location.

7.5 CONCLUSIONS

We have shown that the direction of magnetisation of a dipole source can be derived from ratios of its field components or tensor gradient elements directly above its centre. The NSS transform peaks directly over a dipole and indicates where that analysis can be performed. The ratios of field components and of tensor gradients progressively diverge from their ideal values away from the centre axis of a dipole, and they also diverge for

magnetisations which differ from a dipole. Estimates of magnetisation direction are particularly influenced by plunge. With increasing elevation the magnetic field of a compact source progressively converges towards that of an equivalent dipole, but any advantage of upward continuation to improve magnetisation estimates is restricted by attenuation of the magnetic field and gradients. We present a case study for which the NSS peak closely conforms with the centre of magnetisation derived by parametric inversion, and for which the magnetisation direction from dipole tensor gradient analysis at that peak agrees to within 5° of the inversion model. We only recommend dipole analysis for initial magnetisation direction estimates. If magnetisation directions are required to higher precision they should be obtained by inversion, with a key advantage that inversion can be performed directly on the primary line data, avoiding any problems arising from gridding of the data.

REFERENCES

Beiki M, Clark DA, Austin JR, Foss CA (2012) Estimating source location using normalized magnetic source strength calculated from magnetic gradient tensor data. *Geophysics* **77**, J23–J37. doi:10.1190/geo2011-0437.1

Blakely RJ (1995) 'Potential theory in gravity and magnetic applications'. Cambridge University Press.

Caratori Tontini F, Pedersen LB (2008) 'Interpreting magnetic data by integral moments. *Geophysical Journal International* **174**, 815–824. doi:10.1111/j.1365-246X.2008.03872.x

Clark DA (2012) New methods for interpretation of magnetic vector and gradient tensor data I: eigenvector analysis and the normalised source strength. *Exploration Geophysics* **43**, 267–282. doi:10.1071/EG12020

Clark DA (2013) New methods for interpretation of magnetic vector and gradient tensor data II: application to the Mount Leyshon Anomaly, Queensland. *Exploration Geophysics* **44**, 114–127. doi:10.1071/EG12066

Clark DA (2014) Methods for determining remanent and total magnetisations of magnetic sources – a review. *Exploration Geophysics* **45**, 271–304. doi:10.1071/EG14013

Clark DA, Saul SJ, Emerson DW (1986) Magnetic and gravity anomalies of a triaxial ellipsoid. *Exploration Geophysics* **17**, 189–200. doi:10.1071/EG986189

Dannemiller N, Li Y (2006) A new method for determination of magnetization direction. *Geophysics* **71**, L69–L73.

Fedi M, Florio G, Rapolla A (1994) A method to estimate the total magnetization direction from a distortion analysis of magnetic anomalies. *Geophysical Prospecting* **42**, 261–274. doi:10.1111/j.1365-2478.1994.tb00209.x

Foss CA (2006) 'Evaluation of strategies to manage remanent magnetization effects in magnetic field inversion.' 76th Annual International Meeting, SEG, Expanded Abstracts, 938–942.

Foss CA (2017) 'Resultant-magnetisation based magnetic field interpretation' in V. Tschirhart and M.D. Thomas eds, Proceedings of Exploration 17: Sixth Decennial International Conference on Mineral Exploration, pp.637–648.

Foss CA, McKenzie KB (2011) Inversion of anomalies due to remanent magnetisation: an example from the Black Hill Norite of South Australia. *Australian Journal of Earth Sciences* **58**, 391–405. doi:10.1080/08120099.2011.581310

Fullagar PK, Pears GA (2015) 'Remanent magnetisation inversion.' 24th ASEG International Geophysical Conference, Extended Abstracts. doi:10.1071/ ASEG2015ab188

Helbig K (1963) Some integrals of magnetic anomalies and their relation to the parameters of the disturbing body. *Zeitschrift für Geophysik* **29**, 83–96.

Hood P (1964) The Königsberger ratio and the dipping dyke equation. *Geophysical Prospecting* **12**, 440–456. doi:10.1111/j.1365-2478.1964.tb01916.x

Jackson JD (2007) 'Classical electrodynamics.' John Wiley & Sons.

Lelièvre PG, Oldenburg DW (2009) A 3D total magnetisation inversion applicable when significant, complicated remanence is present. *Geophysics* **74**, L21–L30. doi:10.1190/1.3103249

Li Y (2012) Recent advances in 3D generalized inversion of potential-field data. *SEG Technical Program Expanded Abstracts* **2012**, 1–7.

Lourenço JS, Morrison HF (1973) Vector magnetic anomalies derived from measurements of a single component of the field. *Geophysics* **38**, 359–368. doi:10.1190/1.1440346

McKenzie KB (2020) The magnetic gradient tensor of a triaxial ellipsoid, its derivation and its application to the determination of magnetisation direction. *Exploration Geophysics* **51**, 609–641. doi:10.1080/08123985.2020.1726176

McKenzie KB, Foss CA, Hillan D (2012) 'An improved search for magnetisation direction.' 22nd ASEG Geophysical Conference, Extended Abstracts, 1–4.

Paine J, Haederle M, Flis M (2001) Using transformed TMI data to invert for remanently magnetised bodies. *Exploration Geophysics* **32**, 238–242. doi:10.1071/EG01238

Phillips JD (2005) Can we estimate total magnetisation directions from aeromagnetic data using Helbig's integrals? *Earth, Planets, and Space* **57**, 681–689. doi:10.1186/BF03351848

Phillips JD, Nabighian MN, Smith DV, Li Y (2007) Estimating locations and total magnetisation vectors of compact magnetic sources through combined Helbig and Euler analysis. *SEG Technical Program Expanded Abstracts* **26**, 770–774.

Pratt DA, McKenzie KB, White AS (2014) Remote remanence determination (RRE). *Exploration Geophysics* **45**, 314–323. doi:10.1071/EG14031

Purucker ME (1990) The computation of vector magnetic anomalies: A comparison of techniques and errors. *Physics of the Earth and Planetary Interiors* **62**, 231–245. doi:10.1016/0031-9201(90)90168-W

Roest WR, Verhoef J, Pilkington M (1992) Magnetic interpretation using the 3-D analytic signal. *Geophysics* **57**, 116–125. doi:10.1190/1.1443174

Schmidt PW, Clark DA (1998) The calculation of magnetic components and moments from TMI: a case study from the Tuckers igneous complex, Queensland. *Exploration Geophysics* **29**, 609–614. doi:10.1071/EG998609

Wilson HS (1985) 'Analysis of the magnetic gradient tensor.' Defence Research Establishment Pacific, Technical Memorandum **8**, 5–13.

Wynn WN, Frahm CP, Carroll PJ, Clark RH, Wellhoner J, Wynn MJ (1975) Advanced superconducting gradiometer/magnetometer arrays and a novel signal processing technique. *IEEE Transactions on Magnetics* **11**, 701–707. doi:10.1109/TMAG.1975.1058672

8

Estimation of magnetisation direction by magnetic component symmetry analysis

C.A. Foss, K.B. McKenzie and D.A. Clark

ABSTRACT

We present a novel analysis of magnetic field data to provide estimates of magnetisation direction. Starting with measured total magnetic intensity (TMI) data the analysis requires a wavenumber domain transform to produce a dataset of three orthogonal field components. Each of those three field components can in turn be resolved into three terms of different symmetry which in combination characterise specific components of magnetisation. Relative strengths of the magnetisation components (and thereby magnetisation direction) can be derived from strengths of the isolated field component terms. Analysis is ideally performed on bodies of circular or fourfold symmetry about a vertical axis but is tolerant of moderate departures from this shape assumption (a ratio of 2 in maximum to minimum horizontal extent is typically associated with an error of less than 10° in recovered magnetisation direction). Prior removal of an anomaly from its background field by regional-residual separation is recommended if variation in the background field is significant compared to the anomaly amplitude. We present analyses successfully applied to forward computed fields of complex but symmetric bodies of defined magnetisation and to the Black Hill Norite anomaly in South Australia for which we have palaeomagnetic measurements and alternative magnetic field analyses and inversions.

8.1 INTRODUCTION

Subsurface magnetisations causing local variations in the Earth's magnetic field are generally of unknown direction and this complicates their interpretation and mapping of their distribution. Many methods are used to resolve magnetisation direction from analysis of magnetic field data as reviewed by Clark (2014). The theoretical basis that it is possible to recover valid estimates of magnetisation direction directly from magnetic field data is provided by Helbig (1963) who established that the magnetic moment of a body of magnetisation can be determined from the moment integrals of the B_n, B_e, B_d magnetic field components about a point directly above the centre of magnetisation. The approximate centre of a compact magnetisation can be determined by application of transforms to the total gradient or normalised source strength (Beiki *et al.* 2012). Helbig analysis itself can be adapted to provide magnetisation estimates (Schmidt and Clark 1998; Phillips 2005; Phillips *et al.* 2007; Caratori Tontini and Pedersen 2008; Clark 2014) but the method is highly sensitive to separation of the anomaly from the background field. Approximate relationships based on derivatives of the field components (Phillips *et al.* 2007) reduce this influence of the background field. Other methods to estimate magnetisation direction from magnetic field data include those by Fedi *et al.* (1994), Dannemiller and Li (2006) and McKenzie *et al.* (2012). Magnetic field inversion is a computationally intensive process of estimating magnetisation direction that provides the highest discrimination in resolving magnetisation direction as long as appropriate attention is paid to isolating anomalies of interest from their background fields and expert guidance is applied throughout the inversion process. Both parametric inversion (Foss 2006, 2017; Foss and McKenzie 2011; Pratt *et al.* 2014) and voxel inversion (Paine *et al.* 2001; Lelièvre and Oldenburg 2009; Li 2012; Fullagar and Pears 2015) have been applied to estimation of magnetisation direction from magnetic field data.

Our analysis to determine magnetisation direction shares several key features with Helbig's. Both methods require location of the horizontal centre of magnetisation, both utilise splitting of the field into three orthogonal components, and both recover estimates of the relative strengths of the magnetisation components from integral measures of the strength of magnetic field components. Where the Helbig method uses integral moments of the field components we use the grid standard deviations. Our method also differs from Helbig's in splitting field components into contributions from different magnetisation components according to their symmetry characteristics.

8.2 EQUIVALENCE OF THE GRAVITY GRADIENT TENSOR AND THE MAGNETIC FIELD – MAGNETISATION MATRIX

8.2.1 The gravity gradient tensor for a point mass

The gravity gradient tensor $\mathbf{G(r)}$ at an observation point $\mathbf{r} = (x, y, z)^T$ due to a point mass M positioned at the origin is (see for example Pedersen and Rasmussen 1990; Dransfield 1994):

$$\mathbf{G(r)} = \gamma M \begin{pmatrix} \dfrac{3x^2 - r^2}{r^5} & \dfrac{3xy}{r^5} & \dfrac{3xz}{r^5} \\ \dfrac{3xy}{r^5} & \dfrac{3y^2 - r^2}{r^5} & \dfrac{3yz}{r^5} \\ \dfrac{3xz}{r^5} & \dfrac{3yz}{r^5} & \dfrac{3z^2 - r^2}{r^5} \end{pmatrix} = \gamma M \, \mathbf{T(r)}$$

(Eqn 8.1)

where γ is the universal gravitational constant and $\mathbf{T(r)}$ is a symmetric matrix of Green's functions. Furthermore, when the x, y, z Cartesian axes are aligned in the north, east and vertically downward directions then $\mathbf{T(r)}$ is

$$\mathbf{T(r)} = \begin{pmatrix} T_{nn} & T_{ne} & T_{nd} \\ T_{en} & T_{ee} & T_{ed} \\ T_{dn} & T_{de} & T_{dd} \end{pmatrix} = \begin{pmatrix} \dfrac{3x^2 - r^2}{r^5} & \dfrac{3xy}{r^5} & \dfrac{3xz}{r^5} \\ \dfrac{3xy}{r^5} & \dfrac{3y^2 - r^2}{r^5} & \dfrac{3yz}{r^5} \\ \dfrac{3xz}{r^5} & \dfrac{3yz}{r^5} & \dfrac{3z^2 - r^2}{r^5} \end{pmatrix}$$

(Eqn 8.2)

8.2.2 The 3 × 3 magnetic field–magnetisation matrix for a dipole

The expression for the magnetic field vector $\mathbf{B(r)}$ at an observation point $\mathbf{r} = (x, y, z)^T$ due to a point dipole source possessing a magnetic moment $\mathbf{m}$ located at the origin is given by Blakely (1995, p. 75) as:

$$\mathbf{B(r)} = \frac{C_m}{r^3} \{ 3(\mathbf{m} \cdot \hat{\mathbf{r}}) \hat{\mathbf{r}} - \mathbf{m} \}, \quad \text{(Eqn 8.3)}$$

where $\hat{\mathbf{r}}$ is the unit vector along the line joining the dipole to the observation point, i.e.

$$\hat{\mathbf{r}} = \frac{\mathbf{r}}{|\mathbf{r}|} = \left(\frac{x}{r}, \frac{y}{r}, \frac{z}{r} \right)^T ; r = |\mathbf{r}| = \sqrt{x^2 + y^2 + z^2} \text{ and } r \neq 0.$$

and where C_m is a constant which depends on the system of electromagnetic units used (see Blakely 1995, pp. 67–68). In the International Standard (SI) system of units, C_m = 100 nH/m or 100 nTm/A for magnetic fields expressed in nanotesla (nT), magnetic moments expressed in ampere metre squared (Am^2); and magnetisations expressed in ampere per meter (A/m). The coordinate system adopted here follows the International Geomagnetic Reference Field convention, i.e. x is north, y is east and z is vertically down. The Cartesian components of the magnetic field vector are

$$B_x(\mathbf{r}) = \frac{C_m}{r^3} \left\{ 3(\mathbf{m} \cdot \hat{\mathbf{r}}) \frac{x}{r} - m_x \right\}$$
$$= \frac{C_m}{r^3} \left\{ 3 \left(\frac{m_x x^2}{r^2} + \frac{m_y xy}{r^2} + \frac{m_z xz}{r^2} \right) - m_x \right\}. \text{ (Eqn 8.4.1)}$$

$$B_y(\mathbf{r}) = \frac{C_m}{r^3} \left\{ 3(\mathbf{m} \cdot \hat{\mathbf{r}}) \frac{y}{r} - m_y \right\}$$
$$= \frac{C_m}{r^3} \left\{ 3 \left(\frac{m_x xy}{r^2} + \frac{m_y y^2}{r^2} + \frac{m_z yz}{r^2} \right) - m_y \right\}. \text{ (Eqn 8.4.2)}$$

$$B_z(\mathbf{r}) = \frac{C_m}{r^3} \left\{ 3(\mathbf{m} \cdot \hat{\mathbf{r}}) \frac{z}{r} - m_z \right\}$$
$$= \frac{C_m}{r^3} \left\{ 3 \left(\frac{m_x xz}{r^2} + \frac{m_y yz}{r^2} + \frac{m_z z^2}{r^2} \right) - m_z \right\}. \text{ (Eqn 8.4.3)}$$

For a series of magnetic dipole moments **m** directed along the north, east and downward vertical directions, i.e. $\mathbf{m}_x = m\hat{\mathbf{x}}$; $\mathbf{m}_y = m\hat{\mathbf{y}}$; $\mathbf{m}_z = m\hat{\mathbf{z}}$, or alternatively for a series of magnetisations $\mathbf{J}_n = J\hat{\mathbf{x}}$, $\mathbf{J}_e = J\hat{\mathbf{y}}$, $\mathbf{J}_d = J\hat{\mathbf{z}}$ for a uniformly magnetised sphere of volume v and magnetisation intensity $J = m/v$, the expressions for the north, east and downward vertical components of the magnetic field due to each of the three orthogonal magnetisations may be formulated as a matrix equation as follows:

$$\mathbf{B}_m = \begin{pmatrix} B_n^{Jn} & B_e^{Jn} & B_d^{Jn} \\ B_n^{Je} & B_e^{Je} & B_d^{Je} \\ B_n^{Jd} & B_e^{Jd} & B_d^{Jd} \end{pmatrix}$$

$$= C_m m \begin{pmatrix} \dfrac{3x^2 - r^2}{r^5} & \dfrac{3xy}{r^5} & \dfrac{3xz}{r^5} \\ \dfrac{3xy}{r^5} & \dfrac{3y^2 - r^2}{r^5} & \dfrac{3yz}{r^5} \\ \dfrac{3xz}{r^5} & \dfrac{3yz}{r^5} & \dfrac{3z^2 - r^2}{r^5} \end{pmatrix} = C_m m \mathbf{T}(\mathbf{r}),$$

(Eqn 8.5)

where $\mathbf{T}(\mathbf{r})$ is the (3×3) symmetric matrix of Green's functions specified in Eqn 8.2. The matrix $\mathbf{B}_m$ is known as the magnetic field-magnetisation matrix. Its three rows are the magnetic field components B_n, B_e and B_d for each of the $\mathbf{J}_n$, $\mathbf{J}_e$ and $\mathbf{J}_d$ magnetisations. Importantly, since $\mathbf{T}(\mathbf{r})$ is related to the gravity gradient tensor of a point mass (or uniform sphere with density ρ and mass $M = \rho v$) by the relation $\mathbf{G}(\mathbf{r}) = \gamma M \mathbf{T}(\mathbf{r})$, then the magnetic field-magnetisation matrix $\mathbf{B}_m$ may be expressed in terms of the gravity gradient tensor $\mathbf{G}(\mathbf{r})$ namely,

$$\mathbf{B}_m = \begin{pmatrix} B_n^{Jn} & B_e^{Jn} & B_d^{Jn} \\ B_n^{Je} & B_e^{Je} & B_d^{Je} \\ B_n^{Jd} & B_e^{Jd} & B_d^{Jd} \end{pmatrix} = C_m m \mathbf{T}(\mathbf{r})$$

$$= \left(\frac{C_m m}{\gamma M} \right) \mathbf{G}(\mathbf{r}) = \left(\frac{C_m J}{\gamma \rho} \right) \mathbf{G}(\mathbf{r}). \quad \text{(Eqn 8.6)}$$

The spatial patterns of the gradient tensor elements in Eqn 8.1 and the magnetic components in Eqn 8.5 are identical, as is the symmetry and anti-symmetry of their individual terms about the north and east axes (as is evident from considering the consequence of change of sign in the x and y coordinates respectively). The map representations of these functions are shown in Fig 8.1A and 8.1B.

8.2.3 An expression for magnetisation direction

The direction of magnetisation (or magnetic moment) for a uniformly magnetised body with magnetisation **J** is defined by a declination D_J or azimuthal angle measured positive clockwise from true north to the horizontal projection of the magnetisation vector **J** and an inclination angle I_J which is the angle between the magnetisation vector **J** and its horizontal projection measured positive downwards. An expression for the declination of magnetisation (or of magnetic moment) at any point $\mathbf{r} = (0, 0, -z)^T$ above the point dipole or centre of a magnetised sphere is obtained from the ratio of the horizontal magnetic field components as shown in Chapter 7.

$$D_J = \arctan\left(\frac{J_e}{J_n} \right) = \arctan\left(\frac{-B_e}{-B_n} \right) \quad \text{for } 0 \le D_J \le 2\pi.$$

(Eqn 8.7)

An expression for the inclination of magnetisation I_J is obtained from the ratio of the horizontal field component $B_h = \sqrt{B_n^2 + B_e^2}$ to half the vertical field component B_d as shown in Foss *et al.* (2021)

$$I_J = \arctan\left(\frac{J_d}{J_h} \right) = \arctan\left(\frac{B_d}{2\sqrt{B_n^2 + B_e^2}} \right) \quad \text{for } -\frac{\pi}{2} \le I_J \le \frac{\pi}{2}.$$

(Eqn 8.8)

8.2.4 Poisson's relationship between gravity and magnetic fields

Poisson's relation states that if the boundaries of a gravitational and magnetic source are identical and if the density and magnetisation are uniform within those sources, then the magnetic scalar potential is proportional to the component of gravitational attraction in the direction of magnetisation (Blakely 1995). An expression for the magnetic scalar potential V(**r**) of a uniform body with volume v, density ρ and magnetisation **J** may be derived from the Poisson's relation as follows (Blakely 1995, pp. 91–93):

$$V(\mathbf{r}) = -C_m \left(\frac{1}{\gamma M} \right) \nabla U(\mathbf{r}) \cdot \mathbf{m} = -\left(\frac{C_m}{\gamma \rho} \right) \mathbf{g}^T(\mathbf{r}) \cdot \mathbf{J}, \quad \text{(Eqn 8.9)}$$

where $\mathbf{g}^T(\mathbf{r})$ denotes the transpose of the gravity field vector. For a magnetic dipole with moment **m** or a magnetic sphere with uniform magnetisation **J**, density ρ and equivalent mass M, its magnetic scalar potential V(**r**) is

$$V(\mathbf{r}) = \frac{C_m}{r^3} (m_x x + m_y y + m_z z) \quad \text{(Eqn 8.10)}$$

Expressions for the anomalous magnetic field vector are derived by taking the gradient of the magnetic scalar potential in Eqn 8.9, then

$$\mathbf{B}(\mathbf{r}) = -\nabla V(\mathbf{r}_\mathrm{p}) = \left(\frac{C_\mathrm{m}}{\gamma\rho}\right)\nabla \mathbf{g}^T(\mathbf{r})\cdot\mathbf{J}$$

$$= \left(\frac{C_\mathrm{m}}{\gamma\rho}\right)\mathbf{G}^T(\mathbf{r})\cdot\mathbf{J} = C_\mathrm{m}\,\mathbf{T}^T(\mathbf{r})\cdot\mathbf{J}, \qquad \text{(Eqn 8.11)}$$

where $\mathbf{T}^T(\mathbf{r}_\mathrm{p})$ is the transpose of matrix $\mathbf{T}(\mathbf{r}_\mathrm{p})$ at an external measurement point.

For a northerly magnetisation $\mathbf{J} = J\hat{\mathbf{x}}$ and from the expressions for $\mathbf{T}(\mathbf{r}_\mathrm{p})$ in Eqn 8.5, the components of the magnetic field vector obtained from Eqn 8.11 agree with the expressions for $B_n^{Jn}, B_e^{Jn}, B_d^{Jn}$ in the first row of matrix Eqn 8.6. Similar agreements with $B_n^{Je}, B_e^{Je}, B_d^{Je}$ in the second row of matrix Eqn 8.6 are obtained from Eqn 8.11 for

an easterly magnetisation $\mathbf{J} = J\hat{\mathbf{y}}$ and also with $B_n^{Jd}, B_e^{Jd}, B_d^{Jd}$ in the third row of matrix Eqn 8.6 for a downward vertical magnetisation $\mathbf{J} = J\hat{\mathbf{z}}$. These results are in agreement Pedersen and Bastani (2016). In conclusion the results shown in Sections 2.1 and 2.2 are consistent with and proven by an application of Poisson's relation.

8.3 MAGNETIC COMPONENT SYMMETRY (MCS) ANALYSIS

Grids of total magnetic intensity (TMI) can be resolved into grids of three orthogonal components J_e (east), J_n (north) and J_d (down) (Lourenço and Morrison 1973). We have established that for a dipole the 3×3 matrix of magnetic field components due to orthogonal magnetisation components of equal strength (magnetisation direction of

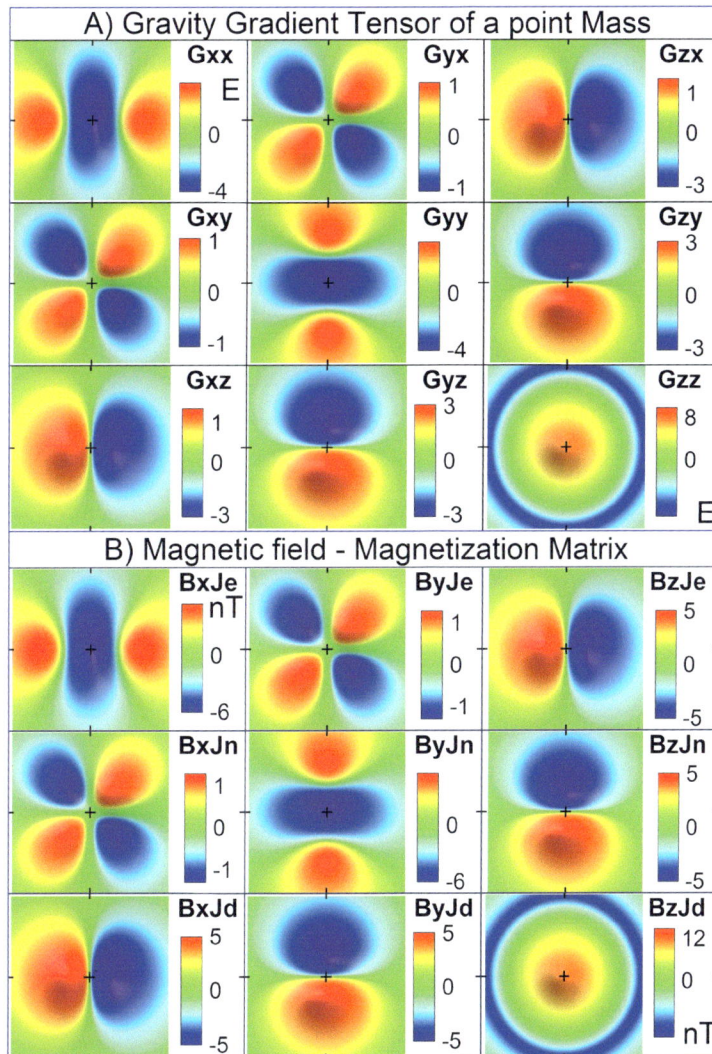

Fig. 8.1. A) Gravity gradient tensor of a point mass, and B) matrix of magnetic field components M_iJ_k where i = x,y,z and k = e,n,d are indices of the field and magnetisation respectively.

declination 45°, inclination +35.3°) is equivalent to the gravity gradient tensor of a point mass, with one-to-one equivalence between individual elements. The individual terms of the gravity gradient tensor and the magnetic field–magnetisation matrix are characterised by their symmetry patterns in the horizontal plane, with individual elements belonging to one of three symmetry classes: symmetric and antisymmetric about two orthogonal planes, antisymmetric about two orthogonal planes, or symmetric about two orthogonal planes (including full circular symmetry) as expressed in the individual terms of Eqns (8.1) and (8.5). The vertical magnetic field component B_d comprises two contributions with single axis anti-symmetry (B_d^{Jn} and B_d^{Je}) and a third contribution (B_d^{Jd}) with full circular symmetry. The two horizontal components: B_n and B_e are comprised of one contribution of single axis anti-symmetry (B_n^{Jd} and B_e^{Jd} respectively), one contribution of two axes symmetry (B_n^{Jn} and B_e^{Je} respectively), and one contribution of two axes anti-symmetry (B_n^{Je} and B_e^{Jn} respectively). In all three cases the magnetic field components can be uniquely deconstructed into these three contributions of different symmetry and the polarities and relative strengths of those separate field contributions reveal the polarities and relative strengths of the causative magnetisation components.

Figure 8.2 shows a selection of magnetisations suitable for MCS analysis, each with its central vertical axis: A) a dipole (magnetisation component field contributions shown in Fig. 8.1B), B) ellipsoids with two equal horizontal axes, C) circular-section vertical cylinders, D) a square ring (TMI anomaly shown in Fig. 8.4B), E) a NE-SW and SE-NW cross (TMI anomaly shown in Fig. 8.4A), and F) multiple horizontal symmetry bodies on a common vertical axis (TMI anomaly shown in Fig. 8.7A). The ellipsoids and cylinders have horizontal diameter to depth extent ratios of 1:5 and 5:1 to detect any sensitivity to depth extent not tested in the dipole analysis.

MCS analysis is performed by transforming a measured TMI anomaly into its orthogonal field components and then resolving those field components into contributions from magnetisation components assigned according to their symmetry patterns. The analysis is performed by estimating the centre of magnetisation (usually by applying an NSS or total gradient transform to highlight the anomaly centre), flipping the grids about east–west and north–south axes through this point, and variously summing or differencing the original and flipped grids. Comparison of the standard deviations of grids with different symmetry pattern requires

correction that is dependent on the horizontal extent of the grids. A convenient methodology to avoid this problem is to operate on pairs of component grids with identical symmetry pattern. The sequence of operations is presented in Table 8.1 leading to estimation of declination and inclination of magnetisation from ratios of the standard deviations of the grids (steps 15 and 16). Summing flipped grids tends to cancel regional field differences of that orientation extending across the area of analysis, but differencing the flipped grids retains any parallel regional field variation in the original ratio to the differenced anomaly component. A regional-residual separation should be performed before the MCS analysis for any grids superimposed on a regional field gradient significant in relation to the anomaly amplitude.

Comparison of magnetic field components across a horizontal plane best suits investigation of the horizontal components of magnetisation and the declination of magnetisation. We derive three declination estimates from grid standard deviation ratios of: (i) horizontal field components due to co-parallel magnetisations B_e^{Je} and B_n^{Jn}, (ii) horizontal field components due to cross-horizontal magnetisations B_e^{Jn} and B_n^{Je}, and (iii) vertical field components due to horizontal magnetisations B_d^{Je} and B_d^{Jn}. Note that these three estimates are by no means independent but each is influenced differently by imperfection in the analysis and the mean of the three estimates is more robust than their individual values. The declination of magnetisation is derived from the arctangent of the ratio of the various grid standard deviations as listed in step 15 of Table 8.1.

Estimation of the inclination of magnetisation is more complex and requires multiple grid operations. The vertical component of magnetisation related to inclination angle has a different pattern of influence on the magnetic field measured across a horizontal surface than do the

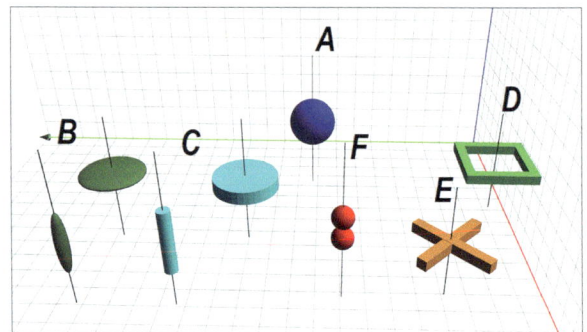

Fig. 8.2. Selection of magnetisation bodies suitable for MCS analysis.

Table 8.1. Sequential processing steps of MCS analysis.

Step	Operation
Initial processing	
1	Regional-residual anomaly separation (recommended)
2	FFT phase transform TMI to B_n, B_e, B_d
3	Centre the anomaly and clip B_n, B_e, B_d to square grids
Grid flip + compute	
4	Flip B_d E-W, $B_d^{Je} = (B_d - B_d$ E-W flip)/2, fd = $(B_d + B_d$ E-W flip)/2
5	Flip fd N-S, B_d^{Jd} = (fd + fd N-S flip)/2
6	Flip Bd N-S, $B_d^{Jn} = (B_d - B_d$ N-S flip)/2
7	Flip B_e E-W, $B_e^{Je} = (B_e + B_e$ E-W flip)/2, fe = $(B_e - B_e$ E-W flip)/2
8	Flip B_n N-S, $B_n^{Jn} = (B_n + B_n$ N-S flip)/2, fn = $(B_n - B_n$ N-S flip)/2
9	Flip fe N-S, Flip fn E-W (generating grids flipped in 2 directions)
10	B_e^{Jd} = (fe + fe N-S flip)/2, B_e^{Jn} = (fe - fe N-S flip)/2
11	B_n^{Jd} = (fn + fn E-W flip)/2, B_n^{Je} = (fn - fn E-W flip)/2
Calculate from grid standard deviations	
12	$B_h^{Jd} = \mathrm{sqrt}(B_e^{Jd} \wedge 2 + B_n^{Jd} \wedge 2)$, $B_d^{Jh} = \mathrm{sqrt}(B_d^{Je} \wedge 2 + B_d^{Jn} \wedge 2)$
13	$B_e^{Jh} = \mathrm{sqrt}(B_e^{Je} \wedge 2 + B_e^{Jn} \wedge 2)$, $B_n^{Jh} = \mathrm{sqrt}(B_n^{Je} \wedge 2 + B_n^{Jn} \wedge 2)$
14	$B_h^{Jh} = \mathrm{sqrt}(B_e^{Jh} \wedge 2 + B_n^{Jh} \wedge 2)$
15	Declination 1 = arctangent (B_e^{Jn}, B_n^{Je}) Declination 2 = arctangent (B_n^{Jn}, B_e^{Je}) Declination 3 = arctangent (B_d^{Jn}, B_d^{Je})
16	Inclination 1 = arctangent $(B_d^{Jd}, B_d^{Jh} *\mathrm{sqrt}(2))$ Inclination 2 = arctangent $(B_h^{Jd}, B_d^{Jh} *\mathrm{sqrt}(2))$ Inclination 3 = arctangent $(B_h^{Jd}, B_h^{Jh} *\mathrm{sqrt}(2))$

horizontal components and therefore recovery of the inclination angle faces different challenges. To investigate fidelity in recovering estimates of inclination of magnetisation we generated magnetic fields from a dipole located at a depth of 100 m with the field computed over a grid of side length 1 km for magnetisations with a range of declinations and inclinations in a −60° geomagnetic inclination field. MCS analysis was applied to component grids derived by FFT of the computed TMI. The resulting inclination estimates are independent of declination, which for this noise-free data was recovered closely even at inclinations up to 80°. The ratio of vertical to horizontal components of magnetisation from which we derive inclination angle climbs steeply at high inclinations (Fig. 8.3A) but the error in the resulting inclination estimate is maximum at mid inclinations (Fig. 8.3B). Errors in individual inclination estimates were in all cases less than 2° and errors in the mean of the three estimates were in all cases less than 1°. We also applied the analysis to fields forward computed from the ellipsoids and circular cylinders shown in Figs 8.2B and 8.2C with depth extent to diameter ratios of 5:1 and 1:5 and reliably recovered both declination and inclination

estimates for magnetisations of shallow, mid and steep inclinations of 15°, 45° and 75° respectively with similar precision to recovery of magnetisation direction for the dipole models. The relationships to derive inclination of magnetisation from ratios of grid standard deviation values are listed in step 16 of Table 8.1.

Extension to more complex but still symmetric magnetisations is illustrated for the rectangular ring and diagonal cross magnetisation distributions shown in Fig. 8.2D and 8.2E. Note that for bodies of symmetry about two orthogonal planes but where those planes are not coincident with the east–west and north–south coordinate axes, that ideal symmetry can be achieved by a suitable rotation of the coordinate system to be coincident with the grid reflection axes and a subsequent reverse rotation of the recovered declination angle (but we show later that this adjustment is not necessary). Forward computed TMI anomalies for the diagonal cross and square ring magnetisations in a geomagnetic field of inclination −60° are shown in Fig. 8.4. The cross body is given a magnetisation of declination 60°, inclination +30°. For this low-inclination magnetisation in a moderately steep geomagnetic field the anomaly pattern is

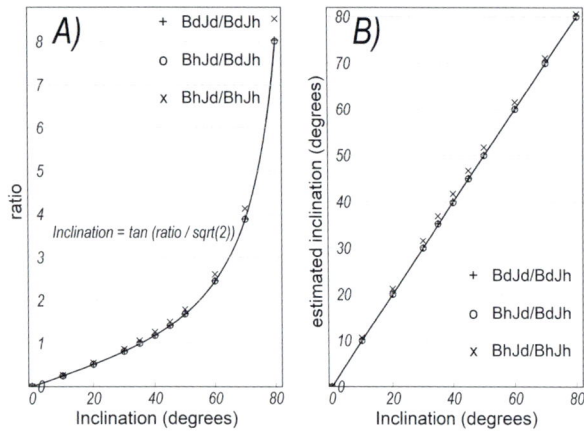

Fig. 8.3. A) ratio of magnetic field component grid standard deviations and B) corresponding inclination angles against angle of inclination of magnetisation.

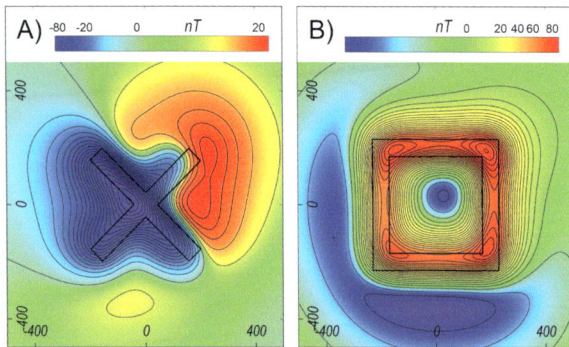

Fig. 8.4. TMI images in a 60° south geomagnetic field over A) a diagonal cross body of magnetisation 60°, +30° and B) a square ring body of magnetisation 120°, −70°.

determined primarily by the declination of magnetisation. The square ring body is given a magnetisation of declination 120°, inclination −70°. For this steep inclination the anomaly pattern is dominated by the spatial distribution of the magnetisation. The two bodies have horizontal widths of 400 to 450 m, depth to the top of the magnetisation of 100 m, and depth extent of 60 m. The grids have a side length of 1 km. The computed TMI anomalies were transformed to B_n, B_e and B_d grids which were then used in MCS analysis following the operations listed in Table 8.1. For both bodies each individual estimate of declination was recovered to within one degree. Individual inclination estimates were recovered to within two degrees, with the mean of the three estimates within one degree. The magnetic component-magnetisation matrix for the diagonal cross body is shown in Fig. 8.5. The body outline is added to the grid images for reference but is not used in the analysis. Figure 8.6 shows the MCS analysis form we use to collate

results. The top left section is a map of grid polarities expected for +ve magnetisation components. In this case there is a complete match with polarity of the recovered grids showing that the magnetisation has +ve inclination and is in the first quadrant (declination 0 to 90°). The grids are quantified by their standard deviation values (annotated in blue in Fig. 8.6) which provide more reliable measures than the peak-to-trough ranges which for measured data are more susceptible to local distortion. The three declination estimates (annotated in red in Fig. 8.6) are derived from ratios of grid standard deviation values together with the grid polarities. Estimation of inclination requires further processing as specified in steps 12 to 14 of Table 8.1. We take the mean declination and inclination values as the best estimated magnetisation direction.

Because magnetic fields are additive MCS analysis can be extended to constructions of symmetric magnetisations centred on a common vertical axis (as for Fig. 8.2 example F). In the case of bodies with identical magnetisation direction the analysis returns that direction as it does for an individual body. The case of bodies with different magnetisation direction is more complex. Figure 8.7A shows a TMI anomaly computed over two dipoles with a difference in magnetisation direction of 150°. The deeper dipole is at a depth of 4,000 m. The shallower dipole directly above it at a depth of 3,200 m has a 50% smaller magnetic moment. By visual inspection the TMI image is more complex than expected for a single dipole but there is no obvious indication that it is due to two magnetisations of such different direction. Figure 8.7B shows the $B_d{}^{Jn}$ grid from MCS analysis. This image clearly reveals the contrasting J_n component polarities of the two differently directed magnetisations with the centre of the image expressing the shallower magnetisation and the outer section of the image expressing the deeper magnetisation. Full MCS analysis reveals the polarity of all three components of both magnetisations. The magnetisation direction recovered from MCS analysis is close to the vector sum of the two magnetisations in a ratio of shallower to deeper of 0.85:1 (compared to 0.5:1 for their magnetisation moments). This ratio represents a preferential 1.7 weighting for the field of the shallower magnetisation. On the central axis there is an inverse cubed weighting of 1.95 for the field of the shallower source and this value decreases continuously with horizontal distance from the axis. For the fields measured 1,000 and 2,000 m shallower over the same magnetisation model MCS analysis

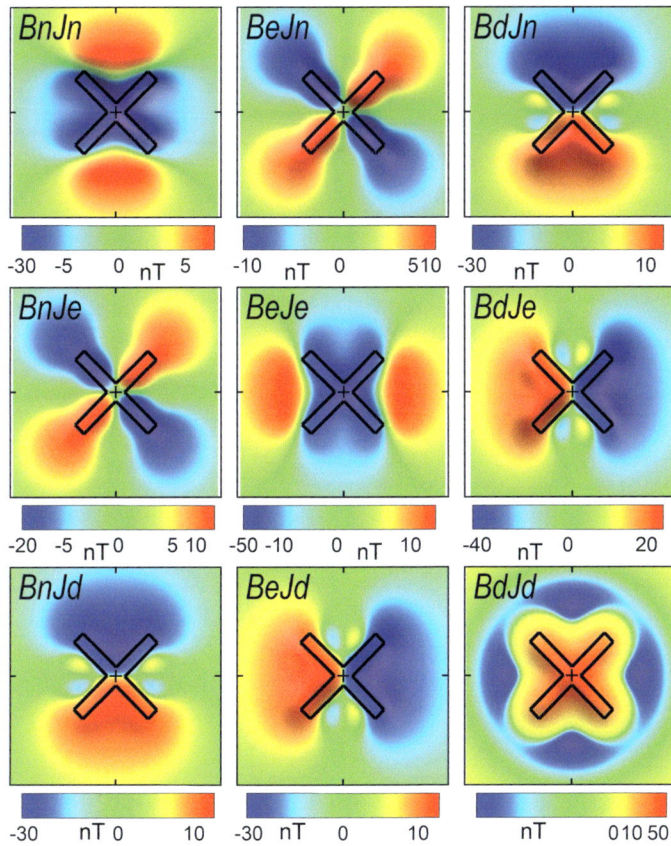

Fig. 8.5. Magnetic field–magnetisation matrix from MCS analysis of the diagonal cross anomaly in Fig. 8.4A.

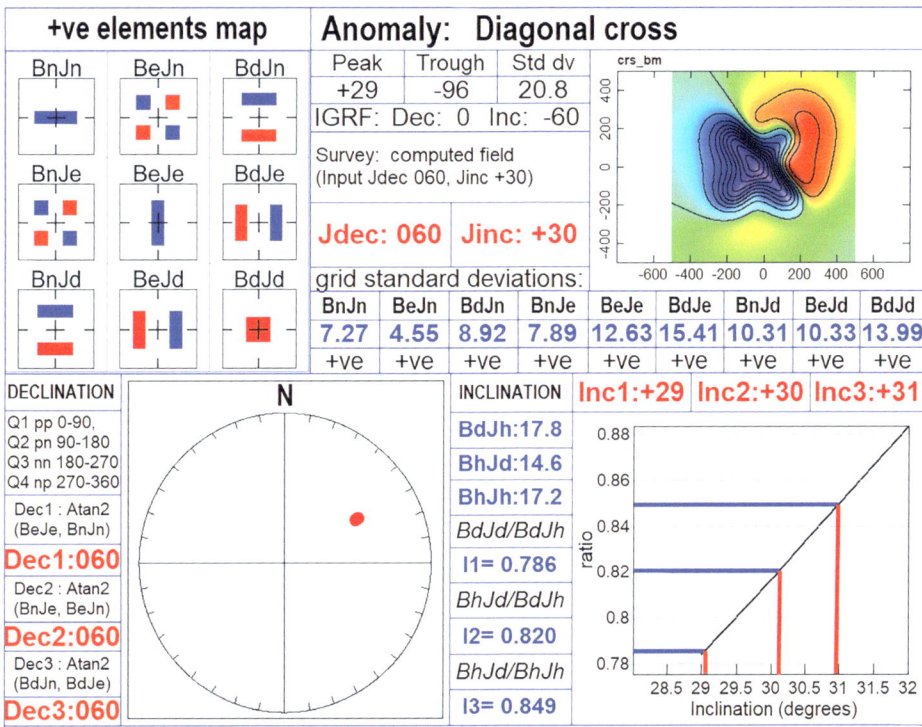

Fig. 8.6. MCS analysis form for the diagonal cross anomaly of Fig. 8.4A.

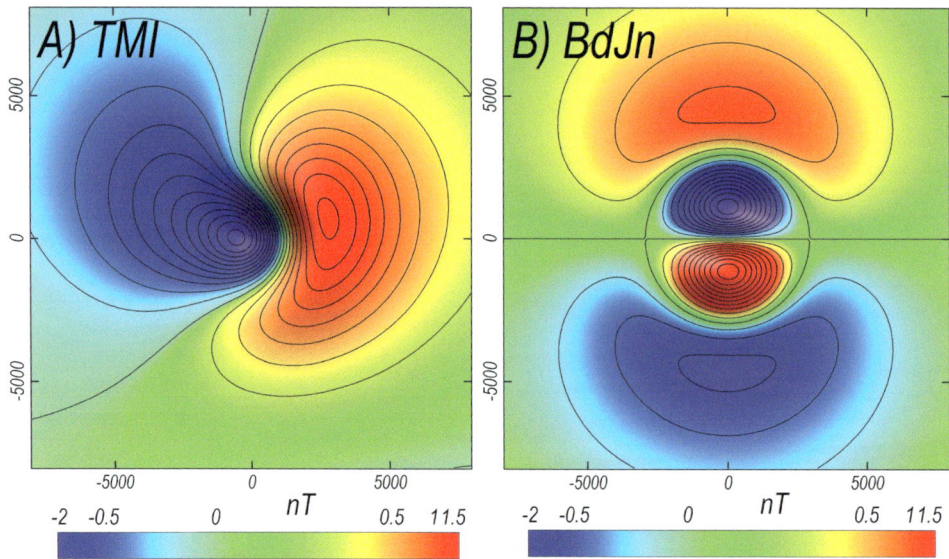

Fig. 8.7. A) TMI image over two vertically stacked magnetisations (see text for details) and B) BdJn image from MCS analysis.

magnetisation direction estimates change by 17° and 43° respectively, moving towards the direction of the shallower magnetisation as the weighting of that magnetisation increases substantially in the lower elevation fields. In these cases MCS analysis does not recover the individual magnetisation directions. However, from fields computed at all three elevations above the model the analysis detects the double magnetisation direction and both the deeper and shallower directions can be assigned declination quadrants and inclination polarities to provide a good starting model for any subsequent inversion studies.

8.4 A CASE STUDY OF THE BLACK HILL NORITE NORTH-WEST ANOMALY, SOUTH AUSTRALIA

The Black Hill Norite also known as the Black Hill Gabbroic Complex is an undeformed, post-tectonic Ordovician intrusion into Cambrian Kanmantoo Group metasediments (Turner 1991). The main lithologies include gabbro, anorthosite, norite, troctolite, peridotite, and diorite. At least some of these rocks have retained a substantial remnant magnetisation that gives rise to distinctive magnetic field anomalies with an equally prominent peak to the south-west and trough to the north-east. Figure 8.8A shows the isolated northwestern TMI anomaly of this pattern which we term the BHN-NW anomaly.

Figure 8.8B shows the outline of the parametric inversion model of Foss and McKenzie (2011) over an

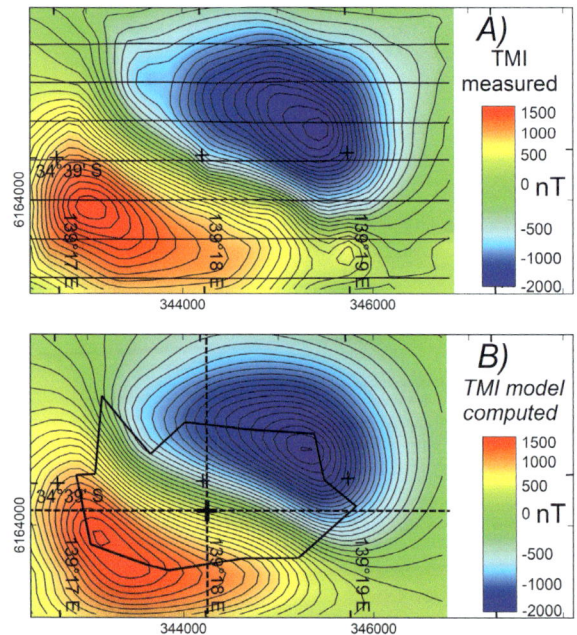

Fig. 8.8. BHN-NW anomaly: A) measured TMI with 400 m spaced E–W flightlines and B) model computed anomaly with the model centre and E–W and N–S grid flip axes shown.

image of TMI computed from the model. The model is a horizontal polygonal sheet of homogeneous magnetisation and its centre was used to locate the N-S and E-W grid flip axes for the MCS analysis. MCS analysis of the BHN-NW anomaly utilises an FFT phase transform of TMI to provide the B_n, B_e and B_d components. As shown in Fig. 8.9 these FFT-derived components are consistent with components forward computed from the TMI inversion model.

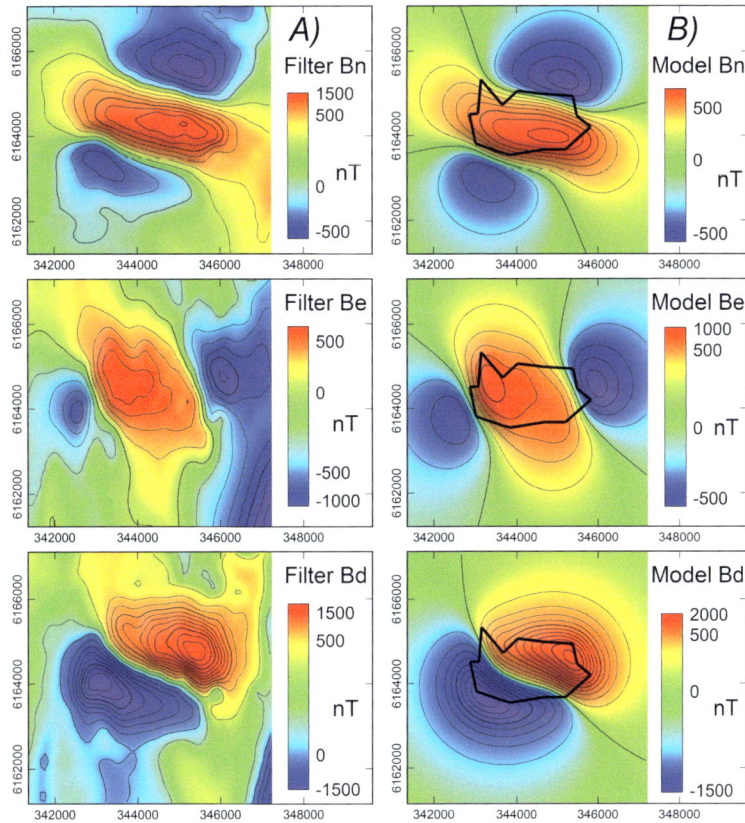

Fig. 8.9. Magnetic field component maps Bn (north), Be (east) and Bd (down): A) derived by FFT analysis of the measured TMI grid and B) forward computed from the inversion model.

MCS analysis was performed on the FFT-derived component grids following the steps listed in Table 8.1. Four steps of the analysis shown graphically in Fig. 8.10 reveal how grids variously symmetric and anti-symmetric are generated by differencing and summing a grid with its reflection. The resulting grids for the nine magnetisation component separations are plotted in Fig. 8.11 using a common linear colour stretch to highlight relative strengths of the grids. The three grids due to the easting magnetisation component (B_n^{Je}, B_n^{Je} and B_d^{Je} in the middle row of Fig. 8.11) and the three grids due to the northing magnetisation component (B_n^{Jn}, B_n^{Jn} and B_d^{Jn} in the top row of Fig. 8.11) have opposite polarity to those for an 'n-e-d' positive magnetisation, and the three grids due to the vertical magnetisation component (B_n^{Jd}, B_n^{Jd} and B_d^{Jd} in the bottom row of Fig. 8.11) have low amplitude consistent with a low positive-inclination magnetisation. The grid standard deviation values are plotted in blue on the MCS analysis form in the lower part of Fig. 8.12 and the resulting declination and inclination estimates are plotted in red. The three declination estimates range between 221° and 237° with a mean of 228°, and the three inclination estimates

range between +11° and +13° with a mean of +12°. The MCS analysis best estimate direction is 228°, +12°.

Figure 8.13 is a stereonet plot of estimates of resultant and remnant magnetisation direction obtained by various authors and methods (note that the resultant and remnant directions cannot be directly compared without consideration of the role of induced magnetisation). Palaeomagnetic studies (Rajagopalan *et al.* 1993; Schmidt *et al.* 1993; Schmidt and Clark 1997) used samples recovered from fresh outcrop in a quarry east of the anomaly within an area of similar but more complex magnetic field variation The first resultant magnetisation estimate derived from the BHN-NW anomaly itself by Schmidt and Clark (1997) using Helbig analysis is in good agreement with earlier remanent and induced magnetisation measurements. There has been a subsequent Helbig study of the anomaly (Phillips 2005), a Helbig and parametric inversion study (Foss and McKenzie 2011) and an inversion study (Pratt *et al.* 2014). The resultant magnetisation estimates derived from the anomaly (points B, C, D, E, G in Fig. 8.13) are well grouped about a mean of 226°, +7° which is within 5° of

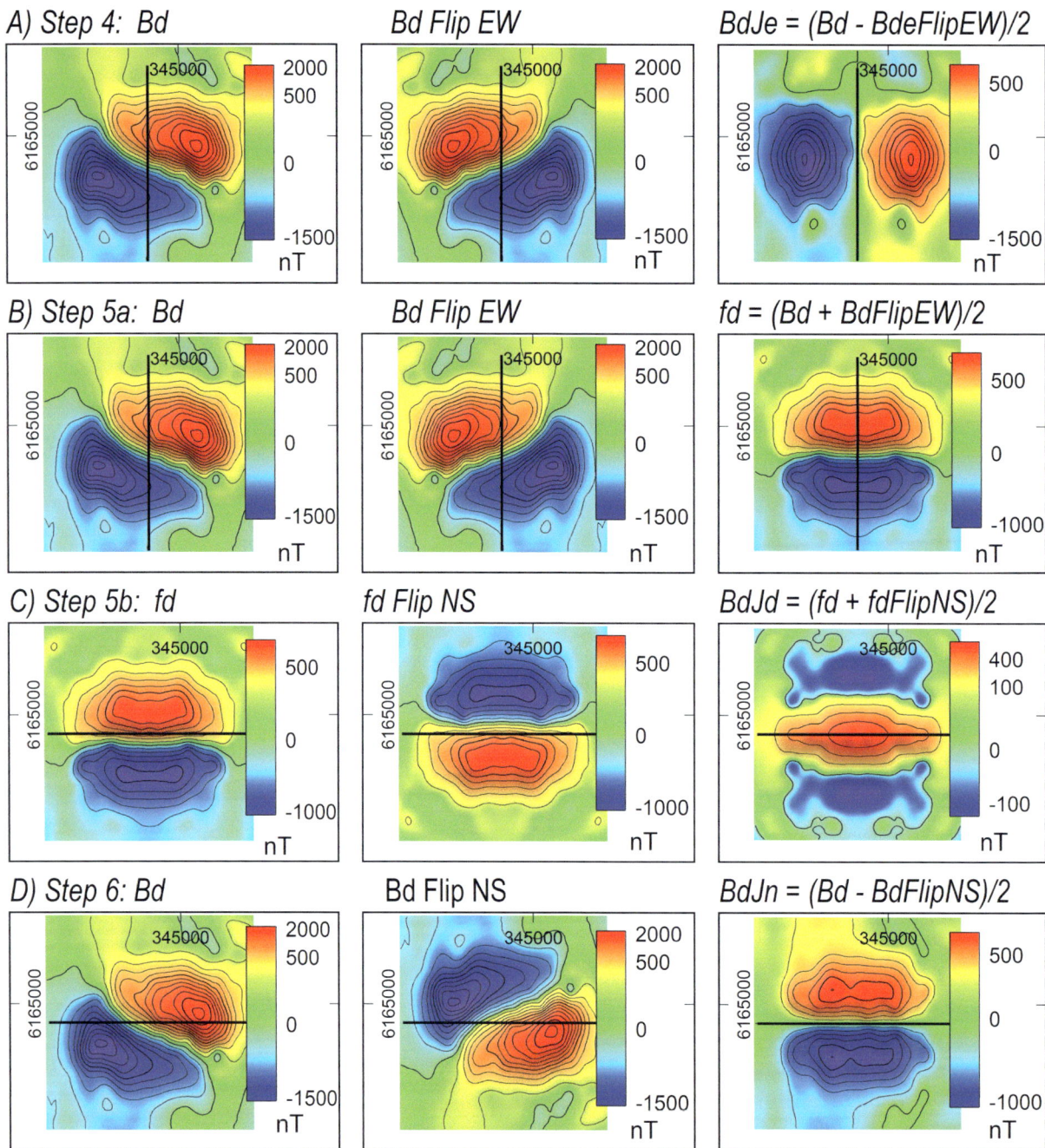

Fig. 8.10. MCS analysis steps 4 to 7 (see Table 8.1) applied to the BHN-NW measured TMI anomaly.

the mean of the three reported remnant magnetisation directions (points I, J, K in Fig. 8.13) of 225°, +12°. This suggests that the magnetisation of the source of the BHN-NW anomaly has a Koenigsberger ratio substantially higher than the value of 2.1 directly measured on samples from the nearby quarry. The MCS analysis direction falls within the population of resultant magnetisation directions from other studies and only 5° from the mean of the other estimates.

The true magnetisation direction of the source of the BHN-NW anomaly is unknown. To evaluate the performance of the MCS method we applied the analysis to the field forward computed from the known magnetisation inversion model of Foss and McKenzie, 2011. The result is presented in Fig. 8.14. The inversion model (body a in Fig. 8.14A) has an average east–west length of 2700 m, average north–south length of 1,500 m and depth to the top of magnetisation below the

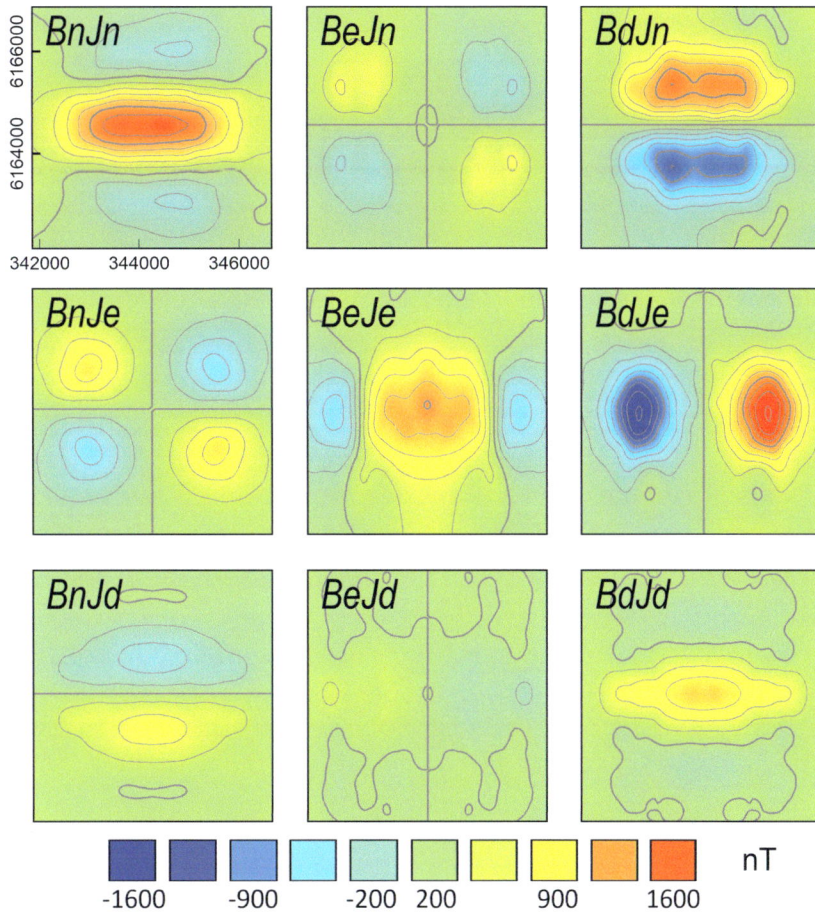

Fig. 8.11. MCS analysis steps 8 to 11 (see Table 8.1) applied to the BHN-NW measured TMI anomaly.

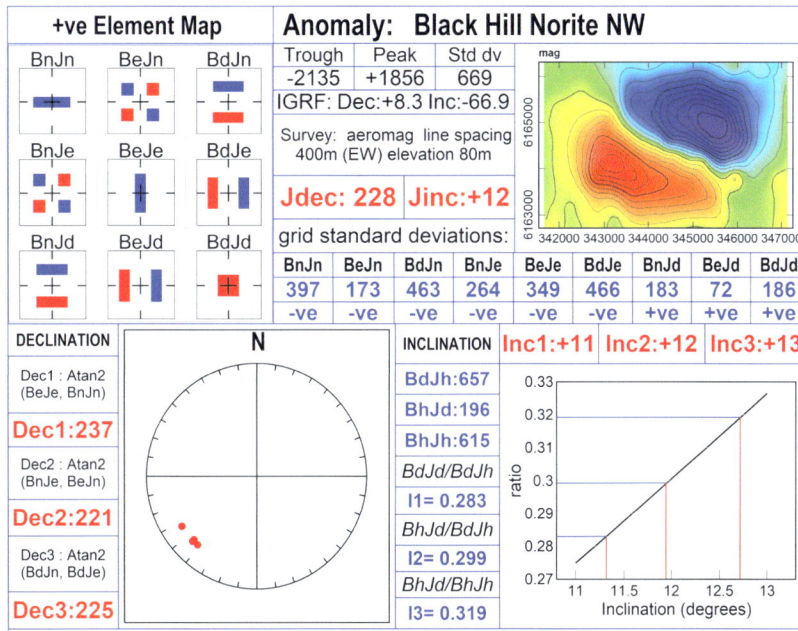

Fig. 8.12. Top) Magnetic field – magnetisation component matrix from MCS analysis of the BHN-NW measured TMI anomaly (all images with a common linear stretch) and Bottom) MCS analysis form with annotated results.

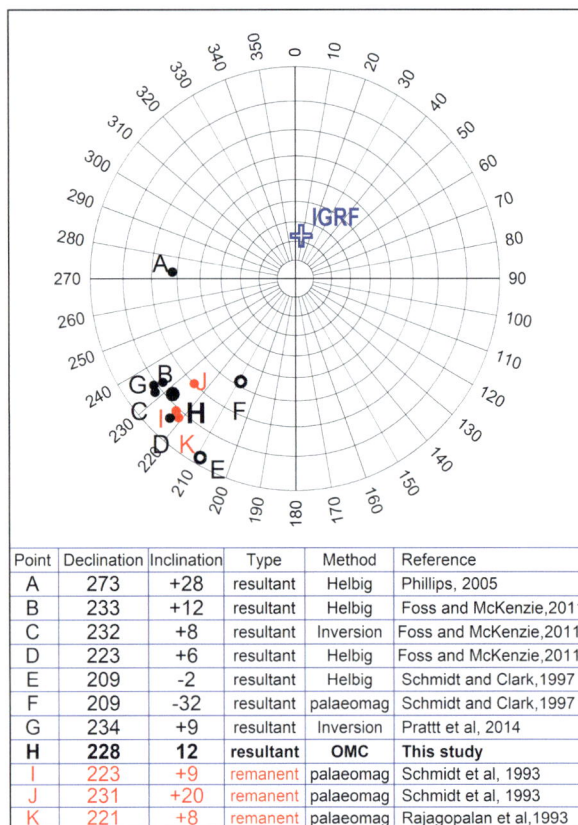

Point	Declination	Inclination	Type	Method	Reference
A	273	+28	resultant	Helbig	Phillips, 2005
B	233	+12	resultant	Helbig	Foss and McKenzie,2011
C	232	+8	resultant	Inversion	Foss and McKenzie,2011
D	223	+6	resultant	Helbig	Foss and McKenzie,2011
E	209	-2	resultant	Helbig	Schmidt and Clark,1997
F	209	-32	resultant	palaeomag	Schmidt and Clark,1997
G	234	+9	resultant	Inversion	Prattt et al, 2014
H	**228**	**12**	**resultant**	**OMC**	**This study**
I	223	+9	remanent	palaeomag	Schmidt et al, 1993
J	231	+20	remanent	palaeomag	Schmidt et al, 1993
K	221	+8	remanent	palaeomag	Rajagopalan et al,1993

Fig. 8.13. Measured remnant (red) and estimated resultant (black) magnetisation directions. Solid symbols are +ve (northern hemisphere) inclination and open symbols are –ve (southern hemisphere) inclination.

measurement plane of 543 m. The error in recovering the input magnetisation direction of the model from MCS analysis of the computed field is 6°. We generated a rectangular prism of the same dimensions and depth (body b in Fig. 8.14A) and also rotated that body by 90° (body c in Fig. 8.14A). For both bodies the magnetisation direction was recovered with an error of only 7°. We also rotated the body clockwise and anticlockwise by 45° (bodies d and e in Fig. 8.14B) and recovered magnetisation direction even more closely with errors of only 4° and 1° (possibly because minimum and maximum extension are not coincident with the axes of grid reflection). These results confirm that MCS analysis is suitable for bodies with horizontal elongation at least up to a factor of 2.

8.5 CONCLUSIONS

We have shown that the 3 × 3 matrix of magnetic field–magnetisation components for a dipole with positive unit magnetisation along each principal axis is identical

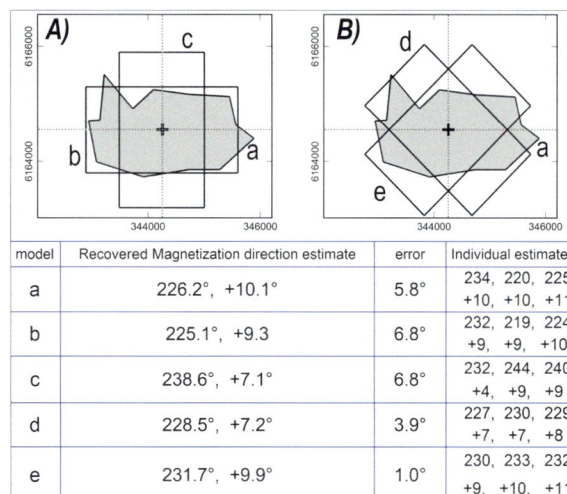

model	Recovered Magnetization direction estimate	error	Individual estimates
a	226.2°, +10.1°	5.8°	234, 220, 225 +10, +10, +11
b	225.1°, +9.3	6.8°	232, 219, 224 +9, +9, +10
c	238.6°, +7.1°	6.8°	232, 244, 240 +4, +9, +9
d	228.5°, +7.2°	3.9°	227, 230, 229 +7, +7, +8
e	231.7°, +9.9°	1.0°	230, 233, 232 +9, +10, +11

Fig. 8.14. Magnetisation directions recovered by MCS analysis of fields computed from the bodies shown in A) and B) with input magnetisation direction of 232°, +8°.

to the gravity gradient tensor of a point mass. For bodies with a vertical axis of fourfold symmetry the direction of magnetisation can be estimated by analysis of field components with specified symmetry characteristics ascribed to specific magnetisation components. The fields of contributions from each orthogonal magnetisation component are isolated by a series of operations on field component grids (transformed from the TMI grid) reflected about north–south and east–west planes through the horizontal centre of magnetisation. The analysis has been successfully applied to computed magnetic fields of complex symmetric bodies and provides an acceptable magnetisation estimate for the measured BHN-NW aeromagnetic anomaly. Analysis of model-computed fields shows that the analysis is reasonably tolerant of departure from ideal symmetry. Further studies are required to establish sensitivity to estimation of the centre of magnetisation but MCS analysis appears well suited for first-pass rapid automated studies.

REFERENCES

Beiki M, Clark DA, Austin JR, Foss CA (2012) Estimating source location using normalized magnetic source strength calculated from magnetic gradient tensor data. *Geophysics* **77**, J23–J37. doi:10.1190/geo2011-0437.1

Blakely RJ (1995) 'Potential theory in gravity and magnetic applications'. (Cambridge University Press).

Caratori Tontini F, Pedersen LB (2008) Interpreting magnetic data by integral moments. *Geophysical Journal International* **174**, 815–824. doi:10.1111/j.1365-246X.2008.03872.x

Clark DA (2014) Methods for determining remanent and total magnetisations of magnetic sources – a review. *Exploration Geophysics* **45**, 271–304. doi:10.1071/EG14013

Dannemiller N, Li Y (2006) A new method for determination of magnetisation direction. *Geophysics* **71**, L69–L73. doi:10.1190/1.2356116

Dransfield MH (1994) Airborne gravity gradiometry. Doctor of Philosophy, The University of Western Australia. doi:10.26182/5c89d3c7789f1

Fedi M, Florio G, Rapolla A (1994) A method to estimate the total magnetisation direction from a distortion analysis of magnetic anomalies. *Geophysical Prospecting* **42**, 261–274. doi:10.1111/j.1365-2478.1994.tb00209.x

Foss CA (2006) 'Evaluation of strategies to manage remanent magnetisation effects in magnetic field inversion.' 76th Annual International Meeting, SEG, Expanded Abstracts, 938–942.

Foss CA (2017) 'Resultant-magnetisation based magnetic field interpretation' in V. Tschirhart and M.D. Thomas eds, Proceedings of Exploration 17: Sixth Decennial International Conference on Mineral Exploration, pp. 637–648.

Foss CA, McKenzie KB (2011) Inversion of anomalies due to remnant magnetisation: an example from the Black Hill Norite of South Australia. *Australian Journal of Earth Sciences* **58**, 391–405. doi:10.1080/08120099.2011.581310

Fullagar PK, Pears GA (2015) 'Remanent magnetisation inversion.' 24th ASEG International Geophysical Conference, Extended Abstracts. doi:10.1071/ASEG2015ab188

Helbig K (1963) Some integrals of magnetic anomalies and their relation to the parameters of the disturbing body. *Zeitschrift für Geophysik* **29**, 83–96.

Lelièvre PG, Oldenburg DW (2009) A 3D total magnetisation inversion applicable when significant, complicated remanence is present. *Geophysics* **74**, L21–L30. doi:10.1190/1.3103249

Li Y (2012) Recent advances in 3D generalized inversion of potential-field data. *SEG Technical Program Expanded Abstracts* **2012**, 1–7.

Lourenço JS, Morrison HF (1973) Vector magnetic anomalies derived from measurements of a single component of the field. *Geophysics* **38**, 359–368. doi:10.1190/1.1440346

McKenzie KB, Foss CA, Hillan D (2012) 'An improved search for magnetisation direction.' 22nd ASEG Geophysical Conference, Extended Abstracts, 1–4.

Paine J, Haederle M, Flis M (2001) Using transformed TMI data to invert for remanently magnetised bodies. *Exploration Geophysics* **32**, 238–242. doi:10.1071/EG01238

Pedersen LB, Bastani M (2016) Estimating rock vector magnetisation from coincident measurements of magnetic field and gravity gradient tensor. *Geophysics* **81**, B55–B64. doi:10.1190/geo2015-0100.1

Pedersen LB, Rasmussen TM (1990) The gradient tensor of potential field anomalies: some implications on data collection and data processing of maps. *Geophysics* **55**, 1558–1566. doi:10.1190/1.1442807

Phillips JD (2005) Can we estimate total magnetisation directions from aeromagnetic data using Helbig's integrals? *Earth, Planets, and Space* **57**, 681–689. doi:10.1186/BF03351848

Phillips JD, Nabighian MN, Smith DV, Li Y (2007) Estimating locations and total magnetisation vectors of compact magnetic sources through combined Helbig and Euler analysis. *SEG Technical Program Expanded Abstracts* **26**, 770–774.

Pratt DA, McKenzie KB, White AS (2014) Remote remanence determination (RRE). *Exploration Geophysics* **45**, 314–323. doi:10.1071/EG14031

Rajagopalan S, Schmidt PW, Clark DA (1993) Rock magnetism and geophysical interpretation of the Black Hill Norite, South Australia. *Exploration Geophysics* **24**, 209–212. doi:10.1071/EG993209

Schmidt PW, Clark DA (1997) Directions of magnetisation and vector anomalies derived from total field surveys. *Preview* **70**, 30–32.

Schmidt PW, Clark DA (1998) The calculation of magnetic components and moments from TMI: a case study from the Tuckers igneous complex, Queensland. *Exploration Geophysics* **29**, 609–614. doi:10.1071/EG998609

Schmidt PW, Clark DA, Rajagopalan S (1993) An historical perspective of the Early Palaeozoic APWP of Gondwana: new results from the Early Ordovician Black Hill Norite of South Australia. *Exploration Geophysics* **24**, 257–262. doi:10.1071/EG993257

Turner SP (1991) Late-orogenic, mantle-derived, bimodal magmatism in the southern Adelaide Foldbelt, South Australia. PhD thesis, University of Adelaide (unpubl.).

9

B$_{z,z}$ peak and trough analysis for rapid mapping of magnetisation direction

C.A. Foss and K.B. McKenzie

ABSTRACT

We present a simple, approximate method to estimate magnetisation direction from analysis of the vertical gradient of the vertical component of the magnetic field B$_{z,z}$ that can be derived from an FFT phase transform from TMI to B$_z$ and an FFT gradient transform from B$_z$ to B$_{z,z}$. The method can be applied semi-automatically (with supervision) to complete survey datasets to quickly detect distribution patterns of magnetisation from scattered anomalies. A key advantage is that the method is independent of the local geomagnetic field direction. We illustrate the method with synthetic data computed from ellipsoid models of different orientations, magnetisation directions and depths and then apply the method to survey data from a test area in eastern Queensland. The current limitations are principally due to challenges in isolation of the magnetic field variations from background fields and resolution limitations of regional aeromagnetic datasets.

9.1 INTRODUCTION

The direction of natural remanent magnetisation (NRM) of a rock is determined by ages, post-acquisition rotations and relative strengths of all its remanent magnetisation components. The resultant or 'total' magnetisation is the vector sum of NRM and an induced magnetisation approximately equal to the product of magnetic susceptibility and magnetic field strength and directed parallel to the local geomagnetic field. Direct determination of remanent and induced magnetisation strengths and directions is based on measurement of oriented rock samples but the resultant magnetisation strength and direction can also be estimated by analysis of magnetic field data. Some recent methodologies used to estimate direction of magnetisation over a compact source include the use of magnetic moments (Helbig 1963; Schmidt and Clark 1998; Phillips *et al.* 2007; Caratori Tontini and Pedersen 2008; Foss and McKenzie 2011; Clark 2012, 2013a, 2014); by inversion of magnetic moments (Medeiros and Silva 1995); from parametric inversion (Foss and McKenzie 2011; Pratt *et al.* 2014) or directly from the magnetic gradient tensor and its eigenvectors (McKenzie 2020).

The most reliable method of estimating magnetisation direction from magnetic field data is by focussed inversion of individual anomalies. However, inversions are time consuming. We would like to find more rapid methods to scan a complete magnetic survey dataset, even if the results are less reliable. In this chapter we propose such a method based on analysis of B$_{z,z}$ the vertical gradient of the vertical component of the field.

The total gradient (TG) and normalised source strength (NSS) transforms of measured TMI highlight sharp magnetic field variations due to shallow magnetisations and show only weak influence of source magnetisation direction. Outputs from the transforms are simpler and more radially symmetric than the input TMI field and they peak over or close to horizontal centres of compact magnetisations (Beiki *et al.* 2012; Clark 2012, 2014; Foss and Austin 2023). Curvature analysis of surfaces (Blakely and Simpson 1986; Phillips *et al.* 2007) applied to gridded magnetic field data locates elongate ridges and troughs as well as equidimensional peaks and wells of those functions. For either TG or NSS transforms we only need to map positive curvature features, and for analysis of compact magnetisations we select equidimensional peaks. Estimation of magnetisation direction for sheet-like magnetisations that produce elongated ridges in the transform outputs is more complex and must be treated separately. Success of the TG and NSS transforms in locating magnetisations is in large part due to their restricted sensitivity to magnetisation direction. To subsequently estimate magnetisation direction from those located magnetisations we need either to return to the TMI data or to select an alternative transform that benefits recovery of magnetisation direction rather than suppresses it.

In an early study of three-dimensional TMI modelling of compact magnetic bodies, Zietz and Andreasen (1967) noted that in high inclination northern hemisphere geomagnetic fields the declination of magnetisation is indicated by the azimuth of the TMI peak-to-trough vector and the inclination of magnetisation is indicated by the amplitude ratio of the TMI peak and trough. In the northern hemisphere a TMI peak to trough ratio much greater than 1 indicates a steep positive inclination of magnetisation, a ratio close to 1 indicates a low-inclination magnetisation and a ratio much less than 1 indicates a steep negative inclination. In the southern hemisphere the roles of the TMI peak and trough are reversed because the sign of the field vector relative to the ground surface is reversed (the convention is that positive down describes the northern hemisphere field and the upward directed southern hemisphere field is negative). TMI is mostly perceived as a consistent magnetic field parameter but its orientation varies between survey locations with different geomagnetic inclination. Resultant magnetisation directions typically vary with geomagnetic inclination because of variation in the direction of their induced magnetisation components. However, even identical magnetisations produce different TMI anomalies in fields of different geomagnetic inclination because of differences in the vector addition with the local background field. The direction of the anomalous field varies substantially around and above a magnetisation and this dictates the anomaly pattern. To reduce this complication we use the field component B$_z$ of fixed (vertical) orientation. Vertical orientation provides advantage because most magnetic fields are measured in a horizontal plane perpendicular to that direction. This is a similar advantage to that derived from use of the reduced to pole (RTP) FFT of TMI data. However, the standard RTP transform also attempts to rotate the magnetisation direction to vertical on assumption that it is initially parallel to the geomagnetic field (Baranov and Naudy 1964; Blakely 1995). The phase change on FFT from TMI to B$_z$ or B$_{z,z}$ is independent of magnetisation direction. In unusual cases that the anomalous field is so strong that it significantly rotates the TMI direction across the region of the magnetic anomaly an intermediate step is required to iteratively adjust the field to a consistent direction as devised by Lourenço and Morrison (1973) and further developed by Schmidt and Clark (2006) and Clark (2013b). Use of the B$_z$ component extends the relationships noted by Zietz and Andreasen (1967) in steep inclination fields. Figure 9.1 shows TMI anomalies (in the upper half of the figure) and B$_z$ anomalies (in the lower half of the figure) due to a dipole with a magnetisation of declination 45° and inclination −45°. The patterns of the TMI anomalies of Figs 9.1A and 9.1C in steep downward (north pole) and steep upward (south pole) fields respectively are as predicted by the relationships of Zietz and Andreasen (1967). The TMI anomaly in the low-inclination equatorial field shown in Fig. 9.1B is quite different. All three B$_z$ anomalies (Figs 9.1D, 9.1E and 9.1F) are identical, resulting from the scalar summation of anomalous and background B$_z$ field components. The interpretational advantages of a single anomaly pattern for one magnetisation direction in all geomagnetic fields is considerable. Derivation of B$_z$ from transformation of TMI data is not especially problematic, even in low geomagnetic inclination fields, provided the data quality and sample spacing are adequate. If these conditions are not met then analysis of the data by any method will be uncertain.

Definition of a B$_z$ anomaly by separation from a background field is of similar difficulty and concern as for TMI data. A common and effective solution for both field isolations (although with escalation of other challenges) is through use of the vertical gradient. For the B$_z$ field component this gradient is B$_{z,z}$. The background field is commonly assumed to contribute only longer wavelength variations and these produce significantly weaker gradients than shorter wavelength field variations of shallow anomalous magnetisations of interest. Conversely,

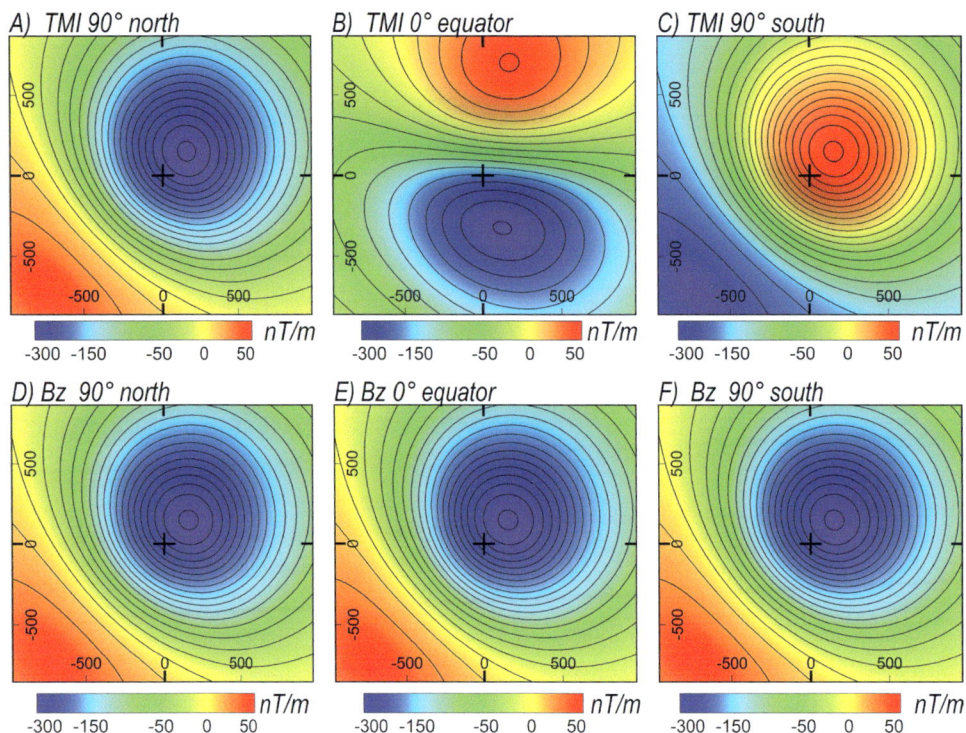

Fig. 9.1. Anomalies in different background fields over an 800 m radius sphere of magnetisation 1 A/m, declination D_M = 45°, and inclination I_M = −45° at a depth of 800 m. The top row are TMI anomalies, the bottom row are Bz anomalies.

investigation of the vertical gradient is problematic where there are significant superimposed shorter wavelength field variations or measurement imperfections. In these cases it can be of advantage to pre-condition the data with a mild upward continuation before estimating the vertical gradient. Use of the gradient can still be of advantage despite the contrary operations of upward continuation (a smoothing operator) and application of the gradient (a sharpening operator). Upward continuation is also more versatile and easily comprehended than the alternative use of fractional derivatives (Cooper and Cowan 2003). The sign of the vertical gradient with respect to the ground surface does not change with geomagnetic hemisphere and so the sign difference of B_z in southern and northern hemispheres persists in $B_{z,z}$ data.

9.2 THE VERTICAL GRADIENT OF THE VERTICAL FIELD COMPONENT: $B_{z,z}$

Figure 9.2 shows six $B_{z,z}$ anomalies for dipoles of different magnetisation direction. As already established, the pattern of these anomalies is independent of the direction of the field in which they are measured. The anomaly for a vertical upward magnetisation (Fig. 9.2A) has a trough to peak amplitude ratio of almost 20:1. The equivalent $B_{z,z}$ anomaly of a vertical downward magnetisation (Fig. 9.2B) is identical other than for a change of polarity. Unlike TMI anomalies, the prominent polarity of $B_{z,z}$ and B_z anomalies reveal if the magnetisation is upward directed (negative polarity) or downward directed (positive polarity) regardless of geomagnetic hemisphere. Anomalies of compact, horizontal, north–south magnetisations are symmetric with equal amplitude peak and trough. For a southerly directed (declination = 180°) horizontal magnetisation the anomaly has a peak to the north and trough to the south (Fig. 9.2E) and for northerly directed (declination = 0°) horizontal magnetisation the anomaly pattern is reversed with the trough to the north and peak to the south (Fig. 9.2F). For north or south directed intermediate inclinations the anomalies have a form intermediate between the anomalies due to vertical and horizontal magnetisations. These patterns are shown in Figs 9.1C and 9.1D for a pair of mid-range northerly directed magnetisations with inclination I_M = −45° and +45° respectively. The anomaly patterns of corresponding southerly directed magnetisations are rotated about a vertical axis by 180°.

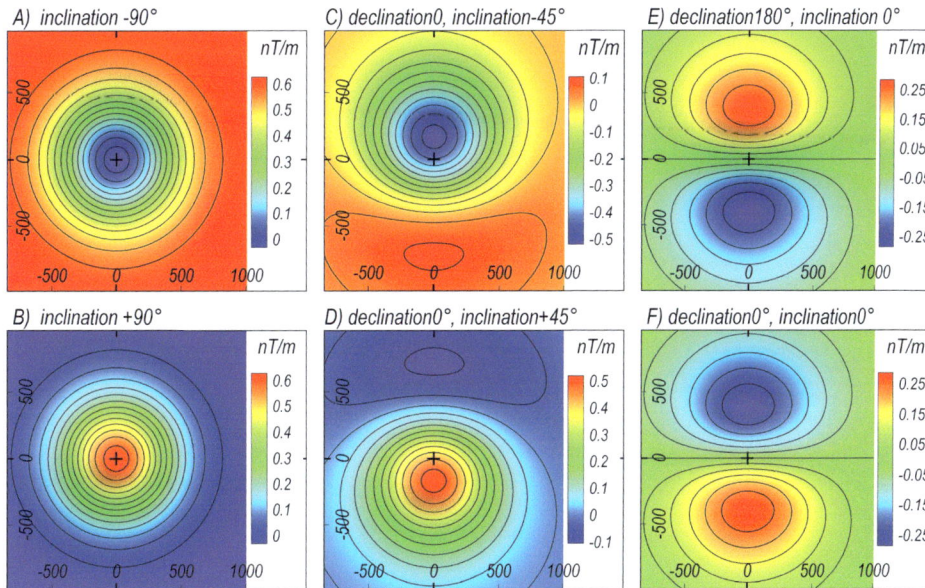

Fig. 9.2. Dipole B$_{z,z}$ anomalies of different magnetisation direction.

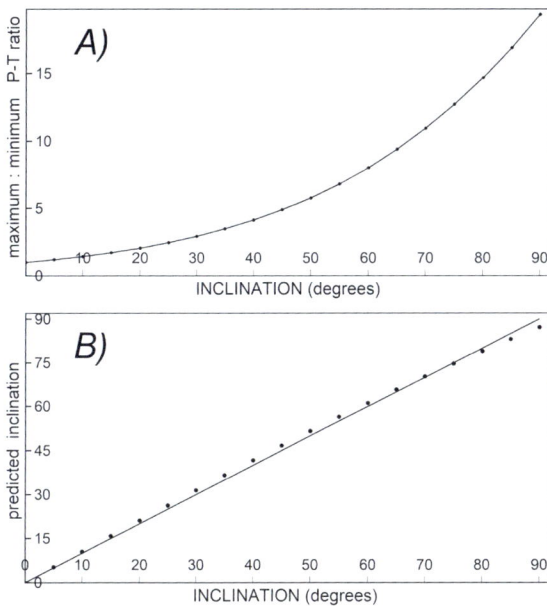

Fig. 9.3. A) peak to trough ratio against inclination of magnetisation, B) the same data (points) on a logarithmic scale with a linear predictor.

As illustrated in Figs 9.1 and 9.2, the declination of magnetisation is recorded by the peak to trough orientation of the B$_z$ or B$_{z,z}$ anomalies. The dipole B$_{z,z}$ ratio of the prominent to weaker feature (the peak to trough ratio for a positive-inclination magnetisation or the trough to peak ratio for a negative-inclination magnetisation) is plotted against inclination in Fig. 9.3A. The ratio increases smoothly from 1 for a horizontal

inclination to almost 20 for a vertical inclination. Figure 9.3B shows the same data plotted as points on a logarithmic amplitude scale, together with a straight-line approximation. This linear prediction of inclination (value = 0 at 0° inclination and 1.34 at 90° inclination) with an average departure from the true inclination of only 0.5° and a maximum departure of less than 2° is adequate for approximate estimation of the inclination of magnetisation.

For a dipole source, the B$_{z,z}$ peak and trough also provide an estimate of depth to the dipole (the centre of a spherical source). The distance between peak and trough varies with inclination of magnetisation. The peak and trough are closest for a zero-inclination magnetisation with a separation of 78% of the depth to centre of magnetisation. For steep inclinations the separation increases to 115%. The conversion factor to estimate centre depth from peak and trough separation is plotted in Fig. 9.4. The centre point of a dipole magnetisation lies on a line joining the B$_{z,z}$ peak and trough of the anomaly with a relative distance from each that is approximately inversely proportional to their amplitudes. For a zero-inclination magnetisation the horizontal centre of magnetisation is midway between the equal amplitude peak and trough. The departure from this approximation is plotted in Fig. 9.5. For an inclination of magnetisation of ± 35° the predicted centre of magnetisation is just over 1% of the peak to trough separation closer to the higher

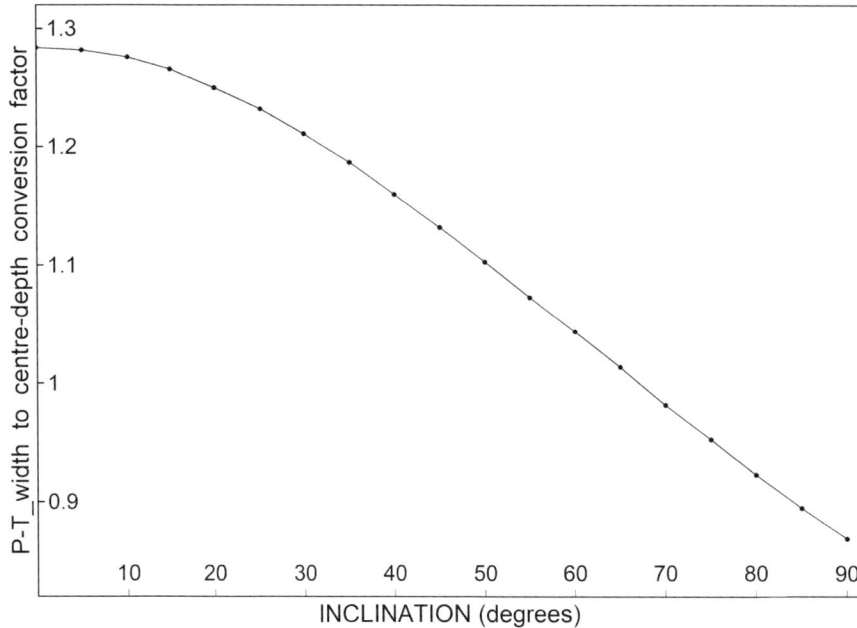

Fig. 9.4. Conversion factor to estimate depth to a dipole centre from the $B_{z,z}$ peak and trough separation as a function of the inclination of magnetisation.

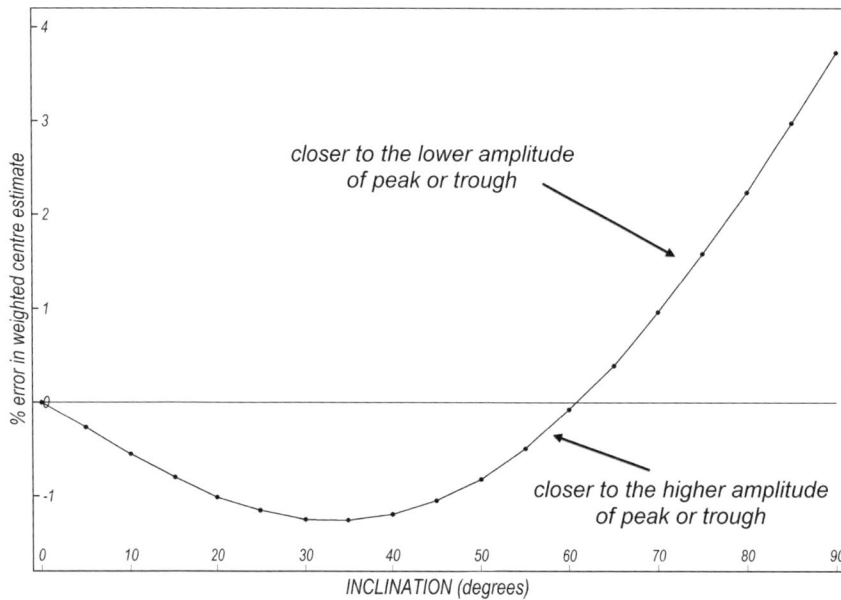

Fig. 9.5. Percentage error in weighted centre prediction of the horizontal centre of magnetisation.

amplitude feature. For an inclination of 61° the prediction is correct and at 90° the prediction is almost 4% closer to the lower amplitude feature than the true centre (for particularly steep magnetisations the distant, weak lower-amplitude feature is difficult to reliably locate and its exact inclination is imprecisely estimated).

9.3 A SYNTHETIC TEST STUDY USING ELLIPSOID SOURCES

A major concern for the magnetisation directions estimated by the proposed $B_{z,z}$ analysis is the validity in representing the distribution of magnetisation with a dipole model. Figure 9.6 illustrates a set of six prolate ellipsoids with axis radial lengths of 800 and 300 m (a

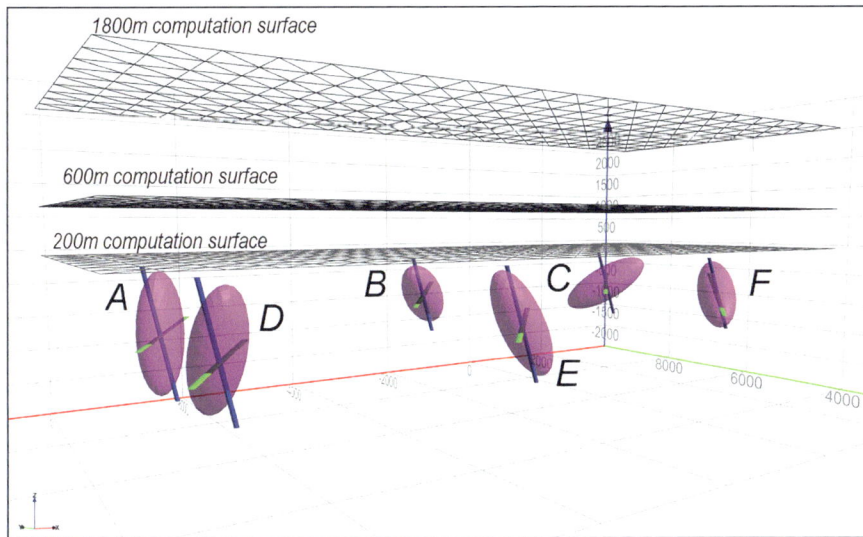

Fig. 9.6. Plunging ellipsoid magnetisation models (magenta) with upward directed magnetisation vectors (blue – steep inclination, green – low inclination). The spatial properties are listed in Table 9.1.

Table 9.1. Ellipsoid body properties list. X_0, Y_0, Z_0 are centre coordinates, the azimuth and dip describe the major axis.

BODY	X_0	Y_0	Z_0	Azimuth°	Dip°
A	−7000.0	7000.0	800.0	0	90
B	−3000.0	7000.0	709.0	0	60
C	1000.0	7000.0	478.0	45	30
D	−7000.0	4500.0	709.0	90	60
E	−3000.0	4500.0	478.0	315	30
F	1000.0	4500.0	478.0	225	30

2.7:1 ratio) and different plunge and azimuth. Expressions for the gradient tensor elements (including B$_{z,z}$) of a uniformly magnetised triaxial ellipsoid are given in McKenzie (2020) and for a right circular cylinder by McKenzie (2022). The prolate ellipsoids we use in this study are a special case of triaxial ellipsoids with two identical length minor axes and one major axis. We selected the six ellipsoids to investigate relationships between their plunge and azimuth angles with two magnetisation directions: A) a high-inclination magnetisation of −75° with a declination of 315° and B) a low-inclination magnetisation of −15° and declination of 45°. B$_{z,z}$ analysis is independent of declination of magnetisation angles relative to the magnetic field but the angle between direction of magnetisation and any spatial extension of the magnetisation has a significant influence on the anomaly shape. We adjusted the body centre

depths to give a consistent depth to the tops of magnetisation of 200, 600 and 1,800 m respectively below three elevation surfaces at which we computed B$_{z,z}$. The purpose of computation at different elevations is to investigate attenuation of the influence of shape with increasing distance from the sources. In proximal fields shape influence is considerable but in distal fields the anomalies more closely approximate to those of dipoles of the same total magnetisation (see also Chapter 4).

Figure 9.7 shows images of B$_{z,z}$ for each of the six ellipsoids in Table 9.1 computed at the shallowest elevation of 200 m above the tops of the magnetisations for the steep inclination (I_M = -75°) magnetisation in Fig. 9.7A and for the low-inclination magnetisation (I_M = −15°) in Fig. 9.7B. Even at this proximity to the top of magnetisation (200 m compared to the long-axis length of the magnetisation of 1,600 m) the direction of magnetisation is the dominant influence on anomaly orientation as measured by the peak to trough azimuth (shown by the white arrows in Fig. 9.7). Peak to trough amplitude ratio is best estimated from contour overlays over the map images. The high trough to peak ratio of all the anomalies in Fig. 9.7A reveals the steep negative inclination of the source magnetisations for those anomalies and the near-unity peak to trough ratios of Fig. 9.7B reveal the low inclination of those source magnetisations. However, the peak to trough amplitude ratio is also influenced by body orientation relative to magnetisation direction. To perform the B$_{z,z}$ dipole analysis

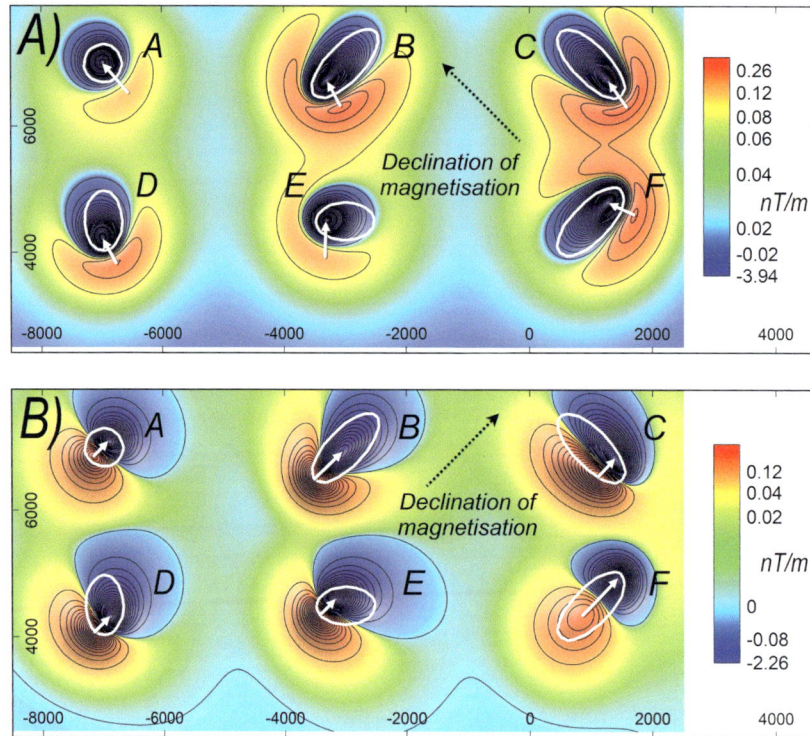

Fig. 9.7. $B_{z,z}$ anomalies for the ellipsoids in Fig. 9.6 with magnetisations of: A) inclination −75°, declination 315°, and B) inclination −15°, declination 45° computed at an elevation 200 m above the tops of the magnetisations. The white arrows are from anomaly maxima to minima.

Table 9.2. $B_{z,z}$ dipole analysis for anomalies of magnetisation: declination 315°, inclination −75°.

Elevation	Body	PT ratio	PT azimuth°	PT inclination°	error°	Centre offset	Centre azimuth°
1,800	A	13.77	314.6	−77.1	2.1	50.5	133.9
1,800	B	12.70	317.7	−74.7	0.8	161.9	168.2
1,800	C	12.07	321.3	−73.2	2.5	157.7	203.1
1,800	D	12.07	321.7	−73.2	2.6	158.6	203.2
1,800	E	11.89	315.4	−72.8	2.2	181.7	135.4
1,800	F	12.07	308.7	−73.2	2.5	155.5	67.1
600	A	17.50	315.0	−84.1	9.1	14.6	135.2
600	B	12.69	322.9	−74.7	2.1	232.5	177.6
600	C	10.16	330.2	−68.1	8.3	292.4	218.7
600	D	18.23	347.4	−85.3	11.3	234.3	263.6
600	E	9.94	315.0	−67.5	7.5	298.4	135.0
600	F	10.16	299.8	−68.1	8.3	292.4	51.3
200	A	27.48	314.9	−97.4	22.4	0.6	109.3
200	B	12.82	328.1	−75.0	3.4	299.4	179.6
200	C	8.21	333.4	−61.9	14.6	432.1	222.2
200	D	23.98	381.0	−93.4	16.7	312.4	268.9
200	E	7.85	315.1	−60.6	14.4	437.4	135.0
200	F	8.21	296.6	−61.9	14.6	432.1	47.8

Table 9.3. B$_{z,z}$ dipole analysis for anomalies of magnetisation: declination 45°, inclination −15°.

Elevation	Body	PT ratio	PT azimuth°	PT inclination°	error°	Centre offset	Centre azimuth°
1,800	A	1.74	45.0	−16.3	1.3	33.6	44.5
1,800	B	1.65	44.9	−14.7	0.3	91.8	168.7
1,800	C	1.60	45.0	−13.8	1.2	95.9	224.8
1,800	D	1.65	45.2	−14.7	0.4	92.4	280.9
1,800	E	1.70	45.0	−15.5	0.5	125.6	122.3
1,800	F	1.88	44.9	−18.5	3.5	166.4	45.0
600	A	1.81	45.0	−17.4	2.4	16.7	44.8
600	B	1.50	44.5	−11.9	3.1	200.6	179.4
600	C	1.27	45.0	−7.1	7.9	257.9	225.0
600	D	1.50	45.4	−11.9	3.1	200.8	270.7
600	E	1.63	44.9	−14.3	0.7	260.7	133.5
600	F	2.60	45.0	−28.1	13.1	304.6	45.0
200	A	1.91	45.0	−19.1	4.1	9.5	44.4
200	B	1.31	44.4	−8.0	7.1	286.6	180.8
200	C	0.90	45.0	3.1	18.1	425.2	225.0
200	D	1.31	45.5	−8.0	7.0	287.1	269.2
200	E	1.55	45.0	−12.8	2.2	408.4	135.0
200	F	4.87	44.9	−46.5	31.5	455.7	45.0

Table 9.4. Mean errors in magnetisation direction and centre point estimation of the B$_{z,z}$ analysis and dipole inversions at the three computation elevations.

Inclination of magnetisation	Elevation	B$_{z,z}$ ANALYSIS RESULTS			INVERSION RESULTS		
		Inclination error	Angular error	Centre offset	Inclination error	Angular error	Centre offset
High (A)	200	14	14	319	15	20	280
High (A)	600	7	8	227	7	11	198
High (A)	1,800	2.1	2.1	144	2.1	3.30	106
Low (B)	200	12	12	312	9.0	9.2	257
Low (B)	600	5	5	207	3.0	3.10	157
Low (B)	1,800	3.5	3.6	101	0.7	0.7	74

we computed B$_{z,z}$ grids for the bodies in Table 9.1 with both the steep-inclination and low-inclination magnetisations detailed above. We then applied algorithms based on Blakely and Simpson (1986) and Phillips *et al.* (2007) to locate the peaks and troughs of these grids and from those results determined distance and azimuth between adjacent peak and trough pairs and their amplitude ratio. We submitted these statistics to the B$_{z,z}$ dipole analysis to obtain the magnetisation direction estimates listed in Table 9.2 for the steep-inclination magnetisation and in Table 9.3 for the low-inclination

magnetisation (in each case 18 estimates for the six bodies at the three elevations). Table 9.4 summarises the mean magnetisation errors and centre offsets from the analyses at each elevation. There is no significant difference in the performance of the analysis in recovering magnetisation direction or centre location from anomalies of the high- or low-inclination magnetisations (there may be sensitivities to directions outside the range we investigated of 15° to 75°). Consistent with reduction of shape effect with elevation, there is significant improvement of the B$_{z,z}$ magnetisation direction estimates with increasing distance from the magnetisation. In practice it is challenging to fully realise this advantage because of the reduction of amplitude with increased distance from the magnetisation (particularly as we are using field gradients). If upward continuation is used to reduce source shape effects it is best to calculate the field gradients beforehand because at higher elevation the ratio of source gradients relative to the background field gradients is reduced.

The $B_{z,z}$ analysis presented above is based on simple dipole models of the distribution of magnetisation. From Table 9.4 we can see that for the 2.7:1 radial ratio ellipsoid models used, the errors in estimated magnetisation direction are less than 10° except at low elevations close to the magnetisations. We consider this acceptable, particularly as geological complexities and measurement imperfections limit the significance of analyses. To better understand the nature and limitations of the dipole basis of our analysis we also inverted the anomalies using dipole models. Figure 9.8A is a composite illustration of the six different ellipsoid models co-centred and with superimposed spherical (dipole) models from individual inversions of the model fields of those ellipsoids. The blue spheres are from inversion of B_z and the red spheres are from inversion of $B_{z,z}$. All bodies are given identical intensity of magnetisation (inversion of dipole sources is completely insensitive to independent values of volume and intensity of magnetisation and there is only low sensitivity for the ellipsoid models). The spherical B_z inversion models have centres of magnetisation that are shallower than the magnetisation

centres of the input ellipsoid models and horizontally displaced in the up-dip direction. Because they are shallower, they also have a reduced magnetic moment to best-fit the deeper-centred ellipsoids. The $B_{z,z}$ inversions have a higher weighting to the shallowest magnetisation and migrate further from the centres of the input models. As a result they have even shallower and more horizontally displaced centre locations with further reduction of magnetic moments. It seems that the $B_{z,z}$ inversion models are less acceptable representations of the input magnetisations than the B_z inversion models. However, in application to measured field data the advantages of improved anomaly separation provided by use of gradients generally outweigh this disadvantage. This is particularly relevant in design of automated or semi-automated analysis to be applied to multiple anomalies across large distances with different background fields.

Figure 9.8B shows (in mesh view) an ellipsoid model from which an input field is computed and a co-centred spherical model of identical volume and magnetic moment. The red spherical model is derived from a $B_{z,z}$ inversion at an elevation of 200 m above the top of the magnetisation, the green model is from a $B_{z,z}$ inversion at 600 m above the top of the magnetisation and the blue model is from a $B_{z,z}$ inversion at 1,800 m above the top of magnetisation. The (red) smallest and most horizontally and vertically displaced spherical model from the shallowest field inversion has the highest error in estimated magnetisation direction. The higher and more distal models have progressively smaller offsets from the centre of the

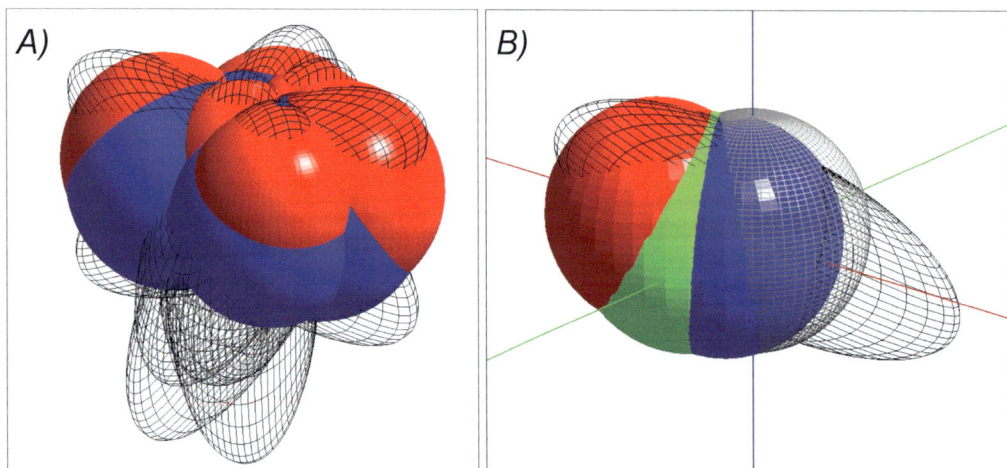

Fig. 9.8. A) co-centred plunging ellipsoid source models with dipole inversion models from individual inversion of the fields of those bodies: B_z inversions (in blue)and Bz,z inversions (in red) and B) Mesh models of a plunging ellipsoid and co-centred, equal volume sphere, with $B_{z,z}$ inversion models of fields 200 m (in red), 600 m (in green) and 1,800 m (in blue) above the top of magnetisation.

Table 9.5. Model statistics derived from inversion of the high-inclination magnetisation I$_M$ = −75°.

Elevation (metres)	Body	Declination (degrees)	Inclination (degrees)	Error (degrees)	Centre offset (metres)	Centre azimuth (degrees)
1,800	A	315.0	−75.0	0.0	0	225
1,800	B	322.1	−72.2	3.4	119	178
1,800	C	329.2	−72.6	4.6	135	216
1,800	D	330.7	−77.4	4.4	130	268
1,800	E	315.0	−72.4	2.6	115	135
1,800	F	300.8	−72.5	4.7	136	55
600	A	315.3	−75.0	0.1	1	211
600	B	331.9	−65.8	10.7	211	179
600	C	343.6	−63.2	15.3	257	215
600	D	21.4	−77.5	15.1	234	269
600	E	315.0	−66.5	8.5	229	135
600	F	286.8	−62.9	15.4	257	56
200	A	316.2	−75.0	0.3	1	221
200	B	338.6	−58.5	18.7	284	179
200	C	343.0	−49.5	28.0	370	215
200	D	53.6	−69.4	27.1	306	269
200	E	316.5	−60.1	14.9	351	135
200	F	287.8	−49.0	28.4	369	55

Table 9.6. Model statistics derived from inversion of the low-inclination magnetisation I$_M$ = −15°.

Elevation (metres)	Body	Declination (degrees)	Inclination (degrees)	Error (degrees)	Centre offset (metres)	Centre azimuth (degrees)
1,800	A	45.0	−15.0	0.0	0	0
1,800	B	45.3	−14.4	0.7	82	186
1,800	C	45.0	−14.7	0.3	88	225
1,800	D	44.7	−14.4	0.7	82	264
1,800	E	44.8	−14.3	0.8	83	140
1,800	F	45.0	−16.8	1.8	111	45
600	A	45.0	−15.0	0.0	0	0
600	B	46.0	−12.7	2.5	160	185
600	C	45.0	−13.4	1.6	189	225
600	D	44.0	−12.8	2.4	160	265
600	E	44.1	−12.3	2.8	185	140
600	F	45.0	−24.1	9.1	246	45
200	A	45.0	−15.0	0.0	0	0
200	B	47.1	−10.2	5.2	239	184
200	C	45.0	−9.6	5.4	316	225
200	D	42.9	−10.3	5.2	239	266
200	E	43.5	−9.7	5.5	304	138
200	F	45.0	−48.8	33.8	443	45

input model and their magnetisation directions converge towards its magnetisation direction. If the anomaly could be detected and successfully inverted at great distance (in a truly distal field) the inversion dipole model should be a spherical equivalent of the ellipsoid model with identical centre coordinates and magnetic moment (including magnetisation direction) as represented by the mesh view sphere in Fig. 9.8B.

The key model statistics for inversion of the high-inclination and low-inclination magnetisations are listed in Tables 9.5 and 9.6 respectively. These results are similar to the dipole-based $B_{z,z}$ analysis results in Tables 9.2 and 9.3. The inversion summary statistics presented in Table 9.4 are also close to the equivalent $B_{z,z}$ analysis statistics, confirming that the dipole-based $B_{z,z}$ analysis performs as expected for fields of dipole sources.

The statistics for estimation of the centre of magnetisation listed in Tables 9.2, 9.3, 9.5 and 9.6 are plotted in map form in Fig. 9.9. The key features recognisable in this figure are that:

1) the $B_{z,z}$ analysis (Fig. 9.9A) and dipole $B_{z,z}$ inversions (Fig. 9.9B) perform almost identically in estimating the centres of magnetisation

2) the plunge azimuth of each body causes a systematic displacement of its estimated centre. There are only minor differences between the displacements for the two magnetisation directions despite their wide angular separation

3) the absolute displacements of the estimated centres of magnetisation are a minimum for $B_{z,z}$ analyses and inversions at the highest elevation (plotted as symbols in blue in Figs 9.9 to 9.11). This is because the more distal field is least influenced by body shape, including plunge. $B_{z,z}$ analyses and inversions of the field at the shallowest elevation (plotted as symbols in red in Figs 9.9 to 9.11) are most strongly displaced towards the shallow up-dip extents of the plunging input models

4) the minimum displacements of the estimated centre of magnetisation from both the $B_{z,z}$ analyses and inversions are for the body that has a vertical axis with no plunge (body 'A').

Figure 9.10 plots the magnetisation direction estimates resulting from the $B_{z,z}$ analyses (Fig. 9.10A) and inversions (Fig. 9.10B). The symbols are identical to those used in Fig. 9.9. Figure 9.10 shows different dispersions of estimated magnetisation directions for the low-inclination magnetisation (plotted as points) and the steep-inclination magnetisation (plotted as crosses). For the low-inclination magnetisation almost all the error in magnetisation direction is in estimation of inclination. For the steeper-inclination magnetisation there is a more equal spread of directions in both declination and inclination. This difference exceeds the reduced significance of declination at high inclinations. The larger rotations of magnetisation direction away from the input

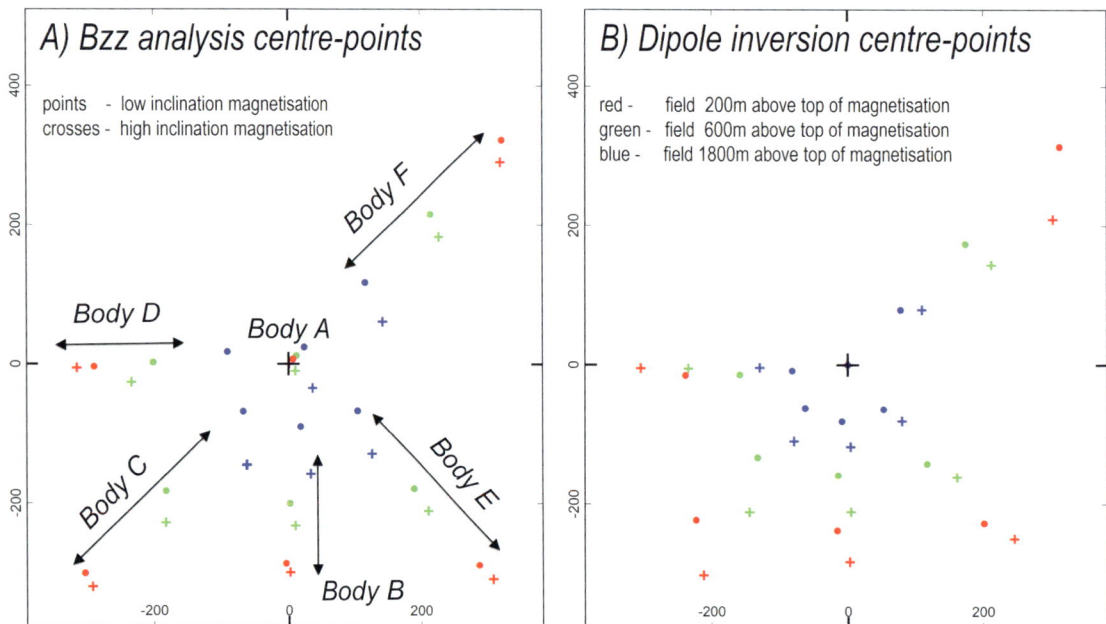

Fig. 9.9. Displacements of estimated centres of magnetisation: A) from the Bz,z analysis and B) from the dipole inversions.

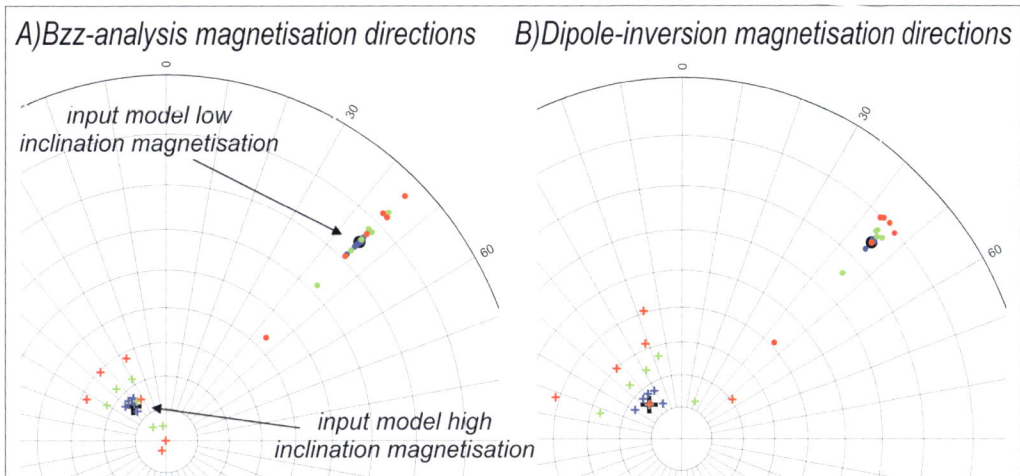

Fig. 9.10. B$_{z,z}$ analysis (A) and inversion model (B) magnetisation directions. Symbols are the same as in Fig. 9.9.

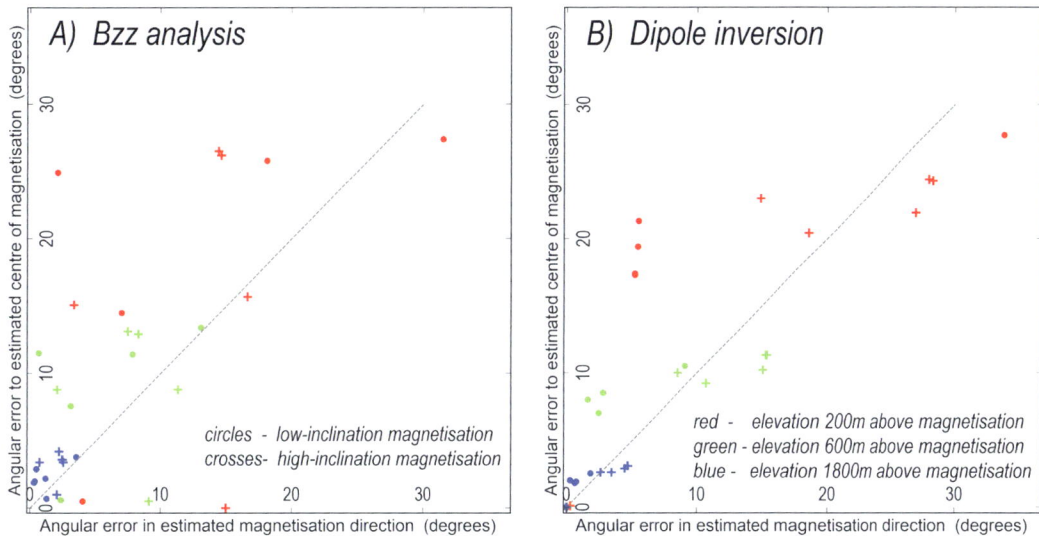

Fig. 9.11. Cross-plots of vertical axis: angular error to the centre of magnetisation (from the computation elevation) against horizontal axis: error in magnetisation directions. Symbols are the same as in Figs 9.9 and 9.10.

magnetisation direction are for B$_{z,z}$ analyses and inversions at the lowest elevation (the red symbols) consistent with those results also having the most substantial displacements of the estimated centres of magnetisation. This correspondence between displacement of estimated centre of magnetisation and error in estimated direction of magnetisation is highlighted in Fig. 9.11 which shows the angular error in positioning the centre of magnetisation (at the elevation of the computed magnetic field) cross-plotted against the angular error in estimated magnetisation direction. This figure emphasises the significant difference between strong influence of source plunge on the magnetic field proximal to the shallowest magnetisation (recorded with red symbols) and the much weaker influence on the more distal field (recorded with blue symbols).

9.4 A FIELD TEST OF THE ANALYSIS APPLIED TO REGIONAL TMI DATA FROM THE WHITE MOUNTAINS NATIONAL PARK AREA, EASTERN QUEENSLAND

Figure 9.12 shows the location of the test area in the White Mountains National Park of north-east Queensland. The area is covered by the Drummond and Galilee aeromagnetic survey (Geoscience Australia survey

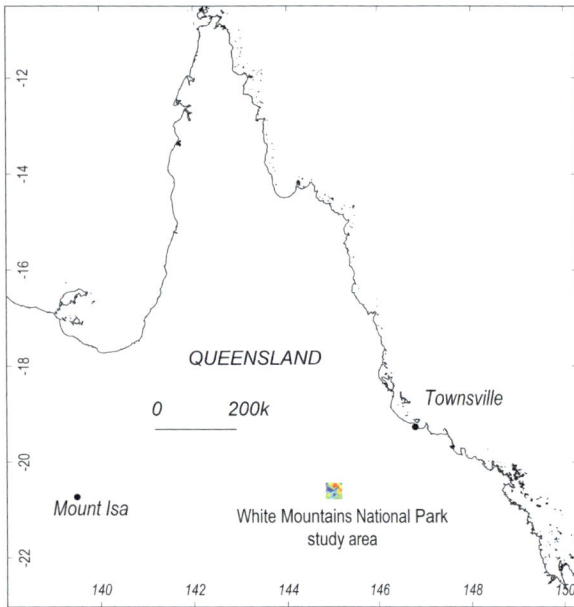

Fig. 9.12. Location of the study area in the White Mountains National Park, eastern Queensland.

sharp, sub-circular anomalies of quite variable pattern, revealing that they are caused by shallow magnetisations of different orientation. Most of the anomalies are defined by measurements on a minimum of three lines, with some anomalies defined by measurements on up to five lines. The area lies within the Hughenden 1:250,000 geological map (Vine and Paine 1974). This is an area of limited outcrop but the most feasible source of the magnetisations are Cenozoic volcanics. Figure 9.13B shows the $B_{z,z}$ transform of the TMI grid that highlights the local anomalies of interest which are of dual polarities and have a wide range of patterns, indicating a wide range of magnetisation directions.

We individually inverted the 11 anomalies highlighted in Fig. 9.14 using clips of the flightline data as shown in the figure. For each anomaly a dedicated background field was defined using a second order polynomial surface. The model-computed field added to this regional was matched by inversion to the tie-line levelled data channel. The inversions used faceted body approximations to plunging pipes of elliptic cross-section with horizontal top and bottom faces. For each single body inversion there are 13 model free parameters: reference point easting, northing and elevation, two cross-section radii, long-axis azimuth, plunge, plunge azimuth, resultant magnetisation intensity, declination and inclination and background field slope and two planar gradients. Several of the anomalies that are immediately

number P793) flown on flightlines with azimuths of 070°–250° at a nominal ground clearance of 80 m and line spacing of 400 m. The data are available for download from the Geoscience Australia Geophysical Archive Data Delivery (GADDS). Figure 9.13A shows TMI variation across a part of that survey area. Broad, medium amplitude magnetic field variations are punctured by

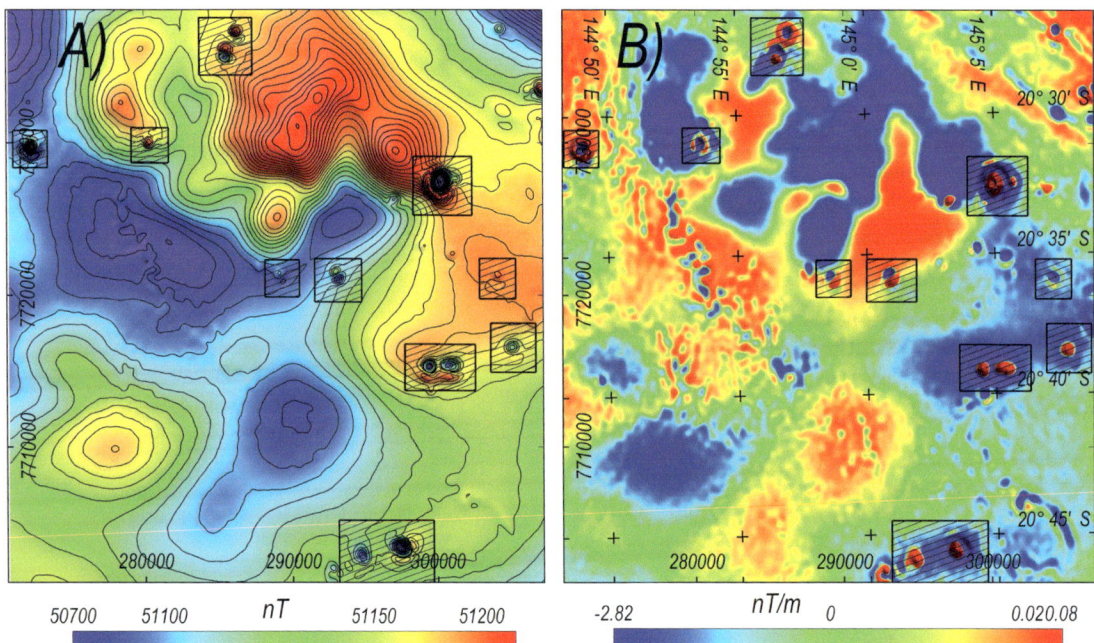

Fig. 9.13. A) TMI and B) the $B_{z,z}$ transform of TMI in the test area.

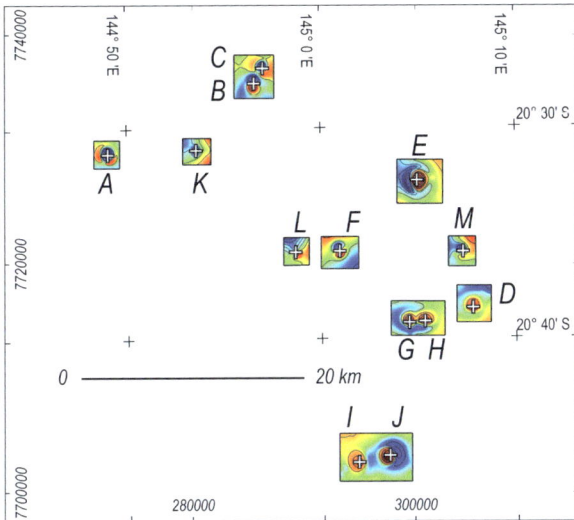

Fig. 9.14. Selected anomalies B$_{z,z}$.

adjacent to each other were investigated in combined double-body inversions because their magnetic fields overlap to the extent that it is difficult to analyse them separately. The magnetisation directions recovered from the inversions were checked by repeating the inversions with substituted triaxial ellipsoid bodies using the same data selections and initial background field definitions but completely independent forward modelling algorithms. In all cases the magnetisation directions from

these check inversions were within 4 degrees of the directions for the pipe models. We report the pipe model results in Table 9.7 because these models provide estimates of depth to the top of magnetisation (a statistic that has little meaning or reliability for the ellipsoid models). Table 9.7 lists depth below sensor to the top of magnetisation and the major diameter for each model. The high ratio of the width of magnetisation to its depth below sensor reveal that the magnetic field has been measured in moderate proximity to the magnetisations.

Figure 9.15 shows a contour plot of the measured TMI field and of TMI forward computed from the post-inversion two-body inversion model for anomalies B and C. The goodness of fit is very high but the input data defining the anomaly is sparse. The peak and trough of each anomaly are defined on only single flightlines, with the adjacent lines serving mostly to constrain the extents of the anomalies. It is most likely that the peak and trough values of these anomalies, and the azimuths between those points would be different if the survey had been flown on lines with an offset of half the flightline spacing, or with infill lines flown between the existing lines. The B$_{z,z}$ analysis results are dependent on only those key statistics of peak and trough values and their locations but hopefully this sampling influence is diluted in inversion results derived for the complete population of anomalies.

Table 9.7. Pipe model spatial and magnetisation parameters.

Anomaly	Intensity of magnetisation (A/m)	Depth below sensor (metres)	Major X-section diameter (metres)	Magnetic moment (A.m^2.10^6)	Plunge angle (degrees)	Plunge azimuth (degrees)
A	2.624	87	422	118	65	7
B	1.458	44	546	80	63	178
C	1.326	85	312	182	69	175
D	2.357	410	478	49	45	0
E	1.550	64	1,188	725	50	176
F	4.881	243	370	19	44	0
G	0.845	180	564	93	89	180
H	1.120	262	1,068	144	73	180
I	1.410	487	942	120	60	142
J	2.985	514	682	274	84	177
K	2.205	225	130	10	85	148
L	1.017	107	236	7	50	3
M	.825	226	806	6	26	192

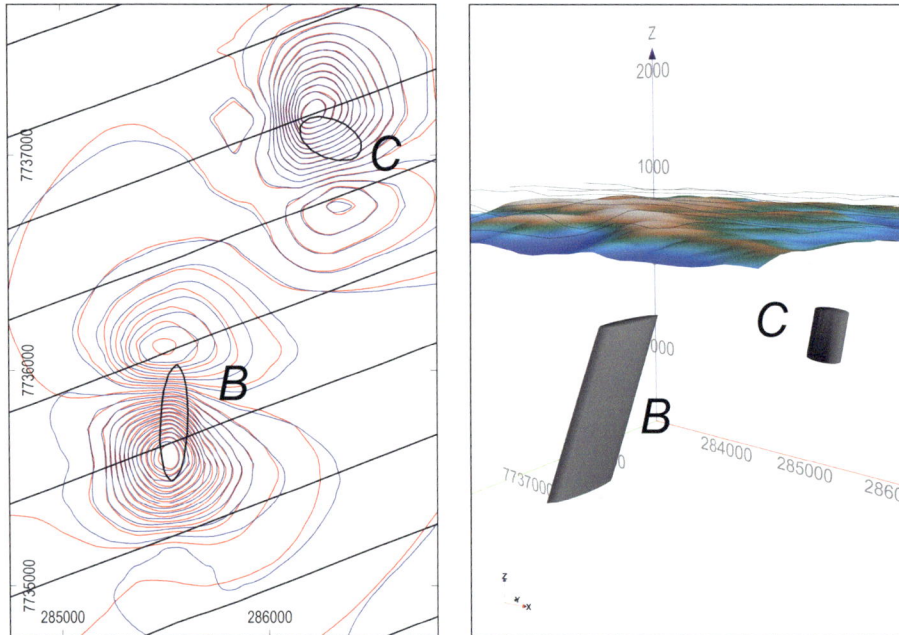

Fig. 9.15. A) measured (blue) and model-computed (red) TMI contours at 10 nT intervals for anomalies B and C with outlines of the source models, and B) perspective view of the models with flightlines and the ground surface.

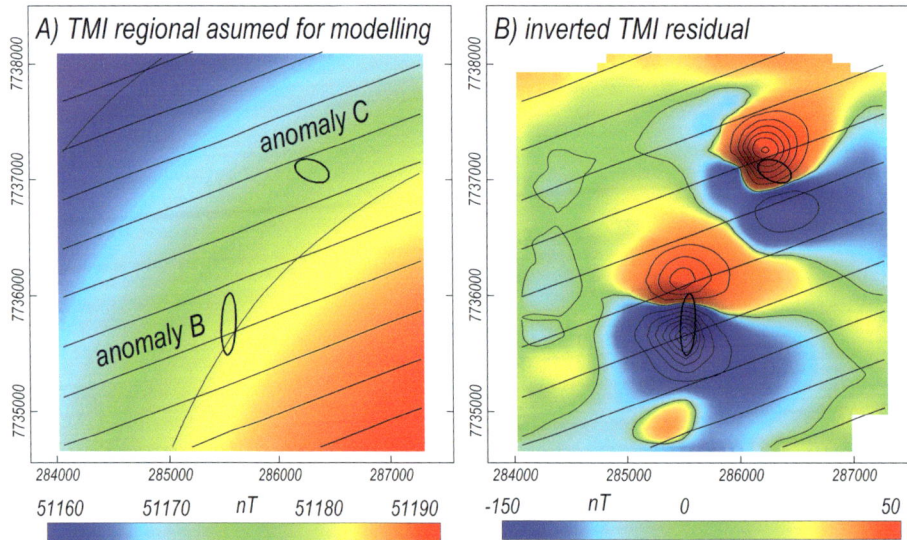

Fig. 9.16. A) regional and B) residual separations of the B and C anomalies (contour intervals 20 nT).

Figure 9.16A images the regional field variation assumed in inversion of anomalies B and C, and Fig. 9.16B images the residual field separation from that regional field. The two fields are contoured at the same interval, showing the regional field variations are subdued in comparison with the amplitudes of the residual anomalies. Superimposed regional fields change the asymmetry of the anomalies, and if this influence is not suppressed it leads to miscalculation of magnetisation direction. The major concern with the regional field is that its influence is amplified if upward continuation is applied to supress shape effects of the magnetisations. Figure 9.17A shows contours of $B_{z,z}$ derived from the TMI field of Fig. 9.15A after it had been upward continued by 400 m (in blue) and the corresponding contours of $B_{z,z}$ (in red) derived after 400 m upward continuation of the residual field separation shown in Fig. 9.16B. These contours are very similar and suggest that for these two

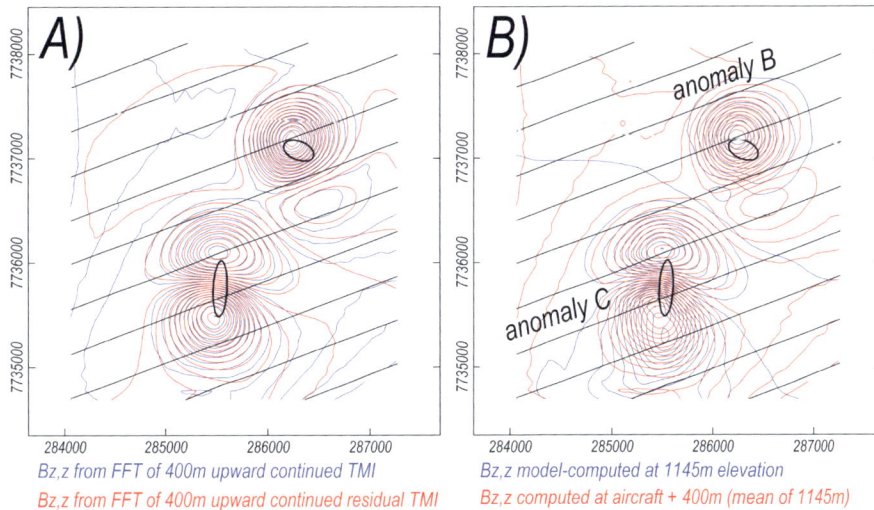

Fig. 9.17. B$_{z,z}$ at elevation 400 m above the survey elevation: A) from FFT upward continuation of measured and residual TMI, and B) model-computed at constant elevation and on survey lines +400 m.

anomalies, regional field separation before upward continuation is not critical (however we anticipate that this may not be the case for other anomalies). Upward continuation should ideally be applied to samples of the field equidistant from the magnetisation but this is not true of the field measurements. Of greater concern are sharp variations in elevation, particularly in areas of rugged relief where the sensor height above the magnetisation is variable both along flightlines and between them. Figure 9.17B shows contours of B$_{z,z}$ derived from the 400 m FFT upward continued residual TMI (in red) and (in blue) of B$_{z,z}$ directly computed from the pipe model after addition of 400 m to the flightline sensor elevation. Differences between these two estimates of B$_{z,z}$ vary with position (controlled by the terrain and its influence on flying height). They are highly disruptive in estimation of the detail of depth to the top of a magnetisation (see for instance Chapter 8) but fortunately have less effect on the more robust estimation of magnetisation direction. For all but the smallest and weakest of the anomalies, upward continuation is beneficial in estimation of magnetisation direction by the B$_{z,z}$ method proposed here. Over a particular area there is likely to be an optimal height for the population of anomalies that compromises attenuation of magnetisation shape effects and introduction of distortion from the upward continuation. Discovery of that 'best' upward continuation height will require testing with inversions of selected anomalies and accumulation of experience.

Table 9.8 lists differences between magnetisation directions recovered from B$_{z,z}$ analysis and dedicated TMI inversions. The mean difference is 10° and the maximum difference is 25°. Individual attention in developing the inversion solutions should result in higher reliability of those directions and we assume that differences between the inversion and B$_{z,z}$ analysis directions are predominantly due to higher error of the B$_{z,z}$ analysis. Inversions of even the most prominent anomalies are expected to have directional uncertainties of at least 5–10° (see Chapter 10) so the mean B$_{z,z}$ analysis error for this population of solutions may be of the order of 15° or more.

The magnetisation direction estimates are plotted in stereographic projection in Fig. 9.18. Dispersion of the individual anomaly inversion and B$_{z,z}$ analysis directions is much greater than expected for the apparent uncertainty in the estimates and is consistent with the wide range of anomaly patterns. The Cenozoic volcanics in this area are related to hot spot activity and may be extruded over a long period. The individual small and shallow bodies also cool quickly and may not faithfully record a stable geomagnetic field direction. Combination of these two factors may explain the wide scatter of magnetisation directions. Independent confirmation of magnetisation direction requires magnetic susceptibility and remanent magnetisation measurement on oriented samples that in this area of sparse (and probably weathered) outcrop would mostly require availability of drillhole cores. Alternatively, the individual anomalies could be measured at lower elevation and higher spatial resolution using drones or ground magnetic surveys to support more diagnostic inversion and B$_{z,z}$ analysis.

Table 9.8. Comparison of $B_{z,z}$ analysis and pipe model inversion magnetisation directions.

Anomaly	Peak-trough amplitude ratio	$B_{z,z}$ analysis declination (degrees)	$B_{z,z}$ analysis inclination (degrees)	Inversion declination (degrees)	Inversion inclination (degrees)	Angular difference (degrees)
A	14.2	334	−78	1	−80	5
B	1.7	3	16	7	16	4
C	4.7	334	−45	11	−46	25
D	13.6	206	77	192	76	3
E	4.0	259	41	255	48	8
F	1.0	340	1	335	−8	10
G	6.6	213	55	188	54	14
H	5.0	215	48	189	47	18
I	8.8	84	64	95	64	5
J	5.1	74	48	67	50	5
K	8.7	61	−64	36	−62	11
L	1.2	329	−5	334	8	14
M	3.3	350	−35	346	−42	8

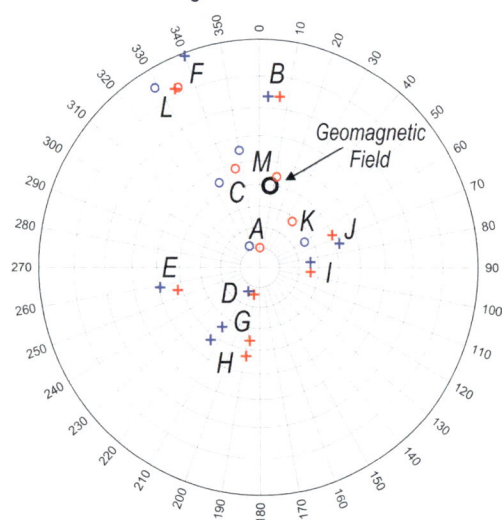

Fig. 9.18. White Mountains magnetisation directions. Red – from $B_{z,z}$ analysis and Blue – from model inversions. Circles are negative inclination, cross are positive inclination.

9.5 CONCLUSIONS

We have established that over compact magnetisations, $B_{z,z}$ anomalies derived by FFT analysis of moderate- to high-resolution TMI data can provide approximate estimates of magnetisation direction. The declination of magnetisation is indicated by the azimuth of the $B_{z,z}$ anomaly and the inclination of magnetisation is indicated by its peak to trough amplitude ratio. The analysis is almost independent of the geomagnetic field direction. Magnetisation estimates derived from measurements close to a magnetisation are corrupted by the influence of the distribution of magnetisation. In more distal fields, magnetisation shape effects are less significant but reduced anomaly amplitudes cause problems in reliable anomaly separation from background and overlapping fields. We have demonstrated recovery of known magnetisation directions from $B_{z,z}$ analysis of synthetic data and shown that magnetisation directions consistent with parametric inversion results are recovered from $B_{z,z}$ analysis of magnetic field data measured above a series of small Cenozoic volcanic bodies in Queensland.

REFERENCES

Baranov V, Naudy H (1964) Numerical calculation of the formula of reduction to the magnetic pole. *Geophysics* **29**, 67–79. doi:10.1190/1.1439334

Beiki M, Clark DA, Austin JR, Foss CA (2012) Estimating source location using normalised magnetic source strength estimated from gradient tensor data. *Geophysics* **77**, J23–J37. doi:10.1190/geo2011-0437.1

Blakely RJ (1995) Potential theory in gravity and magnetic applications. Cambridge University Press.

Blakely RJ, Simpson RW (1986) Approximating edges of source bodies from magnetic or gravity anomalies. *Geophysics* **51**, 1494–1498. doi:10.1190/1.1442197

Caratori Tontini F, Pedersen LB (2008) Interpreting magnetic data by integral moments. *Geophysical Journal International* **174**, 815–824. doi:10.1111/j.1365-246X.2008.03872.x

Clark DA (2012) New methods for interpretation of magnetic vector and gradient tensor data I: eigenvector analysis and the normalised source strength. *Exploration Geophysics* **43**, 267–282. doi:10.1071/EG12020

Clark DA (2013a) New approaches to dealing with remanence: magnetic moment analysis using tensor invariants and remote determination of *in-situ* magnetisation using a static tensor gradiometer. ASEG-PESA Conference, Melbourne, Extended Abstract 1–7.

Clark DA (2013b) New methods for interpretation of magnetic vector and gradient tensor data II: application to the Mount Leyshon anomaly, Queensland, Australia. *Exploration Geophysics* **44**, 114–127. doi:10.1071/EG12066

Clark DA (2014) Methods for determining remanent and total magnetisations of magnetic sources – a review. *Exploration Geophysics* **45**, 271–304. doi:10.1071/EG14013

Cooper G, Cowan D (2003) The application of fractional calculus to potential field data. *Exploration Geophysics* **34**, 51–56. doi:10.1071/EG03051

Foss CA, Austin JR (2023) Distal, proximal and sweet spot limitations in source information content of magnetic field data. 4th AEGC Conference, 13–18 March 2023, Brisbane, Australia, 6p.

Foss CA, McKenzie B (2011) Inversion of anomalies due to remanent magnetisation: an example from the Black Hill Norite of South Australia. *Australian Journal of Earth Sciences* **58**, 391–405. doi:10.1080/08120099.2011.581310

Helbig K (1963) Some integrals of magnetic anomalies and their relation to the parameters of the disturbing body. *Zeitschrift für Geophysik* **29**, 83–96.

Lourenço JS, Morrison HF (1973) Vector magnetic anomalies derived from measurements of a single component of the field. *Geophysics* **38**, 359–368. doi:10.1190/1.1440346

McKenzie KB (2020) The magnetic gradient tensor of a triaxial ellipsoid, its derivation and its application to the determination of magnetisation direction. *Exploration Geophysics* **51**, 609–641. doi:10.1080/08123985.2020.1726176

McKenzie KB (2022) The magnetic field and magnetic gradient tensor for a right circular cylinder. *Exploration Geophysics* **53**, 329–358. doi:10.1080/08123985.2021.1951117

Medeiros WE, Silva JBC (1995) Simultaneous estimation of total magnetisation direction and 3-D spatial orientation. *Geophysics* **60**, 1365–1377. doi:10.1190/1.1443872

Phillips JD, Hansen RO, Blakely RJ (2007) The use of curvature in potential field interpretation. *Exploration Geophysics* **38**, 111–119. doi:10.1071/EG07014

Pratt DA, McKenzie KB, White AS (2014) Remote remanence estimation (RRE). *Exploration Geophysics* **45**, 314–323. doi:10.1071/EG14031

Schmidt PW, Clark DA (1998) Calculation of magnetic components and moments from TMI: a case study from the Tuckers igneous complex, Queensland. *Exploration Geophysics* **29**, 609–614. doi:10.1071/EG998609

Schmidt PW, Clark DA (2006) The magnetic gradient tensor: its properties and uses in source characterization. *The Leading Edge* **25**, 75–78. doi:10.1190/1.2164759

Vine RR, Paine AGL (1974) Hughenden, Queensland – 1:250 000 Geological Series. Bureau of Mineral Resources of Australia. Explanatory Notes SF/55-1

Zietz I, Andreasen GE (1967) Remanent magnetisation and aeromagnetic interpretation. *Mining Geophysics* **2**, 569–590.

10

Improved constraint of magnetisation direction from UAV surveys of under-sampled aeromagnetic anomalies

C.A. Foss, M. Takáč and G. Kletetschka

ABSTRACT

Aeromagnetic surveys are a rapid and relatively cheap method of remote sensing to map the distribution of sub-surface magnetisations. They cover large areas but at a line spacing that is a compromise between resolution of the data and cost. Sampling sufficiency of aeromagnetic data is particularly a problem for investigation of small, shallow magnetisations. We investigate limitations in mapping both the spatial distribution of magnetisation and its direction from sparsely sampled data and illustrate this analysis using two anomalies measured in aa aeromagnetic survey flown at 250 m line spacing over parts of New South Wales and the Australian Capital Territory. The anomalies are due to adjacent shallow remanence-dominated magnetisations. We show that from the aeromagnetic data the direction of magnetisation is weakly constrained for the larger body and is even more uncertain for the smaller body. We flew two UAV surveys at 58 and 35 m elevation and 20 m line spacing. From the UAV surveys we recovered more reliable estimates of source magnetisation direction for both bodies. An effective survey strategy is to fly aeromagnetic surveys at line spacings close enough to confidently detect anomalies of interest and then fly UAV surveys over selected areas where estimates of source magnetisation need to be upgraded.

10.1 ESTIMATION OF MAGNETISATION DIRECTION FROM MAGNETIC FIELD ANALYSIS OF COMPACT SOURCES

Helbig (1963) established that the direction of magnetisation for a dipole of known location can be uniquely derived from analysis of its external magnetic field, and Zietz and Andreasen (1967) showed that in a steep inclination northern hemisphere geomagnetic field the declination of magnetisation of a compact source can be estimated from the azimuth of its total magnetic intensity (TMI) peak to trough vector and the inclination of magnetisation can be estimated from the peak to trough (P:T) amplitude ratio. Figure 10.1 shows an induced magnetisation anomaly over a dipole in a geomagnetic field of inclination $-60°$ (Fig. 10.1A) and anomalies due to a 30° clockwise horizontal rotation of magnetisation direction (Fig. 10.1B), 30° shallowing (Fig. 10.1C) and 30° steepening (Fig. 10.1D). In this southern hemisphere geomagnetic field, declination of magnetisation is indicated by the trough to peak azimuth as opposed to the peak to trough azimuth in the northern hemisphere. The TMI anomaly of the northerly directed induced magnetisation in Fig. 10.1A has a trough to peak azimuth of 0° and a P:T ratio of 5.54. The anomaly for the 30° horizontal clockwise rotation of magnetisation (Fig. 10.1B) causes a 31° rotation of the trough to peak azimuth and only a minor adjustment of the P:T ratio to 8.1:0. The 30° shallowing of magnetisation (Fig. 10.1C) does not rotate the trough to peak azimuth but reduces the P:T ratio to 2.10, and the 30° steepening of magnetisation direction (Fig. 10.1D) also leaves the trough to peak azimuth unrotated but increases the P:T ratio to 12.93. These relationships are consistent with the Zietz and Andreasen (1967) observations and predictions for anomalies in a steep geomagnetic field.

For the well-defined anomalies shown in Fig. 10.1 the directions of magnetisation can be reliably estimated from their anomaly patterns. However, if the anomalies are only sampled on one or two flightlines there is insufficient information to reliably map those patterns. For instance, if the anomaly is only measured on a single central east–west profile the anomaly trough is not sampled and neither the trough to peak azimuth or peak to trough ratio can be determined.

Figure 10.2 shows flightline plans for a 400 m line-spaced survey superimposed on an image of the TMI field of a vertical circular cylinder of radius 100 m, depth extent 150 m and depth to top 100 m below the computed field elevation. The magnetisation has an intensity of 5 Amp/metre and is parallel to the local geomagnetic field of inclination −60°, declination 000°. The flightlines in

Fig. 10.2A are centred over the magnetisation and the flightlines in Fig. 10.2B are off-centred by 200 m (50% of the line spacing). Figure 10.3 shows stacked profiles of TMI measured on these flightlines. The top row (Figs 10.3A and 10.3C) shows stacked profiles for body-centred surveys along north–south and east–west flightlines respectively. In each case a single profile records high amplitudes of the main anomaly peak with much lower amplitudes on the adjacent lines displaced 400 m from the anomaly peak.

The bottom row of Fig. 10.3 (Figs 10.3B and 10.3D) shows stacked profiles for the off-centred surveys along north–south and east–west profiles respectively. For both flightline orientations the anomaly is predominantly expressed on two adjacent profiles. For the north–south profiles (Fig. 10.3B) the two profiles are identical and of significantly lower amplitude than the missing central profile through the anomaly peak (Fig. 10.3A). TMI variations on the east–west profiles (Fig. 10.3D) are of opposite polarities. The substantial differences between adjacent profiles in Figs 10.3A and 10.3B and in Figs 10.3C and 10.3D reveal that for either profile orientation, definition of the magnetic field would benefit considerably from closer line-spaced coverage.

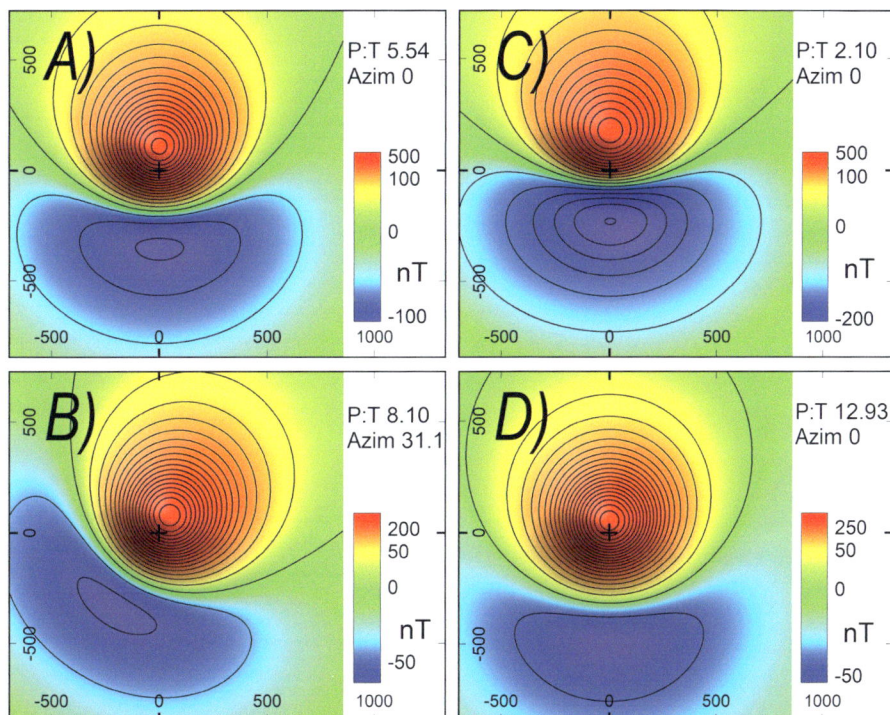

Fig. 10.1. TMI anomalies in a geomagnetic field of declination 0°, Inclination −60° for dipole magnetisations with centre at 400 m depth and magnetisation directions: A) Dec 0° Inc. −60°, B) Dec 30° Inc. −60°, C) Dec 0° Inc. −30°, D) Dec 0°, Inc. −90°.

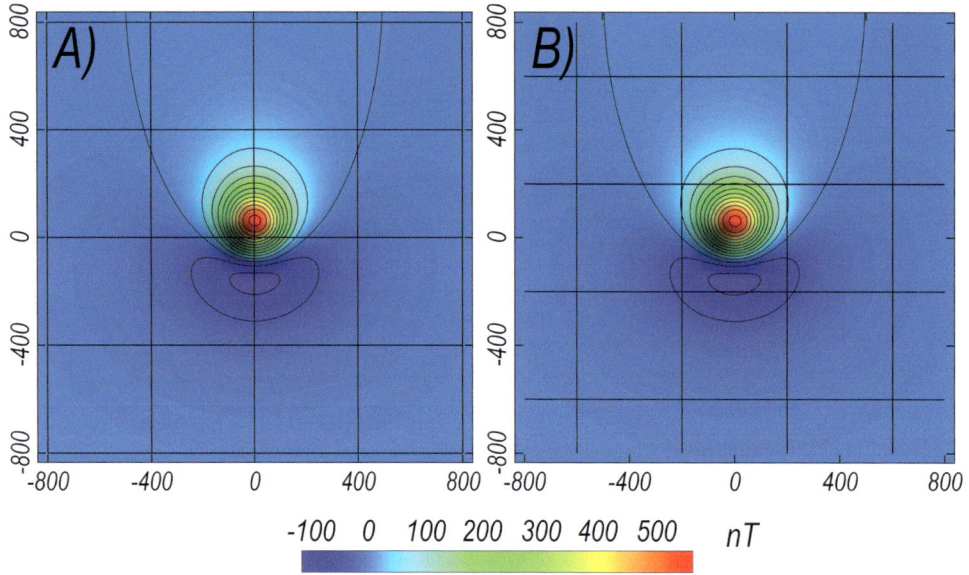

Fig. 10.2. TMI anomalies due to a vertical cylinder of 100 m radius, depth to top 100 m and depth extent 150 m in a geomagnetic field of declination 0°, Inclination −60° with overlays of flightline plans at 400 m spacing: A) centred on the body and B) off-centred.

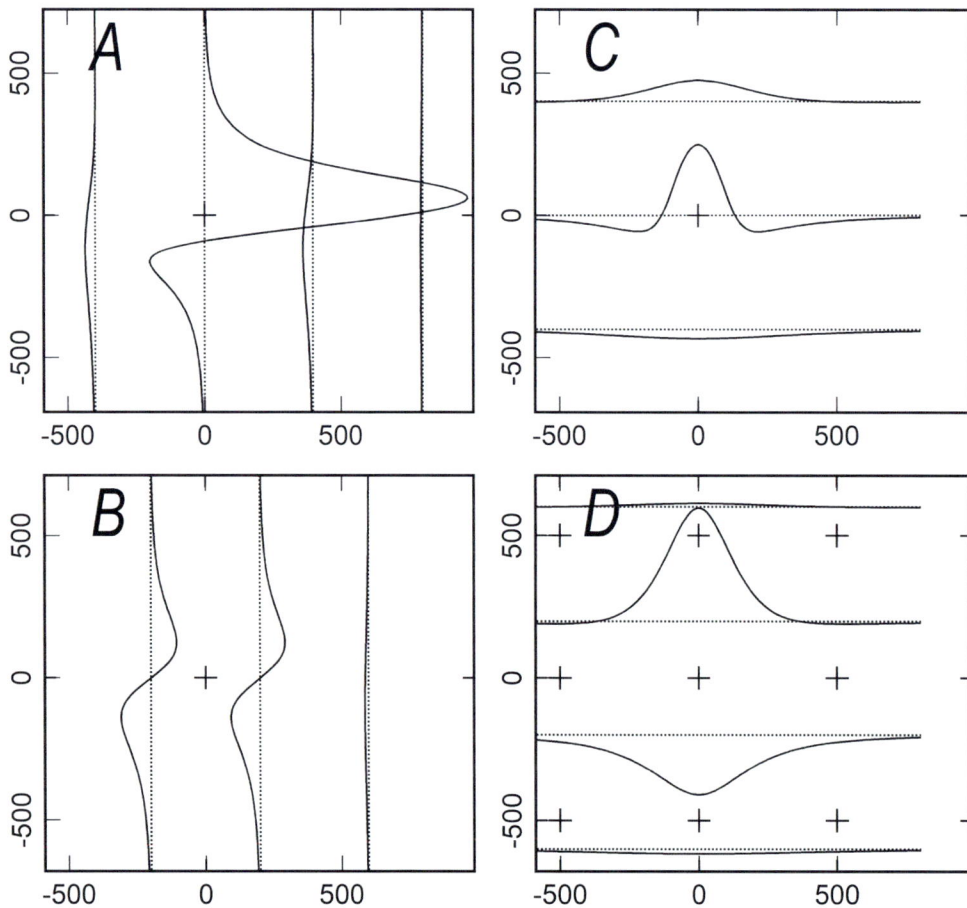

Fig. 10.3. Stacked profiles of TMI for the flightlines shown in Fig. 10.2: A) on-centred N–S lines, B) off-centred N–S lines, C) on-centred E–W lines, D) off-centred E–W lines.

The pattern of TMI variation along each profile in Fig. 10.3 is consistent with the computed TMI images of Fig. 10.2 but the images of Fig. 10.2 cannot be generated from the limited sampling of either 400 m spaced profile set. Anomaly images generated by gridding of each of the four individual 400 m profile sets are shown in Fig. 10.4. All four images are quite different, revealing both the insufficiency of the individual profile sets in sampling the magnetic field and also the substantial influence of flightline orientation. The images from gridding the body-centred profile sets (Figs 10.4A and 10.4C) both peak close to the centre of magnetisation and have peak amplitudes close to the input anomaly peak amplitude. Neither grid image in Figs 10.4A or 10.4C reliably reveals the declination of magnetisation. For the grid in Fig. 10.4A the trough to peak azimuth is constrained by the north–south profile orientation and the grid of Fig. 10.4C is derived from data that do not sample the anomaly trough at all (see Fig. 10.2A). As a consequence, neither declination or inclination of the magnetisation can be reliably determined from these grids.

The grid images of Figs 10.4B and 10.4D are derived from the off-centred profile sets that do not sample the sharp curvatures and high amplitudes at the centre of the anomaly. These grids are even less satisfactory representations of the true field than are the grids from the body-centred profile sets and their lack of sharp curvature suggests that the magnetisation is deeper than it really is.

Total gradient (TG) transforms (Roest and Pilkington 1993; Medeiros and Silva 1995; Wijns *et al.* 2005; Li 2006) and normalised source strength (NSS) transforms (Beiki *et al.* 2012) can be applied to TMI grids to highlight the centre of a magnetisation with reduced dependence on magnetisation direction. However, these gradient dependant enhancements of data are more demanding of sampling sufficiency (Reid 1980). TG transforms of the four TMI grids in Fig. 10.4 are imaged in Fig. 10.5. For the anomaly detected on the body-centred surveys (Figs 10.5A and 10.5C) the TG anomalies are point anomalies centred on the highest sampled TMI value and with minor elongation in the along-line direction (north–south in Fig. 10.5A and east–west in Fig. 10.5C). For the off-centred surveys (Figs 10.5B and 10.5D) the TG peaks are also centred on the extreme sampled TMI values. Gradient enhancement images with peaks centred only on the flightlines is one indication that a magnetic field is most likely under-sampled by the available measurements. For the off-centred north–south flightline survey

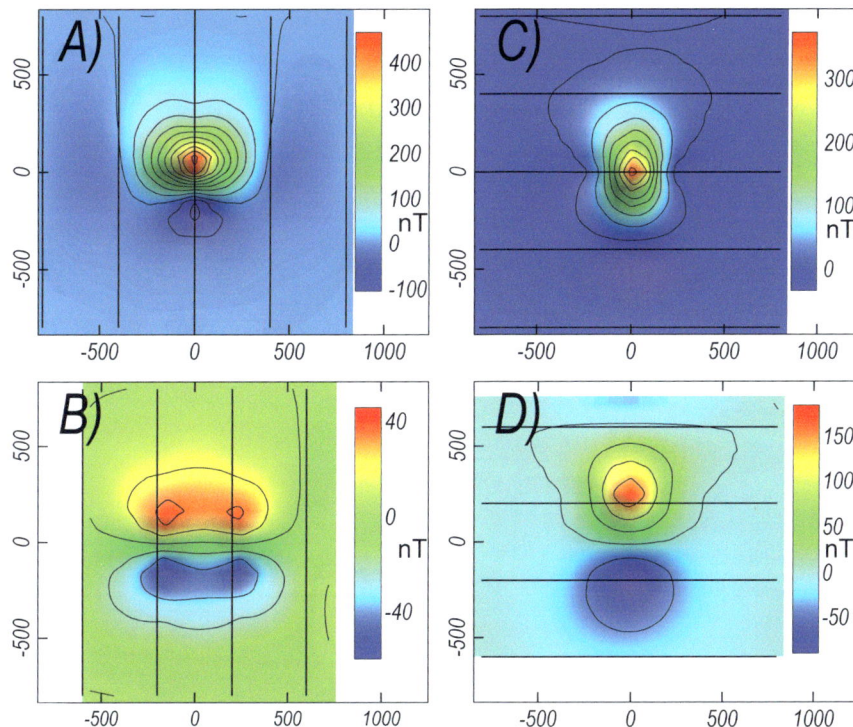

Fig. 10.4. TMI images from gridding of flightline data at 400 m spacing over a vertical cylinder of radius 100 m and depth to top 100 m: A) on-centred N–S lines, B) off-centred N–S lines, C) on-centred E–W lines, D) off-centred E–W lines. Magnetic field and magnetisation directions are both declination 0°, inclination −60°.

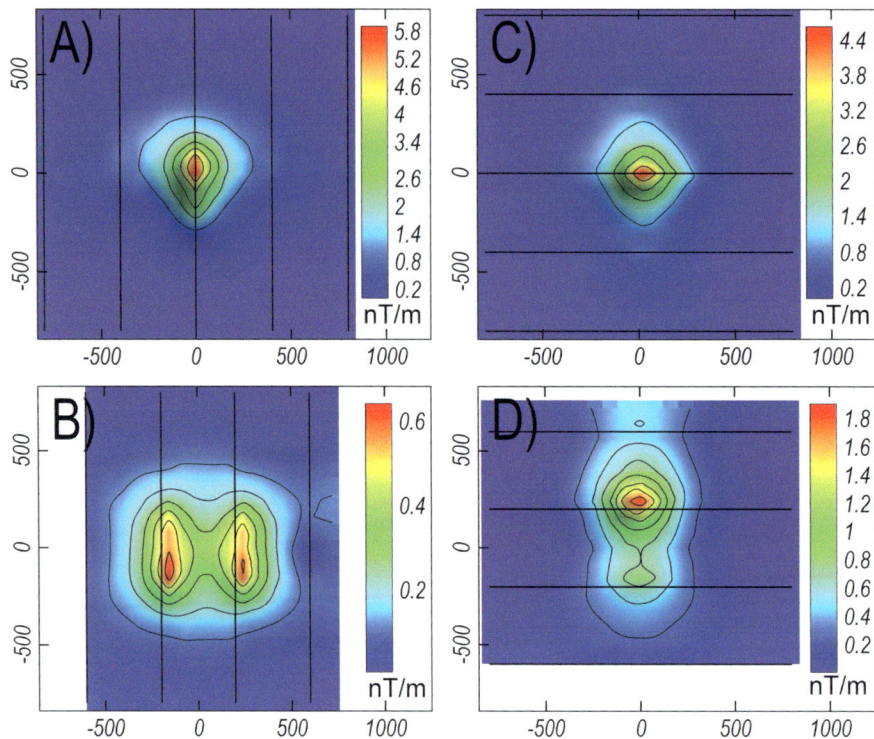

Fig. 10.5. Total gradient (TG) transforms of the grids in Fig. 10.4.

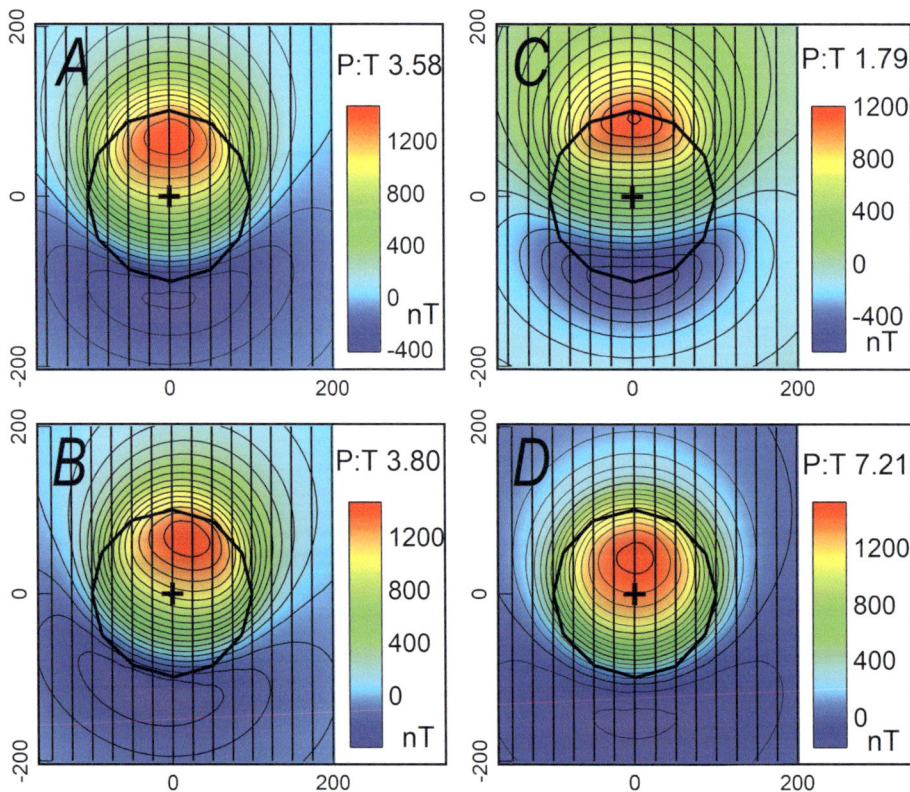

Fig. 10.6. TMI images over the same vertical pipe as for the images of Figs 10.2 to 10.5 but at a reduced elevation of 40 m above the pipe and reduced line spacing of 25 m. Magnetisation directions: A) Dec 0° Inc. −60°, B) Dec 30° Inc. −60°, C) Dec 0° Inc. −30°, D) Dec 0°, Inc. −90°. The geomagnetic inclination is −60°.

(Fig. 10.5C) the grid peaks identically and with strong north–south trend on the two flightlines offset to either side of the anomaly. TG transform of the off-centred east–west flightline survey data also produces two peaks, but in this case the southerly TG peak due to the TMI anomaly trough is less prominent than the northerly TG peak due to the TMI anomaly peak. NSS transforms of the data produce very similar results to the TG transforms and neither transform provides advantage in imaging this under-sampled data other than by highlighting insufficiency of the sampling.

Figure 10.6 shows images of the TMI field computed from the same vertical axis, circular section magnetisation as for Figs 10.2 to 10.5 but in this case with the field computed on more closely spaced lines at 25 m spacing and at the lower elevation of 40 m above the top of the magnetisation. Despite the increased spatial resolution demands of mapping a sharper gradient field at a lower elevation, the substantial reduction in line spacing from 400 m to 25 m provides a far more sufficient sampling of the magnetic field. Figures 10.6A and 10.6B clearly reveal the difference in trough to peak orientation distinctive of the 30° difference in declination of those magnetisations and from both images it is feasible to recover reliable estimates of declination. Figures 10.6B, 10.6A and 10.6C show the change in magnetic fields due to progressive shallowing of a northerly directed magnetisation with inclinations of –90°, –60° and –30° respectively. Consistent with predicted behaviour, the P:T ratio for these three inclinations reduces with reduced steepness of inclination with values of 7.2, 3.6 and 1.8. These values are similar to those for the dipole magnetisations imaged in Figs 10.1D, 10.1A and 10.1C that have values of 12.9, 5.5 and 2.1. The improvements in mapping the magnetic field and recovering source magnetisation information from the lower level, more closely spaced data illustrate the advantage we should expect in flying an unmanned aerial vehicle (UAV) survey to further investigate an anomaly discovered but incompletely specified on a regional aeromagnetic survey.

10.2 THE AEROMAGNETIC ANOMALIES NEAR JINDABYNE, NSW

Figure 10.7 shows the location of an aeromagnetic anomaly south-west of Canberra which we investigated with low elevation and closely spaced UAV surveys. The regional aeromagnetic survey covers part of the Lachlan Orogen (Foster and Gray 2000) and was flown in 2010 by

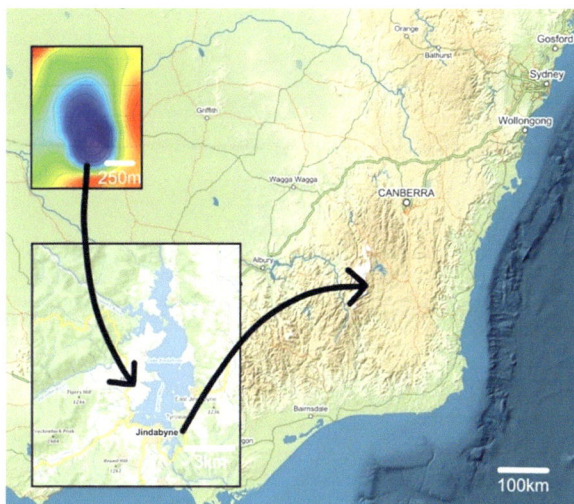

Fig. 10.7. Location map of the main negative aeromagnetic anomaly measured over the west bank of Lake Jindabyne south-west of Canberra.

Fugro Airborne Surveys for the Geological Survey of New South Wales using a fixed-wing Pacific Air Cresco on east–west flight lines at a nominal terrain clearance of 60 m and 250 m spacing. The magnetic field measurements were made with a Scintrex CS-2 magnetometer at a 10 Hz sampling interval giving a point spacing of ~8 m. The aeromagnetic survey is number P1218 in the Geoscience Australia GADDS directory and data can be downloaded from: https://portal.ga.gov.au/persona/gadds.

Figure 10.8 shows a TMI image of a part of the Lachlan aeromagnetic survey. The inset highlights a prominent negative anomaly and a weaker negative anomaly to the south of it. The north–south extent of the main anomaly is 500 m with an amplitude of –470 nT. The smaller anomaly is defined on only one flightline with an amplitude of –190 nT.

As shown by the section of the geological map by Lewis and Glen (1995) in Fig. 10.9, the anomaly location to the west of Lake Jindabyne is mapped as part of the Jindabyne Suite I type Kosciusko Batholith with a main reported lithology of hornblende and biotite bearing tonalite (Lewis *et al.* 1994). However, the small, shallow, reversely magnetised sources of the magnetic anomalies are likely to be small plugs of Cainozoic Monaro Province Volcanics. These feeder pipes and minor relics of basalt flows poorly outcrop but are reported across an area 100 km north–south by 35 km east–west (see fig. 10.35 of Lewis *et al.* 1994) including known occurrences that correlate with similar magnetic anomalies. The geological mapping was

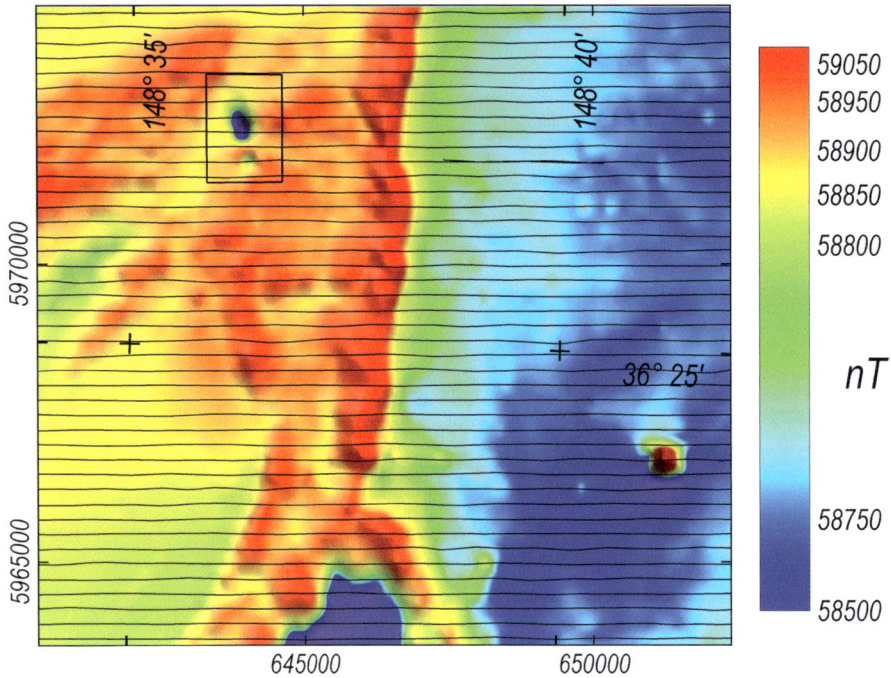

Fig. 10.8. TMI image of a section of the P1218 regional aeromagnetic survey. The inset shows the negative Lake Jindabyne anomaly.

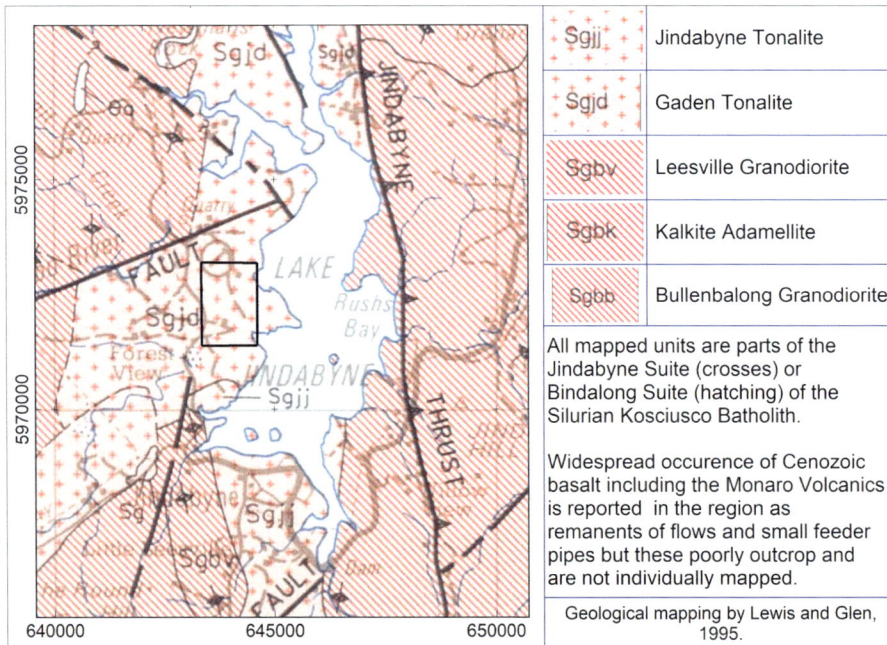

Fig. 10.9. Section of the Bega-Malcoota 1:250,000 geological map (Lewis and Glen 1995).

undertaken before the aeromagnetic survey was flown and had the aeromagnetic and radiometric data been available at the time of the geological mapping these data would most likely have been used to supplement the surface mapping of these strongly magnetic but mostly recessive or covered units.

Figure 10.10A shows TMI variation across the study area (shown as an inset in Figs 10.8 and 10.9). The main anomaly is defined on two east–west flight lines. The line immediately to the south of those is beyond the main anomaly and the line to the north defines the flank of the anomaly but with an amplitude of only

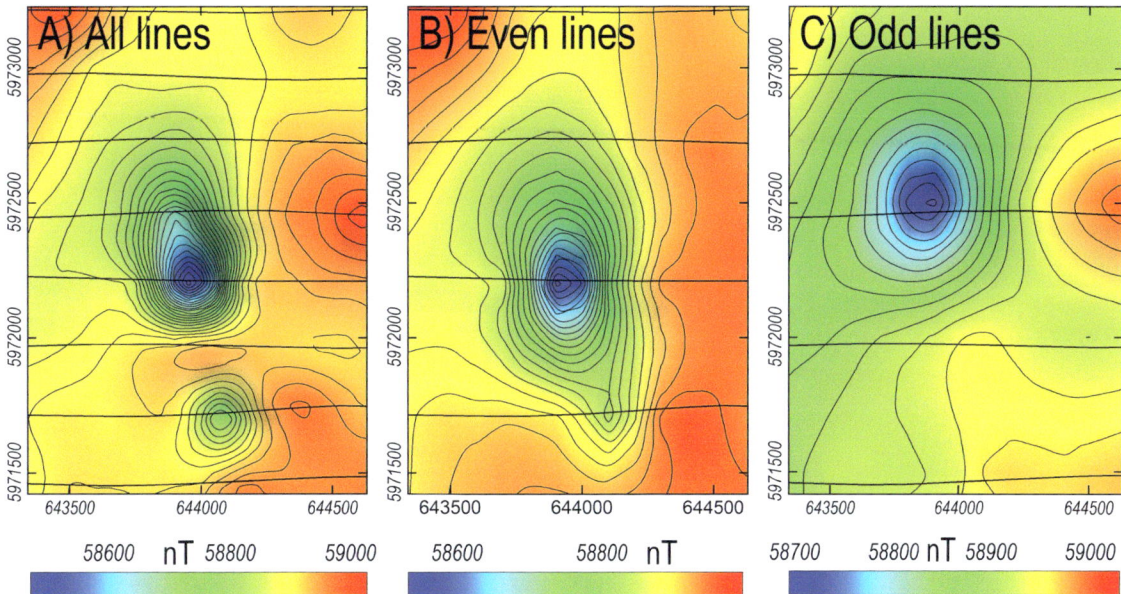

Fig. 10.10. Aeromagnetic survey TMI images of: A) all lines (250 m spacing), B) even lines, C) odd lines.

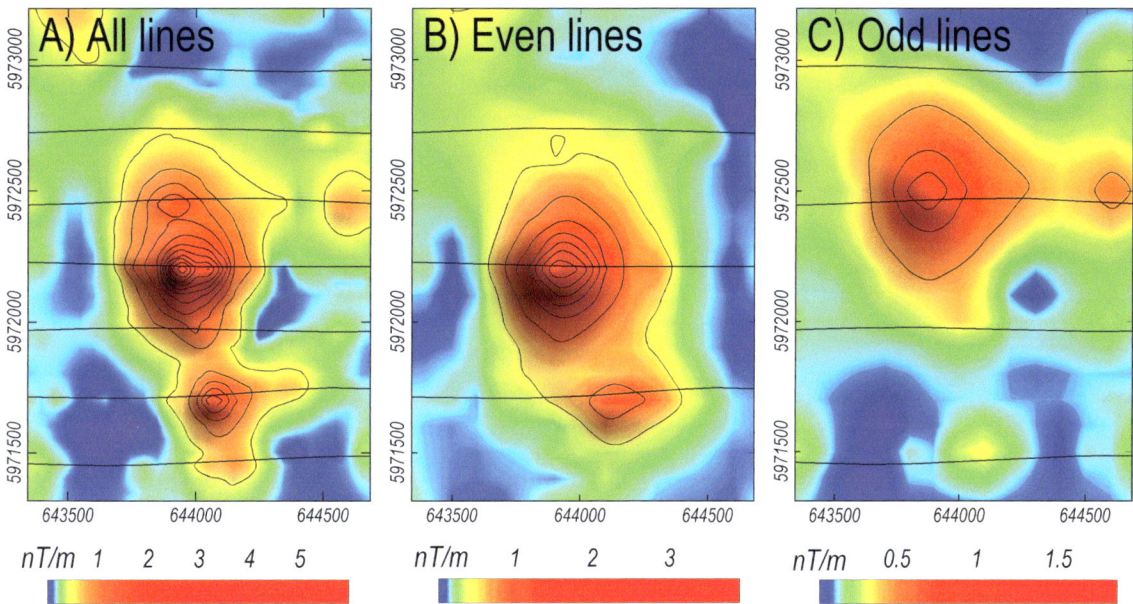

Fig. 10.11. Total gradient (TG) transforms of the TMI grids in Fig. 10.10.

−15 nT (3% of the anomaly trough). The southern anomaly is defined on only a single line. Figures 10.10B and 10.10C show quite different images from gridding of alternate flightlines at 500 m spacing. The image from the even-numbered flightlines in Fig. 10.10B includes the only flightline through the southern anomaly but excludes the flightline between the anomalies, so the gridding joins the two anomalies. The image from the odd-numbered flightlines in Fig. 10.10C excludes the flightline on which the southern anomaly is defined as well as the key flightline through the centre of the main anomaly, producing a mapping of a single anomaly of lower amplitude, displaced to the north. The insufficiency of these 500 m line-spaced data selections revealed by the abrupt differences between them suggests that the 250 m line-spaced survey is itself under-sampled. This insufficiency is further highlighted in Fig. 10.11 that shows total gradient of TMI computed from the survey data (Fig. 10.11A) and from the two alternate-line subsets (Figs 10.11B and 10.11C).

10.2.1 Modelling of the aeromagnetic data

Two inversions of the aeromagnetic survey data were performed, one using a model of two ellipsoids and the other using a model of two elliptic-section pipes. Both inversions allowed a free direction of magnetisation and both models match the input data equally well with no clear justification to favour either model over the other. Sections along the two key flight lines through the main and southern anomalies are shown in Fig. 10.12. The pipe models with horizontal top surfaces and abrupt edges are suitable for estimation of depth to the top of magnetisation but the under-sampled data makes depth estimation unreliable. The two inversions co-locate the magnetisations (see Figs 10.12 and 10.13) and estimate very similar magnetisation directions (see Table 10.1). Figure 10.14 shows measured TMI and TMI forward computed from the elliptic pipes inversion model. The under-sampling of the aeromagnetic anomalies makes it easier to find a close fit to the input data by inversion. However, the inversion models are of low reliability because the sample of data they are derived from is inadequate.

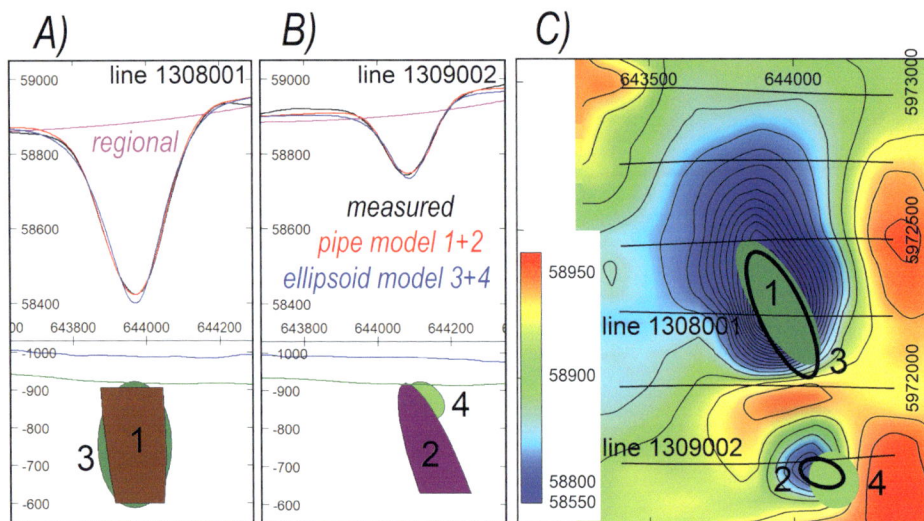

Fig. 10.12. Key aeromagnetic flightline cross-sections through the inversion models and image of model-computed TMI for the elliptic pipe models 1 and 2. The ellipsoid models 3 and 4 produce an almost identical field.

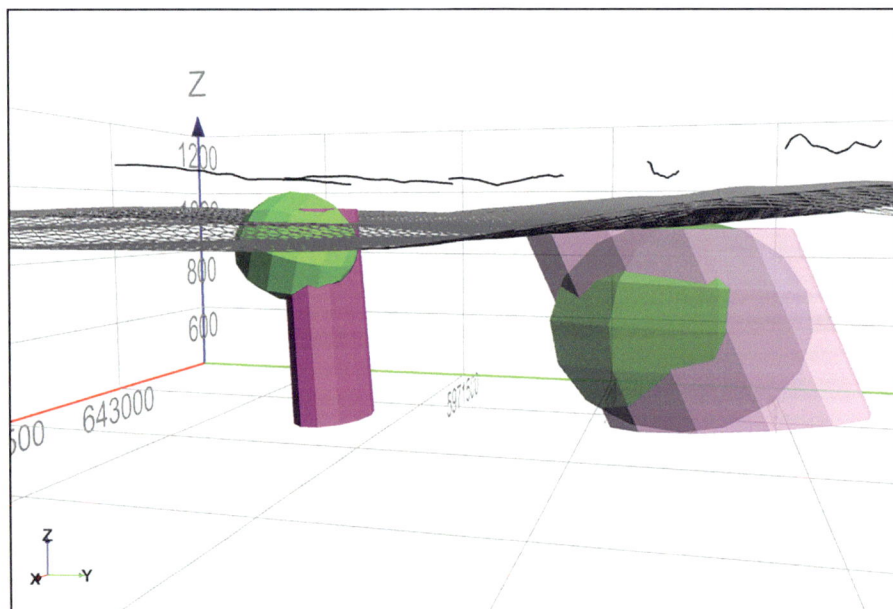

Fig. 10.13. Perspective view of the inversion models shown in Fig. 10.12.

Table 10.1. Inversion model magnetisation details.

Survey	Anomaly	ARAD	Geometry	J_int A/m	J_dec	J_inc	Volume m^3	Moment A.m^2
P1218	Main	309	Ellipsoid	2.11	98	60	29.3×10^6	62×10^6
P1218	Main	309	Elliptic pipe	2.28	80	61	36.3×10^6	83×10^6
P1218	Main	309	Frustum	2.33	82	55	37.2×10^6	87×10^6
P1218	**Main**	**Mean(3)**		**2.24**	**86**	**59**	$\mathbf{34.3 \times 10^6}$	$\mathbf{77 \times 10^6}$
Drone_980	Main	314_377	Ellipsoid	14.7	55	59	1.69×10^6	25×10^6
Drone_980	Main	314_378	Elliptic pipe	6.42	57	68	4.09×10^6	26×10^6
Drone_980	Main	314_379	Frustum	6.76	84	70	4.37×10^6	30×10^6
Drone_980	**Main**	**Mean(3)**		**9.29**	**64**	**66**	$\mathbf{3.38 \times 10^6}$	$\mathbf{27 \times 10^6}$
Drone_980	South	315_380	Ellipsoid	4.05	39	56	0.62×10^6	2.5×10^6
Drone_980	South	315_381	Elliptic pipe	3.21	51	55	1.28×10^6	4.1×10^6
Drone_980	South	315_382	Frustum	2.15	50	55	1.94×10^6	4.2×10^6
Drone_980	**South**	**Mean(3)**		**3.14**	**47**	**55**	$\mathbf{1.28 \times 10^6}$	$\mathbf{3.6 \times 10^6}$
Drone_950	South	316	Ellipsoid	5.79	63	69	1.87×10^3	11×10^3
Drone_950	South	316	Elliptic pipe	4.92	55	60	2.33×10^3	11×10^3
Drone_950	South	316	Frustum	4.92	55	65	2.48×10^3	12×10^3
Drone_950	**South**	**Mean(3)**		**5.21**	**57**	**65**	$\mathbf{2.23 \times 10^3}$	$\mathbf{11 \times 10^3}$
	LARGE	**MEAN(6)**		**5.8**	**76**	**63**	$\mathbf{18.9 \times 10^6}$	$\mathbf{52 \times 10^6}$
	SMALL	**MEAN(6)**		**3.9**	**51**	**60**	$\mathbf{1.9 \times 10^3}$	$\mathbf{7.6 \times 10^3}$

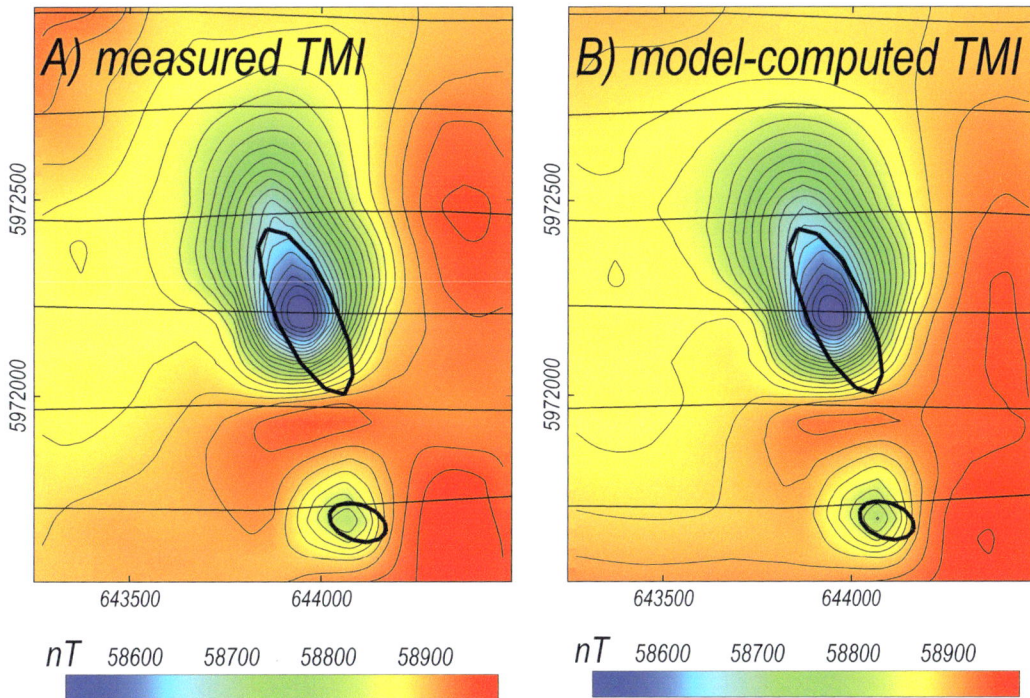

Fig. 10.14. A) measured TMI from the P1218 aeromagnetic survey, and B) TMI forward computed from the elliptic pipe inversion model.

10.3 THE UAV SURVEY OVER THE JINDABYNE AEROMAGNETIC ANOMALY

The UAV/drone used for this magnetometer survey was a DJI Inspire 2, equipped with a fluxgate magnetometer (shown at take-off in Fig. 10.15). The total take-off weight of the drone with the magnetometer system is ~3.75 kg. The theoretical maximum flight speed is 94 km h^{-1} As a safety measure the key system components are duplicated. Modification of the drone for this aerial magnetometer survey application primarily consist of holders for the magnetometer module, a modified chassis, and a sensor cable incorporated into the landing gear.

The drone and control systems carried to site are shown in Fig. 10.16. The drone is controlled by a native DJI interface with dual-band 2.400–2.483 GHz and 5.725–5.850 GHz communication. The flight parameters, setting of profile waypoints and measurement control takes place from the ground station via a tablet and remote controller. The magnetometer is a three-axis vector fluxgate sensor with flat ring core. The measurement range is +-75uT, nonlinearity is < 0.05%, with a noise level < 20 pT rms/√Hz per 1Hz. The three-component orthogonality after calibration is < 0.1°. Sensor location is determined by GPS and GLONASS positioning and altitude control with estimated accuracies of 1.5 m vertical and 3 m horizontal. Data sampling is at 62.5 or 250 samples/sec with the temperature of the sensor and unit and system information data recorded together with the magnetic field data.

The magnetometer has been designed specifically for UAV aerial magnetometry. It consists of a three-axis

Fig. 10.16. The complete mobile UAV system carried to site.

vector fluxgate magnetometer with flat-ring cores. The magnetometer is suspended on a 2.5 m long cable below the UAV. The control electronics of the magnetometer, GPS receiver, and power supply are fixed to the drone body. The complete magnetometer system, including the battery, weighs ~750 g. The magnetometer is independent of the drone and has its own power supply and is independent of GPS and GLONASS positioning. Data stored on the memory card for each measurement point at a frequency of 62.5 Hz are the three orthogonal components of the magnetic field, GPS position coordinates, altitude, temperature, and other system information.

10.3.1 The UAV data acquisition

Two independent surveys were conducted at fixed altitude. Survey 1 (S1) at a mean altitude of 58 m above ground and the smaller Survey 2 (S2) at a mean altitude of 35 m above ground. The Jindabyne survey area has a topographic range of ~80 m. The ground station established in the north-east of the S1 area, near the topographic peak provided a clear view and visual contact with the drone at all times. The survey flight plan had been pre-designed and required only slightly modification for safety considerations after an on-site evaluation.

The drone was assembled on site and the magnetometer was connected. The transmitter was paired with the ground station before the magnetometer and drone were turned on and paired with the controller. The GPS satellite selection was made and the home point was saved. At the same time the magnetometer position was calibrated. The complete survey path with all way-points had been uploaded and stored in the drone's memory.

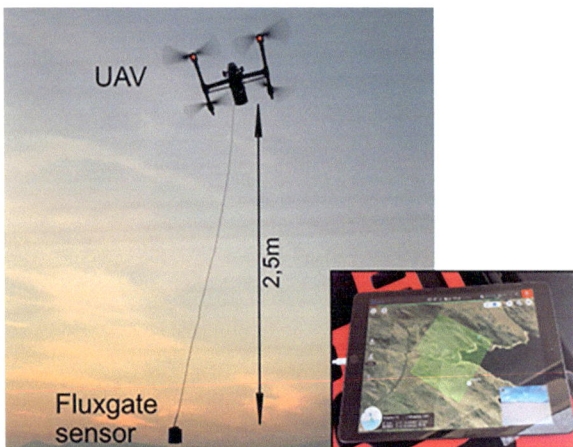

Fig. 10.15. UAV in flight with magnetometer sensor hanging down (left). UAV position during survey is streamed in real time to the ground station (right).

The magnetometer recording was started just before take-off and continued throughout the survey in autonomous mode.

The batteries were changed regularly during the survey as planned in the survey design and choice of ground station. The battery level and the individual cell voltages were monitored in real time during the entire survey both automatically by the drone and also by operator overview. Depending on the distance from the ground station and the estimated remaining available flight time, the drone regularly returned to the ground station where it landed for battery replacements, which was performed in about one minute. The drone then autonomously took off and flew to the position (both vertical and horizontal location) of the last magnetic measurement point to resume magnetic survey measurements. Neither the drone or the magnetometer needed to be turned off during the battery replacement, and the magnetometer data recording was not paused. Data acquired outside the measured profiles during battery replacement are removed from the measurement set in post-survey data processing.

Survey S1 (Fig. 10.17) is a grid of 43 parallel profiles with a spacing of 20 m and a total profile length of 43,974 m. The mean altitude is 58 m above ground. When planning the S1 profiles it was necessary to consider the slope to the north-east. The highest point in the survey area limits the lowest possible height for a fixed altitude survey (if a lower elevation had been required the survey could have been flown on a variable elevation drape surface). The point selected as the ground station was the highest point in the survey area that was suitably free of obstacles for take-off and landing.

The survey includes 86 turning points. The flight speed was set at 54 km/h. Take-off and landing were vertical. All other movements of the UAV over the surveyed area were at fixed altitude. Three battery replacements were required during the survey. These were performed at ends of profiles to both minimise distance to the ground station and reduce disruption of the data acquisition (for locations of the break points see Fig. 10.17). Survey area S2 (Fig. 10.18) is a subset of the S1 area flown with a different profile orientation and at a lower altitude (a mean clearance of 35 m above ground). The lower altitude was possible because of the gentle slopes of this area near the shore of the lake. This survey has 16 parallel profiles and 32 turning points, flown with a single battery set. The profile spacing of 20 m and flying speed of 54 km h^{-1} are the same as for the S1 survey. The total profile length is 11,057 m.

The three mappings of the magnetic field by the aeromagnetic and two drone surveys are shown in the TMI images of Fig. 10.19. The mean elevation of the aeromagnetic data within the study area (not published in the data but recovered by adding the radar altimeter measurements of ground clearance to the SRTM ground elevation) is 1,008 m; 28 m and 58 m higher than the two drone surveys. TMI data ranges for the three surveys are listed in Table 10.2. Over the main anomaly the higher and lower elevation drone surveys have increased data ranges of +50% and +130% compared to the aeromagnetic survey and the increases in amplitude are higher for the smaller southern anomaly that is poorly sampled by the aeromagnetic survey. The principal advantage of the magnetic field mapping from the drone surveys is the closer line spacing rather than the increase in amplitude due to their lower elevations.

Fig. 10.17. Survey S1 flight lines. The red point is the ground station, the green point is the survey start location and the blue points mark survey breaks for battery replacement.

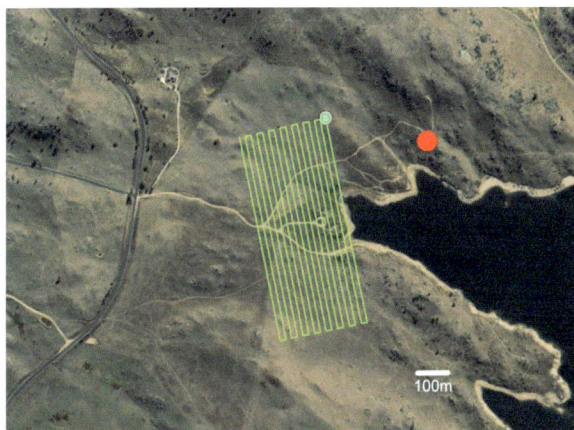

Fig. 10.18. Survey S2 flight lines. The red point is the ground station, the green point is the survey start location.

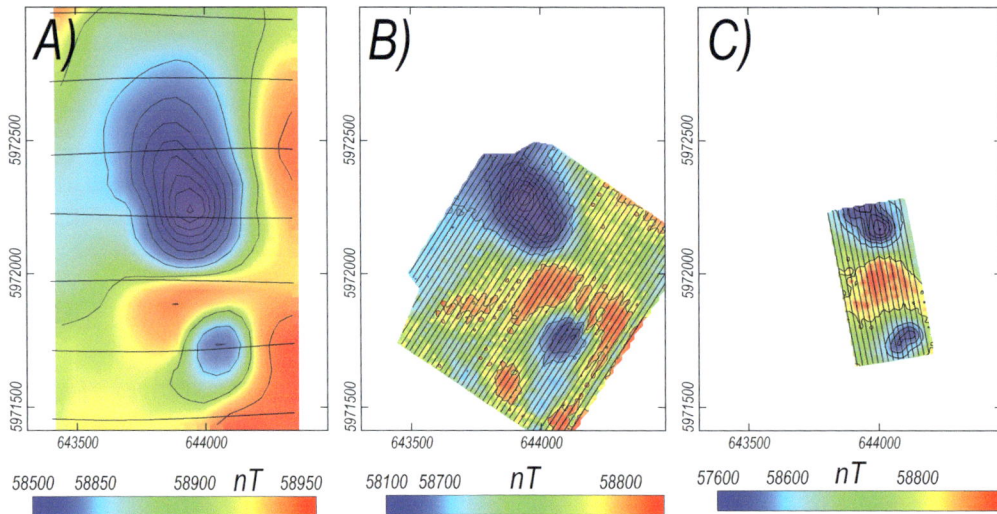

Fig. 10.19. TMI images: A) aeromagnetic survey, B) drone survey S1, C) drone survey S2.

Table 10.2. Magnetic survey data ranges.

Anomaly	Aeromagnetic range (nT)	Survey S1 range (nT)	Survey S2 range (nT)
Main	470	710	1,100
Southern	190	260	640

10.3.2 Modelling of the UAV survey results

The UAV data was inverted using the same ModelVision software and methodology as was used for the aeromagnetic data inversions. Figure 10.20A shows TMI measured on the larger, higher elevation S1 UAV survey over the main aeromagnetic anomaly. The corresponding field computed from a polygonal pipe model derived

from inversion of this data is shown in Fig. 10.20B. The trough of the anomaly is the combination of a dominant minimum to the north-west and a secondary minimum to the south-east. These may be expressions of a single complex magnetisation or two smaller, close magnetisations producing anomalies with extreme overlap. A ground survey would be required to reliably discriminate between these model options. We choose to model the anomaly as due to a single complex body. Inversion of this model permits horizontal movement of the individual vertices as well as changes to the elevation of the top surface, magnetisation intensity and direction, depth extent and plunge. We also inverted the anomaly with models of simpler geometry. Each model occupies almost

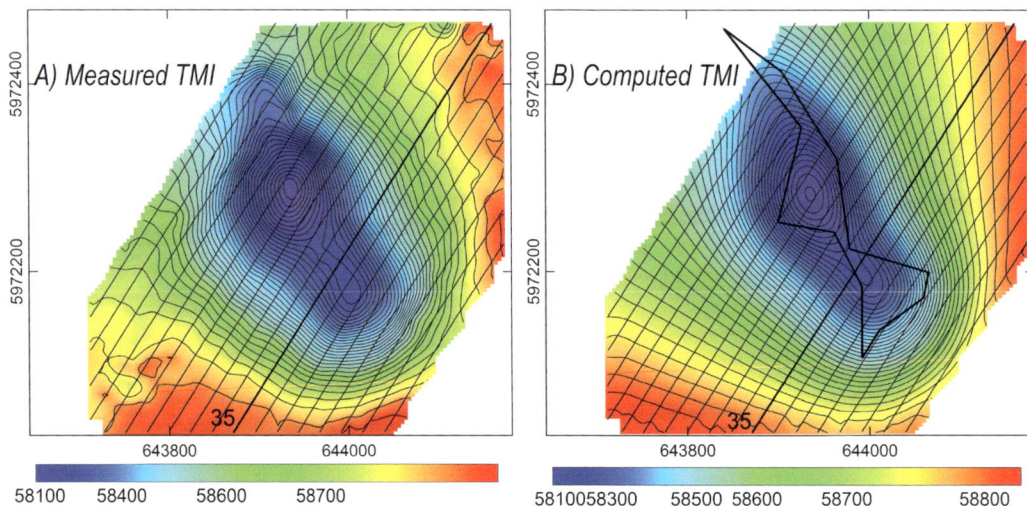

Fig. 10.20. S1 survey Larger Anomaly TMI: A) measured, B) model computed.

the same space (Fig. 10.21) and all have similar magnetisation directions (Table 10.1). All models acceptably match the data but the additional complexity of the polygonal pipe model provides the best fit with a model that has two lobes, one over each of the trough minima. Figure 10.21 shows a section through the model along one of the central flightlines and a perspective view of the overlap of the alternative models.

Figure 10.22A shows an equivalent image of TMI over the smaller southern anomaly measured from the lower

elevation S2 UAV survey and Fig. 10.22B shows the field forward computed from a polygonal pipe inversion mode derived from inversion of that data. This anomaly is more compact but of equivalent complexity to the S1 Larger Anomaly. The polygonal pipe inversion of this anomaly also develops a model with two nodes (Fig. 10.22B). A flightline traverse through this and the alternative models is shown in Fig. 10.23A. The different models again occupy almost the same space (Fig. 10.23B) and have similar magnetisation directions (Table 10.1).

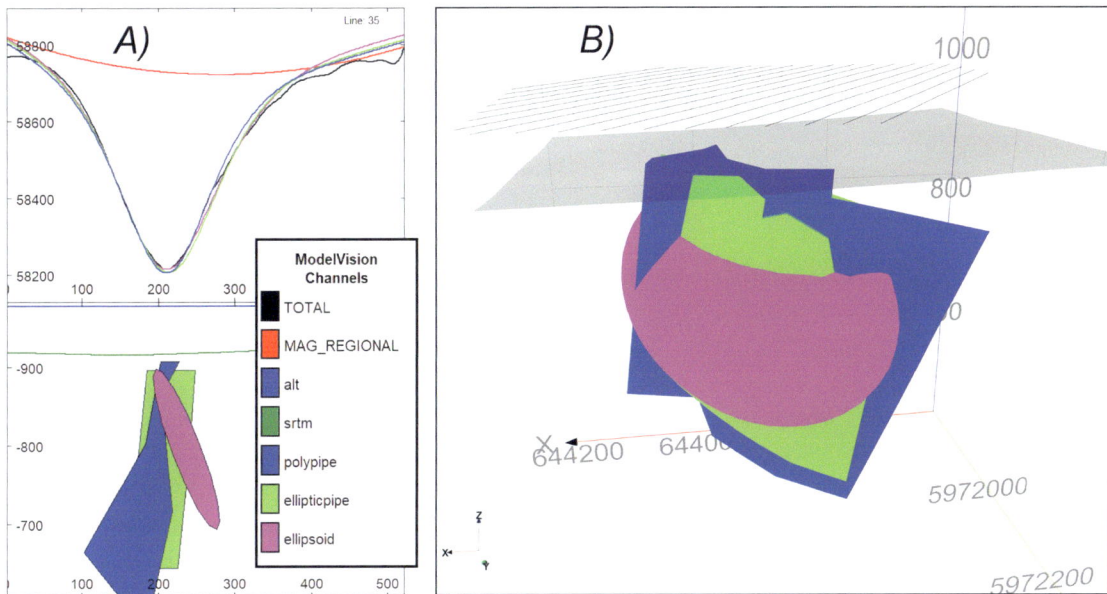

Fig. 10.21. S1 survey Larger Anomaly: A) central flightline section, B) model perspective view.

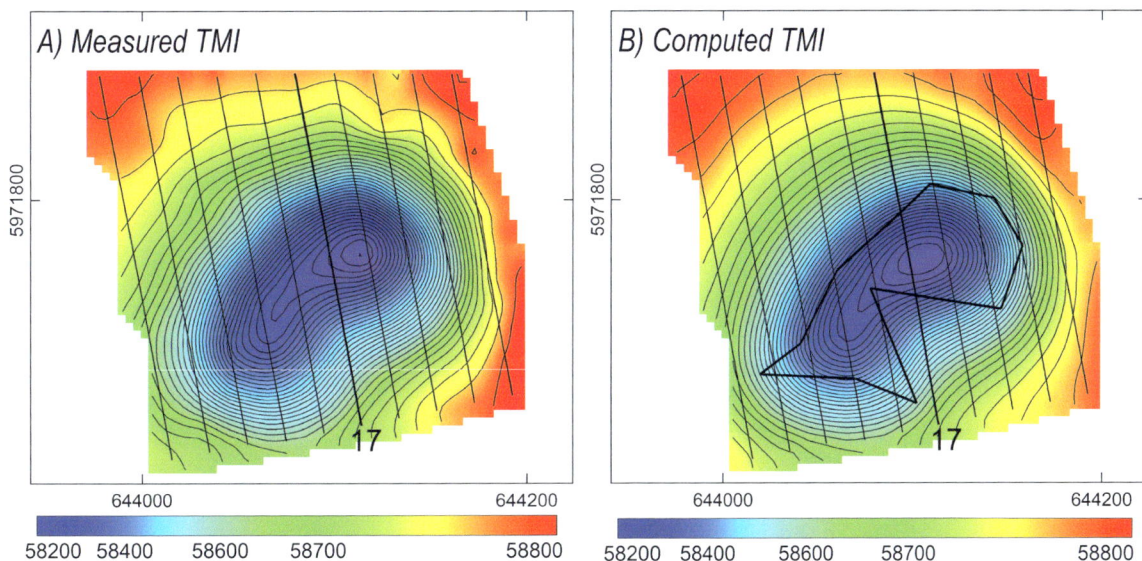

Fig. 10.22. S2 survey Smaller Anomaly TMI: A) measured, B) model computed.

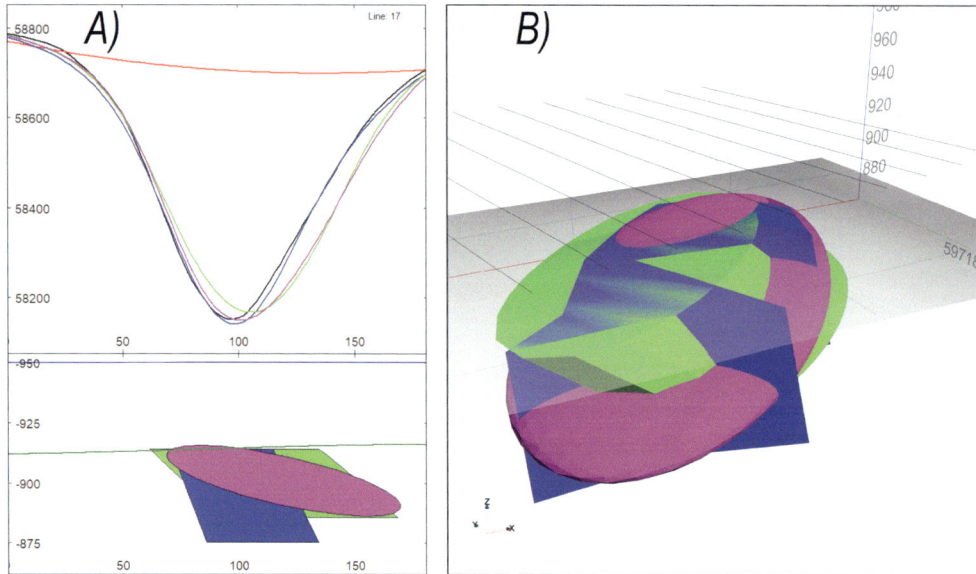

Fig. 10.23. S2 survey Smaller Anomaly: A) central flightline section, B) model perspective view.

10.4 ESTIMATION OF MAGNETISATION DIRECTION FROM THE AEROMAGNETIC AND UAV DATA

The model magnetisation directions listed in Table 10.1 are plotted on a stereonet in Fig. 10.24. Four datasets were inverted: the main anomaly measured on the aeromagnetic survey and on the higher UAV survey, and the southern anomaly measured on both UAV surveys. The southern anomaly is insufficiently sampled on the aeromagnetic survey for meaningful modelling, and the main anomaly was only partially covered on the lower UAV survey. The difference between the mean directions of the three models from each of the two surveys of the main anomaly is 12° and the difference between the mean directions of the three models from each of the two surveys of the main anomaly is 11°. The difference between the mean direction of the six inversions of the main anomaly data (blue symbols in Fig. 10.24) and of the six inversions of the southern anomaly data (red symbols in Fig. 10.24) is 12°. Visual inspection of the directions in Fig. 10.24 suggests that the directions for the two anomalies (the red and blue symbols) may belong to two slightly different populations and the difference between them would be consistent with secular variation between slightly different emplacement and (rapid) cooling ages for the two bodies.

In this study the estimated magnetisation direction for the main anomaly has not changed appreciably from analysis of the aeromagnetic data to analysis of magnetisation data from the UAV survey, but the UAV survey has substantially increased confidence in this direction. The UAV survey has also provided a magnetisation direction estimate for the southern anomaly, that was insufficiently sampled by the aeromagnetic survey to justify estimation of its magnetisation direction. These results from the surveys at Jindabyne are consistent with the synthetic modelling results presented earlier.

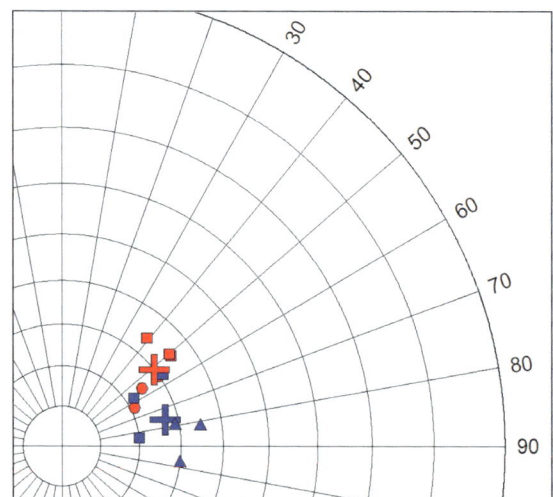

Fig. 10.24. Model magnetisation directions, blue – main anomaly, red – southern anomaly, triangles – aeromagnetic survey, squares – higher UAV survey, circles – lower UAV survey.

ACKNOWLEDGEMENTS

Marian Takáč and Gunther Kletetschka were partially supported from the Czech Science Foundation 20–08294S, and MEYS, LTAUSA 19141.

REFERENCES

Beiki M, Clark DA, Austin JR, Foss CA (2012) Estimating source location using normalized magnetic source strength calculated from magnetic gradient tensor data. *Geophysics* **77**, J23–J37. doi:10.1190/geo2011-0437.1

Foster DA, Gray DR (2000) The structure and evolution of the Lachlan Fold Belt (Orogen) of eastern Australia. *Annual Review of Earth and Planetary Sciences* **28**, 47–80. doi:10.1146/annurev.earth.28.1.47

Helbig K (1963) Some integrals of magnetic anomalies and their relation to the parameters of the disturbing body. *Zeitschrift für Geophysik* **29**, 83–96.

Lewis PC, Glen RA (1995) 'Bega-Mallacoota 1:250 000 Geological Sheet SJ155–4 SJ155–8. Second edition.' Geological Survey of New South Wales, Sydney.

Lewis PC, Glen RA, Pratt GW, Clarke I (1994) 'Bega-Mallacoota 1:250 000 Geological Sheet SJ155–4 SJ155–8: Explanatory Notes.' Geological Survey of New South Wales, Sydney.

Li X (2006) Understanding 3D analytic signal amplitude. *Geophysics* **71**, L13–L16. doi:10.1190/1.2184367

Medeiros WE, Silva JBC (1995) Simultaneous estimation of total magnetization direction and 3-D spatial orientation. *Geophysics* **60**, 1365–1377. doi:10.1190/1.1443872

Reid AB (1980) Aeromagnetic survey design. *Geophysics* **45**, 973–976. doi:10.1190/1.1441102

Roest W, Pilkington M (1993) Identifying remanent magnetization effects in magnetic data. *Geophysics* **58**, 653–659. doi:10.1190/1.1443449

Wijns C, Perez C, Kowlczyk P (2005) Thetamap: edge detection in magnetic data. *Geophysics* **70**, L39–L43. doi:10.1190/1.1988184

Zietz I, Andreasen GE (1967) Remanent magnetization and aeromagnetic interpretation. *Mining Geophysics* **2**, 569–590.

11

Magnetic anomalies of terrain-delimited magnetisations in northern New South Wales as a training set for Martian magnetic field mapping

C.A. Foss and J.R. Austin

ABSTRACT

Total magnetic intensity (TMI) variations measured by an aeromagnetic survey over the Miocene age Lamington Volcanics in northern New South Wales are well explained by inversion of homogeneously magnetised digital terrain models. At different locations these models variously have normal magnetisation and magnetisation dominated by reverse remanence. In consequence of closely matching measured TMI variation the orthogonal magnetic field components forward computed from these terrain models also closely match the equivalent components derived from frequency-domain transform (FFT) of measured TMI. We show that terrain magnetisations can be reasonably recovered from inversion of three-component magnetic field data along single traverses using the constraint of an independently known spatial model. For measurements of limited spatial distribution such as along isolated flight paths, along rover tracks or at dispersed landing sites, multi-component data provide additional information to single-component data. Direct measurement of magnetic vector components in the Earth's strong background field incurs prohibitive errors arising from even slight sensor misorientation. However, these penalties are much smaller where background fields are weak. Furthermore, in weak background fields the direction of the total field varies considerably across and around anomalies, and for magnetic field mapping this makes TMI less suitable than vector components of consistent direction. We assign the normal and reverse magnetisations estimated from the Tenterfield modelling study to an isolated topographic feature of width 23 km and height 800 m to the north of the Argyre Basin region of Mars and are able to reasonably recover those magnetisation directions from multi-component inversion of emulated data on short flight-paths over the feature and sparse station measurements on rover tracks around it. The success of such survey methods will depend on the validity of a homogeneous magnetisation terrain model and the dominance of the field from that magnetisation across the range of measurement locations.

11.1 INTRODUCTION

Many igneous rocks form positive topographic features where they have been emplaced by eruptions over the land surface or where surrounding less resistive materials have been eroded away. We show that in northern New South Wales the magnetic field variation measured by an aeromagnetic survey over the Lamington volcanics (Duggan and Mason 1978) can be well explained by digital terrain models assigned a homogeneous magnetisation. Over parts of the region there is strong positive correlation between the magnetic field and terrain elevation. For these areas the best estimated magnetisation direction is close to that of the present geomagnetic field. Over other parts of the region there is strong negative correlation between the magnetic field and terrain elevation revealing that magnetisation is dominated by reverse remnant components almost anti-parallel to the present field. We show that in both cases there is close agreement between magnetic field components forward computed from the magnetic terrain models and those transformed from the measured total magnetic intensity (TMI) data. For these sufficiently sampled fields there would be limited advantage in having multi-component data. However, multiple components do provide advantage if the magnetic field is sparsely sampled. On Earth, direct measurement of vector components of the magnetic field is penalised by large errors from misalignment of sensors, primarily arising from a background field that is much stronger than the anomalies of interest (almost 54,000 nT in this study area in north-west New South Wales). The Martian magnetic field dynamo has switched off (e.g. Schubert *et al.* 2000; Mittelholz *et al.* 2020) leaving anomalies largely due to remnant magnetisation in a weak background field. Lack of a strong background field causes the TMI vector to change direction within and around local anomalies. TMI measurements owe their value more from being an (approximately) uni-directional vector measurement than being a (pseudo-) scalar measure of the strength of the field. Where unlocked from the consistent direction of a dominant background field, TMI measurements lose some of their appeal for magnetic field mapping. In low background fields such as those found on Mars this provides advantages for multiple vector component or multiple vector gradient (tensor element) measurements. We selected a Martian topographic feature, assigned it the same magnetisations estimated in the Tenterfield study in northern New South Wales, and forwarded computed its vector component anomalies on the assumption that it is surrounded by material of much weaker magnetisation. If Mars has such isolated and terrain-bound magnetisations then it may be possible to derive large-volume estimates of their mean magnetisation from a sparser set of magnetic field measurements than is required for buried sources of unknown position, orientation and extent. Optimisation of magnetic field measurement is of critical advantage in an environment where it is not feasible to measure the field as extensively as we do on Earth.

11.2 TERRAIN-BOUND MAGNETISATIONS AT TENTERFIELD, NEW SOUTH WALES

Figure 11.1 shows the location of part of the 100,000 line kilometre Grafton-Tenterfield regional aeromagnetic survey (Geoscience Australia GADDS survey P1252) flown for the Geological Survey of New South Wales in 2011 using a Cessna 404 fixed wing aircraft on east–west flightlines at 250 m line spacing and with a nominal ground clearance of 60 m. The actual ground clearance is highly variable because of the rugged terrain. The survey extends to the Queensland border and covers an area known as 'the Scenic Rim'. TMI and DEM mapping from the survey data are imaged in Fig. 11.2. The magnetic field strength varies by over 1,600 nT and the variation in elevation is over 800 m. Shown in Fig. 11.2 are the outlines of two detailed study areas. In the north-east area there is a clear negative correlation between ground elevation and TMI, and in the south-west area there is a strong positive correlation between elevation and TMI. The moderately steep geomagnetic inclination of −58° causes a predominantly positive TMI anomaly over a body with magnetisation in the same direction as the field (a 'normal' magnetisation). For terrain-bound magnetisations in contrast against air, any negative correlation between terrain and magnetic field intensity is due to rocks with predominant reverse remanent magnetisation. In the south-west area where there is positive correlation between TMI and terrain either the remanent magnetisation is normal or, if reverse, it is weaker than the induced magnetisation. Our preferred interpretation is that remanent magnetisation and magnetic susceptibilities are similar in both areas, with difference in polarity of the remanent

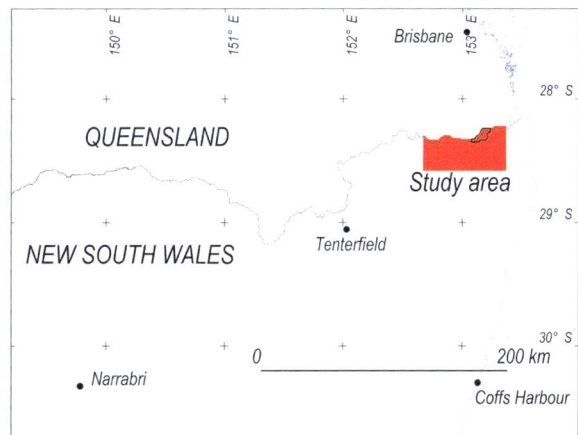

Fig. 11.1. Location of the Tenterfield study area.

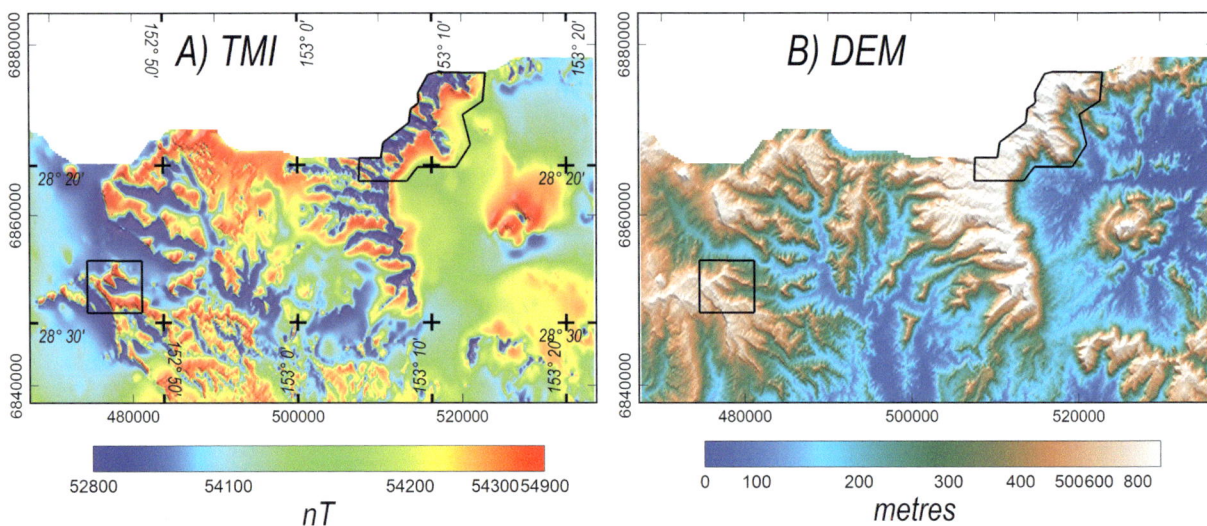

Fig. 11.2. A) TMI and B) DEM images of the Tenterfield area of north-east New South Wales with the two study areas highlighted.

magnetisation due to slight age differences that span a reversal of the Earth's magnetic field.

For the two study areas in Fig. 11.2 we built separate magnetisation models using the terrain surface as the upper surface of magnetisation and a horizontal surface approximately coincident with the base of the Lamington Volcanics as mapped by Brown *et al.* (2007) as the base of magnetisation. The terrain image for the north-east area is shown in Fig. 11.3A and the faceted, closed volume model generated from that surface and the nominated horizontal base level is shown in Fig. 11.3B. For each area we emulated the background magnetic field (the field that would be expected if the terrain magnetisation was absent) using a second-order polynomial of low curvature. Following the recognised correlations of terrain and TMI noted above, we assigned an initial magnetisation parallel to the local geomagnetic field to the model in the south-west area and a magnetisation anti-parallel to the local geomagnetic field to the model in the north-east area. We then trialled several estimates of magnetisation strength to quickly produce modelled field variations of similar amplitude to the measured TMI variation. This provided starting matches between the measured data and model fields that were close enough to ensure rapid and stable convergence of inversion. The inversions were also naturally stable because the only free parameters were their bulk magnetisations. In the initial inversions of each area only three parameters were free to vary: the strength, declination and inclination of magnetisation. Subsequently, the

level of the background field and its north–south and east–west planar gradients (another three parameters) were also set free to search for any reduction of data misfit by adjustment of our initial estimate of the background field. The close fit of measured and post-inversion model computed fields in the north-east area is shown in Fig. 11.4. This is strong testament to the general validity of representing this terrain with a magnetisation that is homogeneous in strength and direction. Non-uniqueness generally imposes considerable constraints on such claims of the validity of inversion models, but because the location and extent of magnetisation are locked in place by the terrain surface there is little scope to propose any broad-scale departure from the model (that undoubtedly deviates considerably from the true ground magnetisation at any discrete location). For the reverse-magnetisation north-east area the optimum magnetisation estimate is 3.8 A/m, declination 188° and inclination +57°. This direction is 178° from the local geomagnetic field direction (only 2° from being anti-parallel to it).

Figure 11.5A shows the terrain image and Fig. 11.5B the corresponding faceted terrain model for the south-west area. The match between measured and model computed TMI achieved by inversion of magnetisation for this model shown in Fig. 11.6 is similar to that for the north-east area shown in Fig. 11.4, revealing no difference in performance of inversion for the different magnetisation directions. The optimum magnetisation estimate for the south-west area model is 4.2 A/m,

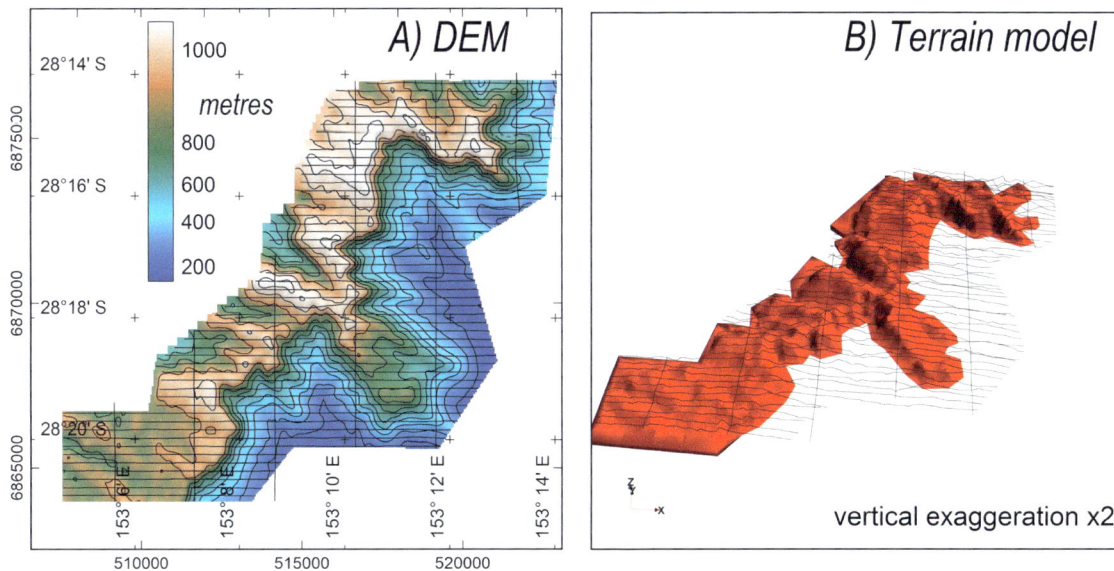

Fig. 11.3. A) terrain elevation and B) the corresponding terrain model and survey flightlines for the north-east area.

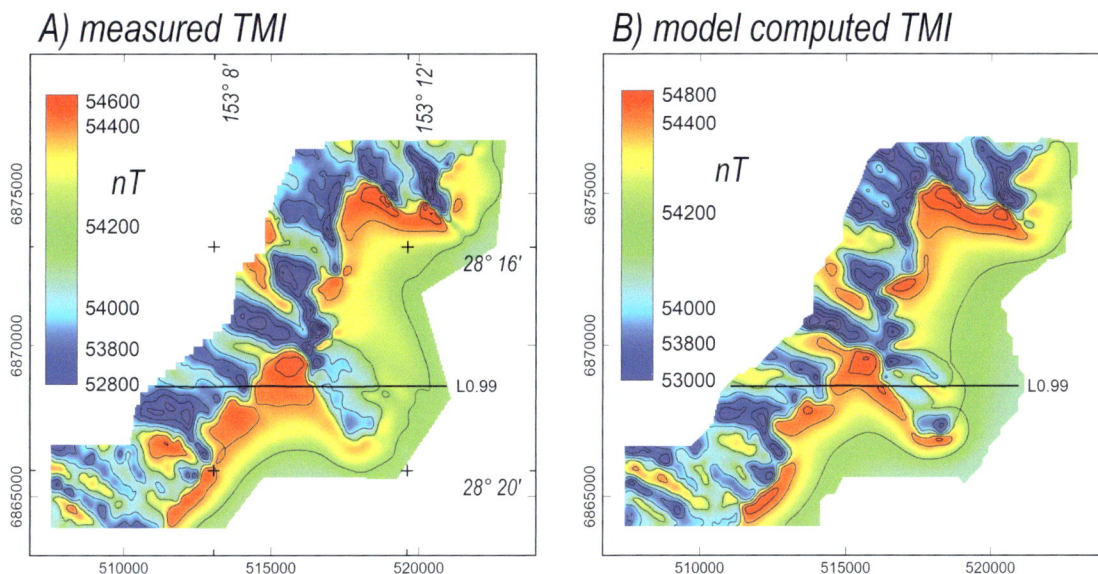

Fig. 11.4. A) measured TMI and B) model forward computed TMI for the (reverse magnetisation) north-east area.

declination 18° and inclination −54°. This direction is 6° from the local geomagnetic field direction and 6° from being anti-parallel to the magnetisation direction in the north-east area. Note that the resultant magnetisation directions for the two areas are not expected to be exactly anti-parallel unless the remanence directions are exactly anti-parallel to each other and also parallel and anti-parallel to the induced magnetisation. Using simple addition and subtraction of the two model magnetisations the best estimates of common magnetic properties for the two areas are: susceptibility 0.014 SI, remanent

magnetisation intensity 4.0 A/m and Koenigsberger ratio 7. However, those estimates are dependant on assumptions of conformity of magnetisation between the two areas and across large model volumes. The model estimates are likely to have little significance to magnetisations that would be measured on single rock samples.

Figure 11.7 shows example flightline sections over the post-inversion models from each of the two areas (for locations see Figs 11.3 to 11.6). Note that the data misfits shown in Fig. 11.7 are not optimised to the data only along those specific lines but to all the lines in each

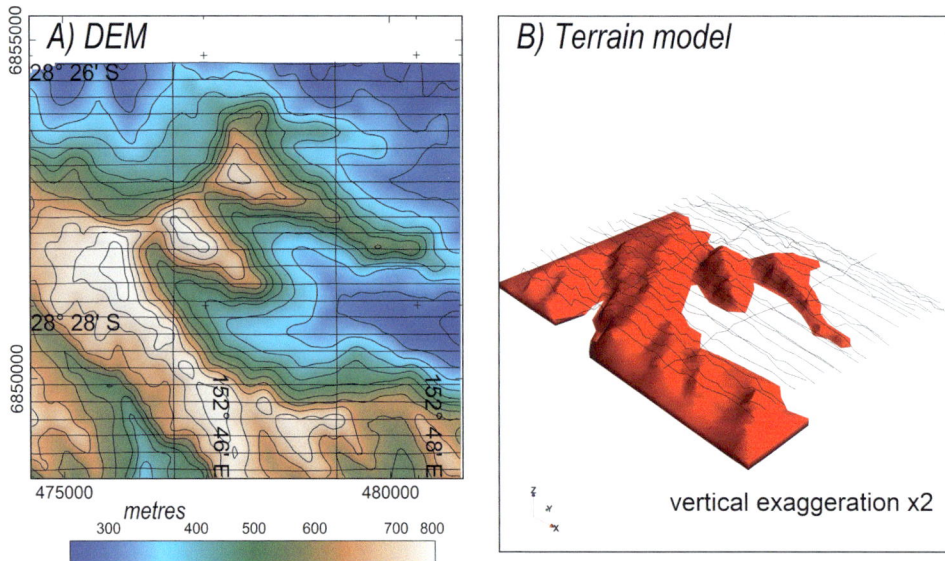

Fig. 11.5. A) terrain elevation and B) the corresponding terrain model and survey flightlines for the south-west area.

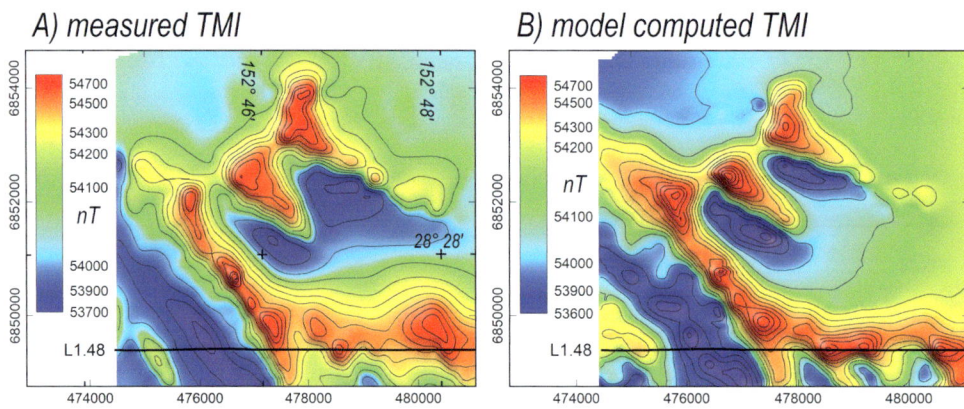

Fig. 11.6. A) measured TMI and B) model forward computed TMI for the (reverse magnetisation) north-east area.

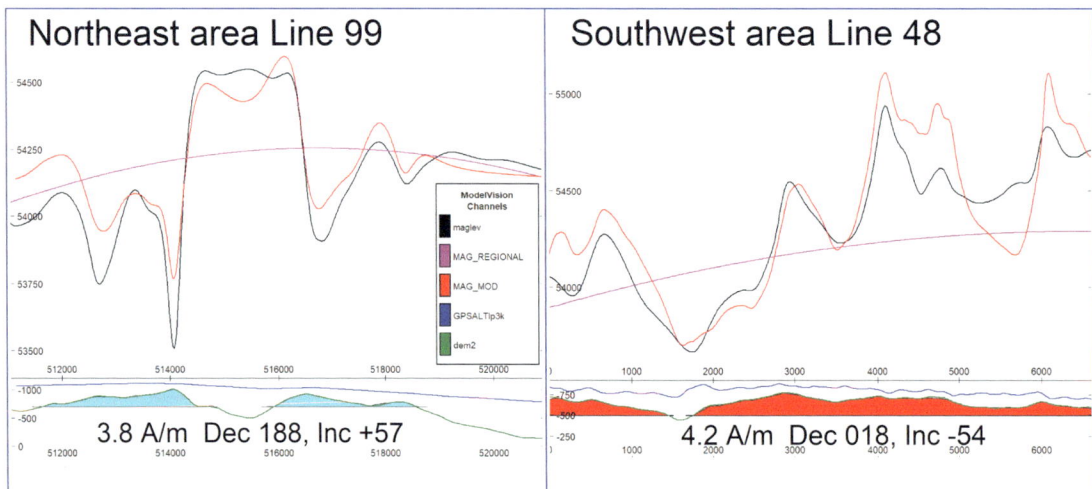

Fig. 11.7. Individual inversion model sections A) from the (reverse magnetisation) north-east area model and B) from the (normal magnetisation) south-west area model.

area. Matching the data on only those single lines would reduce the displayed data misfits on those lines, but the significance of the changes to the models to achieve that would be uncertain. Selection of large model areas with multiple flightlines of considerable length stabilises estimation of the mean magnetisation but it also increases concerns of that magnetisation poorly representing subareas. Further reduction of the misfit between the measured and model computed fields could be achieved by adding local detail to the models but any such changes would require validation.

11.3 TENTERFIELD SURVEY MULTIPLE COMPONENT ANALYSIS AND INVERSION

Magnetic field surveys measure TMI not for interpretational advantage but because of the convenience of low sensitivity to sensor orientation. In almost all surveys, measured TMI values are indistinguishable from the value of the vector in the local geomagnetic field direction (exceptions are for surveys of particularly high-amplitude variations in which there are local rotations of the TMI direction). Assuming that TMI is consistently directed and is therefore a potential field (Blakely 1995) it can be converted to a vector in any other direction using an FFT phase transform (as noted by Blakely 1995 this transform may be unstable for certain combinations of angles if the input data is noisy). TMI can for instance be converted with three different phase transforms to the set of east, north and vertical field components B_x, B_y, B_z. The conditions are that the field is sampled at close spacing and high resolution across a sufficient distance to faithfully treat the range of wavelengths present. These transforms modify the data but do not add information to them. In practice they introduce distortion because of their dependence on the gridding process and (of particular relevance to the Tenterfield survey) on any irregularity in elevation of the measurement surface. Figure 11.8 shows on the top row the B_x, B_y, B_z components forward computed from the TMI inversion model (these three orthogonal components are computed as an intermediate step in calculation of TMI). The bottom row of Fig. 11.8 shows the equivalent components derived by FFT from TMI. Because the model matches

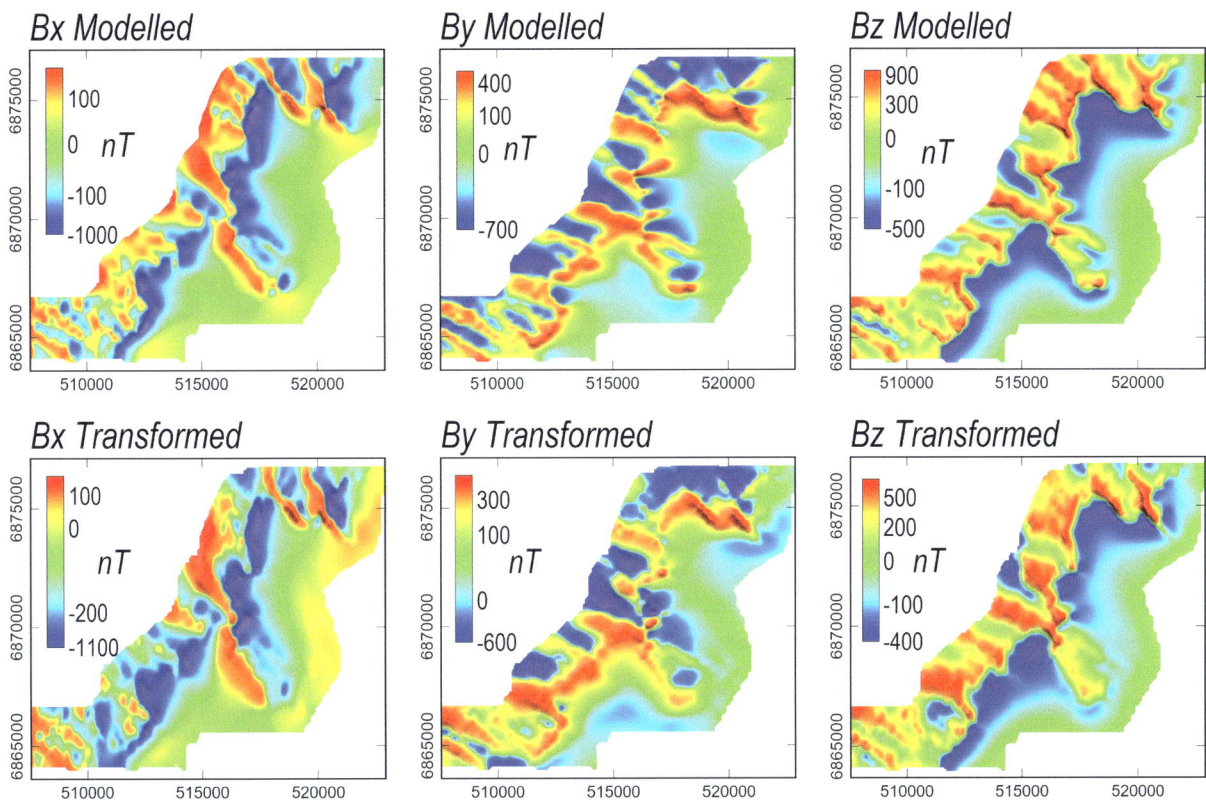

Fig. 11.8. (Left column) Inversion model computed B_x, B_y and B_z components, and (right column) equivalent FFT transformed components for the north-east reversely magnetised area.

the TMI data across a wide area the model components approximately match the same components derived by phase transform of the TMI. Component transforms are independent of magnetisation direction and Fig. 11.9 shows the same predictive success of the FFT components in matching the model components for the reverse magnetisation area.

There is no information gained in transforming the TMI grid data to the different component grids. However, for a restricted distribution of data, such as a single flightline, there is advantage in having multiple magnetic field components and/or gradients. Figure 11.10 shows single flightline magnetisation inversions of the two terrain models. Line 99 is from the south-west model and line 48 is from the north-east model. The top panel in Fig. 11.10 shows measured TMI (solid lines), model-computed TMI from the all-line TMI inversions (dashed lines) and model-computed TMI from the single-line TMI inversions (dotted lines). The dedicated single line inversions fit the data more closely because the data-fit is optimised to only those single lines but the multi-line

inversions also fit most features on those lines quite closely. This stability of the inversions is largely due to constraint of the spatial variation of magnetisation from the terrain-model.

The magnetic field component panels (B_x, B_y, B_z) in Fig. 11.10 show (solid lines) the input data from FFT phase transform of the measured TMI grid interpolated onto the lines, (dashed lines) the component forward computed from the all-line TMI inversion, and (dotted lines) the component forward computed from the single flightline component inversions. The inversion magnetisation direction estimates are plotted in Fig. 11.11. The all-line inversions produce the most consistent magnetisation estimates (plotted with triangular symbols in Fig. 11.11). Note that the closed symbols (normal magnetisations) are on the opposite hemisphere to the open symbols (reverse magnetisations). The smaller symbols in Fig. 11.11 are magnetisation directions from the single line, single component (including TMI) inversions. These directions are reasonably clustered with mean departures from the average values of 11° and 13°

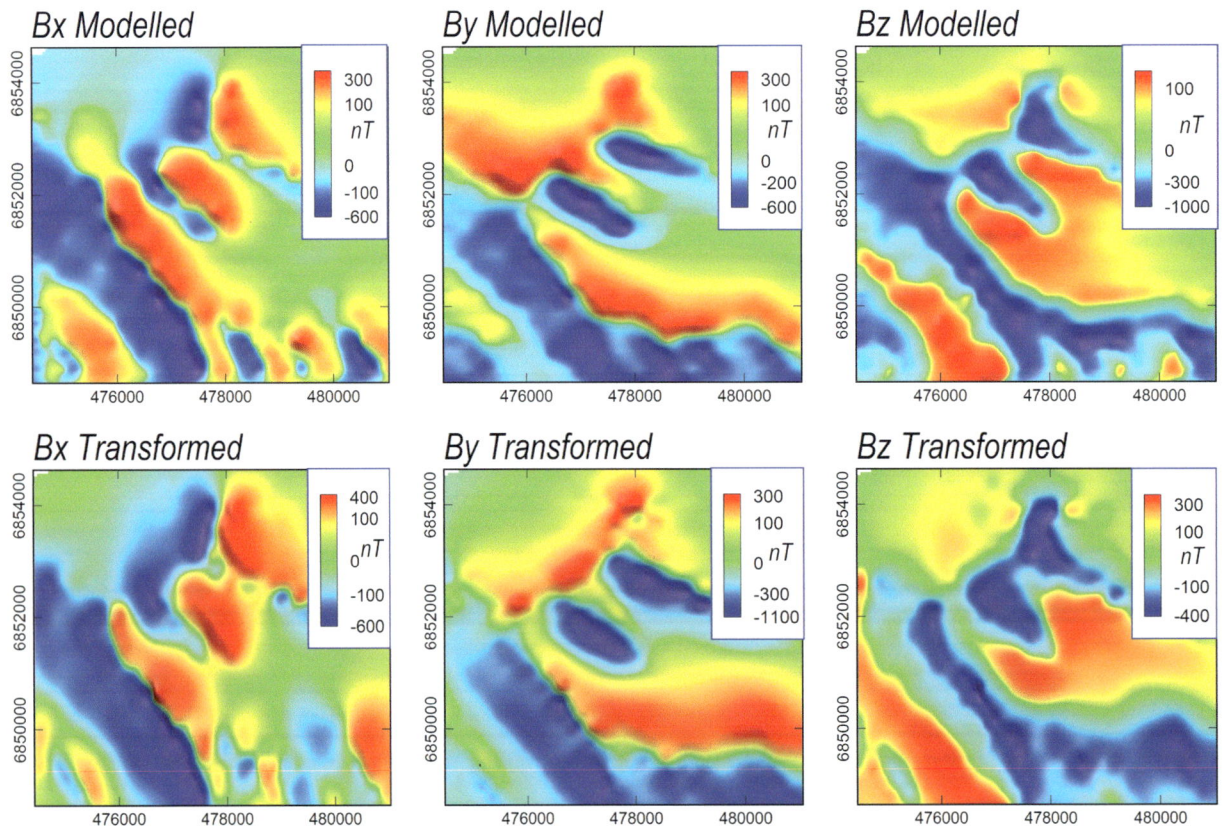

Fig. 11.9. (Left column) Inversion model computed Bx, By and Bz components, and (right column) equivalent FFT components for the south-west normally magnetised area.

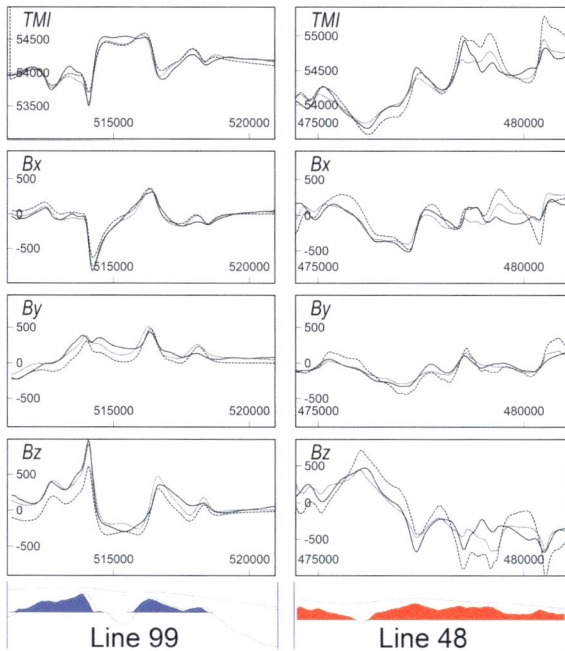

Fig. 11.10. Single flightline inversions of TMI and Bx, By and Bz components. The solid line is the input channel measured TMI and FFT transformed components, the dashed line is the output channel of the complete areas TMI inversions and the dotted lines are the output channels of the different component single flightline inversions.

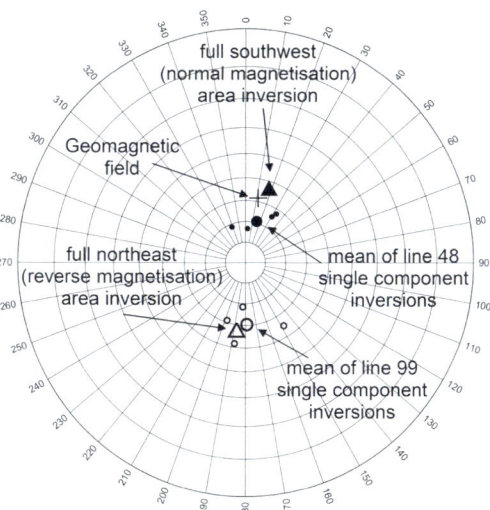

Fig. 11.11. Inversion magnetisation directions (negative inclination 'normal' directions as closed symbols and positive inclination 'reverse' directions as open symbols). Triangles are all-area TMI inversion directions, the larger circles are the means of the invididual component single-line inversions (small symbols).

for the normal and reverse magnetisations respectively, and those means are 6° and 15° from the all-lines inversion directions. These results indicate that using the terrain-model constraint a reasonable magnetisation

estimate can be obtained from inversion of an individual magnetic field component on a single flightline. Note that if the constraint of the terrain model was not available (for instance if the top of magnetisation was buried by a non-magnetic cover) this would not be possible. Inversion of two or more components along flightlines improves magnetisation direction estimates not just from averaging of results as shown here, but also through joint inversion in which models simultaneously explain the multiple data channels.

11.4 SYNTHETIC MARTIAN TERRAIN MODEL STUDY

Figure 11.12 shows the location (322.2°E, 23.5°S) of an isolated terrain feature north of the Argyre Basin and 160 km south-east of the centre of Roddy Crater. The terrain data is the Mars MGS MOLA DEM (version 2) derived from measurements by the Mars Orbital Laser Altimeter (Smith *et al.* 2001) flown on the Mars Global Surveyor spacecraft (Albee *et al.* 2001) and processed by Neumann *et al.* (2001) and Neumann *et al.* (2003). The feature has a width of 20–25 km and a maximum height above the surrounding ground of 800 m. We select the feature solely as being of convenient size and compactness for this study (small enough to be considered a surface detail but large enough to be well sampled by the 463 m cell-size DEM). From the recent MAVEN magnetic field data acquired at elevations as low as 130 km above surface, Mittelholz *et al.* (2020) report an absence of strong magnetisation over the Argyre Basin but heterogeneous magnetisations in surrounding regions. Derivation of magnetisation from terrain models is scale independent provided it is meaningful to estimate bulk magnetisation values over the selected area. Mostly, we envisage the methods as suitable for surface exploration of small-scale features. In the absence of a known magnetisation we chose the estimates derived at Tenterfield of strength 4.2 A/m, declination 18°, inclination −54° (magnetisation 'A') and the almost anti-parallel magnetisation ('B') of strength 3.8 A/m, declination 188°, inclination +57°. The terrain grid feature is shown in Fig. 11.13A. Figure 11.13B shows the faceted terrain model derived (using ModelVision software) from the grid and a horizontal base surface that we used to forward compute the magnetic field. Figure 11.13C shows a simple three-dimensional polyhedral body (a ModelVision 'frustum' class body) also used as an approximate representation of the terrain.

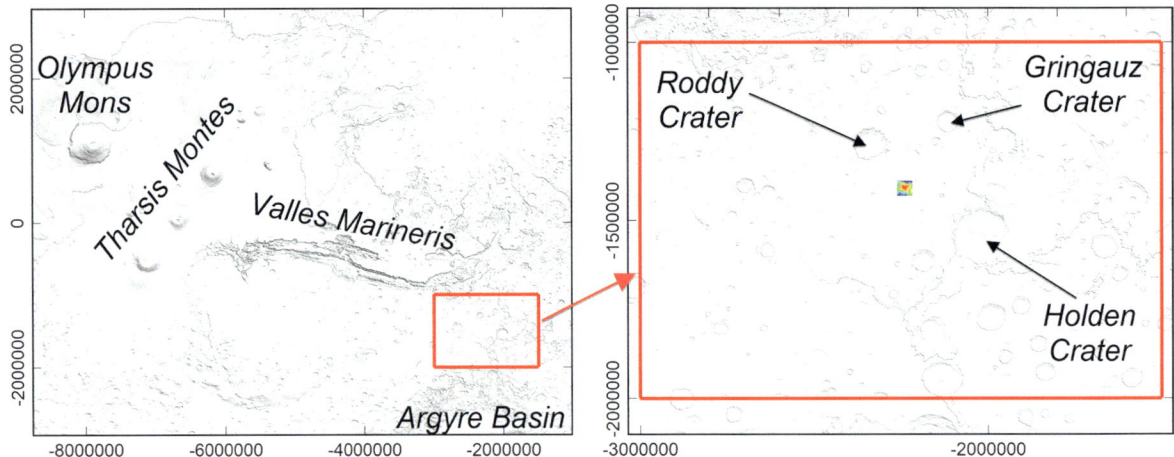

Fig. 11.12. Location of the synthetic terrain magnetic study.

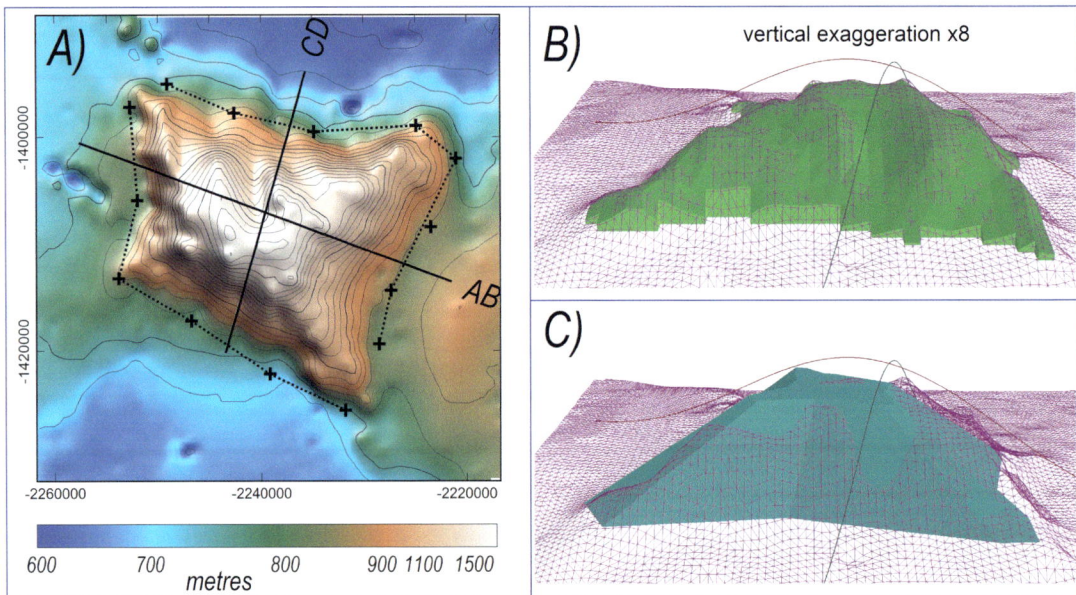

Fig. 11.13. A) terrain image of the isolated topographic feature of study, B) faceted horizon-base model of the terrain feature used for forward field computation, C) polyhedral terrain-approximation model used for inversion.

Figure 11.14 shows magnetic component maps computed just above the ground surface from the detailed terrain model using the two assigned magnetisations. The contour interval is 10 nT and the images have been clipped to exclude the higher amplitude field variations directly above the feature. If there are no overlapping fields and measurements can be made with sufficient resolution, only a sparse distribution of multi-component magnetisations is required to resolve the direction of magnetisation from the mapped fields. Figure 11.14 shows that each magnetic field component switches polarity between fields of the two almost anti-parallel magnetisations (magnetisation A in the top row and

magnetisation B in the bottom row). The decrease in B_z on approaching the terrain feature from any direction as shown in the righthand top frame in Fig. 11.14, reveals that the inclination of magnetisation is negative. The greater challenge of characterising magnetisation direction from magnetic field mapping of TMI is illustrated in Fig. 11.15 in which the azimuth of the magnetic field is imaged as it swings around the terrain feature and there is no consistency as with the B_z component in Fig. 11.14 from which to quickly determine the polarity of magnetic inclination. Vector magnetic field measurements require reference orientations. The vertical reference is easily obtained and there are various options for a

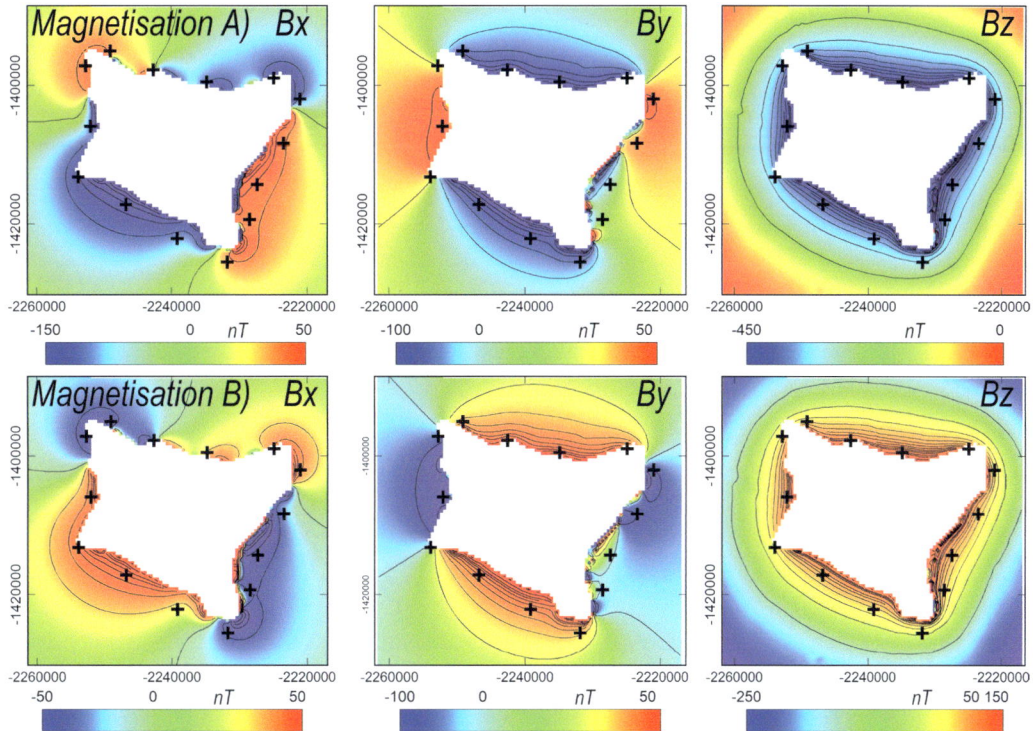

Fig. 11.14. Magnetic field components computed from the terrain model. Upper row for a magnetisation of declination 18°, inclination −54° (magnetisation A) and lower row for a magnetisation of declination 188°, inclination +57° (magnetisation B).

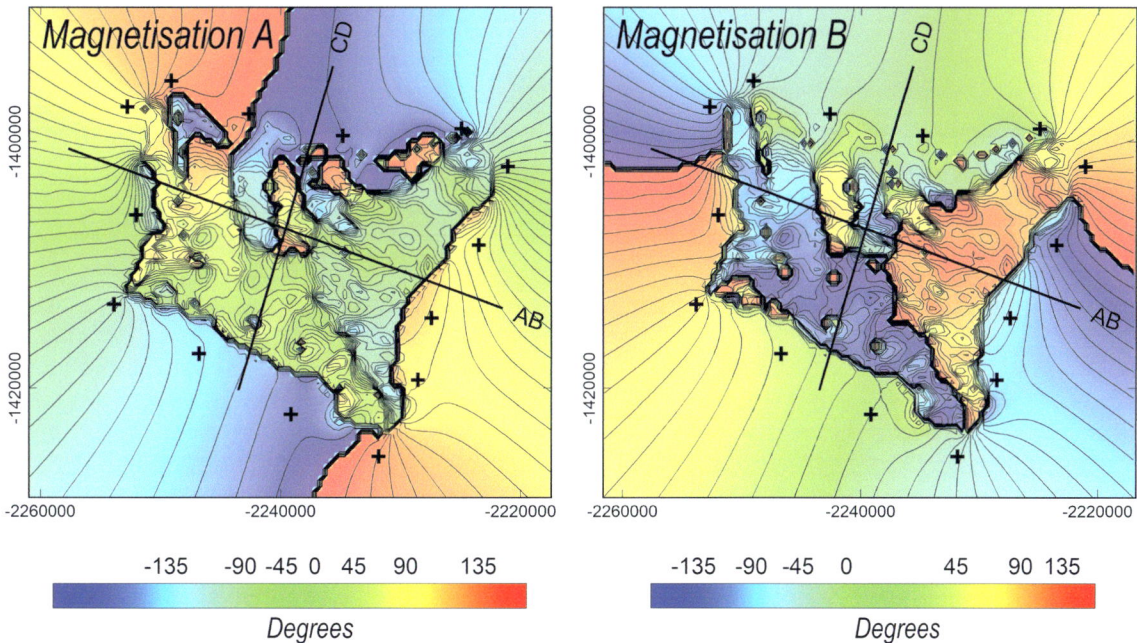

Fig. 11.15. Field declination maps generated by the terrain model fields with (left) magnetisation A and (right) magnetisation B.

horizontal reference, including sun shadows (just as a sun compass is used to orient palaeomagnetic samples).

The horizontal component grid images plotted in Fig. 11.14 all show approximate symmetry about two orthogonal axes. A traverse starting at any point and half-encircling the feature fully samples the component anomaly patterns that express the magnetisation direction. If the anomaly is isolated from other magnetic field variations, four equi-spaced multi-component measurements half-circling the terrain feature are sufficient to recover an estimate of its magnetisation direction. Exact location of the stations is not critical, but stations closer to the feature have stronger signal. We computed magnetic component values 3 m above terrain at six stations on the western track and eight stations on the eastern track (at approximately 8 km intervals) as marked by crosses in Figs 11.13 to 11.15. The traverses start and end near the north-west and south-east corners of the terrain feature. We computed the magnetic component values using the detailed terrain model assigned the magnetisations of ~4 A/m estimated in the Tenterfield study. The average range of the components on each track is 54 nT with a standard deviation of 19 nT. We added noise with a standard deviation of 4 nT to each channel and separately inverted the data along each track with multi-channel (three-component) inversions of the simplified frustum terrain model. Figure 11.16 shows the terrain model computation, model computation with noise and inversion model computation of the

three magnetic field components along the two tracks. The four magnetisations recovered from the inversions (for the two tracks and the two magnetisations) are listed in Table 11.1 and plotted in Fig. 11.18. The mean error of the magnetisation estimates is 7°. These results suggest that it should be possible to recover magnetisation estimates on a rover track around an isolated terrain feature if the terrain model of that feature reasonably represents the distribution of a homogeneous magnetisation of at least moderate strength, if there are no other strong magnetisations nearby and if the measurements can be made to a resolution better than 4 nT.

We also emulated three-component magnetic field data collected on two traverses above the feature (as shown in Figs 11.13B and 11.13C) of length 38.5 km and 27.5 km, at ~350 m above the ground and 500 m spacing along the traverses. Magnetisation estimates derived for the homogeneously magnetised terrain feature would benefit only slightly from more closely spaced measurements, but closer measurement spacing would help resolve any local inhomogeneities in magnetisation.

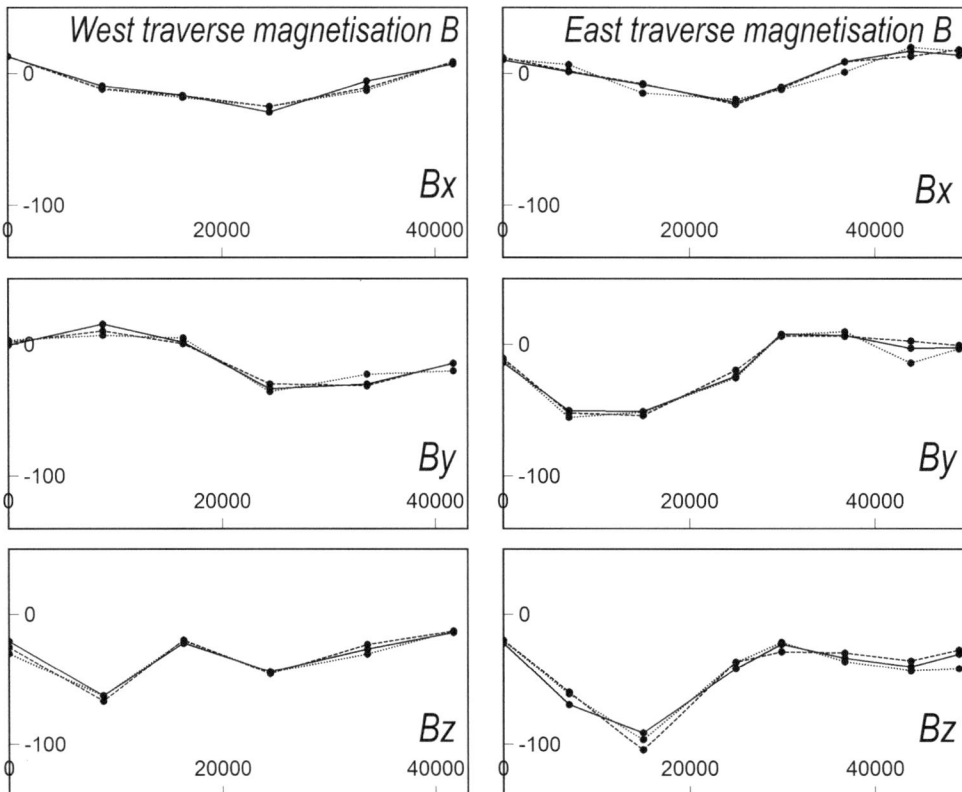

Fig. 11.16. Three-component profiles at stations on west and east tracks with magnetisation B. The three curves of detailed terrain-model compution with and without 4 nT added noise and the simple terrain-model inversion of the noise-added data all form a close envelope.

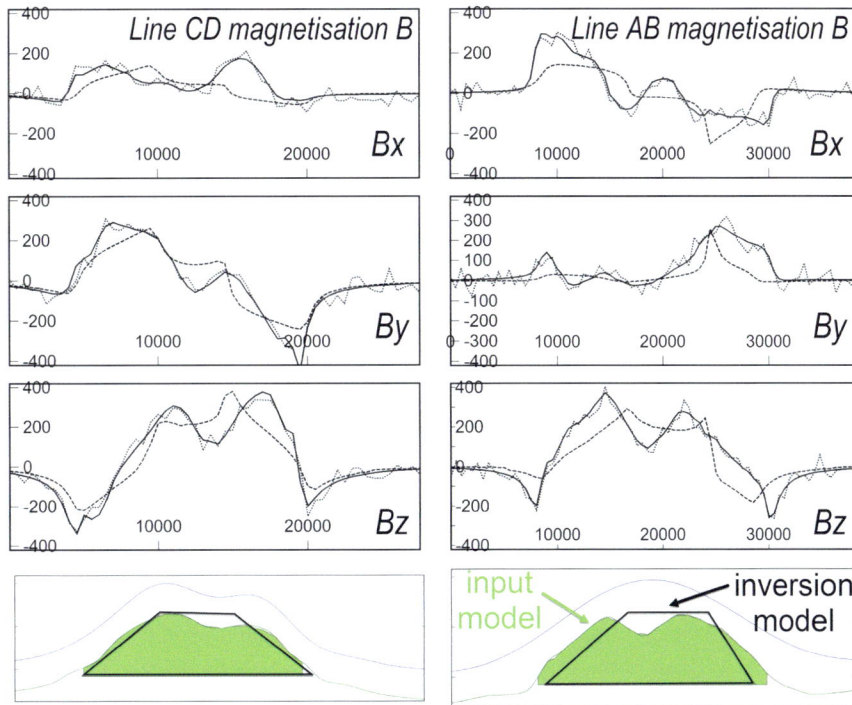

Fig. 11.17 Three component profiles on flightlines AB (right column) and CD (left column) for the terrain model with magnetisation of declination 188°, inclination +57°. Solid lines – forward computed from the terrain model, dotted line – with added noise, dashed line – output from the inversion of the input terrain model data with added noise.

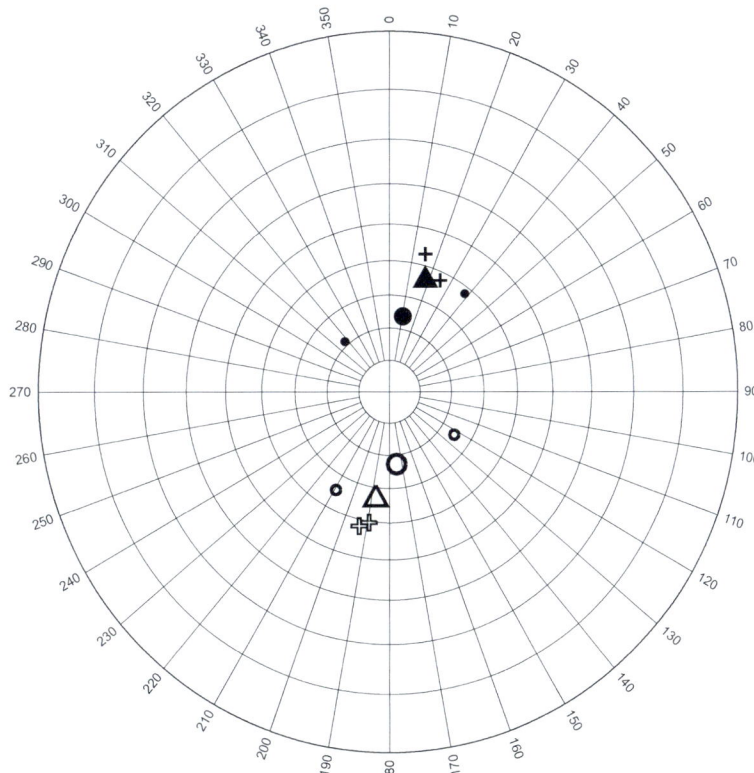

Fig. 11.18. Magnetisation directions: Input magnetisations (triangles), flightline inversions (small circles), average of the two flightline inversions (larger circles) and ground track inversions (crosses). Open symbols are positive inclination, closed symbols are negative inclination.

Table 11.1. Magnetisation direction estimates from inversion of the ground data around the feature and the aerial data above it.

Survey	Magnetisation	Intensity A/m	Declination°	Inclination°	Error°
Ground					
west track	A	4.7	25	−52	5
	B	3.2	193	48	9
east track	A	4.2	15	−47	7
	B	3.8	189	50	7
Aerial					
line AB	A	2.7	318	−69	31
	B	2.5	123	65	31
line CD	A	3.6	38	−52	12
	B	3.3	209	55	12
mean	A		11	−66	4
	B		174	67	12

Figure 11.17 shows model-computed magnetic components with and without noise and the simplified model-computed components derived from inversion of the noise added data. The amplitude of the magnetic field variations at this elevation over the terrain are an order of magnitude higher than on the ground tracks around it, with an average range of 526 nT and average standard deviation of 127 nT. To accommodate the greater challenges of making vector measurements from a moving platform we added a higher noise level of 30 nT standard deviation to the input field data. The magnetisation estimates recovered from inversion of the data on each profile are listed in Table 11.1. In this case there are systematic differences between results for both input magnetisations on each line, with errors of 31° on line AB and 12° on line CD. These errors are not due to the added random noise but to the horizontal top of the simplified inversion model that misrepresents the part of the terrain closest to the overflight measurements. This misrepresentation is more significant on profile AB than on profile CD. In practice it will be the validity of the terrain model in representing the spatial distribution of magnetisation that limits reliability in recovering the magnetisation direction.

11.5 CONCLUSIONS

Over the Miocene age Lamington volcanics in north-east New South Wales there is positive and negative correlation between terrain elevation and TMI, due to magnetisation that is respectively parallel and anti-parallel to the geomagnetic field. We have recovered estimates of these magnetisations from inversion of the magnetic field data using spatial models of magnetisation derived from the digital elevation model (DEM). Recovery of magnetisation direction is more challenging where magnetisation is buried and inversion has to determine its distribution as well as its intensity and direction. We show that the advantage of a known spatial model of magnetisation is sufficient to allow recovery of reasonable estimates of magnetisation direction from inversion of a single profile of TMI (or any other single component) and more tightly constrained estimation of magnetisation direction from inversion of single profiles of multi-component data. Use of vector component data is impractical in the strong background field of the Earth that penalises even slight misorientation of the sensor but is feasible in the weak background fields of Mars where the magnetic field variations are predominantly the signal of interest. We use a local Martian terrain feature of width 20–25 km and height 800 m to generate synthetic three-component magnetic field data and show that input magnetisation directions can be reasonably recovered from analysis of sparse measurements along either rover tracks around that feature or overflights above it. The practicality of these methods depends primarily on the strength of magnetisation and its geological distribution in relationship to terrain features.

REFERENCES

Albee AL, Arvidson RE, Palluconi F, Thorpe T (2001) Overview of the Mars Global Surveyor mission. *Journal of Geophysical Research* 106(E10), 23291–23316. doi:10.1029/2000JE001306

Blakely RJ 1995, Potential Theory in Gravity and Magnetic Applications: Cambridge University Press.

Brown RE, Cranfield LC, Denaro TJ, Burrows PE, Henley HF, Stroud WJ, Brownlow JW (2007) Warwick - Tweed Heads 1:250 000 Metallogenic Map SH/56 2–3. Geological Survey of New South Wales, Maitland and Geological Survey of Queensland, Brisbane.

Duggan MD, Mason DR (1978) Stratigraphy of the Lamington Volcanics in far Northeastern New South Wales. *Journal of the Geological Society of Australia* 25, 65–73. doi:10.1080/00167617808729014

Mittelholz A, Johnson CL, Feinberg JM, Langlais B, Phillips RJ (2020) Timing of the martian dynamo: New constraints for a core field 4.5 and 3.7 Ga ago. *Science Advances* 6, 1–7.

Neumann GA, Rowlands DD, Lemoine FG, Smith DE, Zuber MT (2001) Crossover analysis of Mars Orbiter Laser Altimeter data. *Journal of Geophysical Research* 106(E10), 23753–23768. doi:10.1029/2000JE001381

Neumann GA, Smith DE, Zuber MT (2003) Two Mars years of clouds detected by the Mars Orbiter Laser Altimeter. *Journal of Geophysical Research* 108(E4), 5023. doi:10.1029/2002JE001849

Schubert G, Russell CT, Moore WB (2000) Timing of the Martian dynamo. *Nature* 408, 666–667. doi:10.1038/35047163

Smith DE, Zuber MT, Frey HV, Garvin JB, Head JW, Muhleman DO, Pettengill GH, *et al.* (2001) Mars Orbiter Laser Altimeter – experiment summary after the first year of global mapping of Mars. *Journal of Geophysical Research* 106(E10), 23689–23722. doi:10.1029/2000JE001364

12
Magnetisation deficit anomalies

C.A. Foss

ABSTRACT

Sharp magnetic anomalies most commonly arise due to localised strong ('anomalous') magnetisation in contrast to surrounding weaker magnetisation. However, less commonly, magnetic field anomalies are due to absent or weak magnetisation surrounded by stronger magnetisation. I present two studies to illustrate inversion of these 'deficit' magnetisation anomalies. The first study is of the normally magnetised Kalkarindji basalts in the Northern Territory. There is an extensive magnetic high over a wide sheet of basalt with sharp negative anomalies at what appear to be fracture intersections in the sheet. I interpret these magnetic anomalies as due to destruction of magnetisation where the sheet has been altered by heated waters escaping through those fracture intersections from the underlying Beetaloo Basin. Inversion of these anomalies on the assumption that magnetisation is completely destroyed provides estimates of the magnetisation of the surrounding sheet. In the second example I interpret an elliptic region of approximately 8 km diameter of low and smoothly varying magnetic field values north of Ceduna in South Australia as due to a non-magnetic or weakly magnetic granite intruded into surrounding more strongly magnetised basement. Modelling the granite against a nominated surrounding zero magnetisation recovers a magnetisation estimate that can be reversed in direction and assigned to the basement rocks. In this example the magnetisation of the surrounding basement is variable in strength and to obtain a magnetisation estimate from the contrast against the granite I have excised local high-amplitude anomalies. Independent inversion of some of those anomalies returns magnetisation estimates consistent with the basement magnetisation direction inferred from the granite anomaly inversion. This suggests that although the basement magnetisations are highly variable in strength, they are commonly directed. For both these studies the magnetisation contrasts are rotated from the local geomagnetic field directions, suggesting that remanence contributes to those magnetisations.

12.1 INTRODUCTION

Magnetic field modelling and inversion results are almost invariably discussed in terms of absolute magnetisations (or magnetic susceptibility values) whereas gravity modelling and inversion are commonly discussed in terms of density contrasts. In both cases measured variations in the gravity or magnetic field arise due to contrasts in physical properties. Lateral variations in subsurface density are rarely more than 10% of the total density. However, magnetisation contrasts are commonly a much higher proportion of total magnetisation, with one magnetisation intensity that is inconsequential compared to the other. Another factor discouraging reference to contrasts in magnetic field modelling and inversion are that magnetisation contrasts are vectors rather than scalars and it is more challenging to assign meaning to vector contrasts than to scalar contrasts.

Voxel inversions of magnetic fields are increasingly being used to supposedly map continuous distribution of magnetisation in the subsurface. It is essential to realise that the magnetic field signal only provides information about magnetisation contrasts and that absolute magnetisations cannot be determined by inverse methods. The most reliable inversion results are obtained by focus on only the most discrete field variations (referred to throughout this book as 'sweet-spots') on the assumption that these magnetisations are in contrast with much weaker magnetisation.

One context for which magnetisation contrasts are sometimes discussed is the question of whether a negative amplitude TMI anomaly (in moderate to steep inclination fields) is due to reverse remanent magnetisation or a magnetic susceptibility lower than the surrounding material. This ambiguity is best resolved by searching beyond the immediate anomaly for any other marginal anomalies due to contrasts against what might be a strong enclosing magnetisation. For weak or diffuse anomalies this test is less diagnostic. Figure 12.1 illustrates a synthetic magnetic profile over both a local reverse magnetisation within a non-magnetic background (plotted in blue) and a magnetisation deficit (a hole) in a normal magnetisation background (plotted in red). The local anomalies themselves are almost

identical, and the only chance of discriminating between their sources would be through recognition that the broader magnetisation (with a hole in it) contributes to the background field. Inversion of the magnetic field variation caused by a hole gives an apparent magnetisation direction opposite to that of the surrounding magnetisation, with inclination of opposite polarity and declination rotated by 180°. The two interpretations of a discrete reverse magnetisation or a cavity in magnetisation are end-case models in which one component is assumed to have zero strength relative to the other. Without this assumption, estimation of either magnetisation is highly non-unique. Without multiple constraints, voxel inversions in which the magnetisation of each element of the ground is assigned an absolute magnetisation intensity and/or direction are indefensible.

If the ground beneath a well measured magnetic field consists of only two completely induced magnetisations of different strength (a single contrast in magnetic susceptibility) there should be no bias towards estimation of the directions of those magnetisations. However, many complex distributions of variously directed magnetisations can be misrepresented as suitably distributed zones of high and low magnetic susceptibility in a background of intermediate susceptibility. Figures 12.2A and 12.2D show two TMI anomalies of very different pattern in a −60° inclination field due to magnetisations of declination 30°, inclination −45° and declination 300°, inclination +15° respectively. In this relatively steep southern inclination geomagnetic field the contrast in declination of magnetisation can be estimated from the trough to peak azimuth and the contrast in inclination can be estimated from the peak and trough amplitude ratio (see Chapter 10). However, it is also possible to emulate these anomalies with only induced magnetisations. In this high-inclination field, TMI field variations are predominantly positive over high magnetic susceptibilities and negative (weaker than surrounding values) over low magnetic susceptibilities. The anomaly peaks and troughs due to the two single magnetisations as imaged in Figs 12.2A and 12.2D can be approximately matched with a distribution of magnetisation consisting of higher susceptibility towards the anomaly peak and lower susceptibility towards the trough as shown in the perspective views of Figs 12.2C and 12.2F. A simple magnetisation such as the single magnetisation dipole models used to generate the anomalies in Figs 12.2A and 12.2D can always be

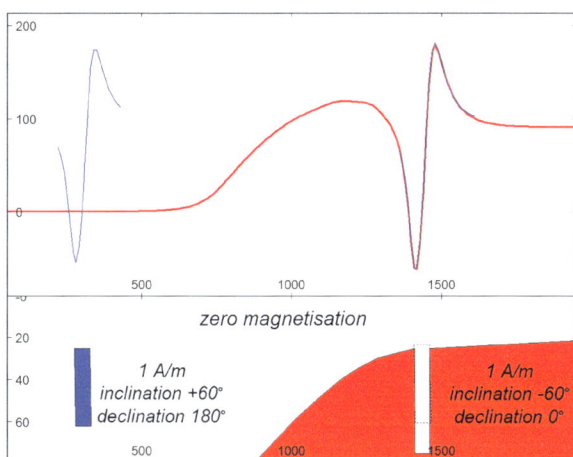

Fig. 12.1. Equivalent anomalies produced by a body with reverse magnetisation (in blue) and a hole in a body of normal magnetisation (in red).

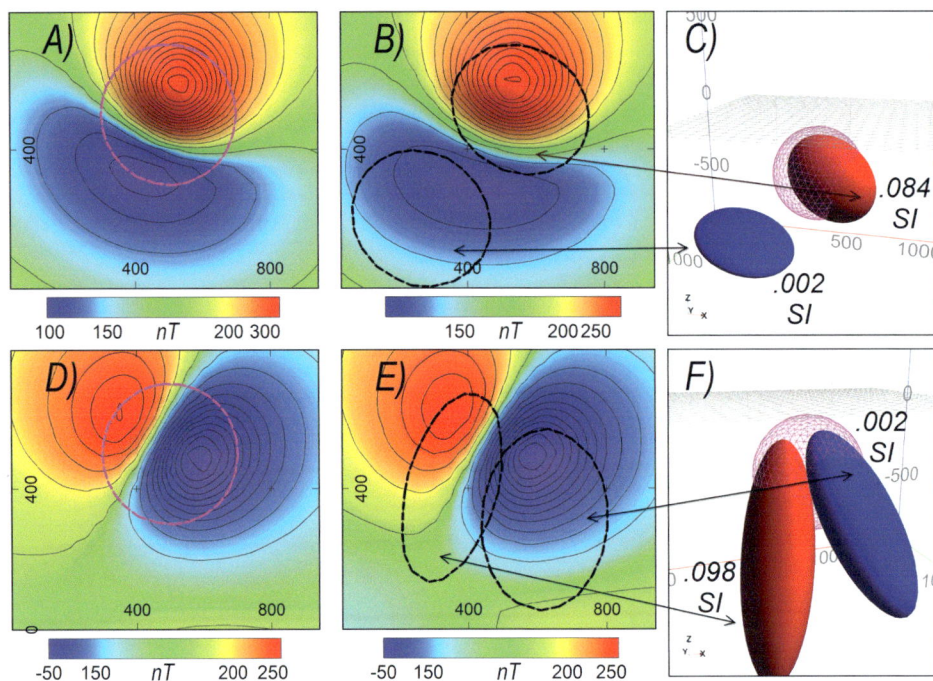

Fig. 12.2. TMI anomalies in a −60°inclination field due to remanent magnetisations: A) and D) and almost identical anomalies: B) and E) due to pairs of low and high susceptibility magnetisations. The remanent magnetisations are isolated and the induced magnetisations are in a background of 0.05 SI. C) and F) show the model perspectives: magenta – remanently magnetised sphere, red and blue – high and low susceptibility ellipsoids.

matched by more complex magnetisations given sufficient degrees of freedom. Conversely, few magnetisations due to complex distributions of high and low magnetic susceptibility can be closely matched by an alternative model of a single, compact magnetisation. Although the anomalies shown in Fig. 12.2 can be generated by either a simple single magnetisation model or a model of both high and low magnetic susceptibility, should such an anomaly be encountered in measured data, in the absence of any alternative information the more justified interpretation is that it is due to a single magnetisation. Complexity should only be added to magnetisation models as required. Note that transforms such as total gradient (TG) or the normalised source strength (NSS) sometimes applied to highlight the location of magnetisation, primarily highlight location of curvature in the magnetic field. If the TMI fields due to the alternative magnetisation models in Fig. 12.2 are identical then any enhancement transforms of those fields are also identical. Where there are minor differences between the alternative model fields the transforms at best emphasise those differences. Enhancements may assist in discriminating between models but they rarely provide diagnostic discrimination where that was not evident in the primary data.

12.2 MAGNETISATION DESTRUCTION ANOMALIES OVER THE KALKARINDJI BASALTS IN THE NORTHERN TERRITORY

Figure 12.3 shows the location of the study area in the Betaloo Sub-basin of the Northern Territory, Australia. The Beetaloo Sub-basin (Williams 2019) is a southern component of the PalaeoProterozic to MesoProterozoic McArthur Basin that contains up to 9 km of sediments and volcanics (Ahmad *et al.* 2013). Large areas of the Beetaloo Sub-basin are unconformably overlain by the Cambrian Kalkarindji Suite of volcanics (formerly known as the Antrim Plateau Volcanics) mostly in the form of extensive basaltic sheets, and the Kalkarindji is in turn completely covered by weakly or non-magnetic Phanerozoic Carpenteria and Georgina Basin and younger sediments. There has been considerable investigation of the petroleum prospectivity of the Beetaloo Sub-basin with multiple wells drilled and over 9,000 km of two-dimensional reflection seismic lines (Williams 2019; Markov *et al.* 2021). There is also complete regional aeromagnetic coverage of the area by the Northern Territory Geological Survey. The study area is covered by the 2014 Dunmarra survey (P1268) flown on north–south flightlines at 400 m spacing and a nominal 80 m terrain clearance.

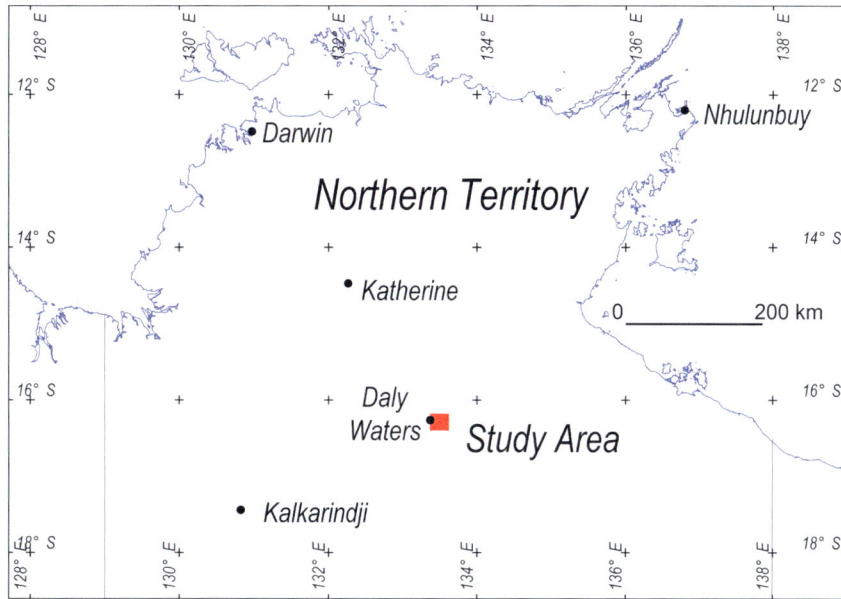

Fig. 12.3. Location of the study area over the Kalkarindji basalts in the Beetaloo Basin of the Northern Territory.

The TMI image in Fig. 12.4 shows contrast between the rough image texture of short-wavelength field variations due to shallow magnetisation contrasts across the top of the Kalkarindji basalts (covering much of the area in Fig. 12.4) and the smoother, longer-wavelength field variations over deeper magnetisation contrasts where the sheet is absent (the eastern section of Fig. 12.4). The highest amplitude magnetic field variations in Fig. 12.4 are over the sharp eastern edge of the sheet of Kalkarindji basalts. Consistent with magnetisation parallel to the southern hemisphere geomagnetic field, northern edges of the sheet have predominantly positive anomalies and southern edges have predominantly negative anomalies. This is illustrated in Fig. 12.5 that shows modelling of a grid traverse over a north-eastern tip of the sheet with the major field variations marking its edges. As shown in Fig. 12.5, the field variation can be well explained by a sub-horizontal sheet of

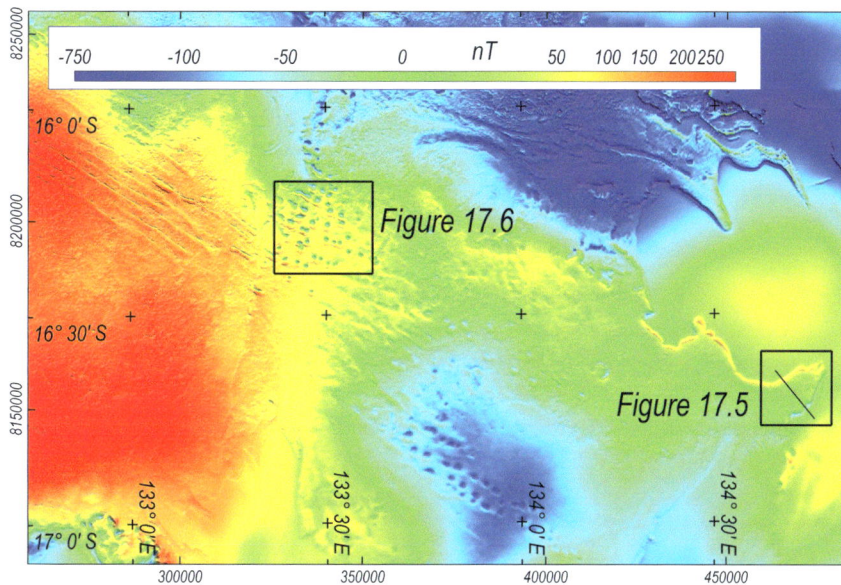

Fig. 12.4. TMI over the study area shown in Fig. 12.3.

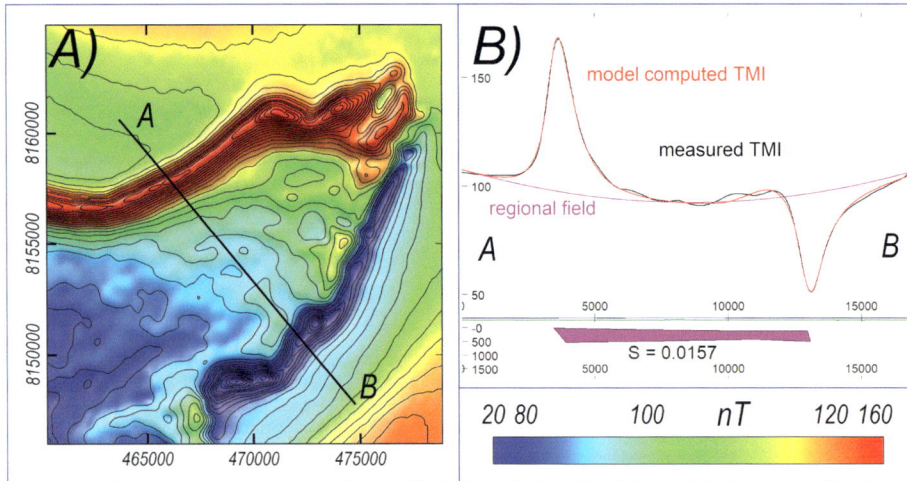

Fig. 12.5. A) TMI image and line of section over a north-east corner of the sheet of Kalkarindji basalts and B) a model along the line of section.

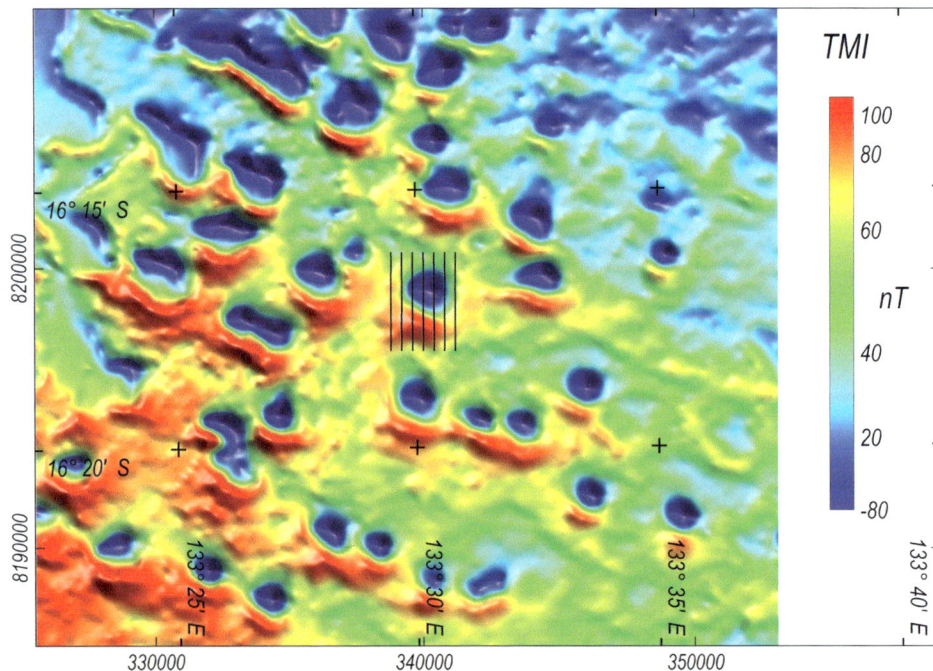

Fig. 12.6. TMI detail of the area located in Fig. 12.4. Modelling of the central anomaly marked by the north–south flightlines (400 m spacing) is shown in Fig. 12.7.

homogeneous magnetic susceptibility of between 0.01 and 0.02 SI and ~500 m thickness. There is little sensitivity to the specific thickness or magnetic susceptibility values of the best-fit model, that almost certainly represents a far more complex distribution of magnetisation (possibly including multiple layers). In modelling a single line of data there is also little sensitivity to magnetisation direction.

Over substantial areas of the Kalkarindji basalts (including the north-west area of Fig. 12.4) there are sharp local, parallel and equi-spaced linear magnetic field variations marking vertical displacements of the top of the sheet across steeply dipping faults that are imaged in the seismic sections (Markov *et al.* 2021). In the study area this pattern of magnetic field variation is mostly replaced by series of elliptic negative anomalies aligned in rows along the dominant east-south-east to west-north-west faulting direction. This pattern is only seen in areas that are clearly within the sub-crop of the Kalkarindji basalts and is interpreted as due to local destruction of the magnetised sheet by hot fluids flowing along the intersections of fault and/or joint patterns that cut the sheet (Foss and Dhu 2016). This interpretation is supported by discovery of bitumen within and above the Kalkarindji volcanics (Glass *et al.* 2013) that may be hydrocarbons from a petroleum system in the underlying basin that were entrapped in the fluid flow. Detail of an area of these TMI anomalies (located in Fig. 12.4) is shown in Fig. 12.6. The anomalies are predominantly negative with a weaker positive to the south. Many are elongated parallel to the interpreted major faulting direction and most have peak to trough ranges of 100 to 200 nT.

If the area of Fig. 12.6 was underlain by a homogeneous smooth-topped horizontal sheet of Kalkarindji basalt with distant edges, the sheet would not have any obvious magnetic field expression. The pattern of the individual anomalies in Fig. 12.6 is consistent with magnetisations oppositely directed to the local geomagnetic field (reverse remanent magnetisations) but with knowledge of the magnetised Kalkarindji sheet they are interpreted as magnetisation contrasts due to absence of the normal magnetisation of the sheet. The section of north–south flightlines plotted in Fig. 12.6 highlight an anomaly that I have inverted using a polygonal-section body of horizontal top and bottom

and free magnetisation direction (see Fig. 12.7). Inversion places the body at approximately the same depth and with approximately one-half of the depth-extent as the induced-magnetisation model of the sheet shown in Fig. 12.5 (neither of these parameters are well constrained). The contrast magnetisation of the model is 1.0 A/m with an inclination of +37°, declination 195°. On assumption that the body represents a hole of much lower magnetisation than the surrounding sheet, the sheet magnetisation can be inferred as ~1.0 A/m with an inclination of −37° and a declination of 15°. This direction is rotated by 12° from the local geomagnetic field, suggesting that there is some contribution from remanent magnetisation. The apparent magnetisation of the hole is approximately twice that of the apparent induced magnetisation of the sheet estimated in the model of Fig. 12.5. This gives the two models similar products of magnetisation and depth extent, consistent with the assumption that the reverse anomaly is due to a hole in the sheet filled with material of low intensity of magnetisation. Each of the inferred holes in the sheet provide an opportunity to estimate its local magnetisation. This is most conveniently achieved by inverting for the magnetisation contrasts of the holes and then assigning the reverse of those magnetisations to the sheet.

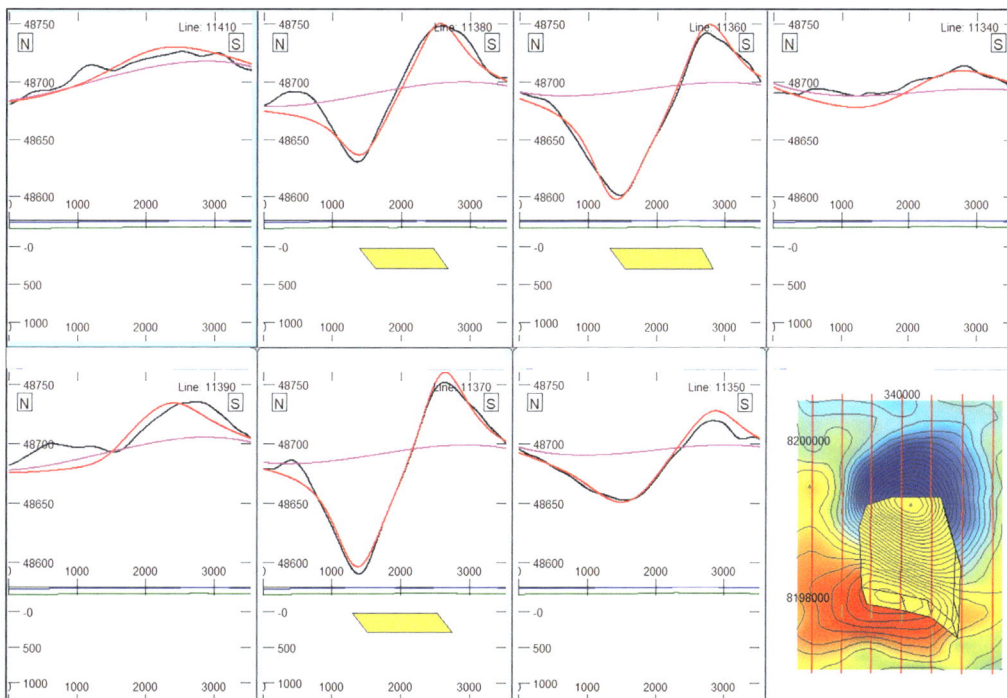

Fig. 12.7. Flightline sections through the model derived from inversion of the anomaly located in Fig. 12.6. The black traces are the measured field, magenta is the assigned background field and the red trace is the model-computed field.

12.3 A NEGATIVE MAGNETISATION CONTRAST ANOMALY OVER A PRESUMED BURIED GRANITE NORTH OF CEDUNA, SOUTH AUSTRALIA

To further illustrate recovery of estimates of magnetisation from anomalies that are the expression of negative magnetisation contrasts I present and extend a study by Foss *et al.* (2020) over a buried granite near Ceduna, South Australia in the southern Gawler Craton (Fig. 12.8). The TMI image in Fig. 12.9 is generated from data acquired on the 2014 South Australian PACE (Plan for Accelerating Exploration) Streaky Bay survey. The survey was flown on east–west flightlines at 200 m spacing and a nominal 60 m terrain clearance. Within the study area there is a range of almost 1,000 nT due to variable magnetisation of the southern Gawler Craton basement that underlies a thin cover of poorly consolidated Eucla Basin sediments. The basement rocks assigned to the Saint Peter Suite of the Nuyts Domain (Ferris and Fairclough 2007; Pawley *et al.* 2016; Reid *et al.* 2019) are undifferentiated within the study area because of the lack of outcrop and are mostly known from coastal exposures to the south. The rocks are dominantly granitic but with a range of inter-banded felsic to mafic and ultramafic rocks, interpreted to have been generated from magma mixing. The nearest basement outcrop, 10 km to the north-east, is assigned to the Hiltaba Granite as shown in the geological map by Blissett (1977) in Fig. 12.10.

The anomaly of interest in Fig. 12.9 is the broad magnetic low that has only minor internal field variation and a sharp, almost elliptic outline. The mostly simple shape of the anomaly, cross-cutting the complex magnetic field expressions of the surrounding basement suggests that the body may be a relatively late-stage intrusion of homogeneous, weak magnetisation as is commonly associated with late or post-orogenic granites. The magnetic low could be explained as due to a region of increased depth to the magnetised basement underlying an essentially non-magnetic cover, but the elliptic, smoothly curved sharp field change defining the edge of the anomaly is more consistent with a granite contact than with basement downthrow across faults that would be expected to consist of linear segments. Gravity variation coincident with the magnetic anomaly does not support confident interpretation because of the wide, 7 km spacing of the gravity stations, but the negative c. 100 μm/sec^2 gravity variation (based largely on a single station towards the centre of the magnetic anomaly) is also consistent with a granite emplaced into a more mafic, denser basement (Foss *et al.* 2020).

There are two major, broadly bifold classifications of granite as 'I-type' and 'S-type' (Chappell and White 1974) based on whole-rock geochemical and mineralogical inference of the igneous or sedimentary nature of the original melt zone, and of 'ilmenite series' and

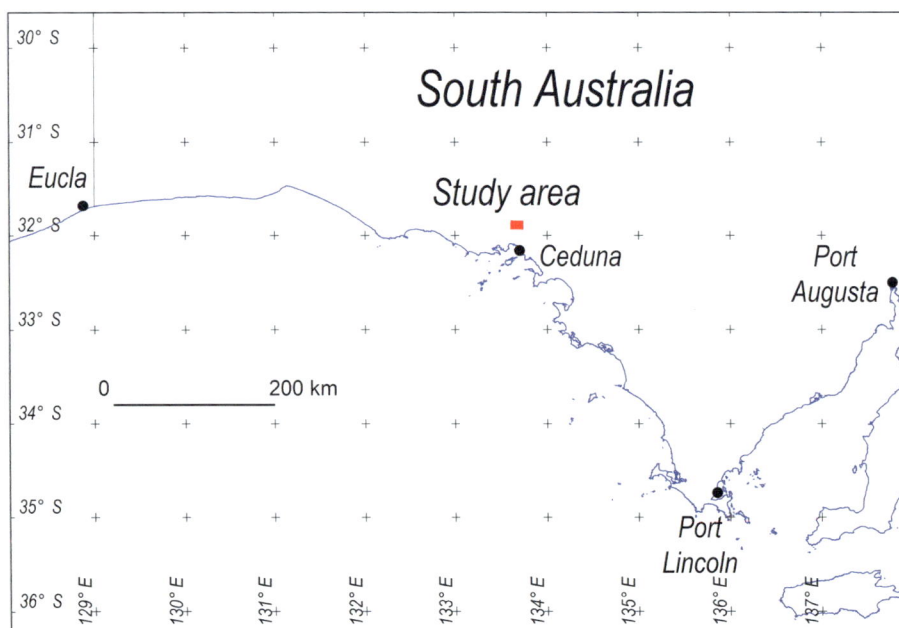

Fig. 12.8. Location of the study area near Ceduna in the southern Gawler Craton, South Australia.

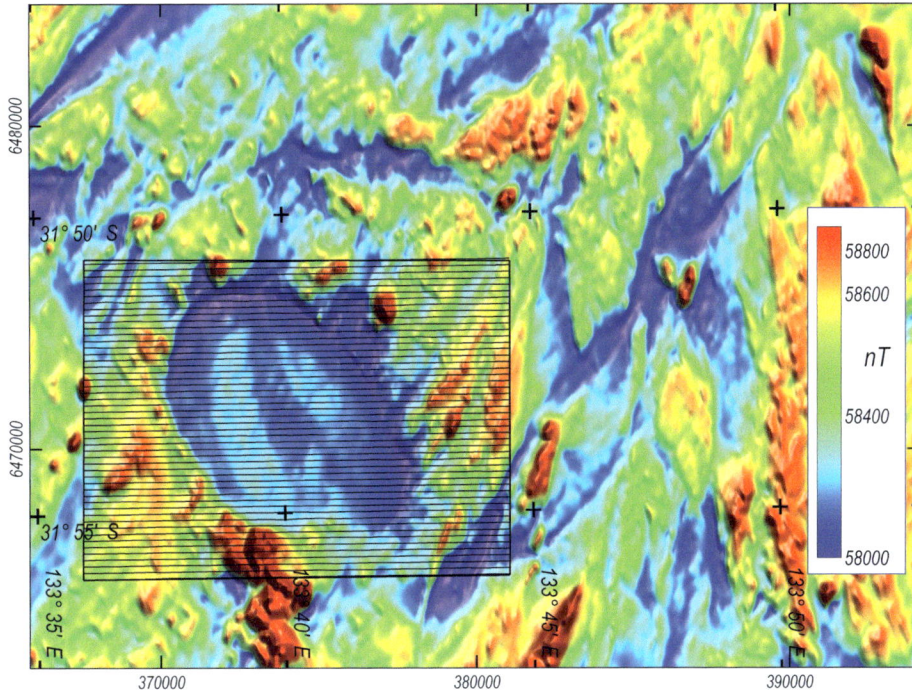

Fig. 12.9. Ceduna area TMI. The east–west flightlines show the area of the main inversion.

Fig. 12.10. Ceduna area 1:25,000 geology map (Blissett 1997) with TMI contours (100 nT interval). There is a single basement outcrop with the rest of the area covered by various Quaternary units.

'magnetite series' (Ishihara 1977) based on their minor iron and titanium oxide opaque minerals that ties directly to magnetisation. There is approximate correspondence between (generally) more strongly magnetised I-type and magnetite series granites and (generally) weakly magnetised S-type and ilmenite series granites.

In the southern Gawler Craton, samples of the Saint Peter Suite have variously been classified as I-type, S-type and A-type (anorogenic or anhydrous) (Frost *et al.* 2001) possibly because of complexities of magma mixing or because many of the granite samples are extensively altered.

In the study of the Kalkarindji volcanics the magnetisation contrast was against a sheet of apparently homogeneous magnetisation, but in this study the magnetic field variations are more complicated. Within and around the study area complex magnetic field variations include several small, relatively simple and isolated anomalies (Fig. 12.11) each defined on three to five east–west flightlines suitable for approximate determination of those local source magnetisation contrasts. These discrete anomalies highlighted in Fig. 12.11 have been independently inverted using plunging elliptic-section pipe homogeneous magnetisation models (Foss *et al.* 2020). Figure 12.12 shows measured and model-computed TMI for the anomaly located in Fig. 12.11 (anomaly 'L' in Foss *et al.* 2020). The complexity of this anomaly is not completely revealed by the 200 m flightline spacing, as is evident from the sharp

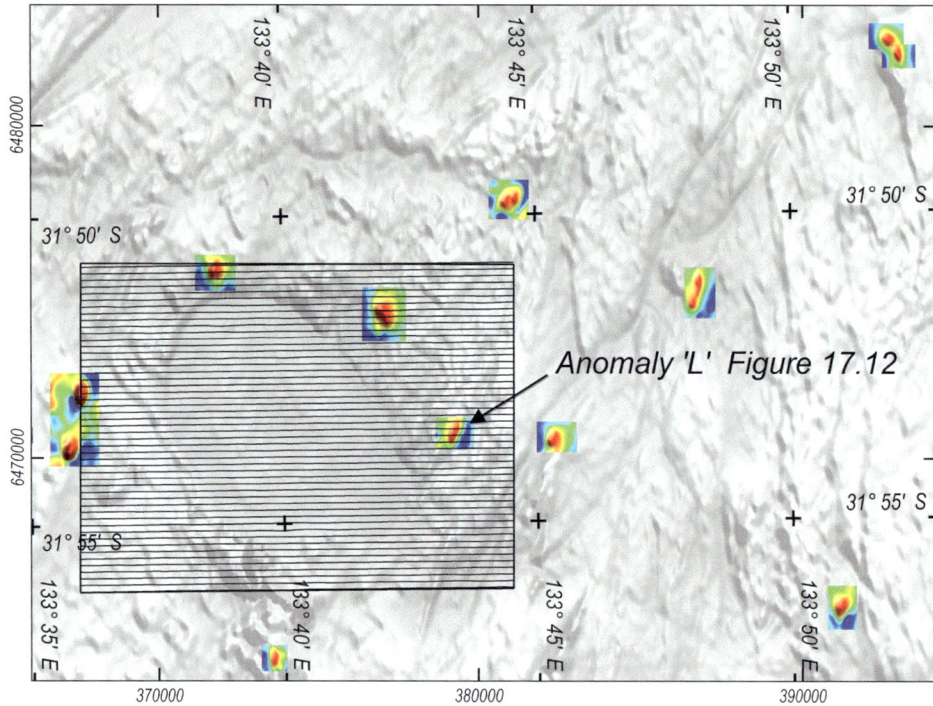

Fig. 12.11. Distribution of selected anomalies that have been individually inverted, shown over a grey-scale image of TMI. Most anomalies have a peak-to-trough range of 100 to 200 nT.

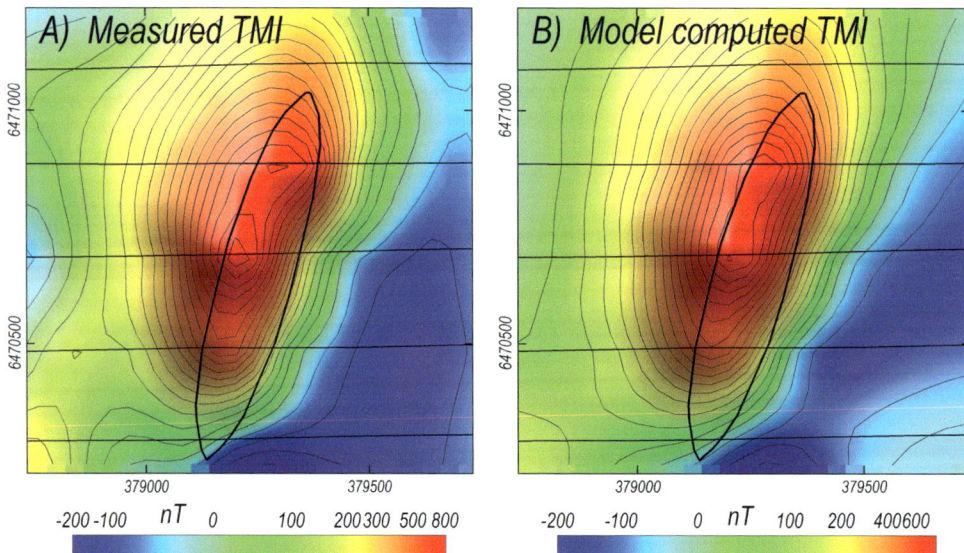

Fig. 12.12. A) measured and B) model-computed TMI over the anomaly located in Fig. 12.11.

differences in amplitude of the three adjacent flightlines across its centre. The model matches the field well on the available flightlines but may not well represent the (unknown) field between them. However, uncertainty of each inversion mostly arises from ambiguity in anomaly separation from complex background fields.

To study magnetisation contrasts between the granite and basement without edge effects from truncation of the models, I assigned magnetisation to the granite in the centre of the model and zero-magnetisation to the bounding basement, selecting 'standard' basement by excising sharp magnetic highs over interpreted strong intra-basement magnetisations (the image gaps in Fig. 12.13). Following the inversions I reassigned magnetisation by setting the standard basement magnetisation (that had been held at zero in the inversions) to the reverse of the granite body magnetisation and setting the granite body magnetisation to zero. A north–south range of 10 km for the model-computed field (50 flightlines at 200 m spacing) covers the variation of

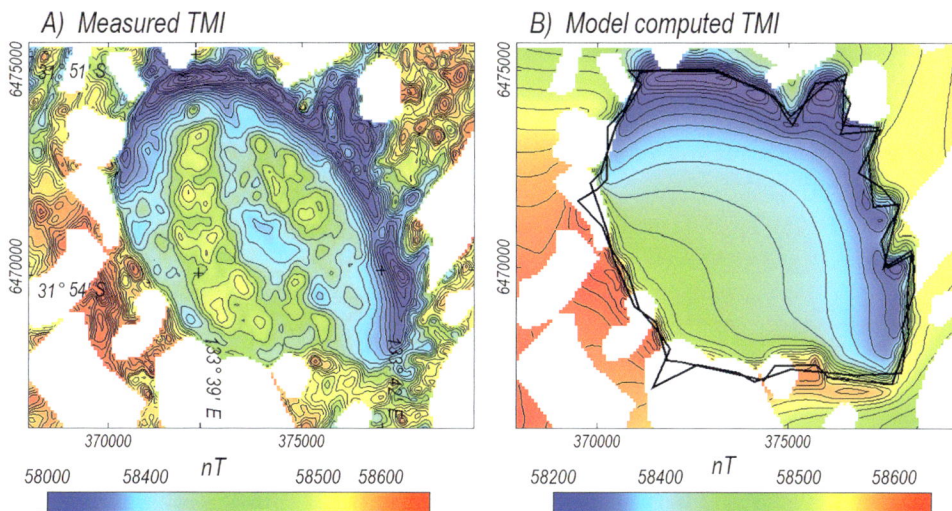

Fig. 12.13. A) measured TMI with high-amplitude basement anomalies excised and B) model-computed TMI (from inversion of even-numbered flightlines) with outline plans of the tops of the odd- and even-numbered inversion models.

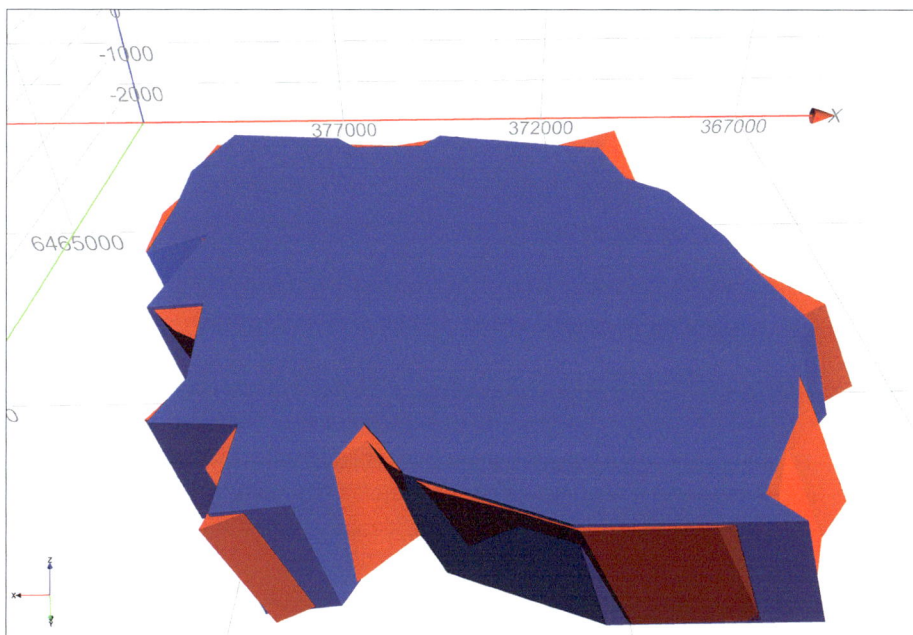

Fig. 12.14. Perspective of the granite inversion models: Blue from the odd-numbered and Red from the even-numbered flightlines.

interest (see Fig. 12.9). I split the flightlines into odd-numbered and even-numbered sets for independent inversion to investigate model repeatability. The study by Foss *et al.* (2020) first represented the granite body with elliptic-section horizontal topped and bottomed sheets, and after an approximate fit to the data had been achieved further inverted these bodies allowing independent movement of vertices defining their horizontal sections. In this study, I digitised the outline of a starting granite model from the TMI image and again used a staged inversion to first solve for the bulk properties of horizontal and vertical position, depth extent and magnetisation contrast. I then further inverted the bodies adding independent movement of the digitised vertices. To encourage stability, these inversions were performed in series of iterations, each allowing only short-distance movement of the vertices from their initial positions. In each iteration series the permitted ranges of the vertices were reset around their final positions in the previous series. Allowing larger single displacements of the vertices might help the models escape local data-misfit minima through exploration of a wider model space, but at the practical cost of a larger number of wasted computations and less effective search of the model space. Figure 12.13B shows the field computed from the odd-numbered line inversion model together with the outlines of the tops of the models from both the odd- and even-numbered line inversions. Comparison of the input data (Fig. 12.13A) and the field computed from an inversion model (Fig. 12.13B) shows that the inversions successfully explain the major long-wavelength features of the input data but lack the complexity to reproduce the short-wavelength features. The objective of inversion is not to match the input data (the task of its objective function) but to support speculation about the geology and magnetisation. These inversions clearly establish that key characteristics of the data are compatible with the model concept that the observed anomaly is due to a single homogeneous weak magnetisation surrounded by stronger magnetisation (also possibly homogeneous at broad scale after local anomalous magnetisations are excised).

Figure 12.14 shows a perspective view of the odd-line and even-line granite inversion model bodies (blue and red respectively) and statistics of the model magnetisations are reported in Table 12.1. The two bodies in Fig. 12.14 and the corresponding models of Foss *et al.* (2020) are all similar. The tops of three of the four model bodies are between 200 and 210 m below the mean ground level

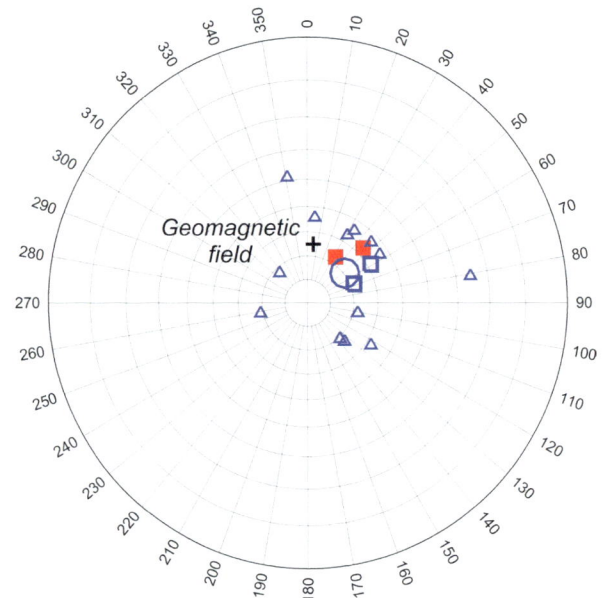

Fig. 12.15. Model magnetisation directions (all negative inclination): blue triangles – small anomalies, blue circle – small anomalies mean magnetisation, blue squares – basement estimates Foss et al. (2020), red squares – basement estimates this study.

(the fourth is 240 m). More reliable spot estimates of depth to magnetisation would best be recovered from single flightline inversions over the western and eastern edges of the body that are almost perpendicular to the flightlines (and on tie-line intersections with the northern and southern edges of the body). Estimated magnetisation contrasts for the four bodies are between 0.42 and 0.59 A/m (Table 12.1) and variations in depth extent are between 1,160 and 1,630 m, with considerable trade-off between the two parameters, such that the magnetic moments of the bodies are all between 32 and 35 $A.m^2.10^9$ as reported in Table 12.1 (the cross-sectional areas of the four granite model bodies are very similar).

Figure 12.15 shows magnetisation directions of the models. The 13 individual high-amplitude positive anomaly models have a range of estimated magnetisation directions, possibly as a combination of true variation in magnetisation direction and errors arising from incorrect anomaly separation (Foss *et al.* 2020). The α95 of the population is 14° (Foss *et al.* 2020) and the angular separation between the mean direction and the geomagnetic field is 18° (Table 12.1) suggesting that the magnetisations include a rotated remanent component. The two inversions by Foss *et al.* (2020) of the main granite anomaly gave a mean magnetisation estimate only 9° different to the mean of the positive anomaly

Table 12.1. Model magnetisation details (*after reversal of direction from the granite-body inversion model).

Study	Inversion	Volume (km3)	Magnetic moment (A.m2.109)	Magnetisation Intensity (A/m)	Dec- lination	Inc.- lination	ARRA
Foss et al. 2020	Mean of Thirteen anomalies			13.2	52°	−70°	18°
Foss et al. 2020	Odd-lines	77	32	0.422	68°*	−68°*	24°
Foss et al. 2020	Even-lines	75	33	0.440	59°*	−59°*	25°
this study	Odd-lines	62	32	0.520	32°*	−67°*	11°
this study	Even-lines	59	35	0.593	46°*	−57°*	21°

magnetisation directions, supporting an interpretation that the anomalous and 'standard' basement magnetisations are similar in direction and only differ in strength. The mean of the two granite-anomaly basement magnetisation estimates in this study is 11° different to the estimate by Foss *et al.* (2020) but is still only 10° from the mean of the positive anomaly directions, so the conclusion that these are the same direction is also supported by these inversion results.

The inversion models do not address magnetisation of the complex high-amplitude anomalies excised as shown in Fig. 12.13. It would be possible to restore that data and introduce additional model bodies to explain the anomalous magnetisations. However, those field variations were excluded because they were considered too complex to provide reliable inversion models. Addressing those anomalies in inversion of the complete dataset would result in a more complete representation of the magnetisation around the granite but would be unlikely to provide new insights into the geology or magnetisation. If those complex anomalies themselves are the focus of interest they could be inverted individually, but with necessarily lower reliability than inversion of the better-defined large granite anomaly.

12.4 CONCLUSIONS

With appropriate selection of the background field, the vector contrast between two magnetisations can be found by magnetic field modelling and inversion with either magnetisation set to zero. This includes the case of a weaker magnetisation enclosed within a stronger one (a magnetisation deficit anomaly). In the case that one magnetisation is much stronger but is set to zero in modelling or inversion, its value is the reverse (of equal strength and opposite direction) to the estimate obtained for the other magnetisation. Without

constraints such as for magnetisations that form part of a terrain or bathymetric surface with contrast against a known zero magnetisation of air or water, it is not possible to derive either absolute magnetisation without knowing or assuming the other. Selection of which magnetisation to set to zero to obtain the contrast value can be decided by convenience – generally depending on the distribution of the magnetisations. In both the case of the holes in the Kalkarindji basalt sheet and of the buried granite near Ceduna it is more convenient to assign magnetisation to the included, more weakly or non-magnetised confined body and then reverse the vector to give the interpreted magnetisation of the surrounding material.

REFERENCES

Ahmad M, Dunster JN, Munson TJ (2013) 'Chapter 15: McArthur Basin': in Ahmad M and Munson TJ (compilers). Geology and mineral resources of the Northern Territory. Northern Territory Geological Survey, Special Publication 5.

Blissett AH (1977) 'Childara, Sheet SH/53–14, 1:250000 geological series'. Geological Survey of South Australia, Adelaide.

Chappell BW, White AJR (1974) Two contrasting granite types. *Pacific Geology* **8**, 173–174.

Ferris GM, Fairclough MC (2007) 'Explanatory Notes for the CHILDARA 1:250 000 Geological Map'. South Australia. Department of Primary Industries and Resources. Report Book 2007/8.

Foss CA, Dhu T (2016) The bark without a dog – magnetic anomalies over holes in a volcanic sheet in the greater McArthur Basin, NT. *ASEG Extended Abstracts* **2016**(1), 1–5. doi:10.1071/ASEG2016ab276

Foss CA, Gouthas G, Katona LF, Hutchens MF, Reed GD, Heath PJ (2020) 'Gawler Craton Airborne Geophysical Survey Region 5, Streaky Bay – Enhanced geophysical imagery and magnetic source depth models'. Report Book 2020/00020. Department for Energy and Mining, South Australia, Adelaide.

Frost RB, Barnes CG, Collins WJ, Arculus RJ, Ellis DJ, Frost CD (2001) A geochemical classification for granitic rocks. *Journal of Petrology* **42**, 2033–2048. doi:10.1093/petrology/42.11.2033

Glass LM, Ahmad M, Dunster JN (2013) 'Chapter 30: Kalkarindji Province': in Ahmad M and Munson TJ (compilers). 'Geology and mineral resources of the Northern Territory'. Northern Territory Geological Survey, Special Publication 5.

Ishihara S (1977) The Magnetite-series and Ilmenite-series Granitic Rocks. *Mineria y Geologia* **27**, 293–305.

Markov J, Foss CA, Swierczek E, Delle Piane C (2021) 'Kalkarindji through the structural lens: structural characteristics of the Kalkarindji basalt from non-seismic geophysical data'. AGES Proceedings, Northern Territory Geological Survey, pp. 43–47.

Pawley MJ, Reid AJ, Dutch RA (2016) Magmatic systems of the Paleoproterozoic St Peter Suite, Western Gawler Craton: insights from reconnaissance mapping. *Mesa Journal* **81**, 4–12.

Reid AJ, Pawley MJ, Wade C, Jagodzinski EA, Dutch RA, Armstrong R (2019) Resolving tectonic settings of ancient magmatic suites using structural, geochemical and isotopic constraints: the example of the St Peter Suite, southern Australia. *Australian Journal of Earth Sciences* **67**, 31–58. doi:10.1080/08120099.2019.1632224

Williams B (2019) 'Definition of the Beetaloo Sub-basin'. Northern Territory Geological Survey, Record 2019–015.

13

Magnetisation excavation anomalies over the Tasmanian dolerites and application to magnetic modelling of small Martian impact craters

C.A. Foss and J.R. Austin

ABSTRACT

Magnetic field variations arise due to contrasts in magnetisation. Where part of a sheet of magnetisation is removed in the excavation of valleys the resulting magnetic field variations can alternatively be modelled as due to the contrast of the remaining sheet against air or the contrast of air against the sheet. Where regions of removed magnetisation are localised there are advantages in modelling the volume from which magnetisation has been removed rather than the more extensive surrounding volume of remaining magnetisation. We present a model study of magnetisation excavated in valleys incised through volcanic sheets in south-east Tasmania and then apply the same methodology in a model emulation of magnetic fields over a Martian impact crater where crustal magnetisation has been excavated in an impact event. On Earth, where there are only limited exposures of homogeneous strongly magnetised rocks and where terrain is continually modified by surface processes, many impact craters have weak or complex magnetic field expressions. However, on Mars many small impacts into an igneous crust might better suit investigation of crustal magnetisation direction. The initial flights of NASA's Ingenuity with over 10 km of distance covered at heights of up to 20 m suggest that future developments may make feasible the measurement of magnetic profiles across minor terrain features. We show that it should be possible to recover magnetisation estimates from small impact craters even if there is associated demagnetisation beneath and around them.

13.1 INTRODUCTION

Magnetic field variations are mostly modelled as departures from a background field postulated to be the field that would be present if the investigated magnetisation was absent. In this chapter we investigate magnetic field variations that we consider due to contrasts between a zone from which magnetisation has been removed by excavation and an adjacent remaining magnetisation. These are specific cases of magnetisation deficit anomalies considered in the Chapter 12, in which a weak or non-existent magnetisation is surrounded by a stronger magnetisation. As discussed in Chapter 11, there are significant advantages in modelling terrain-related magnetisations because the known top surface of the magnetisation constrains estimation of other magnetisation parameters. Advantages of modelling excavated (missing) magnetisation are most evident where the excavations are of limited extent and the field needs only to be modelled to short distances

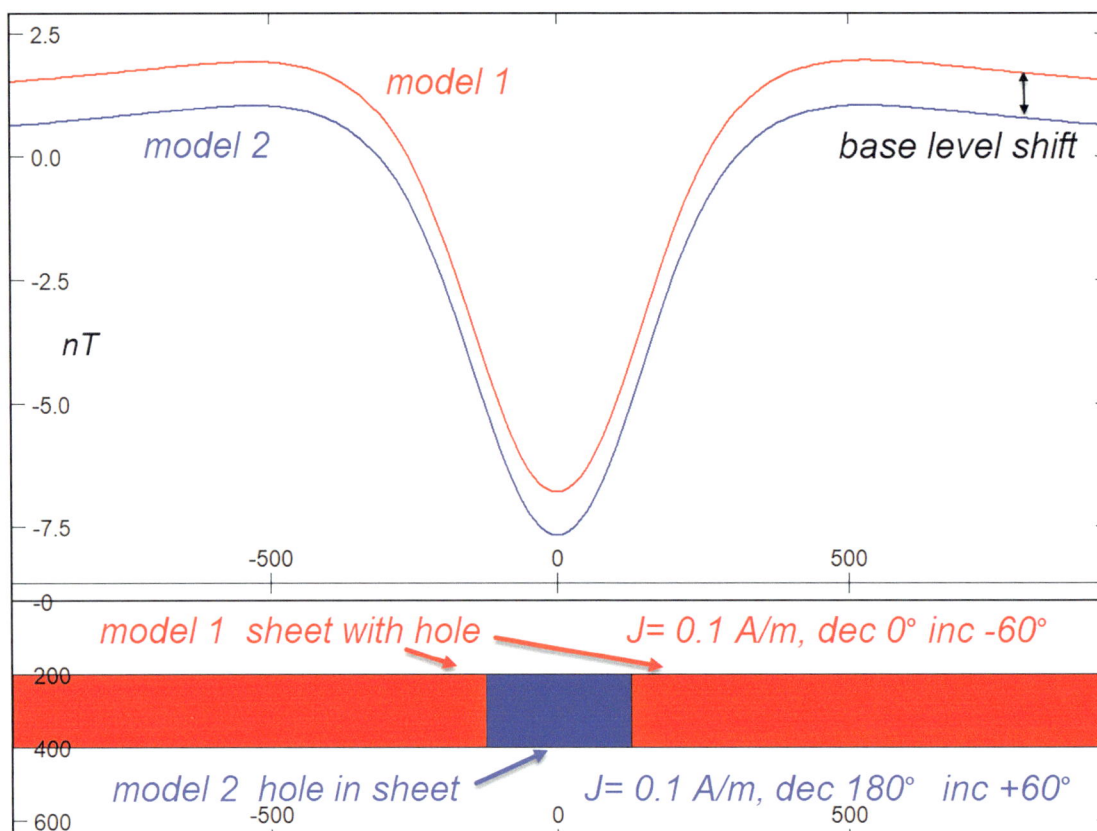

Fig. 13.1. East–west traverse in a 60° south magnetic field of a sheet with a gap in it (in red) and of an equal strength oppositely directed magnetisation occupying the gap (in blue).

away from their edges. The alternative of directly modelling the surrounding magnetisation requires either extending the model much further or facing concerns about edge-of-model termination artefacts extending into the region of interest. Two differences in modelling excavation magnetisations are in the background field levels and the apparent directions of magnetisation. In modelling an excavation magnetisation the magnetic field of the (estimated) complete sheet of magnetisation is included in the background field, and the excavation magnetisation has equal strength and anti-parallel direction to the magnetisation that has been removed.

Figure 13.1 shows the close equivalence between the magnetisation of a sheet with a gap in it and a magnetisation of equal strength and opposite direction occupying the gap. The difference between the two is the base shift due to the difference in magnetisation present in the two models. These different base levels have little significance in modelling or inversion of magnetic field

data because the absolute value of the background field is unknown and is empirically set to the smoothed value of the field surrounding the variation of interest ('the regional field'). The issue of selecting the appropriate background field can be significantly reduced by modelling gradients rather than fields.

We investigated excavation anomalies over a sheet of Jurassic dolerite in south-east Tasmania where the sheets and valleys cut through them dominate the terrain and have associated sharp, high-amplitude magnetic field variations. The dolerites are part of the Antarctic-Australia Ferrar flood-basalt volcanism with trace element compositions that suggest crustal contamination of a mantle source (Brauns *et al.* 2000; Hergt and Brauns 2001). Leaman (2002) suggests a default effective susceptibility value of 0.07 SI for modelling magnetic field variations due to these rocks. Measured susceptibilities are highly variable with a mean of less than 0.02 SI (Leaman 2002) but remanent magnetisation is strong, with Koenigsberger ratios in some cases greater than 10. The remanent magnetisation is mostly sub-parallel to the induced magnetisation or has slightly higher inclination, and stratigraphically lower zones of the dolerites include layers of reverse magnetisation. Jaeger and Joplin (1954) reported magnetisation of samples from borehole cores through several sheets, including one with an estimated thickness of 600 m. They found very variable intensity of magnetisation with steep inclinations and reversal towards the base of the sheet. Irving (1956, 1963) presented paleomagnetic NRM (natural remanent magnetisation) measurements from 30 dolerite sites in central to eastern Tasmania. The closest sampling site to our study area is site 28 that has a reported NRM direction of declination 68° and inclination −80° from two samples with an angular difference of 18°. A study by Schmidt (1976) extended sampling of the dolerites and included alternating field and thermal demagnetisation. These laboratory studies confirmed that the NRM directions measured on the dolerites are stable. A subsequent study by Schmidt and McDougall (1977) with both palaeomagnetic investigations and potassium-argon dating reconfirmed a split in magnetisation directions (including an anomalous easterly direction in our magnetic field study area) but found no systematic correlations between age and magnetisation direction.

We believe that there may be future opportunities to exploit excavation anomalies to estimate crustal magnetisations on Mars where impact craters of a wide range of sizes have excavated crustal material. Robbins and Hynek (2012) have created a database of over 384,000 craters with diameter > 1 km. Of these more than 300,000 have diameters < 3 km. There have been many studies of large Martian impact structures to investigate crustal-scale magnetisation that can be mapped at satellite elevations (Arkani-Hamed 2005; Shahnas and Arkani-Hamed 2007; Louzada *et al.* 2011; Lillis *et al.* 2013; Vervelidou *et al.* 2017). Reduction of magnetisation by excavation is further enhanced by the influence of shock demagnetisation in the rocks surrounding and beneath the craters. Demagnetisation effects depend on the unknown magnetic coercivity of the impacted crust, but Mohit and Arkani-Hamed (2004) and Shahnas and Arkani-Hamed (2007) suggest that at least for craters of several hundred kilometres diameter, demagnetisation may extend from the crater by as much as 80% of the crater radius. Demagnetisation is primarily due to shock effects with a secondary influence of thermal effects. Mittelholz *et al.* (2020) detected magnetic field variations on low-level MAVEN tracks close to a 35 km diameter crater in the region of Apollinaris Patera that include a reduction in the local magnetic field strength by 75%. They modelled the crater with a vertical cylindrical hole with surrounding demagnetisation extending the diameter by 50% and found that demagnetisation accounted for a reduction in magnetic field strength of 60%. However, we are primarily interested in smaller craters (including secondary craters formed by return of ejecta from a primary impact). Reliable mapping of the magnetic field signatures of these smaller craters of only several kilometres diameter or less are at or below resolution limits of the current satellite data but could be suitable for surface exploration by rovers or UAVs. For a vertical cylinder of vertical magnetisation contrast of 1 A/m and height to diameter ratio 0.1, the vertical magnetic field component B_z is 400 nT at a clearance that is 10% of the cylinder height. At a clearance equal to the cylinder height B_z falls to 150 nT. The scale independence of these magnetisation-plus-measurement distributions suggests that craters across a wide range of sizes might be suitable for magnetisation analysis.

13.2 MAGNETISATION EXCAVATION ANOMALIES IN SOUTH-EAST TASMANIA

Figure 13.2 shows the location of the Lost Falls Forest in south-east Tasmania. The East Tasmania Aeromagnetic and Radiometric Survey was flown over this area in 2022 for Mineral Resources Tasmania (MRT) on east-west flightlines at a spacing of 200 m. The data are available for download from the Geoscience Australia GADDS (Geophysical Archive Data Delivery System) as survey P5020. Figure 13.3 shows digital terrain elevation (DEM) and total magnetic intensity (TMI) images of part of the area. Much of the relief is due to thick sub-horizontal layers of basalt. Where rivers cut through these layers they incise deep and often steep-sided valleys. The magnetic field variation in the study area is over 2000 nT and the dendritic drainage pattern seen in the terrain image in Fig. 13.3A is also visible as magnetic lows in the TMI image in Fig. 13.2B. We selected the area outlined in Fig. 13.3 because it is a plateau of subdued elevation changes except where cut by valleys. The associated TMI image suggests that magnetisation is

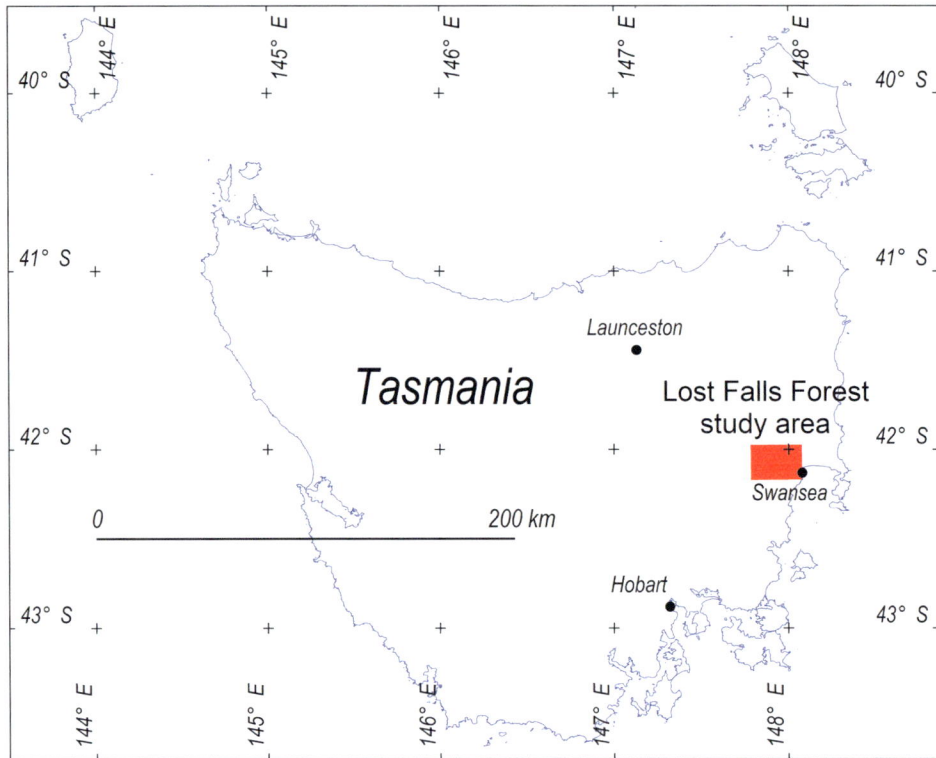

Fig. 13.2. Location of the Lost Falls Forest study area in south-east Tasmania.

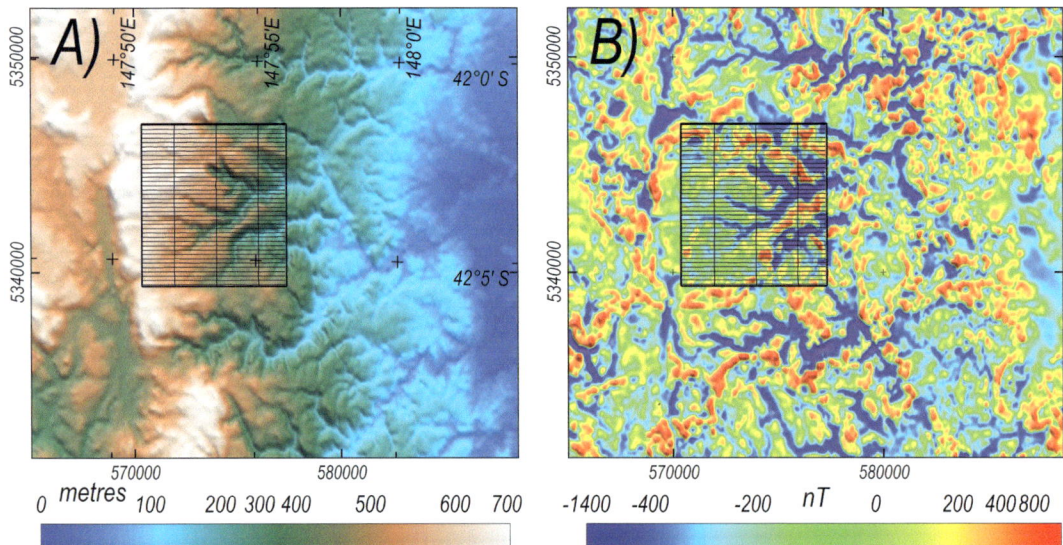

Fig. 13.3. A) DEM and B) TMI images of a section of the survey area over the Lost Falls Forest Reserve.

relatively homogeneous. Many surrounding areas have more pronounced magnetic highs, possibly marking individual plugs or volcanic feeders. Across the inversion area the DEM has a 390 m range from 206 to 596 m with a mean of 471 m. This surface is mapped from radar altimeter data and over the highly vegetated area these data may include some returns from the tree canopy. Over the rugged survey area it is impractical and undesirable to maintain a constant terrain clearance in flying the survey. For data continuity between adjacent flightlines a smooth drape surface is superior. The radar elevation values measuring distance above ground vary from 48 to 197 m with an average of 90 m. We use the DEM grid to generate the top surface of a terrain model and the base surface of the excavation model.

Figure 13.4 shows the magnetisation models used in this study. Figure 13.4A is a terrain magnetisation model generated with a horizontal base and a top surface tessellated from the terrain grid. For this model we selected an initial magnetisation estimate parallel to the local geomagnetic field of declination 15.5° and inclination −72.6°. We then ran inversions to adjust the intensity and direction of magnetisation and the base level and slopes of the background regional field. The optimum magnetisation found by the inversions was 2.7 A/m, declination 73.4° and inclination −68.5°. This direction is 19° from the geomagnetic field and starting model direction, with an inclination 4° shallower than the field. The inclination of remanent magnetisation reported by Irving from remanent magnetisation measurements on samples from the nearby palaeomagnetic site was 11° steeper than our model resultant-magnetisation direction, but the model-estimated declination of resultant-magnetisation of 73° is only 5° different.

In magnetic field inversions the primary trade-off against magnetisation direction is the horizontal

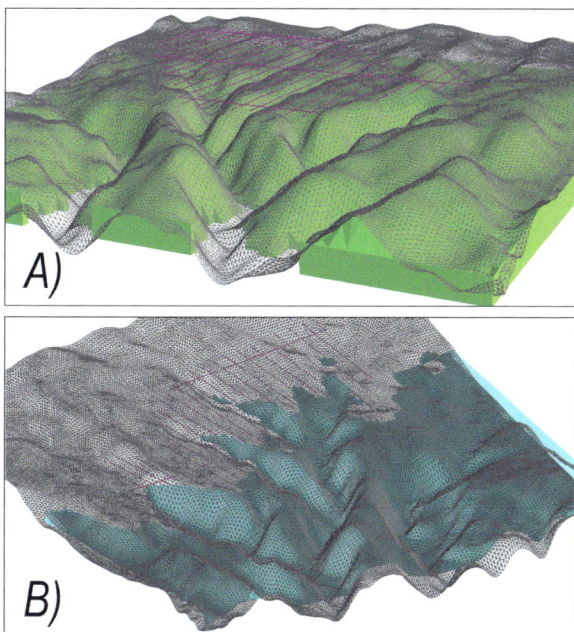

Fig. 13.4. A) terrain model with a tessallated top derived from the DEM grid and a horizontal base and B) an excavation model of the space between a horizontal top surface and the terrain base.

position of the magnetisation. However, in this study both the horizontal and vertical position of the magnetisation are locked in place by constraint of the terrain model. In running the inversions there is limited sensitivity to local features in the measured magnetic field because these effects tend to cancel over the complete model. We believe that the inversion model provides a reasonable estimate of resultant magnetisation across the volume of ground investigated. However, there are likely to be substantial local departures in intensity of magnetisation due to local inhomogeneities in the sheet, consistent with the palaeomagnetic and rock magnetic studies of Jaeger and Joplin (1954), Irving (1956, 1963), Schmidt (1976), Schmidt and McDougall (1977) and Leaman (2002). There may also be extreme local magnetisations resulting from lightning strikes, although the ground clearance of the magnetic field measurements attenuates the expression of those magnetisations substantially.

The close fit between input measured TMI and TMI forward computed from the inversion model is shown in Fig. 13.5B and an example model cross-section (the north–south tie line 190260 located in Fig. 13.5) is plotted in Fig. 13.6. Coincidence of the peaks and troughs of measured and model-computed TMI is controlled by the magnetisation direction because the model is locked in place by constraint of the terrain data. Imperfections in matching the input data are primarily misfits in amplitudes of individual peaks and troughs. These misfits could be reduced with local adjustments of the model magnetisation, particularly towards its top surface. However, such local adjustments would be mostly cosmetic and unlikely to significantly improve reliability of the bulk magnetisation estimate.

Figure 13.4B shows the alternative excavation model of the magnetisation removed from the sheet by erosion of the valleys. This model has a horizontal top and its tessellated base is generated from the terrain grid. This excavation model of free-space fits with the terrain model to create a volume bounded by a horizontal top and base, the edges of which are sufficiently removed from the field measurements that they do not cause field variations within the region of computation. We cannot use the excavation model to directly invert the measured data because, as shown in Fig. 13.6, the survey was flown on a tight drape for this rugged area and in the valleys the sensor elevation passes within the volume of the excavation model. The significant differences in sensor elevation across the region prevent reliable upward

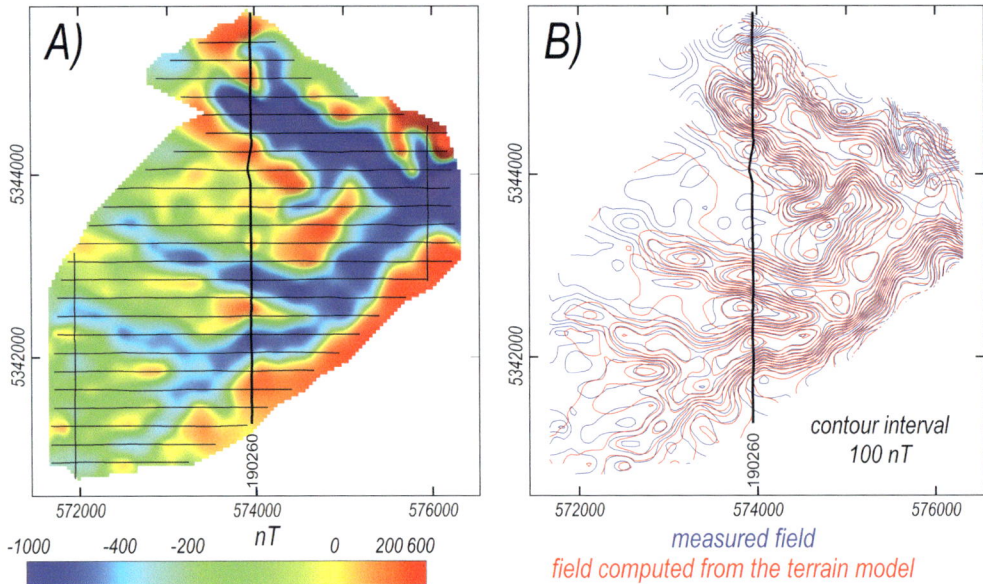

Fig. 13.5. A) measured TMI and B) contours of measured TMI (in blue) and TMI computed from the terrain model (in red).

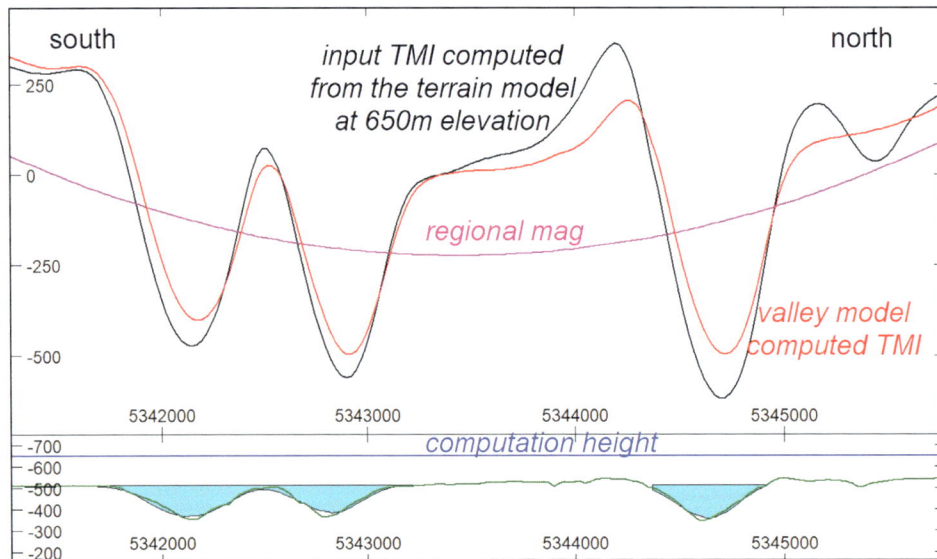

Fig. 13.6. Example south to north tie-line through the terrain model.

continuation of the data and the TMI grid required for upward continuation is itself is of questionable validity given this range of measurement elevation. For this study, since we had a close representation of the field along the flightlines from the terrain model computations (derived using the sensor elevation channel) we were able to use that model as an equivalent source to re-compute the field above and outside the volume of our excavation model. For this we chose a constant elevation just above terrain. This vertical translation to a

virtual measurement set would not have been necessary if the survey aircraft had not flown below the top of the basalt sheet. Because in this case we used the terrain model as an intermediate step to generate the dataset for the excavation-model inversion, the excavation model cannot improve on the magnetisation estimate derived from the terrain model. We do, however, have the considerable advantage that we know the input magnetisation that the excavation-model inversion should replicate.

Figure 13.7 shows the same north–south tie-line as imaged in the lower-elevation terrain-model inversion in Fig. 13.6, with the field recomputed at a constant elevation of 650 m. Inversion of this data has an easier task than the initial inversion of the terrain model because the greater separation from the magnetisation at the higher elevation supresses some of the detail measured at the lower elevation, and because the input data is computed from a homogeneous-magnetisation model. Nevertheless, there are still imperfections in matching the two models with their different regional background

fields. Figure 13.8A shows the field forward computed from the excavation model and Fig. 13.8B shows the close match between the computed fields of both the terrain model and the excavation model. The best estimate of apparent (contrast) magnetisation of the excavation model is intensity 5.03 A/m, declination 233° and inclination +72°. This magnetisation is 173° (rather than the ideal 180°) different to the input terrain model magnetisation. The discrepancy of 7° compared to the difference of 19° between the terrain-model magnetisation and the geomagnetic field direction gives us confidence that if

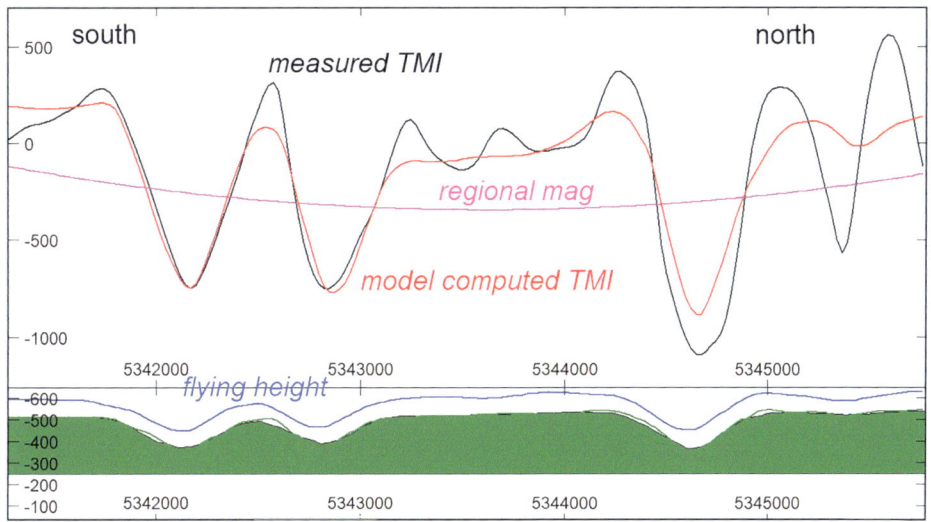

Fig. 13.7. Excavation model cross-section along south to north tie line 190260. The base of the model is derived from the DEM grid, the top surface is horizontal and the magnetisation is inverted from the computed field of the terrain model.

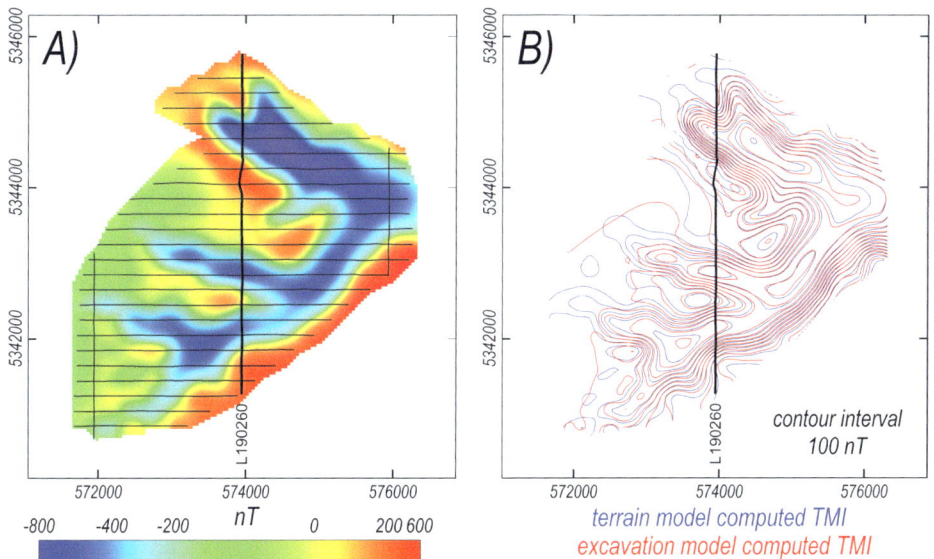

Fig. 13.8. A) TMI forward computed from the terrain model at a constant elevation of 650 m and B) contours of that input field (in blue) and the field computed from the inverted excavation model (in red).

the survey had been flown higher and we could have derived the excavation model directly from inversion of that measured data, that we would most likely have come to the same conclusion that the resultant-magnetisation includes expression of a steep-inclination easterly directed remanent magnetisation.

13.3 SYNTHETIC MAGNETIC MODELLING OF A MARTIAN IMPACT CRATER

The methodology of modelling excavation magnetic anomalies in south-east Tasmania may be suitable for future surface exploration of Mars, particularly for investigation of crustal magnetisation surrounding small impact craters. We anticipate that a terrain model is likely to be the starting point for magnetic field modelling of these small craters and that complexity will only be added if that initial model proves inadequate. To test our ability to recover the crustal magnetisation surrounding an impact crater we selected a simple crater of approximately 8 km diameter from the Mars MGS MOLA DEM (version 2) measured by the Mars Orbital Laser Altimeter (Smith *et al.* 2001) flown on the Mars Global Surveyor spacecraft (Albee *et al.* 2001) and processed by Neumann *et al.* (2001) and Neumann *et al.* (2003). The location of the crater is shown in Fig. 13.9 and the terrain variation is imaged in Fig. 13.10A. The depth of the crater is ~800 m with a depth to diameter ratio of 0.1. This

particular crater has only a minor rim that does not require independent treatment in the modelling. We assume that the crater wall is the bounding surface of a magnetisation that is homogeneous at the crater scale. For surface exploration on Mars we anticipate that greater use could be made of smaller features, but this 8 km diameter crater allows us to generate an acceptable terrain model from the available 463 m cell size grid. In any surface exploration we expect that a more detailed terrain grid would be available for magnetic field modelling. As already mentioned, there is a constant amplitude of magnetic field variations for surveys with measurement elevation and extent proportional to model size, so our findings are size independent.

Figure 13.10A shows the terrain grid over the chosen impact crater and Figs 13.10B and 13.10C show two total magnetic intensity (TMI) images computed by assigning different magnetisations to the model generated from that grid. There is a difference of 116° between these two magnetisation directions but it is challenging to predict those directions from the TMI anomalies. On Earth, anomalous magnetic fields sum with the background field to in some places reinforce it giving positive TMI features and in other places oppose it giving negative TMI features. With experience of magnetic fields measured in a particular geomagnetic inclination the approximate magnetisation direction of compact anomalies can be predicted by visual inspection (see Chapter 5) but on

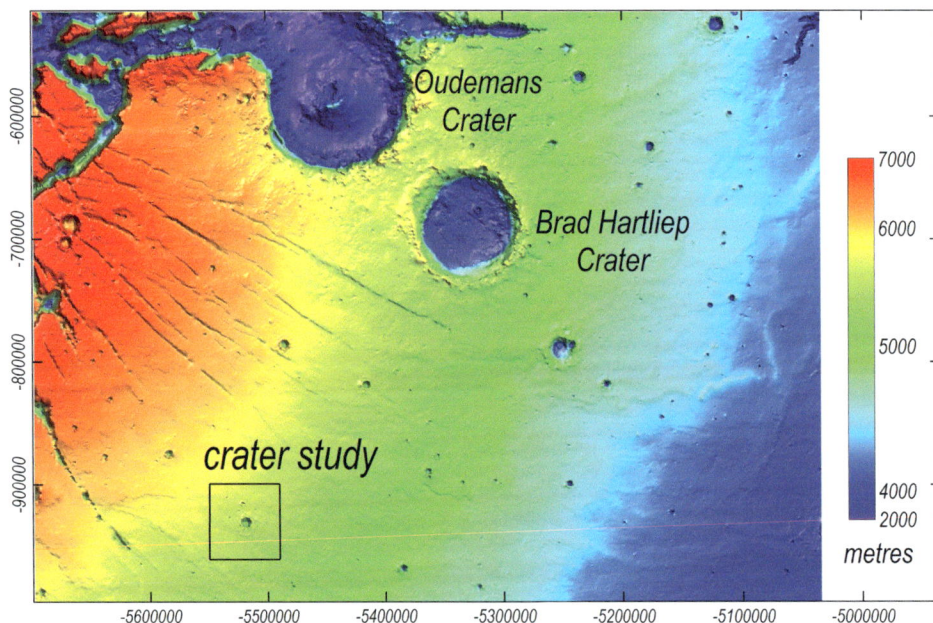

Fig. 13.9. Martian terrain model detail from the MARS MGS MOLA DEM (version 2) showing the location of the crater selected for study.

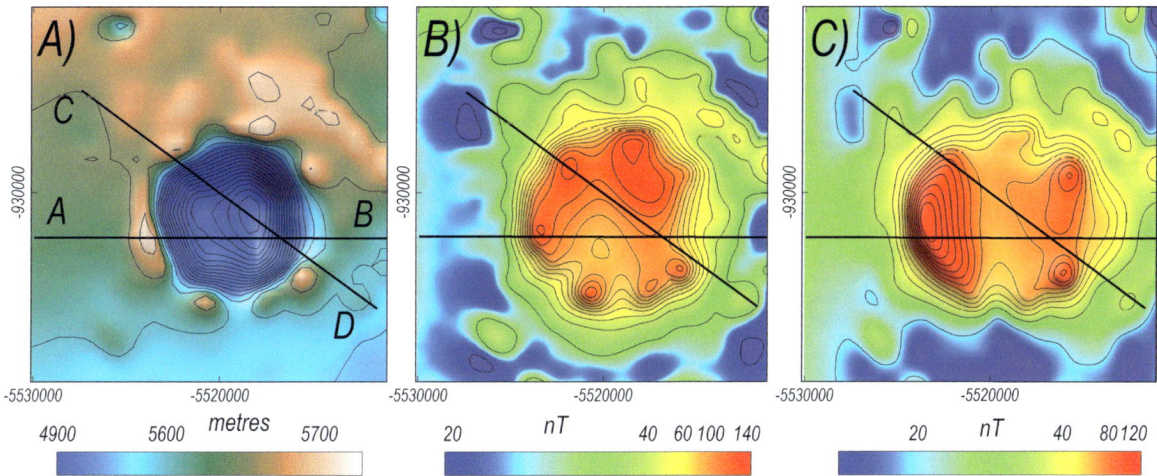

Fig. 13.10. A) terrain image of the selected crater and B) and C) TMI anomalies over the crater for magnetisations in the surrounding crust of B) declination 0°, inclination −60° and C) declination 90°, inclination +30° respectively.

Mars in a very weak background field TMI anomalies are almost universally positive-only and their pattern is less diagnostic of magnetisation direction. In Chapter 16 we discussed the advantages of multicomponent vector data for magnetic field studies in weak background fields and those advantages also apply to excavation anomalies. Figure 13.11 shows images of the three orthogonal vector component anomalies corresponding to the TMI anomalies in Figs 13.10B and 13.10C. For Martian surface exploration it may be impractical to make sufficient measurements to obtain such imagery but we can use these images to understand the component variations along the two profiles shown, and the multi-component data on those two profiles is sufficient to resolve the surrounding magnetisation. For the steep inclination (−60°) magnetisation A in the top row of Fig. 13.11 the two horizontal components show dipole anomalies of similar positive and negative amplitudes. The vertical component is predominantly positive and maps the crater extents more conveniently. Note that the polarity of B_z is opposite to that of the magnetisation because the crater is a deficit magnetisation. For the less steep magnetisation B with an inclination of +30° shown in the lower row of Fig. 13.11, the B_z component is mixed polarity but

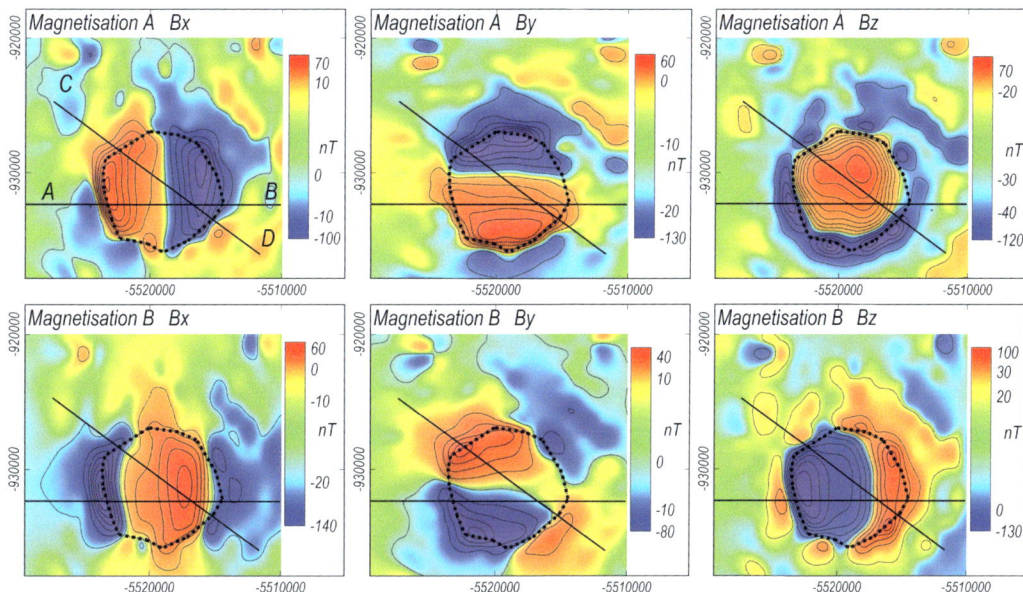

Fig. 13.11. Magnetic field Bx, By, Bz component anomalies over the crater (with dashed outline) for surrounding magnetisations A) declination 0°, inclination −60° (top row) and B) declination 90°, inclination +30° (bottom row).

with dominant polarity also opposite to that of the magnetisation. The horizontal B_x component parallel to the northerly declination of this magnetisation produces a prominent positive anomaly and the horizontal B_y component perpendicular to the magnetisation produces a weaker quadrupole anomaly.

Figure 13.12A shows the terrain model from which the component grids and profiles in Fig. 13.11 were computed and Fig. 13.12B shows the excavation model that uses the terrain grid as its base and has a horizontal top coincident with the ground elevation around the crater. The excavation model was used to invert the profiles computed from the terrain model at an elevation of

6,000 m (360 m above its top). The inversion results are shown for the magnetisation A model in Fig. 13.13 and for the magnetisation B model in Fig. 13.14. These were multi-component inversions of both survey lines that sought to simultaneously optimise the fit to all three components on both lines. These fits do not match the details of the field computed from the more complex terrain model but for both magnetisation directions all three components show an approximate fit and this corresponds with close recovery of the input model magnetisation direction. For magnetisation A, the excavation model has a magnetisation of declination 175.6°, inclination +59.3° that gives an anti-parallel crustal magnetisation estimate of declination 355.6°, inclination −59.3° (a discrepancy of 2.3° with the input direction). For magnetisation B, the excavation model has a magnetisation of declination 269.8°, inclination −28.8° that gives an anti-parallel crustal magnetisation estimate of declination 89.8°, inclination +28.8° (a discrepancy with the input direction of 1.2°). Depending on magnetisation direction (that is unknown before the survey) a single profile may not have an orientation well suited to investigation of the magnetisation. An effective survey procedure would be to fly two profiles of different orientation as shown in Fig. 13.11. Hopefully suitable terrain features (including but not necessarily limited to impact craters) will be sufficiently common that a regional survey could step from one feature to the next.

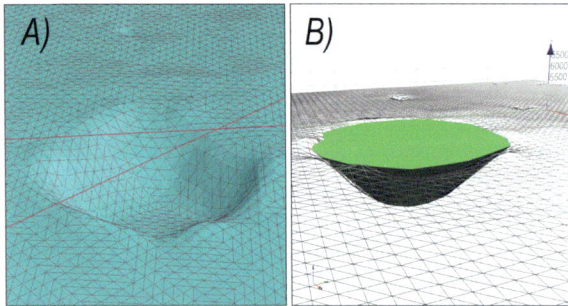

Fig. 13.12. Alternative models of A) terrain used for forward computation of the input fields for this study and B) the space model of the crater excavation used in inversion. The model volumes have magnetisations of equal strength and opposite direction.

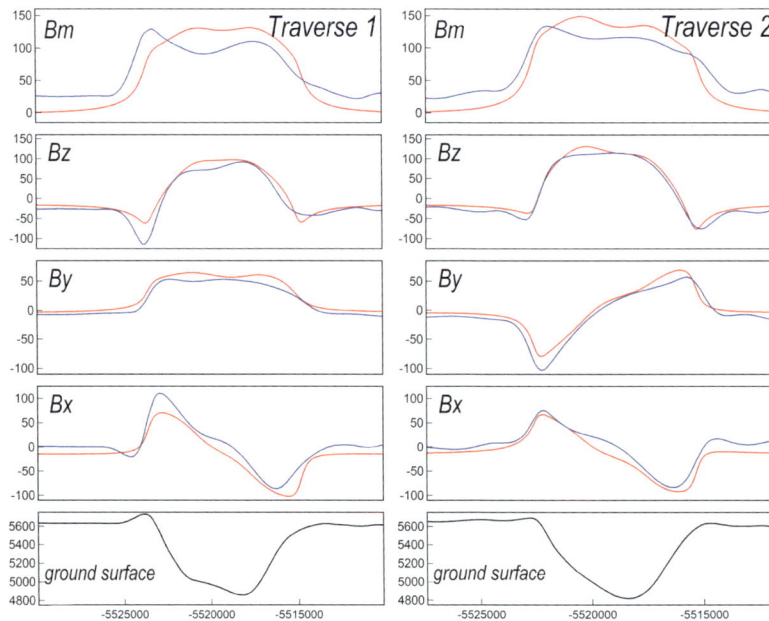

Fig. 13.13. Bx, By and Bz components on traverses A-B (left) and C-D (right): (blue) computed from the terrain model with magnetisation A, and (red) inverted from the excavation model.

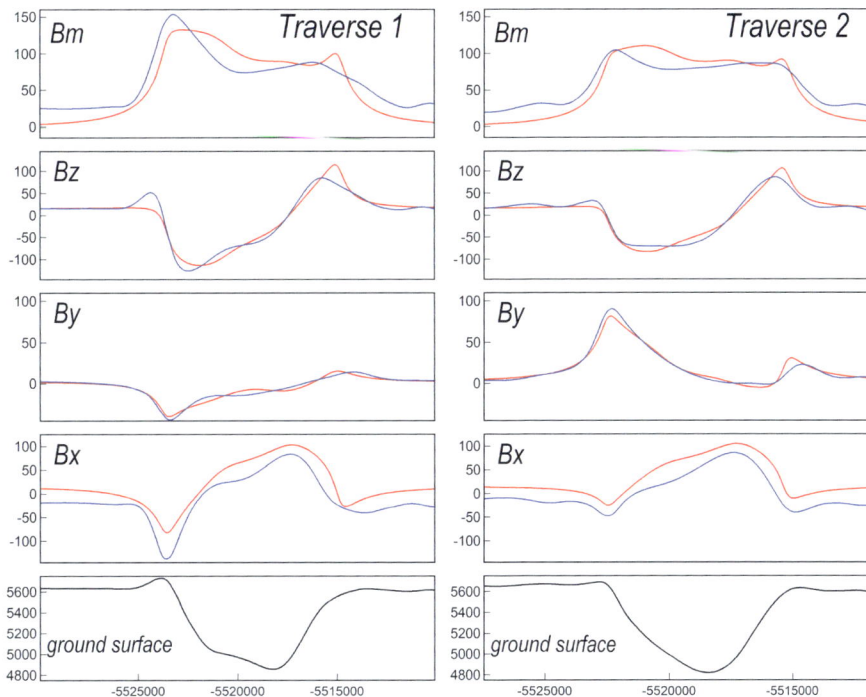

Fig. 13.14. Bx, By and Bz components on traverses A-B (left) and C-D (right): (blue) computed from the terrain model with magnetisation B, and (red) inverted from the excavation model.

13.4 SYNTHETIC MAGNETIC MODELLING OF A MARTIAN CRATER WITH DEMAGNETISATION AND IMPACT MAGNETISATION

We briefly investigate the case that the crater terrain may not be the key bounding surface for crustal magnetisations if there is surrounding demagnetisation caused by the impact. From the crater morphology the expected distributions of stress and temperature generated by the impact can be inferred (Mohit and Arkani-Hamed 2004; Shahnas and Arkani-Hamed 2007) but without knowledge of the pre-impact magnetisation or rock magnetic properties it is not feasible to predict the post-impact magnetisation. We investigate separate cases that an impact into a magnetised crust during a period when the magnetic field is weak or non-existent excavates a crater and causes demagnetisation beneath and surrounding it. In a further example we consider the case that an impact during a period when the magnetic field is strong into a crust that is initially weakly magnetic or non-magnetic excavates a crater and generates an annulus of strong magnetisation around it with a remanent magnetisation acquired on cooling from the impact-generated heating event. These simple conceptual models illustrate aspects of the magnetic field that

might be investigated with advantage in mapping crustal magnetisation.

Figure 13.15 illustrates two east–west multi-component magnetic field profiles acquired at 50 m and 1,000 m above an impact crater for which demagnetisation doubles the depth of missing magnetisation. The crust is assigned a pre-crater, mid-inclination magnetisation with declination 45° and inclination +45°. To model this scenario we create a model of missing magnetisation (either through excavation or demagnetisation) reverse to the crustal magnetisation with declination 225° and inclination −45°. The dashed curves in Fig. 13.15 show the magnetic field computed for the magnetisation contrast due to the excavated crater and the solid curves show the field for the contrast of the crater plus the surrounding demagnetised zone. The crater diameter is 5,000 m, the depth is 500 m and the demagnetisation extends to another 500 m depth. Both curves have similar patterns for each of all the three components, revealing that the additional demagnetisation zone would not cause any significant difference in estimated magnetisation direction. An optimum inversion fit using the crater model would result in overestimation of the intensity of crustal magnetisation as a result of the excluded volume of magnetisation lost in

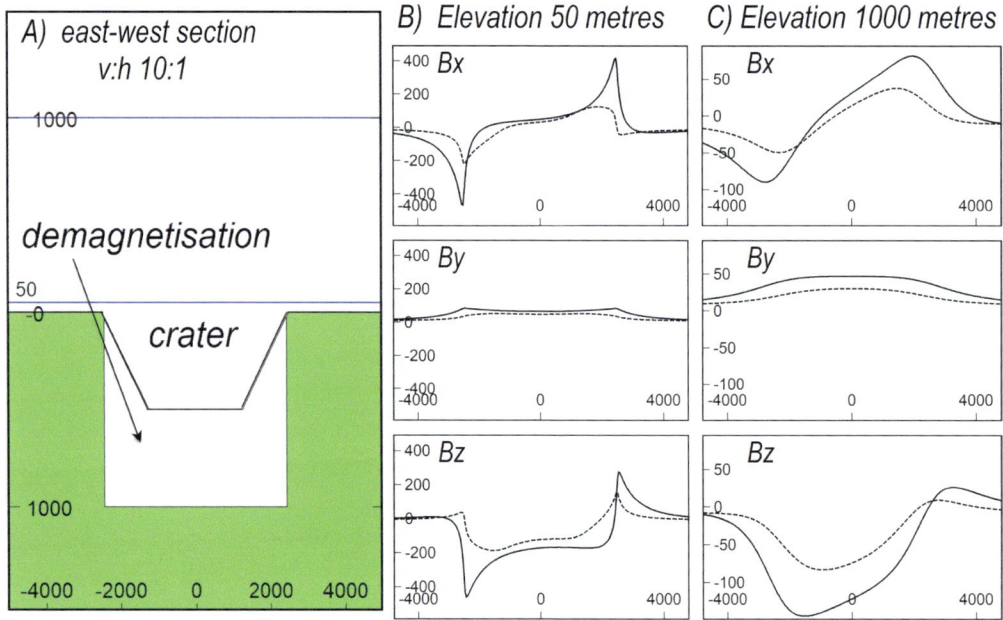

Fig. 13.15. A) 5 km diameter 500 m deep crater with impact demagnetisation extending to an additional depth of 500 m, with a west to east traverse at elevations B) 50 m and C) 1 km. Dashed line for the crater only, solid line for the crater and demagnetised zone.

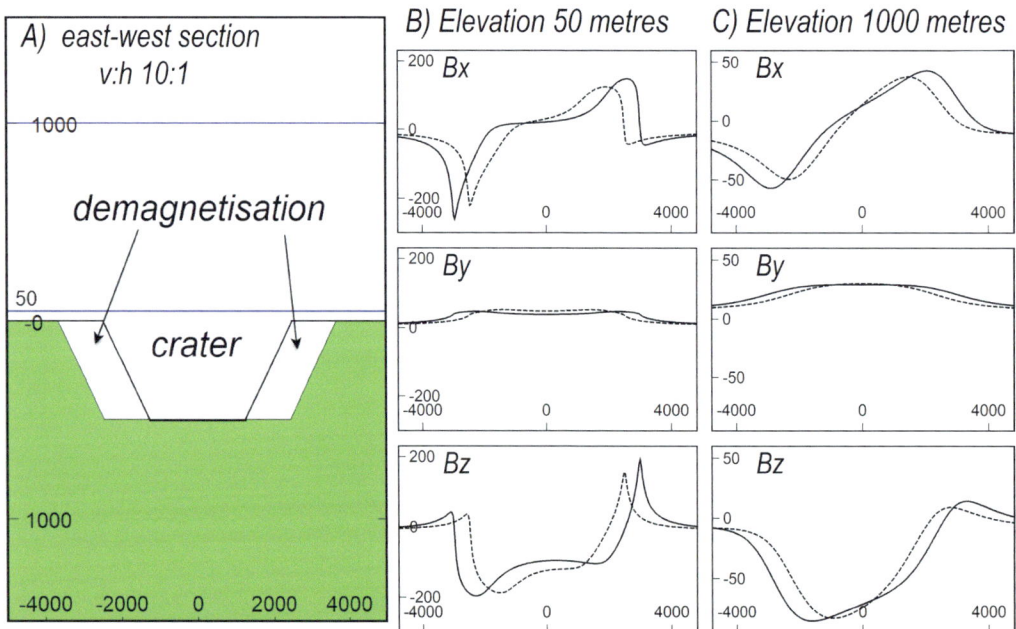

Fig. 13.16. A) Model of a 5 km diameter 500 m deep crater with impact demagnetisation extending 500 around it and magnetic components computed on an east–west traverse at B) elevation of 50 m and C) elevation of 1 km.

the demagnetisation zone. There would be a significant post-inversion data-misfit because the deeper demagnetisation zone has a broader and less sharp magnetic field expression than the shallower crater. This discrepancy is most noticeable in the lower elevation profile (Fig. 13.15B). A discrepancy with the terrain model might occur if the crater is infilled with transported material

or chaotic blocks of random orientation that have a low bulk intensity of remanent magnetisation.

Figure 13.16 shows a model where demagnetisation surrounds the crater. As with demagnetisation beneath the crater, demagnetisation around it also increases the amplitude of the magnetic field variations, but in this case the most prominent change to the anomaly is its

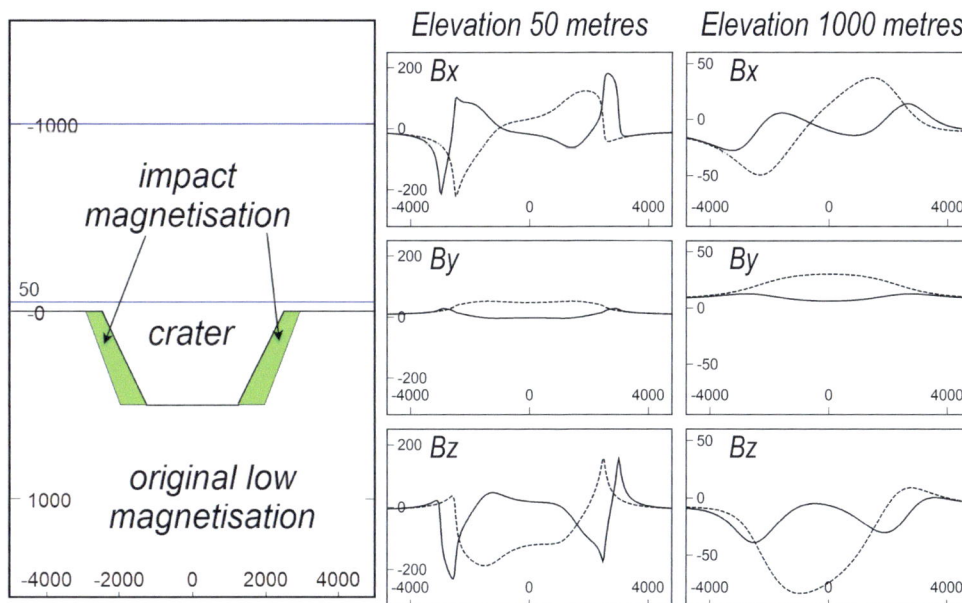

Fig. 13.17. Model with impact magnetisation (in green) extending 500 m around the crater, with magnetic components computed on an east–west traverse at elevations of B) 50 m and C) 1 km. The dashed lines from Figs 13.15 and 13.16 are shown for comparison only.

increased width, with outward migration of the anomaly peaks and troughs from the crater wall to the more distant edge of demagnetisation. The expression of demagnetisation both beneath and around the crater is most evident in the lower elevation measurements. For all three components, at both elevations, peak and trough amplitude ratios of the crater anomalies with and without demagnetisation are broadly similar, giving similar estimates of magnetisation directions. Given the additivity of magnetic fields, combined demagnetisation both around and beneath the crater (with possible error in estimation of its volume) is expected to mostly impact on estimation of the strength of the surrounding magnetisation.

Figure 13.17 shows another simplified concept model where impact into an initially non-magnetic crust occurs during a period of a strong background field. Excavation of the crater does not cause a magnetisation contrast with the originally non-magnetic crust, but a zone of thermo-remanent magnetisation (TRM) is created where the surrounding crust heated by the impact then cools through the blocking temperature of its ferromagnetic minerals. This model of an annular magnetisation surrounding a crater is only marginally more complex than the first two cases we investigated. In the impact magnetisation model of Fig. 13.17 the crater has the same zero magnetisation as the background crust and would not be magnetically visible except for the annulus of

magnetisation surrounding it. In this example we gave the new, annular magnetisation a magnetisation with declination 225° and inclination −45°. This is the same magnetisation direction as the contrast magnetisation for the crater in the previous examples, but the magnetic anomalies are quite different because the annulus magnetisation (20% of the crater diameter) also has an external contact. If the anomaly is well defined it should be straightforward to discriminate between an annular magnetisation surrounding a crater and basement magnetisation extending well beyond it. Figures 13.17B and 13.17C show that the zone of annular magnetisation is most obvious at lower levels. At substantially greater elevations the magnetic field expressions from both sides of the crater overlap and lose their diagnostic characteristics.

We have investigated the case only of impact-induced magnetisation of a previously un-magnetised basement but provided the measurement field extends sufficiently beyond the annular zone it should be possible to individually resolve differently directed (more than ~30° difference) extensive pre-impact and local circum-crater impact-caused magnetisations, even though there is likely to be extensive transition and overlap between them. Nevertheless, this two-magnetisation case (three magnetisations including the zero magnetisation of the crater) is more complex than the single-contrast magnetisations considered previously and would require favourable conditions of strong magnetisations and high-resolution measurements to recover reliable estimates of the magnetisation directions.

13.5 CONCLUSIONS

We have shown both with modelling and inversion of measured TMI anomalies over valleys incised into sheets of the Tasmanian dolerites, and with synthetic modelling and inversion of vector component data computed from the terrain model of a Martian impact crater, that a model of excavated magnetisation can be used to emulate sheets of magnetisation complete other than for that excavation. The surrounding material is then assigned a magnetisation of equal strength and opposite polarity to the removed magnetisation. This method of modelling and inversion has the advantage that models are more compact than would be required to represent the surrounding magnetisation. We hope that with future developments of magnetic surveying on Mars this approach may be of benefit in deriving bulk magnetisation estimates for materials surrounding small impact craters.

REFERENCES

Albee AL, Arvidson RE, Palluconi F, Thorpe T (2001) Overview of the Mars Global Surveyor mission. *Journal of Geophysical Research* **106**(E10), 23291–23316. doi:10.1029/2000JE001306

Arkani-Hamed J (2005) Magnetic crust of Mars. *Journal of Geophysical Research* **110**, E08005. doi:10.1029/2004JE002397

Brauns CM, Hergt JM, Woodhead JD, Maas R (2000) Os isotopes and the origin of the Tasmanian Dolerites. *Journal of Petrology* **41**, 905–918. doi:10.1093/petrology/41.7.905

Hergt JM, Brauns CM (2001) On the origin of Tasmanian dolerites. *Australian Journal of Earth Sciences* **48**, 543–549. doi:10.1046/j.1440-0952.2001.00875.x

Irving E (1956) The magnetisation of the Mesozoic dolerites of Tasmania. *Papers and Proceedings of the Royal Society of Tasmania* **90**, 157–168. doi:10.26749/VKVI2019

Irving E (1963) Paleomagnetism of the Narrabeen chocolate shale and the Tasmanian dolerite. *Journal of Geophysical Research* **68**, 2283–2287. doi:10.1029/JZ068i008p02283

Jaeger JC, Joplin G (1954) Rock magnetism and the differentiation of dolerite sill. *Journal of the Geological Society of Australia* **2**(1), 1–19. doi:10.1080/00167615408728454

Leaman D (2002) The effective magnetisation of the Jurassic dolerites of Tasmania. *Exploration Geophysics* **33**, 166–171. doi:10.1071/EG02166

Lillis RJ, Robbins S, Manga M, Halekas JS, Frey HV (2013) Time history of the Martian dynamo from crater magnetic field analysis. *Journal of Geophysical Research* **118**, 1488–1511. doi:10.1002/jgre.20105

Louzada KL, Stewart ST, Weiss BP, Gattacceca J, Lillis RJ, Halekas JS (2011) Impact demagnetization of the martian crust: current knowledge and future directions. *Earth and Planetary Science Letters* **305**, 257–269. doi:10.1016/j.epsl.2011.03.013

Mittelholz A, Johnson CL, Feinberg JM, Langlais B, Phillips RJ (2020) Timing of the martian dynamo: New constraints for a core field 4.5 and 3.7 Ga ago. *Science Advances* **6**, 1–7.

Mohit PS, Arkani-Hamed J (2004) Impact demagnetization of the Martian crust. *Icarus* **168**, 305–317. doi:10.1016/j.icarus.2003.12.005

Neumann GA, Rowlands DD, Lemoine FG, Smith DE, Zuber MT (2001) Crossover analysis of Mars Orbiter Laser Altimeter data. *Journal of Geophysical Research* **106**(E10), 23753–23768. doi:10.1029/2000JE001381

Neumann GA, Smith DE, Zuber MT (2003) Two Mars years of clouds detected by the Mars Orbiter Laser Altimeter. *Journal of Geophysical Research* **108**(E4), 5023. doi:10.1029/2002JE001849

Robbins SJ, Hynek BM (2012) A new global database of Mars impact craters ≥ 1 km: 2 Global crater properties and regional variations of the simple-to-complex transition diameter. *Journal of Geophysical Research* **117**, E06001. doi:10.1029/2011JE003967

Schmidt PW (1976) The non-uniqueness of the Australian Mesozoic palaeomagnetic pole position. *Geophysical Journal of the Royal Astronomical Society* **47**, 285–300. doi:10.1111/j.1365-246X.1976.tb01274.x

Schmidt PW, McDougall I (1977) Palaeomagnetic and potassium-argon dating studies of the Tasmanian dolerites. *Journal of the Geological Society of Australia* **24**, 321–328. doi:10.1080/00167617708728991

Shahnas H, Arkani-Hamed J (2007) Viscous and impact demagnetization of Martian crust. *Journal of Geophysical Research* **112**, E02009. doi:10.1029/2005JE002424

Smith DE, Zuber MT, Frey HV, Garvin JB, Head JW, Muhleman DO, Pettengill GH, *et al.* (2001) Mars Orbiter Laser Altimeter—Experiment summary after the first year of global mapping of Mars. *Journal of Geophysical Research* **106**, 23689–23722. doi:10.1029/2000JE001364

Vervelidou F, Lesur V, Grott M, Morschhauser A, Lillis RJ (2017) Constraining the date of the martian dynamo shutdown by means of crater magnetization signatures. *Journal of Geophysical Research* **122**, 2294–2311. doi:10.1002/2017JE005410

14

Estimation of magnetisation direction from tripole and quadrupole anomalies in low inclination fields

C.A. Foss and K. Leslie

ABSTRACT

Magnetisations parallel or anti-parallel to a low inclination magnetic field produce tripole anomalies with two flanking lobes and a central lobe of opposite polarity. Magnetisations perpendicular to the field produce petal-shaped quadrupole anomalies with lobes of alternating polarity. We map the distribution of these anomalies as functions of the inclination of the magnetic field and the direction of magnetisation and we present analyses to determine magnetisation direction from the peak and trough azimuths and amplitude ratios of the anomalies. We demonstrate recovery of magnetisation direction from tripole anomalies measured in aeromagnetic surveys in low inclination fields in Brazil and Malaysia. We have not yet encountered natural quadrupole anomalies in aeromagnetic survey data but we have generated quadrupole anomalies with oriented magnetic sources in a low inclination field in Malaysia and in a Rubens coil set. Analysis of these anomalies returns their correct source magnetisation directions. We also present analysis of similarly generated tripole anomalies due to (reverse) magnetisation directed opposite to low inclination fields.

14.1 INTRODUCTION

Most magnetic field surveys measure the total intensity of the magnetic field (TMI) across a near-horizontal plane above a subsurface magnetisation. Patterns of the measured magnetic field variations are determined by this spatial relationship between the magnetisations and measurements, the direction of magnetisation and the orientation of the magnetic field. At high geomagnetic inclinations (Fig. 14.1A) the main field is almost perpendicular to the measurement surface and a compact steep-inclination magnetisation produces a positive anomaly of near-circular symmetry with the morphology almost of a monopole (the anomaly has a central, sharp, high-amplitude peak with a much weaker and diffuse surrounding outer zone of negative polarity). At intermediate latitudes the TMI expression of a compact magnetisation has the morphology of a

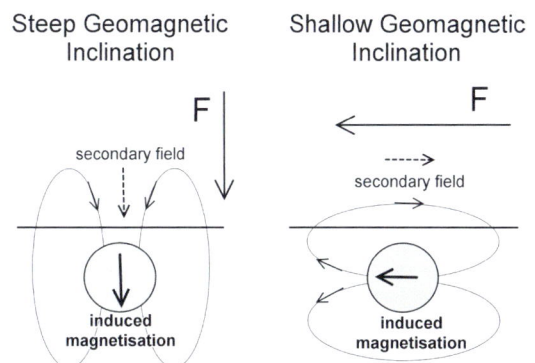

Fig. 14.1. Magnetic field lines for secondary induced fields in an A) polar (vertical) geomagnetic field and B) equatorial (horizontal) geomagnetic field.

dipole with one closed region where the anomalous field sums with the background field to increase TMI and an adjacent closed region where the anomalous field subtracts from the background field to decrease TMI. Close to the equator (Fig. 14.1B) low inclination fields are almost parallel to the measurement surface and field-parallel magnetisations produce oppositely directed secondary fields above them resulting in predominantly negative amplitude TMI anomalies. In low inclination fields anomalies can have more than two closed regions. Throughout this chapter we refer to anomalies by their morphology: tripoles are anomalies with three closed regions and quadrupoles are anomalies with four closed regions.

Figure 14.2 shows a set of anomalies in a zero-inclination field due to magnetisations of north, south, east, west, up and down direction. Horizontal north and south directed magnetisations (Figs 14.2A and 14.2B) that are parallel and anti-parallel to the main field respectively produce tripole anomalies. Of these, field-parallel magnetisations (Fig. 14.2A) have a central trough with peaks to north and south, and magnetisations anti-parallel to the field (Fig. 14.2B) have a central peak with troughs to north and south. Easterly and westerly directed horizontal magnetisations (Figs 14.2C and 14.2D) produce quadrupole anomalies. For easterly directed magnetisations (Fig. 14.2C) the north-east and south-west lobes are positive, and for westerly directed magnetisations (Fig. 14.2D) the north-west and south-east lobes are positive. In steep inclination fields magnetisations are normal or reverse dependant on whether their inclination has the same or opposite polarity to the local geomagnetic field. In low inclination fields normal or reverse magnetisations are those with declination parallel to the field (northerly directed) or anti-parallel (southerly directed).

At high geomagnetic inclinations we have the fortunate situation that the azimuth of a line joining the anomaly peak and trough provides an approximate indication of the declination of magnetisation. Also, the peak to trough amplitude ratio indicates the inclination of magnetisation (Zietz and Andreasen 1967). It is more of a challenge to estimate magnetisation direction from key statistics of a dipole anomaly in moderate to low inclination fields but we show here that in low inclination fields estimates of magnetisation direction can be conveniently recovered from the larger complement of statistics available for more complex tripole and quadrupole anomalies. We restrict definition of tripole and quadrupole anomalies to be those anomalies for which the weakest included lobe has an amplitude no less than 10% of the strongest. This definition reduces concern with anomaly classification in analysis of sparse or noisy measured data or where there are overlapping fields from other magnetisations.

The low-inclination magnetic field expression of magnetisations with northerly declination (field parallel) is summarised in Fig. 14.3. For a horizontal northerly directed magnetisation in a horizontal field (the solid black line in Fig. 14.3B) the north–south profile over the centre of magnetisation is symmetric with two flanking positives. If the magnetisation is not horizontal (e.g. curves 'M30s' and 'M30n' in Fig. 14.3) the flanking anomaly peak in the direction of inclination is reduced and the opposite peak is strengthened. A horizontal magnetisation in a horizontal field produces the most prominent tripole anomaly with the highest ratio of the of the weakest lobe to the strongest. For inclined magnetic fields the maximum symmetry of the anomaly and thereby the strongest tendency towards a tripole anomaly is for magnetisations with inclination of opposite polarity to the field (as seen for the 30° south magnetisation in the 30° north field in Fig. 14.3A and the 30° north magnetisation in the 30° south field in Fig. 14.3C).

The international geomagnetic reference field (IGRF) provides an approximate representation of the geomagnetic field. The absolute inclination of the IGFR is imaged in Fig. 14.4 between inclinations of −30° and +30°. This is the range of the field commonly referred to as low inclination. The IGRF magnetic equator (the line where the model has zero inclination) broadly follows

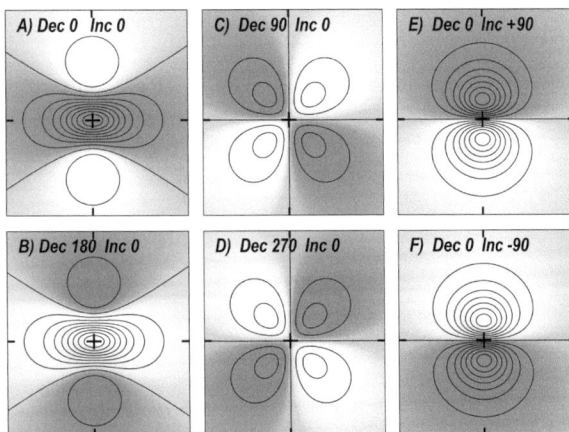

Fig. 14.2. Zero-inclination geomagnetic field TMI anomalies over dipole magnetisations of A) inclination 0° declination 0°, B) inclination 0° declination 180°, C) inclination 0° declination 90°, D) inclination 0° declination 270°, E) inclination +90° and F) inclination −90°.

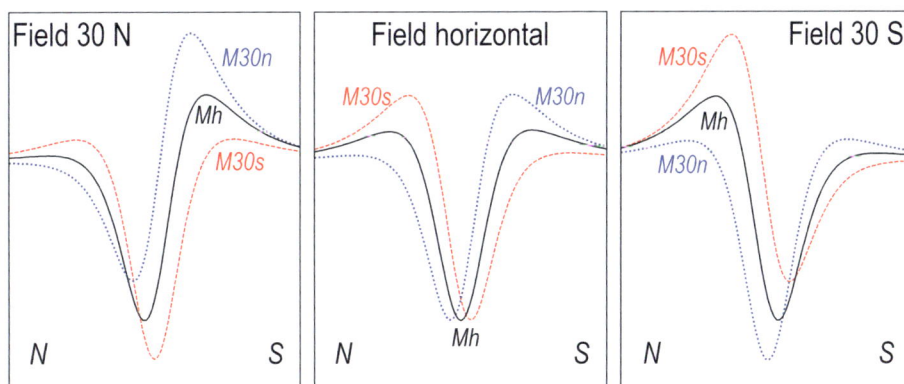

Fig. 14.3. north–south TMI profiles over dipoles with -30° (30°S),0° (horizontal) and +30° (30°N) inclination magnetisations in geomagnetic fields of inclination: A) 30° North, B) horizontal and C) 30° South.

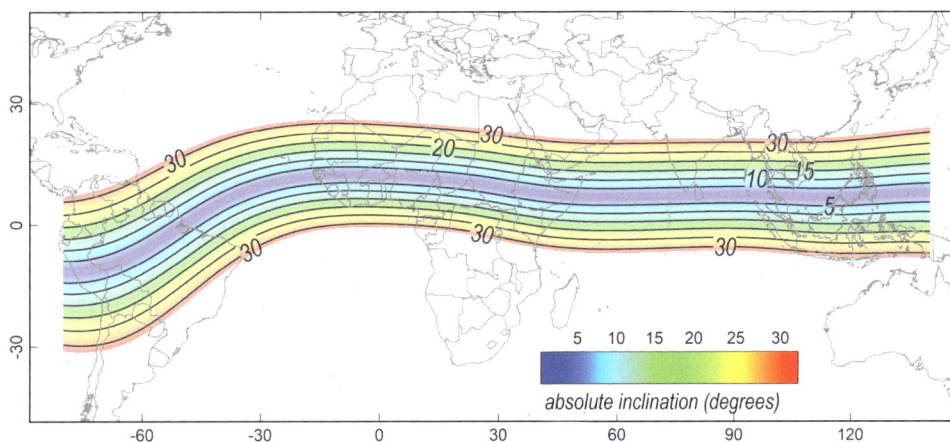

Fig. 14.4. Region of absolute geomagnetic inclination < 30.

the geographic equator but with departures to the north over western central Africa and to the south over the northern part of south and central America. The major land areas with significant regions of low inclination magnetic field are northern Brazil and central America, western Africa to Somalia and Ethiopia, the southern Arabian Peninsula, India and the northern part of South-East Asia and the Philippines. All these regions can be expected to have the tripole anomalies we describe here as well as currently unreported (as far as we are aware) quadrupole anomalies. The average north–south distance between the ± 15° contours is ~1,500 km and between the ± 30° contours is ~3,000 km. The likelihood of finding tripole or quadrupole anomalies is highest closest to the equator. Tripole or quadrupole anomalies can be found at geomagnetic inclinations steeper than 30° but these are mostly due to elongate bodies aligned north–south with polarisation towards their ends.

14.2 THE MAGNETISATION RANGE OF TRIPOLE AND QUADRUPOLE ANOMALIES

To map the range of equidimensional source magnetisations that produce tripole and quadrupole anomalies we made closely spaced computations of their TMI fields and recorded the peak and trough amplitudes. We investigated magnetisations at 10° intervals of declination and inclination (reduced to 5° intervals across ranges of sharp transition). For these synthetic fields anomaly definition is sufficient to reliably determine their morphology, but for measured data anomaly morphology can only be reliably assigned if the field sampling and measurement resolution is adequate and if magnetisations are compact and reasonably isolated.

Figure 14.5 shows the range of tripole and quadrupole anomalies with declination and inclination for compact magnetisations in a zero-inclination field. Bodies of more complex shape must be considered separately. Compact source magnetisations with inclination steeper than 40° generate morphological dipole anomalies (as shown in Fig. 14.5). Anomalies of magnetisations with declination parallel or anti-parallel to the field (declinations close to 0° or 180°) transition between dipole and tripole morphology at inclinations of ± 15°. Anomalies of magnetisations with declination perpendicular to the field (east or west) transition between dipole and quadrupole morphology at steeper angles of ± 35°. Of the 308 results on the 10° × 10° grid plotted in Fig. 14.5, 69% are morphological dipoles, 22% are morphological tripoles and 8% are morphological quadrupoles. On the expectation that many magnetisations are within 15° of the local geomagnetic field direction (because of the contribution of induced magnetisation) tripole anomalies should be common and possibly the dominant anomaly morphology in near-horizontal fields. From our limited experience of such low inclination fields tripole anomalies generally appear to be in the minority (some may go unrecognised if the field is insufficiently sampled). If a survey dataset has anomalies of more than one morphology, a selection of anomalies of each morphology should be analysed or inverted to investigate whether the different magnetic field expressions reveal magnetisations of different direction (possibly generated in geological events of different age).

Figure 14.5 maps the range of magnetisations with declination between 0° and 180°. The pattern is symmetric for magnetisations with declination between 0° and −180°. For low inclination, northerly directed magnetisations of declination within the range of ± 90° the tripole anomalies are composed of a central trough flanked by weaker peaks. For low inclination, southerly directed magnetisations (declinations between 90° and 180° or between −90° and −180°) the anomalies are tripoles composed of a central peak and flanking troughs.

Figure 14.6 shows the distribution pattern of anomaly morphologies in a background field of inclination −30°. Compared with the pattern for the horizontal background field in Fig. 14.5 the proportion of dipoles increases from 69% to 81%, tripoles decrease from 22% to 18% and the quadrupoles have a larger relative decline from 8% to only 1%. The non-zero inclination of the background field also skews the distribution pattern of tripole and quadrupole anomalies. Most of the magnetisations with declination between 0° and 90° that generate tripole anomalies have inclination of similar value but opposite sign to the background field. Most of the reverse magnetisations with declination between 90° and 180° that generate tripole anomalies have inclination of magnetisation similar

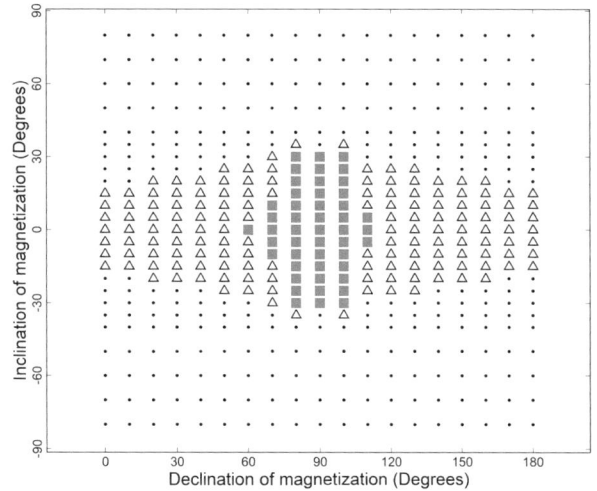

Fig. 14.5. Morphology of compact source anomalies in a horizontal field (points are dipoles, triangles are tripoles and squares are quadrupoles).

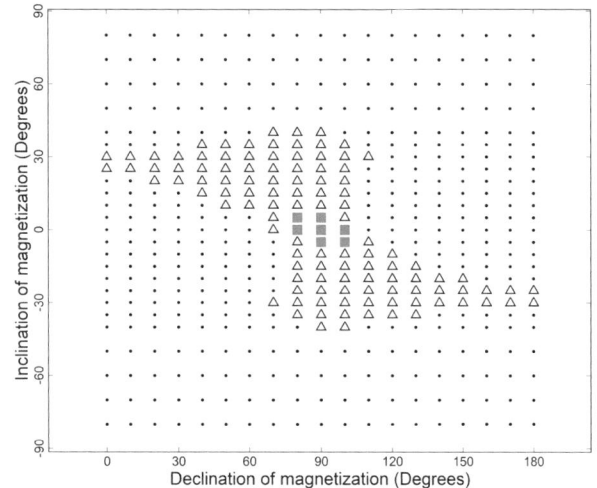

Fig. 14.6. Morphology of compact source anomalies in a 30° S field (points are dipoles, triangles are tripoles and squares are quadrupoles).

in value and sign to the geomagnetic field. In this steeper geomagnetic inclination field quadrupole anomalies are limited to magnetisation inclinations within 10° of horizontal and declination within 10° of east or west.

Figure 14.7A shows three symmetric anomalies generated by dipolar magnetisations. The lowest amplitude anomaly is generated by a horizontal magnetisation in a horizontal field. A reverse magnetisation in a vertical field has double the amplitude and very similar shape. A reverse magnetisation in a 45° inclination field has a similar shape and an intermediate amplitude. In Fig. 14.7B the three curves are scaled to a common trough amplitude. This highlights the decrease in relative amplitude of the flanking peaks with increase in inclination of the background field. Of the three curves shown in Fig. 14.7 the 10% criteria line we use to define tripole or quadrupole anomalies only intersects the profile for the 0° inclination background field.

The disappearance of tripole and quadrupole anomalies with increase in magnetic field inclination is further investigated in Fig. 14.8. Figure 14.8A plots the percentage amplitude ratio of the weakest lobe of tripole and quadrupole anomalies relative to the strongest lobe for declination of magnetisation parallel and perpendicular to the field respectively. At zero inclination this ratio is much higher for quadrupole anomalies (100%) than for tripole anomalies (20%). However, quadrupole anomalies attenuate more rapidly with increasing field inclination to create a crossover between the relative amplitudes of the two anomalies at less than 15° field inclination. The quadrupole anomalies cross the 10% classification threshold at the lower inclination angle of ~20°

compared to tripole anomalies at ~40°. This is consistent with the greater loss of quadrupole anomalies than of tripole anomalies with increase in steepness of inclination of the background field from 0° to 30° as shown in Figs 14.5 and 14.6.

The absolute increase in anomaly amplitude with background field inclination for magnetisations of inclination opposite to the field (optimum candidate magnetisations for tripole and quadrupole anomalies) is plotted in Fig. 14.8B. In a zero-inclination field the anomaly amplitude due to a 0° declination magnetisation is significantly greater than for an identical strength 90° declination magnetisation and it remains higher across the relevant range of field inclinations up to 45°. Combining Figs 14.8A and 14.8B, Fig. 14.8C plots the amplitudes of the weakest anomaly lobes normalised to the anomaly peak for a 0° declination, 0° inclination magnetisation in a 0° inclination background field (point

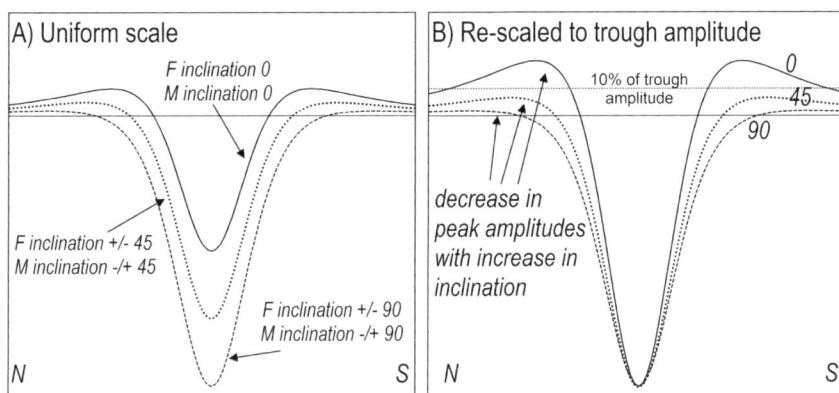

Fig. 14.7. TMI Profile from dipole with inclination of magnetisation 0° in field of inclination 0° and similar profiles of magnetisations +/− 45° and +/− 90° in fields of −/+ 45° and −/+ 90° respectively.

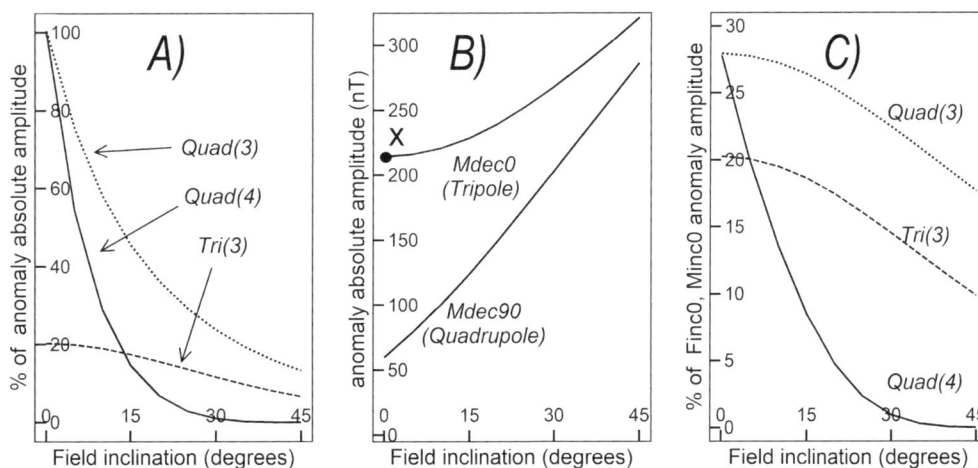

Fig. 14.8. A) decrease in amplitudes of minor lobes as percentage of the total anomaly amplitude with increase in background field inclination, B) increase in absolute anomaly amplitude, C) resulting change in true amplitude of the lobes.

'X' in Fig. 14.8B). Figure 14.8C reveals the considerable discrimination against detection of the weaker fourth lobe of quadrupole anomalies (the curve labelled 'Quad(4)') and also shows favoured detection of the third lobe of those anomalies (the curve labelled 'Quad(3)') compared to the third lobe of tripole anomalies (the curve labelled 'Tri(3)').

14.3 ESTIMATION OF MAGNETISATION DIRECTION FROM A TRIPOLE ANOMALY

We use a similar approach to estimating magnetisation direction from tripole anomalies in low inclination fields that Zietz and Andreasen (1967) used for dipole anomalies in steep northern geomagnetic fields. Figure 14.9 shows TMI maps produced in a zero-inclination field by magnetisations with four rotations, each of 30° (the field of the non-rotated magnetisation is shown in Fig. 14.2A). In Figs 14.9A and 14.9B the magnetisation has 30° rotations of declination, clockwise and anti-clockwise respectively. Consistent with the behaviour noted by Zietz and Andreasen (1967), the primary expression of these rotations of magnetisation direction is a rotation of the peak-trough axis of the anomaly. For tripole anomalies we measure this axial direction of the anomaly from the azimuth linking the two flanking lobes of the anomaly. In Figs 14.9C and 14.9D the magnetisations have 30° rotations of inclination. Also consistent with the high inclination relationships reported by Zietz and Andreasen

(1967), these differences in inclination of magnetisation do not change the azimuth of the anomaly axis but change the anomaly amplitude ratios. In low inclination fields this is most pronounced in the amplitude ratio of the two flanking lobes of the anomaly. Encouraged by these relationships, we investigate estimation of the declination of source magnetisation from tripole anomalies using the azimuth of a line joining the flanking lobes and the inclination of magnetisation from amplitude ratios of the flanking lobes.

If the central (main) lobe of a tripole anomaly is negative the declination of the source magnetisation is in the range of ± 90°. If the central lobe is positive the declination of the source magnetisation is in the range of +90° to +180% or −90° to −180%. This polarity of the anomaly determines the sense in which to determine the azimuth between the two flanking lobes regardless of their relative amplitudes. The estimate of declination of magnetisation is given by adding to 0° (for a central negative anomaly) or to 180° (for a central positive anomaly) twice the angle of departure from the azimuth joining the two flanking lobes. For example, in Fig. 14.9A the central lobe is negative and the azimuth of the line joining the flanking lobes is +15°. The declination of magnetisation is derived by adding twice +15° to 0° to give +30°. The cross-plot of estimated declination against true declination in Fig. 14.10 shows that predictions are very close to the true values with only minor departures for declinations close to 90° (these are the anomalies with a fourth, below-threshold lobe that would otherwise qualify as

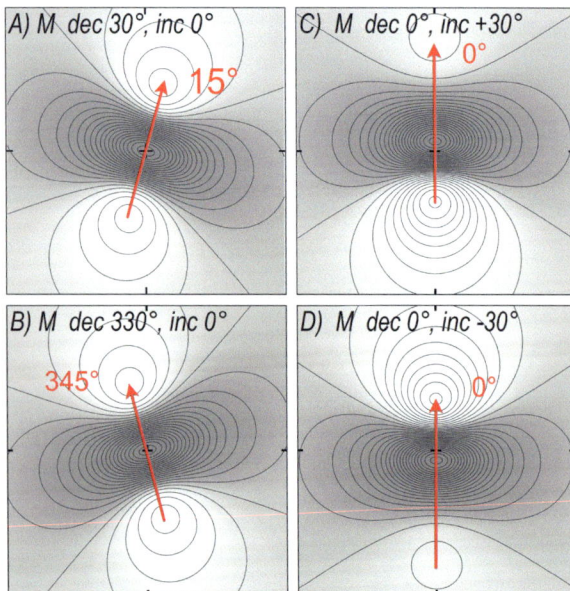

Fig. 14.9. Change of tripole anomaly pattern in a zero-inclination field for four 30° rotations of magnetisation direction.

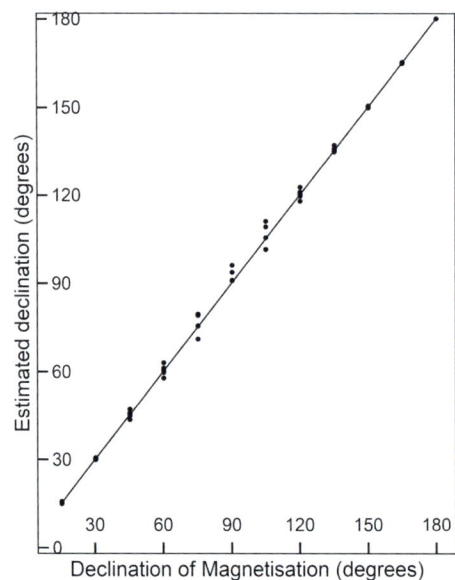

Fig. 14.10. Declination of magnetisation estimates derived from the two flanking lobes of tripole anomalies.

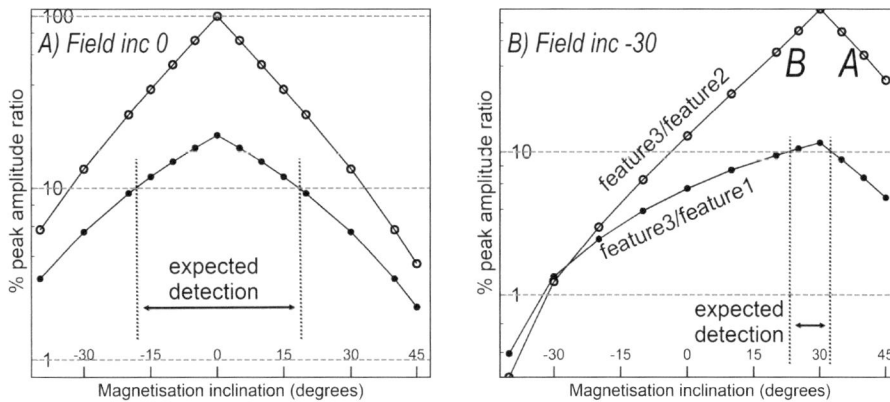

Fig. 14.11. Amplitude ratios of tripole anomalies in A) zero inclination and B) 30°south background fields. Open symbols are ratios of the flanking lobes and closed symbols are ratios of the strongest to weakest lobe (with the dashed line as the 10% threshold for definition of tripole anomalies).

quadrupole anomalies). The points plotted in Fig. 14.10 are for all anomalies detected as tripole anomalies in field inclinations up to 45° at the 15° intervals of field inclination that we investigated.

Estimation of inclination of source magnetisation from tripole anomalies is slightly more complicated than estimation of declination but can be achieved equally effectively. The amplitude ratios of the lobes of the anomaly are sensitive to the inclination of magnetisation and to the inclination of the field. Figure 14.11 shows plots of the percentage amplitude ratio of the two flanking lobes of anomalies due to northerly directed magnetisations in fields of inclination 0° (Fig. 14.11A) and -30° (Fig. 14.11B). Also plotted in those figures is the percentage amplitude ratio of the weakest flanking lobe to the strong central lobe. Following our definition of tripole and quadrupole anomalies, analysis is only required where the weakest to strongest lobe ratio exceeds 10%. In the zero-inclination field (Fig. 14.11A) there is a 40° range of inclination for analysis but this contracts to less than 10° in a field of inclination ± 30° (Fig. 14.11B). Where the inclination of magnetisation is of equal value and opposite sign to the inclination of the background field (the 100% ratio in Figs 14.11A and 14.11B) the flanking lobes of a tripole anomaly have equal amplitudes. The ratio of the minimum to maximum peaks decreases symmetrically away from this central value and on a logarithmic scale the variation is almost linear across the range of analysis. Inclination of magnetisation is to a good approximation given by adding or subtracting to the reverse of the geomagnetic inclination a departure angle from that direction given by:

$$departure\ angle° = 30 * (2 - log10\ (\%amplitude\ ratio))/0.88 \qquad \text{(Eqn 14.1)}$$

This leaves the task of determining whether to add or subtract the departure angle because as shown in Fig. 14.11, an amplitude ratio value has two intersections on the prediction curves (marked 'feature3/feature2' in Fig. 14.11B). In a zero-inclination field the departure angle is positive where the larger flanking lobe is towards the south and is negative where the larger lobe is to the north. For non-zero geomagnetic inclination, the departure angle acts to steepen inclination (increase the absolute inclination) if the larger of the flanking lobes is towards the pole and acts to shallow inclination (decrease the absolute inclination) if the larger of the flanking lobes is towards the magnetic equator. For example, in a −30° geomagnetic inclination field (Fig. 14.11B) with the larger flanking lobe towards the pole (in this case to the south) if the amplitude ratio of the two flanking lobes gives a departure angle of 10°, the inclination of magnetisation is +30° +10° = +40° (point 'A' in Fig. 14.11B). Alternatively, with the larger lobe towards the magnetic equator (in this case to the north), the inclination of magnetisation is +30° −10° = +20° (point 'B' in Fig. 14.11B).

14.4 ANALYSIS OF A MEASURED BRAZILIAN TRIPOLE ANOMALY

To investigate the application of relationships between TMI tripole anomalies and magnetisation direction developed in the previous section we have tested the analyses against field data measured in low inclination geomagnetic fields in Brazil and in Malaysia. Figure 14.12B shows a tripole TMI anomaly measured on an aeromagnetic survey in the Goias area of Brazil. Data was downloaded from the Geological Survey of Brazil CPRM website (https://geoportal.cprm.gov.br) from a

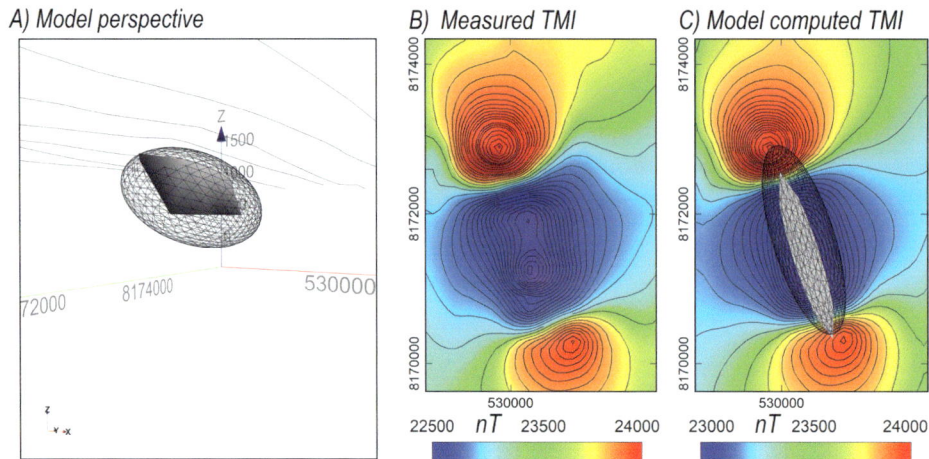

Fig. 14.12. A) perspective view of the ellipsoid (net) and elliptic pipe (solid) models, B) measured and C) model-computed TMI.

survey flown in 2004 with north–south flight lines at 500 m spacing. The mean terrain clearance over the anomaly is 127 m and the IGRF direction at the site is declination 340°, inclination −24°. The anomaly is prominently tripolar with a dominant central trough and two flanking high-amplitude peaks. The peaks are almost 2,800 m apart on an azimuth of 339° (within 1° of the geomagnetic field orientation). We inverted the anomaly using two separate model geometries – a horizontal-top elliptic section pipe and an ellipsoid (Fig. 14.12A) with independent forward modelling algorithms for each body type. The resulting pipe inversion model has a magnetisation direction of declination 344°, inclination +8° and the ellipsoid inversion model has a declination 345°, inclination +11°. The difference between these two directions is only 3°. Both model bodies are elongate in a direction close to the geomagnetic declination. This contributes to their polarisation and may explain the unusually high amplitude of the flanking positive anomaly lobes relative to the central low. The ratio of the three lobes measured from the background field level is (from north to south) 1.34: −1: 0.516. The flanking peak to peak amplitude ratio as used in Fig. 14.12 to estimate the inclination of magnetisation is 39%. This gives a departure angle of 10°. The highest amplitude peak is the northern one, towards the equator for this southern geomagnetic inclination field. Therefore, the reverse of the geomagnetic inclination (+24°) is reduced by 10° to give an estimated inclination of magnetisation of +14°. This inclination is less than 5° from the mean of the two inversion estimates of +10°. The total angular difference between the inversion estimate

of declination 345°, inclination +10° and the analysis estimate of declination 339°, inclination +14° is 7°. The difference between the inversion magnetisation estimate and the geomagnetic field direction is 34° so the analysis method is quite sufficient to support inference from the inversion that the source of this anomaly has a significant component of remanent magnetisation different in direction to the local geomagnetic field.

14.5 ANALYSIS OF A MEASURED MALAYSIAN TRIPOLE ANOMALY

Figure 14.13B shows a tripole anomaly in Malaysia measured in a geomagnetic field of declination 0°, inclination −12.6° on the Central Belt aeromagnetic survey flown for the Geological Survey of Malaysia in the 1980s. One of the authors first investigated this anomaly when based at the University of Malaya. We thank the Department of Minerals and Geoscience Malaysia (JMG) for permission to present this example. The line spacing across the anomaly is 570 m and the mean terrain clearance based on radar altimeter measurement is 134 m (the survey was flown before availability of GPS positioning). The anomaly consists of a central negative and two flanking peaks, revealing that the source magnetisation is of low inclination and directed approximately to the north. We inverted the anomaly using two alternative models of a horizontal-top elliptic pipe and an ellipsoid. The elliptic pipe model gave a magnetisation of declination 348°, inclination +12° and the ellipsoid model gave an almost identical magnetisation of declination 348°, inclination +11°. These two inversions

A) model perspective

B) measured TMI

C) model computed TMI

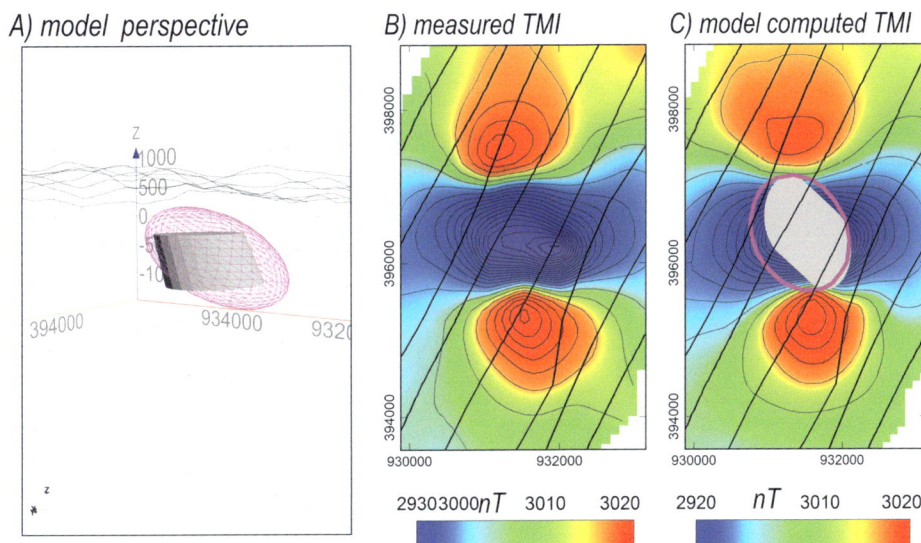

Fig. 14.13. A) perspective of alternative pipe and ellipsoid models, B) the measured TMI anomaly, and C) the anomaly forward computed from the pipe inversion model.

(independent other than for the common input data and regional separation) each closely match the measured field. However, uncertainty in the magnetisation direction is significantly greater than the 1° difference between them because the anomaly peaks and trough are each defined primarily on single and different flight lines. The line joining the two anomaly peaks has an azimuth of 351° (9° from the site IGRF declination) and the relative amplitudes of the three anomaly lobes from north to south are 0.35: −1: 0.45. The 77% ratio of the two flanking peaks gives a departure angle of 4° in the inclination of magnetisation analysis. The stronger southern peak is towards the pole (the survey area lies north of the geographic equator but south of the geomagnetic equator that is the determinant for this magnetic field analysis). For this anomaly orientation the departure angle is subtracted from the reverse field inclination of +13° to give an estimated inclination of +9°. This direction is less than 3° different from the inversion estimates, despite the anomaly being of only modest amplitude and sparsely sampled.

Both tripole anomalies from Brazil and Malaysia conform with the prediction that the most favourable circumstance to find these anomalies is where inclination of magnetisation is of opposite polarity to the inclination of the background field. The results from analyses of both tripole anomalies are also in good agreement with magnetisation directions recovered from inversions that provide the best available bulk estimates of the source resultant magnetisation directions.

14.6 ESTIMATION OF MAGNETISATION DIRECTION FROM QUADRUPOLE ANOMALIES

Figure 14.2C and 14.2D show quadrupole anomalies generated in low inclination fields from magnetisations that are also of low inclination and are directed perpendicular to the field (to east and west respectively). Figures 14.5 and 14.6 show that quadrupole anomalies are generated over a more restricted range of magnetisation directions than tripole anomalies. For a field inclination of 30° (Fig. 14.6) quadrupole anomalies are only found for magnetisations less than 10° from horizontal and less than 20° from magnetic east or magnetic west. We have not yet discovered a convincing example of a natural quadrupole anomaly in aeromagnetic data. The rarity of these anomalies is most likely to be due to the condition that there is a 90° difference between the declination of magnetisation and the geomagnetic field direction. This requires strong dominance of remanent magnetisation and a significant rotation of the resultant magnetisation into either of two small ranges suitable for generation of the anomalies. For young rocks this declination of magnetisation would have to be by a substantial and specific physical (tectonic) rotation. The additional complexity of quadrupole anomalies also places greater demands on the sufficiency of sampling the field, with some quadrupole anomalies possibly unrecognised because they are incompletely sampled.

For analysis of tripole anomalies we utilised statistics that are not available for anomalies of dipole

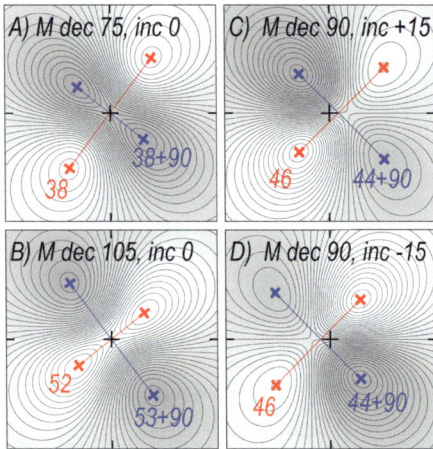

Fig. 14.14. Centres of four quadrupole anomalies with the trends of south-west to north-east positive lobes (in red) and south-east to north-west negative lobes (in blue).

morphology. This advantage of anomaly complexity is further extended with quadrupole anomalies for which several new statistics can be defined. In this study we use statistics derived from the two positive lobes and from the two negative lobes because visual inspection suggests that these features provide the most simple and evident expression of magnetisation direction.

Quadrupole anomalies occur across such small ranges of magnetisation direction that immediately on their recognition, magnetisation directions can be automatically ascribed with reasonable accuracy (an easterly magnetisation for anomalies with north-east and south-west positive lobes and a westerly magnetisation for

anomalies with north-west and south-east positive lobes). We use synthetic data analysis to investigate how these approximate magnetisation directions can be refined. Figure 14.14 shows four quadrupole anomalies in a zero-inclination geomagnetic field. The red crosses mark the positive lobe centres and the blue crosses mark the negative lobe centres. For easterly directed magnetisations (north-east and south-west positive lobe anomalies) the declination of magnetisation is approximately given by:

$$declination = 90° + 2 * (azimuth\ of\ sw\ to\ ne\ positive\ lobes − 45°)$$
$$= 90° + 2 * (azimuth\ of\ nw\ to\ se\ negative\ lobes − 135°) \qquad (Eqn\ 14.2)$$

and for westerly directed magnetisations (north-west and south-east positive lobe anomalies) the declination of magnetisation is approximately given by:

$$declination = 270° + 2 * (azimuth\ of\ se\ to\ nw\ positive\ lobes − 315°)$$
$$= 270° + 2 * (azimuth\ of\ ne\ to\ sw\ negative\ lobes − 225°) \qquad (Eqn\ 14.3)$$

Declination of magnetisation for 52 quadrupole anomalies with easterly directed magnetisation generated in magnetic fields from 0° to 40° inclination are plotted in Fig. 14.15A. Only five estimates of declination in both the peaks and the troughs analyses (Eqns 14.2 and 14.3) are in error by more than 5° and the mean error of each analysis is 2.2°. In almost all cases the errors in the negative and positive lobe analyses for each

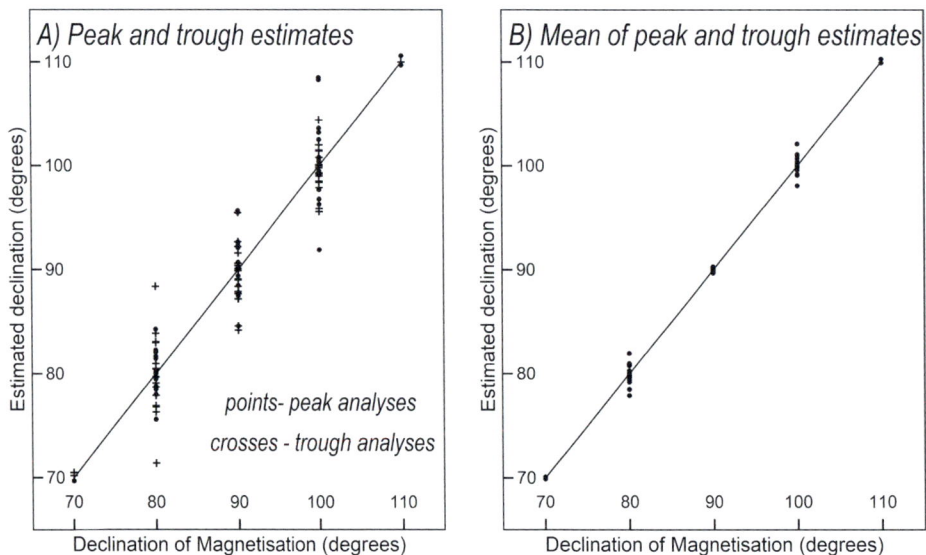

Fig. 14.15. Quadrupole anomaly declination of magnetisation estimates.

magnetisation are of opposite sign and the mean error of the combined analyses (plotted in Fig. 14.15B) reduces further to 0.6°.

We can also use the positive and negative lobe pairs to estimate inclination of magnetisation from the quadrupole anomalies in the same way that we used the single flanking pair of lobes to estimate magnetisation direction from tripole anomalies. Two (generally different) departure angles are estimated from the negative and positive polarity lobe pairs of quadrupole anomalies using Eqn 14.1. There are several combinations and permutations to consider in the search for rules to estimate inclination of magnetisation from quadrupole anomalies. We can expect to find separate rules for the negative and positive lobe pairs, and we need to investigate influence of polarity of inclination for both the magnetic field and the magnetisation, as well as which angle is steeper or shallower than the other. Our systematic tests found a surprisingly simple and condensed set of rules:

1) For analysis of the negative lobe pair the reference inclination (at which the peak to trough amplitude ratio is 1) is that of the background field and for the positive lobe pair the reference inclination is the negative of the field inclination.
2) For analysis of negative lobe pairs the departure angle is added to the reference inclination if the stronger negative lobe is to the north and subtracted if it is to the south. For positive lobe pairs the departure angle is added if the stronger lobe is to the south and subtracted if it is to the north.

We determined these rules from model computed anomalies for which we knew the background field value was zero. However, for analysis of measured anomalies (should they be found) separation from their background fields will be critical (this is required to determine the lobe amplitudes). Determination of the background field level is complicated by the complexity of the anomalies and is best done from field values around their margins. The field value at the saddle point towards the centre of an anomaly can be very different to the background value. Fortunately, there is a considerable advantage that error in estimating the background field value causes opposite errors in estimating inclination from the negative and positive lobes, with the mean of the two estimates far less susceptible to the error in background field separation.

Figure 14.16 shows two quadrupole anomalies due to magnetisations of 90° declination in a background field of inclination −15°. The quadrupole anomaly in Fig. 14.16A has a magnetisation of the same inclination (−15°) as the background field. The negative lobes of this anomaly have identical amplitudes and the minimum to maximum ratio of 100% gives a departure angle of 0° (from Eqn 14.1). The magnetisation inclination estimate (point 'A' in Fig. 14.17B) is the reference direction of −15°. the positive lobe amplitude ratio is 14.6%. This gives a departure angle of 28.5° (from Eqn 14.1). Subtraction of this departure angle from the reference angle of +15° gives an estimate of the inclination of magnetisation of −13.5° (point 'B' in Fig. 14.17B) only 1.5° different from the estimate derived from the negative lobe pairs. Identical results are found from analysis of the anomaly from a magnetisation of opposite inclination shown in Fig. 14.16B. This analysis gives inclination estimates plotted as points 'C' and 'D' in Fig. 14.17B, again with a mean

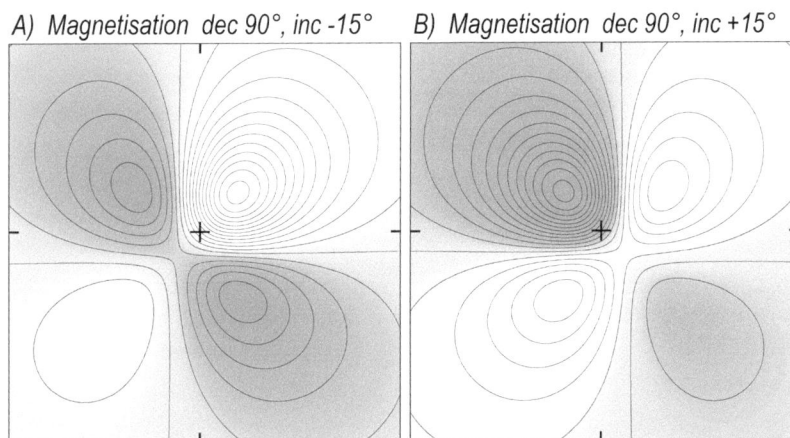

A) Magnetisation dec 90°, inc -15°   B) Magnetisation dec 90°, inc +15°

Fig. 14.16.   Quadrupole anomalies with magnetisations of the same inclination angle A) of identical polarity, and B) of opposite polarity to the 15° south geomagnetic field.

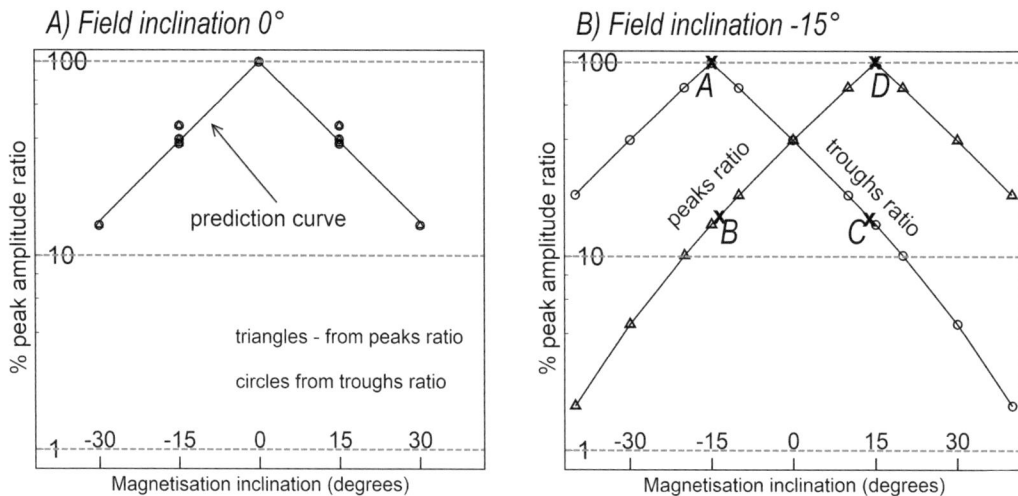

Fig. 14.17. Quadrupole anomaly peaks and troughs amplitude ratios in: A) zero-inclination field and B) inclination 15° south field.

error of less than 1° in estimation of inclination from analysis of the negative and positive lobe pairs.

Figure 14.17A shows the prediction curve for inclination of magnetisation derived from both positive and negative polarity lobe pairs for anomalies in a zero-inclination field. The reference angles of each curve are zero and the two curves overly each other. Figure 14.17B shows corresponding curves for quadrupole anomalies in a −15° inclination field. In this non-zero inclination field the positive and negative lobe curves separate. The apex (the reference point) of the positive lobes curve is at the inclination of opposite sign to the background field and the apex (reference point) of the negative lobes curve is at the same inclination as the background field. It is these reference points from which the departure angles of Eqn 14.1 are measured. Points A, B, C and D plotted in Fig. 14.17B are from analysis of the two anomalies in Fig. 14.16. In the general case most points would be expected to have non-zero departure angles and lie at different distances along the curves. The negative and positive lobe estimates are generally displaced from each other because of their different amplitude ratios, but most importantly the abscissa values for both should be close to provide consistent inclination estimates.

14.7 QUADRUPOLE ANALYSIS TEST ON MEASURED ANOMALIES

We have an analysis with which to estimate source magnetisation direction from quadrupole anomalies – we need measured quadrupole anomalies on which to perform this analysis. We have not yet encountered a natural quadrupole anomaly in aeromagnetic survey data but we are able to generate them. Figure 14.18A shows a quadrupole anomaly measured in a Rubens coil set in Sydney, set to generate a low inclination (+4°) field, and Fig. 14.18B shows a quadrupole anomaly measured in a natural low inclination field of −10° at Kuala Pilah, Malaysia. The sources of the anomalies are magnetite-gabbro cylindrical palaeomagnetic plugs of diameter 2.5 cm and height 2.5 cm. The source of the Kuala Pilah anomaly has a natural (NRM) magnetisation selected because of its high strength and convenient orientation almost along the axis of the cylinder. This plug was laid horizontally with the magnetisation directed to the east. The survey was performed in a mostly wooden 'kampong' (village) house distant from electrical sources other than the winch used to pull the magnetometer along its track and the battery used to power the magnetometer and record the data. The sample has a Koenigsberger ratio > 20 and its induced magnetisation has an almost imperceptible contribution to the anomaly. The source of the anomaly measured in the Rubens coil set is also a strongly magnetic paleomagnetic plug, in this case with an isothermal remanent magnetisation (IRM) of even higher Koenigsberger ratio applied along its axis using a pulse magnetiser. This core was laid horizontally with its magnetisation directed to the west.

We measured the magnetic field data with a three-component fluxgate magnetometer drawn along a 1.3 m track and measured distance along the track with a potentiometer connected to the axis of the winch drawing the magnetometer. We performed the survey by moving the magnetic source at 2 cm intervals along a baseline perpendicular to the track between measurement of each survey 'flight' line. We measured background fields with the source removed before, after and at regular intervals during the survey, and from the averages of those check lines we generated background profiles for each field component. We then subtracted those background field component profiles from the corresponding survey line component profiles as a very effective isolation of the anomaly (unfortunately this is not possible in standard aeromagnetic surveying). This process supresses superimposed field variations, time variations of the field during the survey and long-wavelength variations due to slight bends along the track that disorient the component sensors. We performed repeatability tests and found that we could improve the results by placing the paleomagnetic plugs in three-dimensional printed holders to achieve more precise orientation and positioning before measuring each survey line. In processing the data we added fixed (survey average) background component values back to each survey line to correct for the base shift applied in subtracting the background field. We then gridded each of the three component channels and generated two synthetic 'tie-lines' perpendicular to the survey line and beyond the main

anomaly and interpolated the gridded component data onto those lines. From analysis of these tie-lines we applied base shifts as required to reduce line-to-line level variations. Finally, we generated a TMI channel on each survey line from the three level-adjusted component values. This TMI channel is the data we inverted (we also separately inverted the individual component channels and recovered consistent estimates of magnetisation from each of those inversions).

14.7.1 The Kuala Pilah anomaly

The Kuala Pilah magnetic field has a declination of 0°and inclination −10°. The core sample was oriented with the intention to generate a magnetisation with declination and inclination of ~90° and 0° respectively. Inversion of the resulting TMI anomaly shown in Fig. 14.18B gave an optimum estimated magnetisation direction of declination 85°, inclination 0°. The amplitude ratios of the negative and positive lobes are 47% and 70% respectively for departure angles of 11° and 5°. The inclination of magnetisation estimate from the negative lobe pairs is given by adding the departure angle of 11° to the reference inclination of -10° to give an inclination of +1°. The inclination of magnetisation estimate from the

A)west-magnetisation measured in Sydney

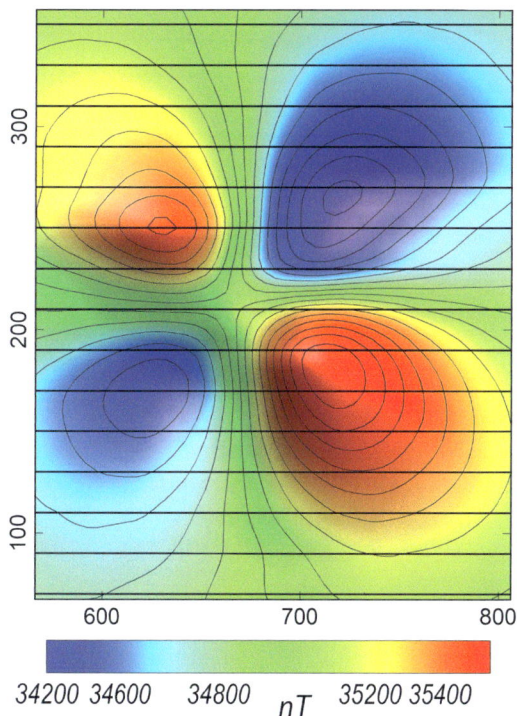

B)east-magnetisation measured in Kuala Pilah

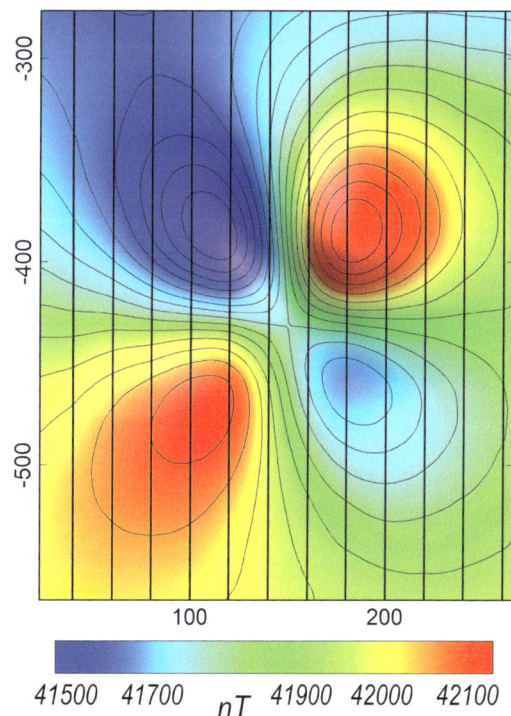

Fig. 14.18.   Measured quadrupole anomalies A) in a +4° inclination field in a Rubens coil set and B) in −10° inclination field at Kuala Pilah, Malaysia. Distance scales are in millimetres.

positive lobe pairs is given by subtracting the departure angle of 5° from the reference inclination for that curve of +10° to give an inclination of +5°. There is a difference of 4° between the two estimates and the mean estimated inclination of +3° is only 2° different to the inversion result.

The declination of magnetisation is estimated from Eqn 14.2 (the anomaly pattern reveals that it is an easterly directed magnetisation). The trend of the positive lobes in Fig. 14.18B is 42° and Eqn 14.2 gives an estimated declination of magnetisation of 84° (90° + 2* (42° – 45°)). The trend of the negative lobes in Fig. 14.18B is 138° and from these lobes Eqn 14.2 also gives an estimate of 84° for the declination of magnetisation (90° + 2* (135° – 138°)). These declination estimates from the quadrupole analysis are within 1° of the estimate provided by inversion.

14.7.2 The Sydney Rubens coil set anomaly

The field in the Rubens coil set in Sydney was set to a declination of 2° and inclination +4°. The optimum estimated magnetisation direction from inversion of the TMI anomaly shown in Fig. 14.18A is declination 269°, inclination +8°. The amplitude ratios of the negative and positive lobes of this anomaly are 75% and 54% respectively, giving departure angles of 9° and 4°. The inclination of magnetisation estimate from the negative lobe pairs is given by adding the departure angle of 4° to the reference inclination of +4° to give an inclination of +8°. The inclination of magnetisation estimate from the positive lobe pairs is given by adding the departure angle of 9° from the reference inclination for that curve of −4° to give an inclination of +5°. There is a difference of 3° between the two estimates and the mean inclination of 6.5° is less than 2° from the inversion result. From inspection of the orientation of the positive lobes of the quadrupole anomaly in Fig. 14.18A we can determine that it is caused by a westerly directed magnetisation and we estimate the declination of magnetisation using Eqn 14.3. The trend of the negative lobes is 227° and of the positive lobes is 314°. Equation 14.3 gives corresponding estimates of the declination of magnetisation of 274° and 268°, a difference of 6°, with a mean value of 271° that is only 2° away from the inversion result.

14.8 TRIPOLE ANALYSIS OF MEASURED REVERSE-MAGNETISATION ANOMALIES

We found several tripole anomalies in Brazilian and Malaysian aeromagnetic survey data but all are due to normal magnetisations. Normal magnetisations are more common than reverse magnetisations – in large part due to the bias from contribution of induced magnetisation. However, it is quite feasible that an extensive search of low-inclination magnetic field data from multiple areas will find reverse-magnetisation tripole anomalies (particularly in areas with young volcanic sequences that span one or more geomagnetic field reversals). Reverse magnetisation tripole anomalies are likely to be more common than quadrupole anomalies. To confirm that the tripole anomaly analysis can also be applied to study of reverse magnetisation we generated these anomalies using the same methodology as for quadrupole anomalies, but with magnetisations oriented anti-parallel to the field rather than perpendicular to it. Figure 14.19 shows tripole anomalies generated in a Rubens coil set in Sydney (Fig. 14.19A) and in the natural field at Kuala Pilah, Malaysia (Fig. 14.19B). The central peak of the tripole anomalies immediately reveals that they are due to low-inclination magnetisations approximately anti-parallel to the field (with declinations close to ± 180°).

The TMI anomaly shown in Fig. 14.18B is due to a near-horizontal reverse magnetisation placed in the Kuala Pilah magnetic field of declination 0°, inclination +10°. Inversion of the anomaly gives a best estimate of the direction of magnetisation of declination 183°, inclination −2°. The azimuth joining the flanking anomaly minima is 180° and the amplitude ratio of the two minima (measured from the background field) is 42%. This ratio gives a departure angle (Eqn 14.1) of 13°. The higher amplitude minima is towards the equator and so the estimated inclination of magnetisation is +10° −13° = −3°. This is only 1° different to the inversion estimate of −2°.

The TMI anomaly shown in Fig. 14.18A is due to a near-horizontal reverse magnetisation placed in a field of declination 2°, inclination +4° generated in a Rubens coil set in Sydney, Australia. The azimuth joining the flanking minima is 0°. The amplitude ratio of the weaker to stronger flanking lobes is 91%, generating a low departure angle of 1.4° (Eqn 14.1). The stronger flanking lobe is towards the pole (to the north for this positive inclination field) and steepens the angle of inclination from the field inclination value of +4° to +5°, only 1° from the inversion result of +4°. For both of these generated anomalies the tripole analysis returns a close estimate of the magnetisation direction.

A) south-magnetisation measured in Sydney   B) south-magnetisation measured in Kuala Pilah

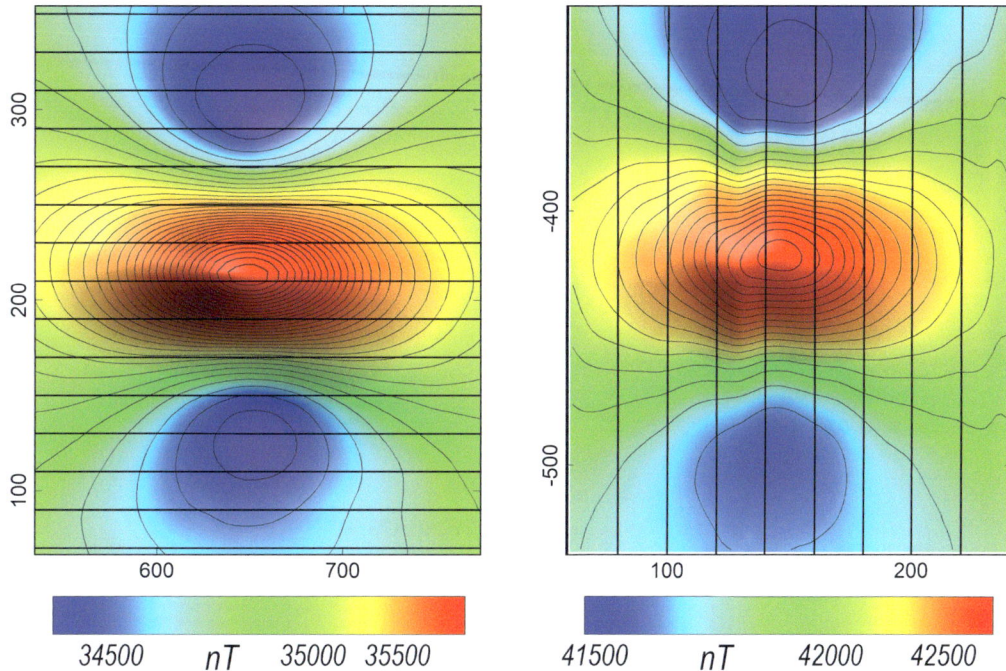

Fig. 14.19.   reverse-magnetisation tripole anomalies measured A) in a Rubens coil set in Sydney and B) in the natural field at Kuala Pilah, Malaysia. Distance scales are in millimetres.

14.9 CONCLUSIONS

Low-inclination magnetisations parallel or anti-parallel to field generate tripole anomalies with two flanking lobes to either side of a central lobe of opposite polarity. Low-inclination magnetisations perpendicular to the field generate quadrupole anomalies with lobes of alternate polarity. We present analyses to recover magnetisation directions from these anomalies. We have demonstrated successful application of the analyses to tripole anomalies in aeromagnetic data measured in Brazil and Malaysia. We have also successfully applied the analyses to quadrupole and reverse-magnetisation tripole anomalies we generated with an oriented magnetic source in a low inclination field in Malaysia and in a Rubens coil set in Sydney. These analyses recover estimates of magnetisation more reliably than would be expected for anomalies of dipole morphology in fields of steeper inclination.

REFERENCE

Zietz I, Andreasen GE (1967) Remanent magnetization and aeromagnetic interpretation. In *Mining Geophysics, Volume II: Theory*. (Eds WE Heinrichs, RC Holmer, RE MacDougall, GR Rogers, JS Sumner and SH Ward) pp. 569–590. Society of Exploration Geophysicists, Houston, TX.

15

The Australian Remanent Anomalies Database as a precursor to a global database

C.A. Foss, P. Warren, P.W. Schmidt and S. Patabendigedara

ABSTRACT

We have developed the Australian Remanent Anomalies Database (ARAD) as a resource of magnetic field data, models and magnetisation direction estimates ascribed in part to remanent magnetisation. As far as we are aware, this is the first national database of magnetisation directions established from magnetic field data. ARAD is deployed in the AuScope Discovery web portal. In Austalia there is restricted outcrop of fresh basement rocks and most rock magnetisations causing variation in the geomagnetic field are only known (with considerable associated uncertainty) from analysis or inversion of that magnetic field data. ARAD is designed as a location to store and search for magnetisation details. It supports download of any magnetic field data, models or magnetisation direction estimates that have been supplied for the database solutions.

Across large parts of Australia, FAIR (findable, accessible, interoperable and reusable) magnetic field data (https://ardc.edu.au/resource-hub/making-data-fair/) have been acquired and are distributed by Federal and State and Territory Governments. ARAD solutions are predominantly derived from these data, and a major objective for ARAD is to support analysis and interpretation of the same data. FAIR magnetic field coverage is increasingly available in many other countries, and we have included example solutions in ARAD from surveys beyond Australia. We have also included solutions derived from NOAA's global EMAG2 database (https://www.ncei.noaa.gov/products/earth-magnetic-model-anomaly-grid-2). Global and regional magnetic field compilations rely substantially on aeromagnetic data and the data quality and sufficiency of those surveys determines to what extent they support reliable recovery of magnetisation estimates from measured magnetic anomalies. Broad anomalies due to large crustal magnetisations produce fields that can be usefully analysed even from sparse and high-elevation magnetic field coverage, but smaller magnetisations that generate the majority of existing ARAD solutions are unresolved in present global magnetic field compilations.

The minimum requirement for an entry in the database is that key details are supplied of the anomaly and the survey in which it was measured. Close to magnetisations (in their 'proximal' fields) anomalies are of high amplitude but generally complex shape that creates challenges in estimation of magnetisation direction. Further from magnetisations (in their mid to far or 'distal' fields) estimation of the direction of magnetisation is less disrupted by the shape of distribution of magnetisation but isolation of these lower-amplitude anomalies is more problematic.

15.1 INTRODUCTION

The direction of magnetisation of a rock, whether known from laboratory measurement or from analysis and inversion of magnetic field data is of limited value without corresponding information from other rock units or magnetic anomalies. A key objective of the remanent anomalies database is to provide a context for analysis of magnetisation directions by comparison with directions from adjacent anomalies and/or palaeomagnetic measurements. Rocks acquire magnetisation during geological events such as the cooling of an igneous intrusion or metamorphic terrain through the blocking temperature range of its ferromagnetic grains or on sediment deposition and diagenesis. Complications arise from post-acquisition remagnetisation events and/or tectonic rotations that reorient magnetisation but under favourable circumstances directions of remanent magnetisation can also provide information about the timing and nature of that tectonism (Irving 1964; McElhinny 1973). In palaeomagnetism the requirement for multiple results that can be compared and combined was recognised long ago and there are published palaeomagnetic databases (Pisarevsky and McElhinny 2003; Pisarevsky 2005) and websites (Jarboe *et al.* 2012; Pisarevsky *et al.* 2022). Schmidt *et al.* (1990), Schmidt (2014) and Schmidt and Clark (2000) have reviewed palaeomagnetic data from Australia, and as more magnetisation directions are recovered from magnetic field analyses it will become increasingly productive to compare those directions with the palaeomagnetic data. Indirect determination of magnetisation from magnetic field data is particularly valuable in Australia where extensive cover and deep weathering restrict access for conventional palaeomagnetic sampling. Building a database of magnetisation directions from analysis of magnetic field data faces several additional challenges to creating a palaeomagnetic database: estimation of magnetisation from magnetic field data is indirect and uncertain, recovered magnetisation directions are the result of both induced and remanent magnetisation and the estimates depend on additional factors such as anomaly separation.

The Earth's magnetic field is a spatially continuous combination of the external fields of many different crustal magnetisations together with time-varying contributions from both the core and ionosphere. The magnetic field includes local variations (anomalies) different from the background field that arise from (anomalous) magnetisation different to the background magnetisation. Magnetic field anomalies are generally detected and defined visually in image, contour, magnetic profile or stacked profile displays. Anomaly separations are non-unique and require interpretive justification. Where anomalies cannot be visually recognised in a well-designed data display it is unlikely that magnetic field analysis or inversion can provide reliable results. Chapters 5 and 6 refer to 'sweet-spots' as segments of data that support analysis or inversion (in these cases for estimation of magnetisation direction). A single, discrete sweet-spot supports estimation of only a single, discrete magnetisation direction. However, multiple analyses or inversions can be applied to obtain different estimates of that direction and each of those alternate results can be entered into the database. All analytic and inversion techniques are based on similar assumptions about the nature and distribution of magnetisation and depend identically on validity of the input magnetic field data. Ability to cross-compensate for variation in estimated magnetisation direction with variation of other magnetisation parameters (particularly its horizontal location) means that even the most appropriate analysis or inversion of a well-defined sweet-spot leaves uncertainty in magnetisation direction greater than 5°. Voxel inversions are reported to recover multiple magnetisation directions across complete volumes of the sub-surface, but that level of detail cannot be justified and only a single magnetisation direction should be reported for each discrete magnetic field feature. Magnetic field variations that are indistinct or overlap extensively do not support meaningful recovery of magnetisation direction. ARAD accommodates quality factors and confidence levels that have been assigned to models and magnetisation direction estimates to guide their subsequent use. However, these statistics are only indicative (e.g. the 5° limit suggested above is an approximate estimate derived from sensitivity and repeatability analyses of measured and synthetic data).

A significant realisation that we came to early in construction of ARAD is that it is not helpful to think of a magnetic anomaly as primarily a feature of the magnetic field. A magnetic anomaly is only known by the measurements that express it and its primary definition must be with respect to those measurements. If a subsequent survey acquires new data at the same location then that new survey defines a new anomaly. The old anomaly remains valid and now has a new relationship to the more recent data with which it shares common sources. Also, an anomaly remains valid even if its source is subsequently removed (e.g. by mining) and is no longer a feature of the present magnetic field. If we consider the hypothetical event of an instant reversal of the geomagnetic field, anomalies in the new field (that would need to be re-surveyed) could then with considerable advantage be related to the corresponding anomalies of quite different shape in the previous field to analyse pre- and post-reversal anomaly pairs with different induced magnetisation components and a common hard remanent magnetisation. This thought experiment highlights the concept that an anomaly belongs primarily to the survey data that defines it rather than being a fixed feature in the (continually changing) magnetic field.

The magnetic field measurements from which anomalies in the database are defined are almost exclusively of total magnetic intensity (TMI) made on aeromagnetic surveys. However, the database can also accommodate anomalies measured in ground or drone surveys and from vector component, field gradient or tensor data. These different magnetic field expressions are all directly related to each other and with sufficient data precision and distribution can be derived from each other (except for the indefinite integral in determination of a field from its gradients). ARAD allows that (with various advantages) models and magnetisation estimates can be derived from transforms and enhancements of the primary measured data. Examples of transforms include upward continuation of the field to higher elevation that reduces the influence of shape details in the distribution of magnetisation, and vertical derivatives to more effectively separate anomalies of interest from their background field.

Magnetic field anomalies are mostly due to local magnetisations stronger than the surrounding material, and in many cases much stronger. Magnetic field analysis and interpretation commonly assumes that the absolute magnetisation is identical to its contrast against surrounding magnetisation (i.e. that the background magnetisation is zero). Magnetisation estimates recorded in ARAD are almost exclusively of this effective (contrast) magnetisation but the values quoted as directions (declination and inclination) and strengths (A/m) or total magnetisations ($A.m^2$) can optionally include correction to absolute values against any independently known background value. From magnetic field images across many different geological settings it is evident that not all magnetic anomalies of interest are confidently separated from other overlapping field variations. In many cases geological processes result in complexity and variation within rock units that give rise to clusters of overlapping field variations. In such cases it is more challenging to isolate anomalies for analysis or inversion and it is also more difficult to justify the assumption that magnetisation contrast is a reliable estimate of absolute magnetisation.

Separation of an anomaly from its background field or overlapping fields of adjacent magnetisations is necessarily interpretive and should be supported by visual inspection to confirm that the results appear reasonable. Automated methods of anomaly detection and separation are convenient but must be supervised. The task is suitable for a comprehensively trained expert system but at the time of writing no such reliable system is available.

Where convenient, database entries include the sample of magnetic field data from which they were derived. These are in most cases of the supplied (processed TMI) data but optionally can be of data after a regional field separation or a transform such as the vertical derivative or upward continuation has been applied. In some cases the data are also supplemented with transforms such as the total gradient (also known as the analytic signal or modulus of the analytic signal) or normalised source strength (NSS). These transforms reduce the influence of magnetisation direction and provide approximate mapping of the horizontal distribution of magnetisation with reduced sensitivity to its direction. Data can be supplied in profile (mostly flightline) and/or grid format. The profile data are the more primary as grids create secondary data away from measurement locations. However, grid images are convenient for initial visual interpretations, and many processing operations are most conveniently performed by grid-based fast Fourier transforms (FFT). Note that magnetisation direction cannot be reliably resolved from analysis of single-channel data on a single profile. ARAD does not require specific data formats, but ESRI (ER Mapper) format with an easily read ASCII header file is encouraged for grids and ASCII or the widely used (but proprietary) Geosoft (GDB) format for profile data. Where data are provided for an anomaly they are packed in a zip file that can optionally include any other supporting information (that should be described by a simple 'readme' text file).

The intention when ARAD was first designed was that it would be made open for public submission of results and would grow by community contribution. Unfortunately, at present there are security issues in providing public access to externally populate ARAD and it is not feasible to allow direct public upload to it.

15.2 DESIGN OF THE DATABASE

Figure 15.1 is a simplified schematic of the database structure built around the central component of the anomaly list. The key relationships are that anomalies are generated by magnetisations controlled by geology and are detected and defined by surveys. The section of the database displayed in the upper half of Fig. 15.1 links geology to rock magnetic and palaeomagnetic studies. The geology is defined by geological events, geological ages and lithologies. This section of the database was designed to establish the fundamental relationships of the rocks carrying the magnetisations but to date it is only sparsely populated. Palaeomagnetic information is already provided in a global database that includes Australian data (Pisarevsky *et al.* 2022). There are benefits in having a single database source for palaeomagnetic data and so we have not proceeded to populate that section of the database. A future option is to link ARAD to an external palaeomagnetic database.

Few of the magnetic anomalies already in the database have samples from outcrop or drillhole available for palaeomagnetic or rock magnetic measurement. In north America and Scandinavia, recent glaciation has resulted in exposure of fresh samples of rock at surface and in those regions it is naturally more convenient to establish links between rock magnetism and the magnetic field. In Australia it is much more difficult to do this using only sparse drilling information. However, this restriction also raises the importance of the database as the only source of magnetisation information in some areas (albeit indirect and only of resultant magnetisation).

Database relationships mapped in the lower half of Fig. 15.1 are at present used much more extensively than those in the upper half. The linkage between anomalies and surveys is essential and the survey from which they are derived is a required statistic of all anomalies. At the time of writing all 300+ anomalies have at least one model. Previously, magnetic modelling results were disseminated in summary form in reports or publications available only to those aware of the studies and with access to the reports. ARAD models available for download from a web portal are much more publicly visible and accessible.

As a guide to considerations in building and using the database, the key elements of the surveys, anomalies, models and magnetisation estimate tables are shown in Table 15.1. Where feasible the element values are populated from drop-down options. Design of the database is a compromise between trying to ensure that all the information that is required and most of the information that is useful is collected, without the inconvenience of having to populate or navigate through rarely used options.

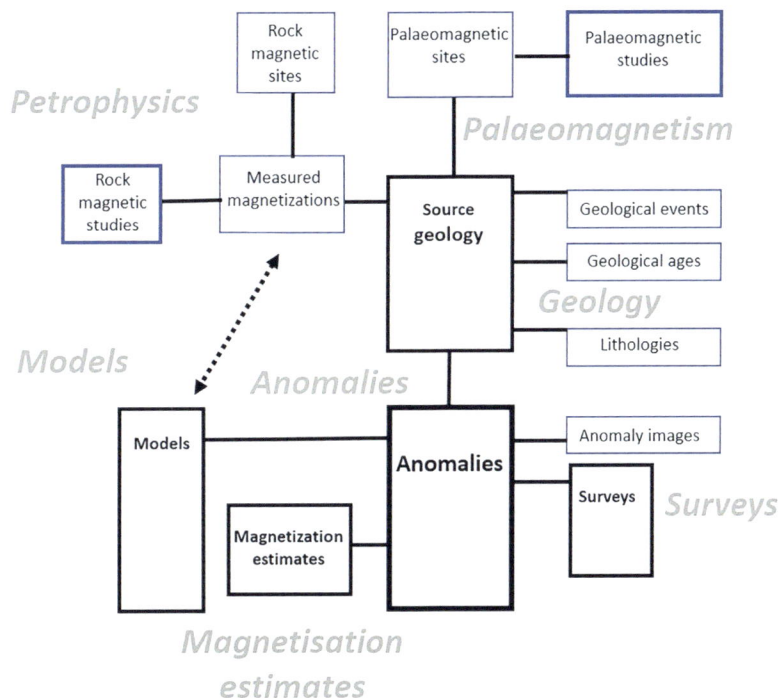

Fig. 15.1. Schematic of the database structure.

Table 15.1. Key anomaly database elements.

Surveys	Anomalies	Models	Magnetisation estimates	Magnetisation*
Survey ID	Anomaly ID	Model ID	Estimate ID	Known susc Y/N
Survey name	Anomaly name	Model name	Anomaly ID	Known remn Y/N
Database name	Survey ID	Anomaly ID	Estimation method	susceptibility
type	Geology ID	data type	Regional removal	Remanence int
year	data type	data derivation	centre	Remanence Dec
elevation type	data derivation	continuation ht	confidence	Remanence Inc.
ground clearance	continuation ht	data distribution	Analysis problem	Resultant int
single/multiline	longitude	model confidence	Continuation height	Resultant Dec
line spacing	latitude	model problem	Moment amplitude	Resultant Inc.
line orientation	IGRF Int/Dec/Inc.	Remanence Y/N	declination	Koenigsberger ratio
point spacing	extent N–S	Inversion Y/N	inclination	ARRA
area	extent E–W	model geometry	ARRA	
line km	+ve amplitude	spatial control	Optimisation stat	
highest elevation	−ve amplitude	elevation datum	Cluster statistic	
lowest elevation	shape	no. of bodies	Upload date	
data type	azimuth	top elevation	author	
E–W extents	data quality	depth below ground	comments	
N–S extents	data problem	volume		
outline filename	entry date	magnetisation*		
author	author	upload date		
comments	comments	author		
		comments		

15.3 DATABASE COMPONENTS

15.3.1 Surveys

The first action to upload an anomaly to the database is to locate the survey in the surveys list or to create a new survey entry. Anomalies cannot be entered into the database without being assigned to a survey. In Australia, Geoscience Australia maintain a survey database for the regional magnetic field data available from their GADDS (geophysical archive data delivery system) web portal (https://portal.ga.gov.au/persona/gadds). If the survey is in the Geoscience Australia survey database then the Geoscience Australia survey identifier (the project or 'P' number) is used but we also allow entry of surveys not in the GADDS register. The various survey parameters such as line spacing, terrain clearance and flightline direction all influence the definition of anomalies and the confidence with which magnetisation estimates can be derived.

15.3.2 Anomalies

The anomalies list is the core of the database. An anomaly is defined by survey data and is synonymous with a specific package of data. The unique specification of an anomaly is defined by the survey, geographic extents of the anomaly, data type, and any data transforms or continuation heights that have been applied. Transforms of the measured data, such as vertical derivatives or upward continuations define new anomalies. Also, if alternate flightlines are selected to investigate repeatability of results then each of the subsets of flightlines defines a separate anomaly. The versatility of this approach is required because an analysis or inversion only recovers information from the data submitted to it. Enhancements and transforms do not create new data but differentially weight the expression in the data of various parts of the magnetisation. By design, transforms and enhancements can focus on a magnetisation of interest but at the cost of possible distortion.

For some anomalies critical data might be missing. Misrepresentation of the magnetic field by undersampling is more problematic than is widely appreciated, possibly because grid images give the misleading suggestion that the magnetic field is continuously defined. Grid representation of a magnetic field is substantially determined by the gridding algorithm that interpolates apparent data where none has been measured. If continuity of field curvature can be reasonably assumed between survey points or lines then advanced gridding algorithms such as those based on anisotropic diffusion (Naprstek and Smith 2019; Davis 2022) can be used with advantage over standard minimum curvature methods (Briggs 1974). These issues are considered in greater detail in Chapter 2. Chapter 10 considers the advantages of drone surveys flown at close line spacing to infill anomalies that have been detected but are insufficiently sampled by aeromagnetic surveys. Although the anomalies in both the aeromagnetic and drone surveys are generated by the same source magnetisation, different surveys provide different information about that magnetisation due to variations in their line spacing and flying heights.

15.3.3 Magnetisation estimates

There are several methods to estimate direction of magnetisation from magnetic field data (Clark 2014) including new methods presented in this book (Chapters 7 to 9). Two of the major approaches to estimation of magnetisation direction are those based on Helbig analysis (Helbig 1963; Phillips 2005; Foss and McKenzie 2011) and those investigating trial directions in correlation between transforms with low and high sensitivity to magnetisation direction (Fedi *et al.* 1994; Stavrev and Gerovska 2000; Dannemiller and Li 2006; Gerovska *et al.* 2009; Li *et al.* 2017; Liu *et al.* 2020). Magnetisation direction estimators are analyses designed to automatically recover estimates of magnetisation direction from magnetic field data without the need for inversion. The various methods each have different strengths and weaknesses but all have similar requirements for suitability of the input data. If data support reliable estimation of magnetisation direction by one method they will generally also support the alternative methods. None of the automated processes to estimate magnetisation direction have the versatility or reliability of a well performed and user-guided inversion. Magnetisation estimates are stored in the database together (optionally) with the data from which they are derived. Estimators of magnetisation direction will become more important when they are developed to the stage that they can provide sufficient resolution and reliability to define populations of anomalies with a common magnetisation direction. To be useful, the lower reliability of the estimates must be compensated by the ease of generating solutions in an automated or at least lightly supervised

process. Ideally the solutions should include one or more quality factors that can be applied as a filter to reject less reliable solutions and to promote the more reliable solutions for verification and upgrade by inversion.

15.3.4 Models

Models, unlike magnetisation direction estimates, provide fully specified solutions from which a magnetic field (hopefully consistent with the input measured field) can be forward computed. A source model for any anomaly is optional and each anomaly can have multiple alternative models. A model is specified by the data from which it is derived, together with its key parameters. Models can be forward computed using a magnetisation direction from rock-magnetic measurement, magnetic field analysis or by specification. Alternatively (as is the case for most ARAD anomalies) models can be derived by inversion that discovers an estimate of magnetisation direction from the input data. Significant attention is paid through other chapters of this book (especially Chapters 1 and 5) to the non-uniqueness of models. Any model that conforms to independent constraints and produces an acceptable match to the input data (a subjective decision) is a candidate representation of the true magnetisation. The fields computed from models are added to associated background fields to match the measured field and this separation of fields is an essential aspect of any model.

To present, most ARAD anomalies have been isolated, modelled and inverted using ModelVision software (Pratt *et al.* 2020). These models are stored in the ModelVision (ASCII) 'tkm' format that includes magnetisation information and enables their further interrogation by modelling or inversion. The models are also stored in purely spatial 3D-DXF and Gocad T-surf formats for inclusion in GIS with other spatial models and data. Models derived using alternative software can also be uploaded to ARAD with the main requirement that each individual model body is characterised by and reported using a single magnetisation direction. Projections of the models are inherited from projection of the data from which they have been derived (that must be reported in the metadata).

15.4 THE DATABASE OF AUSTRALIAN MAGNETISATIONS

Figure 15.2 shows the distribution of ARAD solutions across Australia displayed in the AuScope web portal and

Fig. 15.2. ARAD solutions (300+) over a TMI image of Australia captured from the Geoscience Australia web portal.

Fig. 15.3 shows selection of anomaly 1 (the north-west Black Hill Norite anomaly) with buttons for download of the anomaly grid, models and images. Many of the ARAD solutions have been derived from the same data as the background TMI image or data of similar resolution, but some ARAD solutions are derived from higher resolution (but still publicly available) data with anomalies that may not be evident in the background image.

The 300+ Australian solutions currently in ARAD represent the more emphatic expressions of remanent magnetisation in the Australian magnetic field but these are only a minute selection of the total field variations, most of which have some contribution from remanent magnetisation. Although the prime objective of the database is to detail and map remanent magnetisation, ARAD also includes solutions with magnetisation parallel to the local geomagnetic field. We envisage the ideal future of the database as a record of all magnetisation estimates that can be recovered from the magnetic field. With improved understanding of the expression of magnetisation in the magnetic field data and how to optimise its recovery it is feasible that the number of ARAD solutions can be grown exponentially. We hope that there will also be continual upgrade of the magnetic field coverage from which new and improved magnetisation estimates can be derived. Just as with mining, where upgraded processes enable metal recovery from reprocessing of old mine tailings, it should be feasible to continually upgrade ARAD as analytic techniques and inversion procedures improve.

15.5 EXAMPLE DATABASE ENTRIES BEYOND AUSTRALIA

National borders need not limit the scope of ARAD. The database has been designed specifically for use with FAIR magnetic field data because solutions derived from proprietary data are not open to verification. FAIR magnetic field data is becoming increasingly available across the world and supports extension of ARAD or the

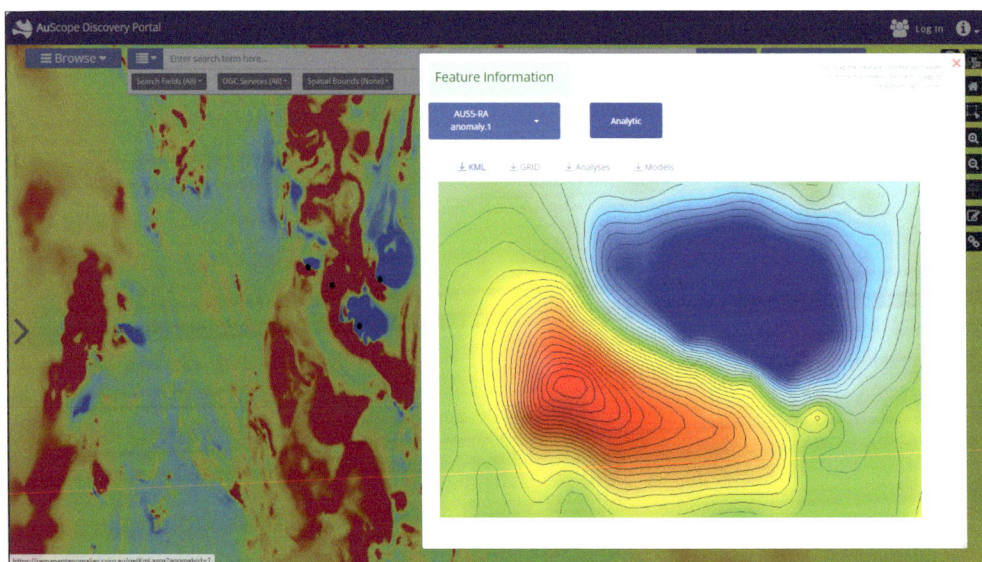

Fig. 15.3. AuScope portal zoom-in of the TMI and solutions in the vicinity of the ARAD001 solution from the Black Hill Norite of South Australia. The pop-up sidebar provide details of the anomaly with buttons to download a kmz image, a grid clip and models.

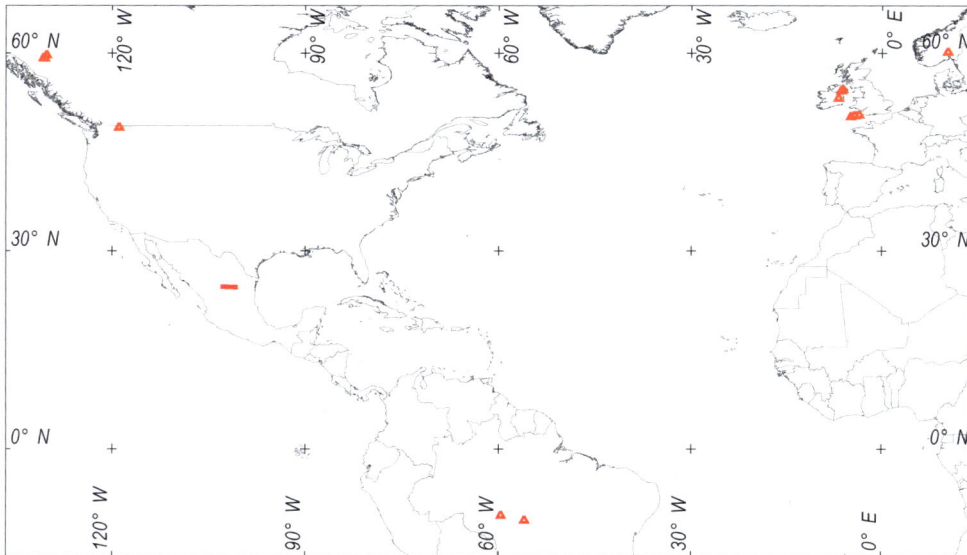

Fig. 15.4. Distribution of solutions beyond Australia as of March 2025.

Table 15.2. List of ARAD solutions outside Australia as of July 2024.

Anomaly	Country	Name	Longitude	Latitude	Description
313	Brazil	Caca_dos_Tapiunas	-59.60	-10.10	Intrusive
331	Brazil	Rio Claro	-55.89	-10.82	Wide circular sheet
332	Canada	Jennings_River ne	-130.18	59.72	Reverse magnetic plug
333	Canada	Jennings_River sw	-130.62	59.31	Reverse magnetic plug
334	Canada	Mathews Tuya	-130.43	59.19	Thin sheet
335	Mexico	El Puerto de la Palma	-102.71	24.58	Volcanic complex
336	Mexico	Los Chiqueros	-102.18	24.55	Volcanic complex
337	Mexico	El Cabresante	-101.46	24.61	Volcanic complex
338	Mexico	El Penuelo	-100.80	24.58	Volcanic complex
339	UK	Hennock	-3.66	50.61	Possible mineralisation
340	UK	Trewinnion	-4.98	50.38	Intrusive?
341	UK	Callington	-4.33	50.50	Intrusive?
342	USA	Clackamas Mtn NW	-118.91	48.73	Plug?
343	USA	Clackamas Mtn SE	-118.87	48.71	Plug?
344	Ireland	Broomhedge	-6.18	54.49	Reverse magnetic plug
345	Ireland	Clare	-6.48	54.32	Reverse magnetic plug
346	Ireland	Grange Common	-6.91	53.21	Two volcanic bodies?
347	Norway	Haugsbygd	10.34	60.19	Intrusive plug?

creation of equivalent databases in other regions. Here we present database entries generated beyond Australia to illustrate the benefits such a database provides in value-adding to the magnetic field data and hopefully ARAD will grow with new solutions both within and outside Australia. Figure 15.4 shows the March 2025 distribution of database solutions beyond Australia. Table 15.2 lists the anomalies and Table 15.3 provides key estimated statistics of the source magnetisations for those anomalies.

Table 15.3. Peak-Trough amplitudes of the anomalies in Table 15.2 and their magnetisation estimates.

Anomaly	Peak-Trough Amplitude (nT)	Model(s)	Intensity (A/m)	Magnetic Moment (A.m^2.10^6)	Declin-ation	Inclination	ARRA
313	730	Elliptic pipe	40.46	49,890	336°	−54°	53°
331	732	Elliptic pipe	1.836	491,288	350°	+5°	11°
332	767	Elliptic pipe	8.29	846	255°	−85°	168°
333	443	Elliptic pipe	3.069	252	132°	−77°	165°
334	1,888	Polygonal pipe	11.20	6,317	300	+85°	14°
335	946	Elliptic pipe	45.49	1,258,400	357°	+64°	27°
336	807	Elliptic pipe	19.05	913,800	339°	+49°	25°
337	1,303	Elliptic pipe	24.42	2,107,500	25°	+52°	24°
338	1,762	Polygonal pipe	21.01	2,231,000	16°	+62°	28°
339	659	Elliptic pipe	6.77	62	178°	−20°	134°
340	91	Polygonal pipe	0.616	290	180°	−22°	137°
341	146	Elliptic pipe	7.87	159	191°	+27°	86°
342	1,736	Elliptic pipe	10.7	501	218°	−87°	164°
343	2,475	Elliptic pipe	54.6	210	165°	−64°	167°
344	745	Elliptic pipe	4.60	144	185°	−45°	156°
345	1,643	Elliptic pipe	19.5	212	224°	−68°	163°
346	267	Elliptic pipe	2.95	291	340°	+47°	21°
347	2,800	Circular pipe	35.95	638	190°	−31°	138°

15.5.1 Anomaly ARAD331, Brazil

Figure 15.5 shows a TMI anomaly measured by the Japuíra survey in the Mato Grosso region of Brazil. The data can be downloaded from the Geological Survey of Brazil geoscience system (https://geosgb.sgb.gov.br/geosgb/about_geosgb_en.html). The survey was flown on north–south flightlines at a line spacing of 500 m and a nominal terrain clearance of 100 m. Figure 15.5B shows the anomaly computed from inversion with a homogeneous horizontal-top, elliptic-section pipe model of average diameter of 14 km and depth extent 1,800 m. As shown by the similarity of the images in Fig. 15.5 A and 15.5B, the model matches the anomaly quite successfully. However, compromises in matching data over such a large area with a simple model mean that at any one location the model is an unreliable representation of the local in-ground magnetisation (e.g. of spot values of depth to the top of magnetisation). Figure 15.6 shows a profile extracted from the model along a central north–south flightline (note the data-fit is not derived exclusively on this flightline but on the complete flightline set). The high-amplitude, short-wavelength variations would cause considerable uncertainty for a model derived from single flightline data but there is effective cancellation of these

irregularities over the large number of flightlines inverted. This stability of the model was confirmed by running inversions with the complete dataset, and with even- and odd-numbered flightlines only. All three inversion models returned very similar magnetisation estimates. The −5° geomagnetic inclination at the site was chosen as the starting model magnetisation direction, and the +5° post-inversion estimate of the inclination of magnetisation (an ARRA of 10°) is consistent with the tri-pole morphology of the anomaly with flanking positive lobes to both the north and south of the central low body as explained in Chapter 14.

15.5.2 Anomalies ARAD342, ARAD343 Clackamas Mountain, Washington State, USA

Figure 15.7 shows a segment of TMI measured by the Republic Survey flown for the United States Geological Survey (USGS) and Washington Department of Natural Resources as part of the Earth Mapping Resources Initiative (EMRI) in Washington State, USA (Staisch *et al.* 2024) (https://www.sciencebase.gov/catalog/item/65959be8d34e3265ab152eed). The survey was flown on flightlines at 200 m spacing and with a mean terrain clearance for the two anomalies

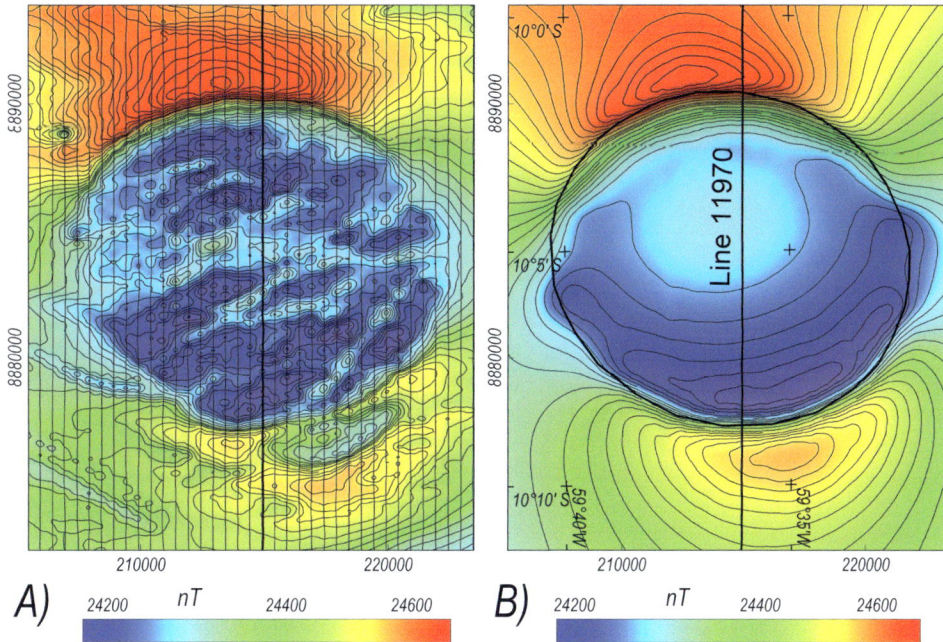

Fig. 15.5. A) measured and B) model-computed TMI for ARAD anomaly 331. The outline of the elliptic-section model is plotted in B). A section through the model along flightline 11970 is shown in Fig. 15.6.

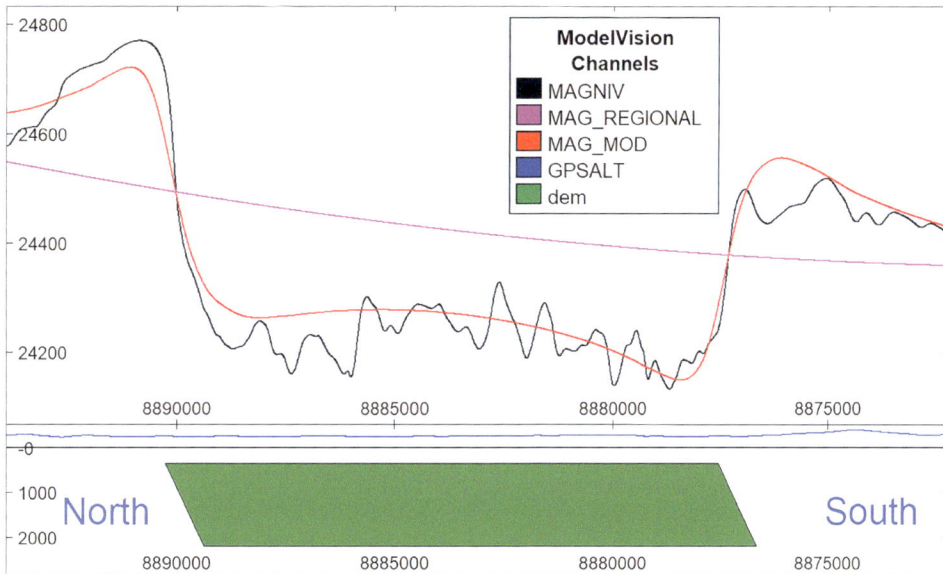

Fig. 15.6. North–south section along central flightline 11970 (for location see Fig. 15.5).

studied of 130 m. Figures 15.8A and 15.8B show anomalies ARAD342 and ARAD343 in more detail with overlays of the top of the elliptic-section pipe models generated to match them. Each anomaly, defined by five to six east–west flightline segments of approximately 1 km length, is composed of a prominent subcircular trough with a lower-amplitude peak to the

north. The polarity of the anomalies is opposite to that expected of induced magnetisations at this latitude and suggests that they are due to reverse magnetisations. The ARRA values for the two anomalies are 167° and 164° (Table 15.3) compared to 180° for an exact reverse magnetisation. The anomalies are modelled with both plunging elliptic-section pipes (shown in red) and

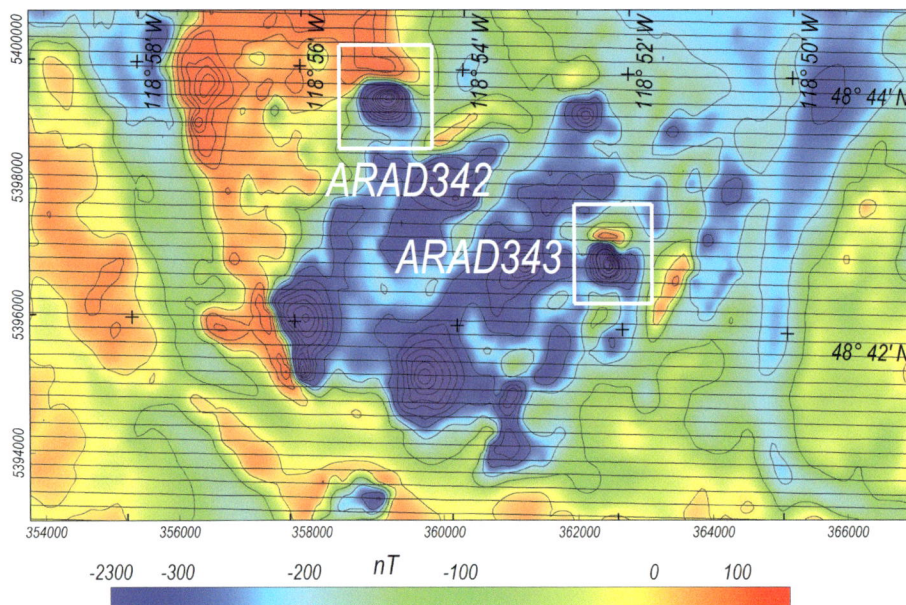

Fig. 15.7. Segment of TMI from the Republic Survey flown over Clackamas Mountain, including anomalies ARAD342 and ARAD343.

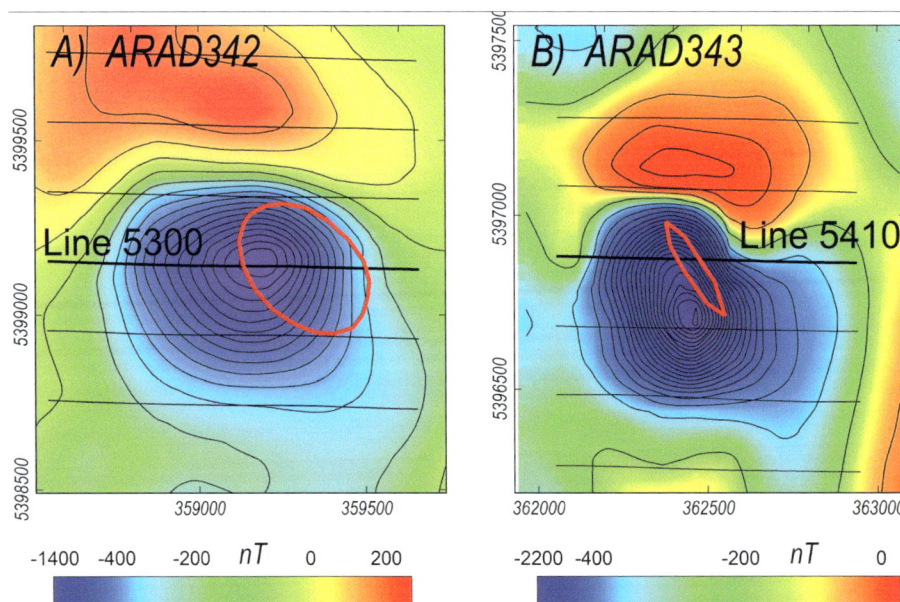

Fig. 15.8. TMI anomalies with east–west flightlines and outlines of the tops of inversion models for: A) ARAD342 and B) ARAD343.

ellipsoids (shown in blue). Sections through the inversion models along the central flightlines shown in Figs 15.9A and 15.9B reveal that the magnetisations are outcropping or very close to surface. Inversions with the two body types (independent other than they share the same input data and almost identical anomaly separations) give magnetisation directions with differences of only 3°and 8° for anomalies ARAD342 and ARAD343

respectively. Unfortunately, even for these probably outcropping magnetisations it is unlikely that palaeomagnetic measurement of remanent magnetisation (NRM) can be applied in modelling the anomalies because these prominent sites are likely to be heavily lightning struck. Nevertheless, it may be possible to determine the direction of remanent magnetisation with thermal or alternating-field demagnetisation.

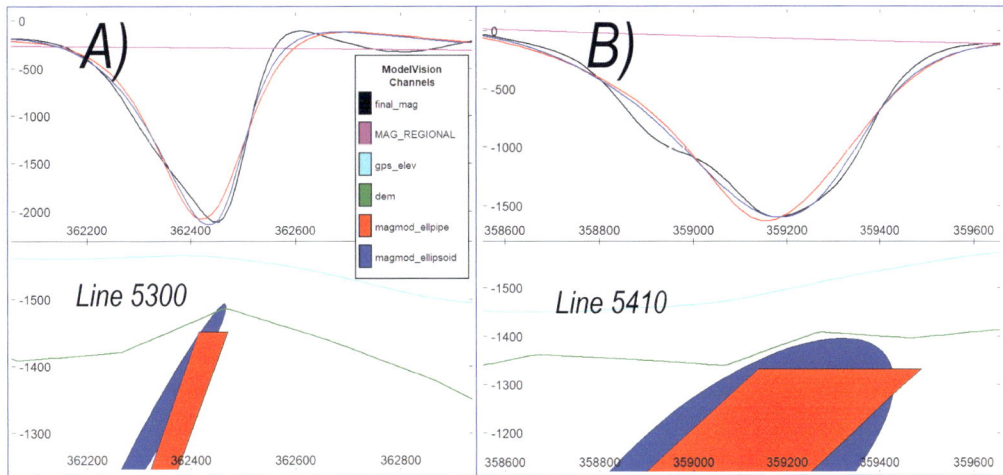

Fig. 15.9. Central west-to-east flightlines for A) Anomaly ARAD 342 and B) Anomaly ARAD343.

15.5.3 ARAD anomaly 334 over Matthews Tuya, British Columbia, Canada

Figure 15.10 shows a segment of TMI measured by the Jennings River Survey in northern British Columbia. The survey was flown on north-west-south-east flight-lines at 500 m spacing and a nominal terrain clearance of 150 m. The data can be downloaded from the Natural Resources Canada website (https://osdp-psdo.canada.ca/dp/en/search/metadata/NRCAN-GEOSCAN-1-223258)

or the Geoscience British Columbia website (https://www.geosciencebc.com/projects/2005-059/). The geomagnetic inclination of +76° at this site is steep and the almost single-signed negative amplitude anomalies ARAD332 and ARAD333 (Fig. 15.10) are due to magnetisations with similarly steep southerly inclinations (inversion results for these two anomalies are listed in Table 15.3). The predominantly positive anomaly over Matthews Tuya (anomaly ARAD334) is clearly due to a

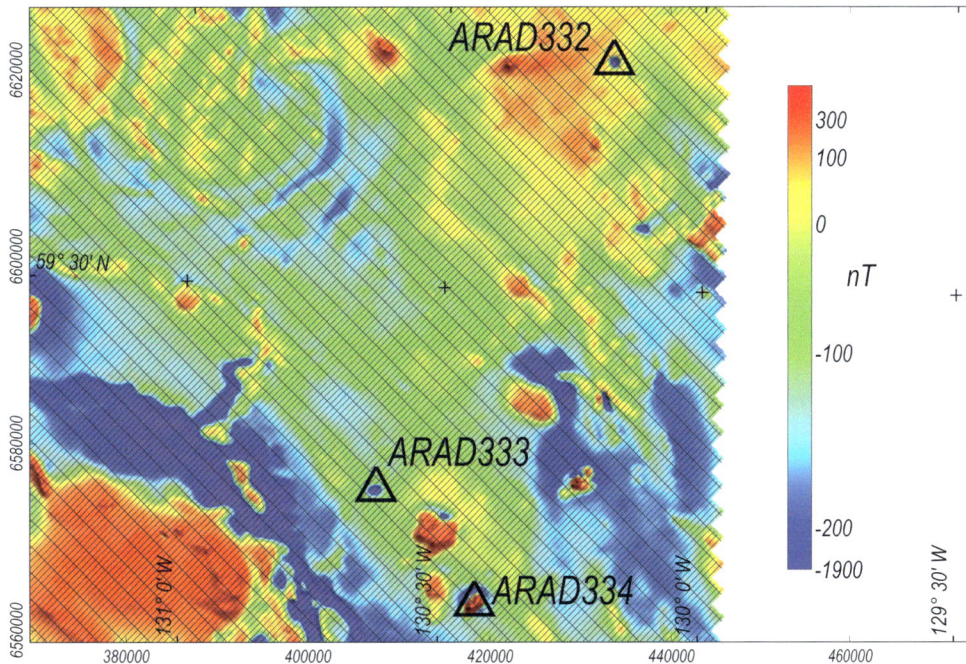

Fig. 15.10. Section of the Jennings River Survey in British Columbia including anomalies ARAD332, ARAD333 and ARAD334.

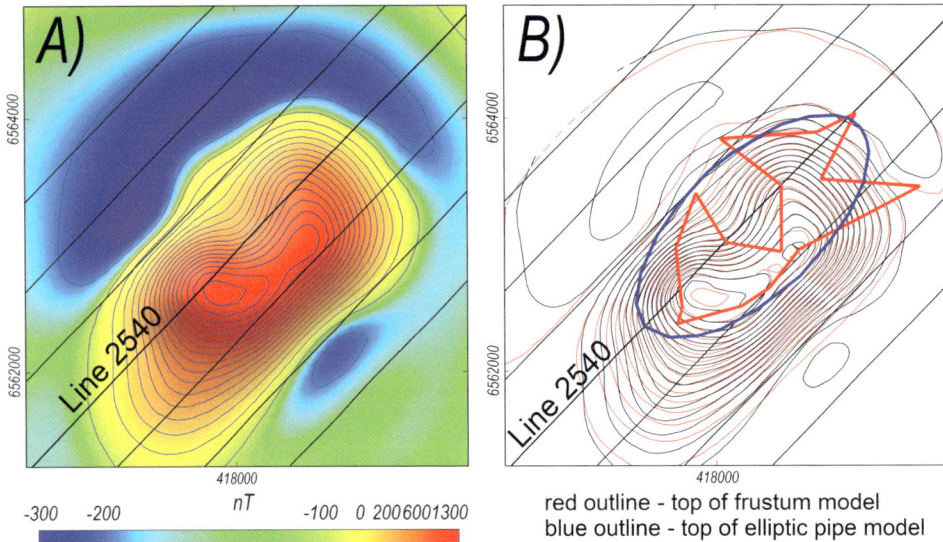

Fig. 15.11. A) TMI image and flightlines over Anomaly ARAD334 and B) measured (black) and computed from polygonal-section model (red) contours with outlines of the tops of the polygonal-section (red) and elliptic-section (blue) models.

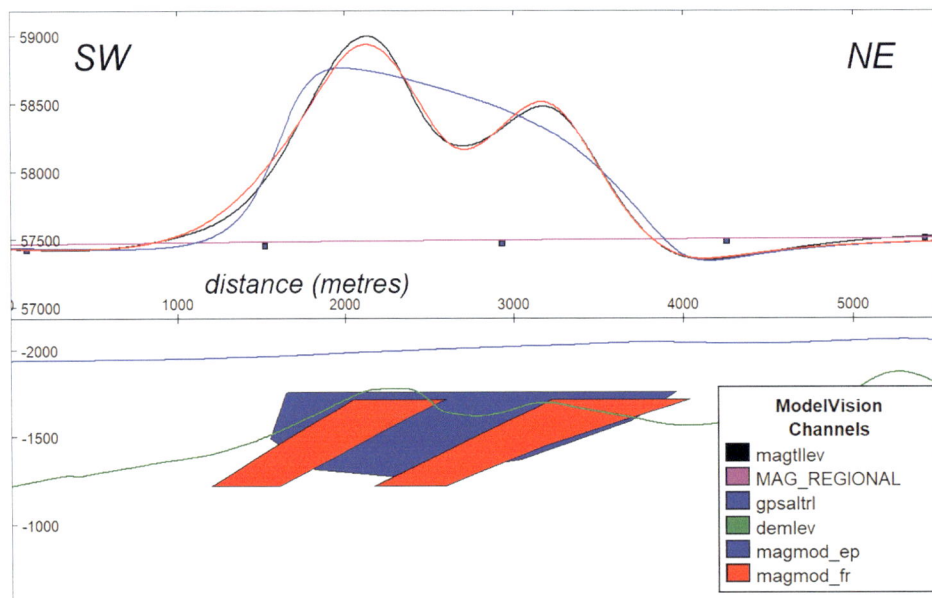

Fig. 15.12. Section along flightline 2540 with polygonal-section (red) and elliptical-section (blue) models.

quite different magnetisation, closer to the geomagnetic field direction. Figure 15.11A shows the measured anomaly over Matthews Tuya (a tuya is a flat-topped, steep-sided volcanic sheet erupted through a glacier or ice sheet). The anomaly is elongate in the south-west-north-east direction parallel to the flightline direction and is sampled on only two lines (each of which has very different expression of what is clearly a complex anomaly). We inverted the anomaly using both a pipe of simple, elliptic section and a body of polygonal section that could adapt to a more complex shape through inversion of individual

vertices. Figure 15.11B shows contours of the measured magnetic field and of the closely matching field computed from the polygonal-section model. Also shown in Fig. 15.11B are the plans of the top section of the elliptic-section (blue) and polygonal-section (red) models. Only the polygonal-section model is complex enough to match the detail of the anomaly and it is possible that the two centres of magnetisation revealed by this model might be separate but adjacent bodies. Both models are of limited depth extent as consistent with a tuya, are broadly coincident and have similar magnetisation directions

(Table 15.3) with a difference of only 7°. Figure 15.12 shows a section through the models on the westernmost of the two lines that pass through the anomaly. The section shows the limited depth extent and that the polygonal model has attempted to match the two anomaly peaks with two segments coincident with two topographic peaks.

15.5.4 ARAD anomaly 346, Grange Common, County Kildare, Ireland

Figure 15.13 shows a segment of TMI over Ireland measured in the Tellus Survey nation programme of the Geological Survey of Ireland. The data can be downloaded from Tellus (https://www.gsi.ie/en-ie/programmes-and-projects/tellus/Pages/default.aspx). The section of the survey including anomalies ARAD344 to ARAD346 was flown on north-north-west to south-south-east flightlines at 200 m spacing and a nominal 60 m terrain clearance (variable in the drape over the topography associated with ARAD346). Anomalies ARAD344 and ARAD345 are compact, almost circular negative anomalies apparently due to reversely magnetised volcanic pipes (for magnetisation details see Table 15.3 or download the models from ARAD). Figure 15.14A shows anomaly ARAD346 over a body of Ordovician andesites at Grange Common similar to those exposed in the Hill

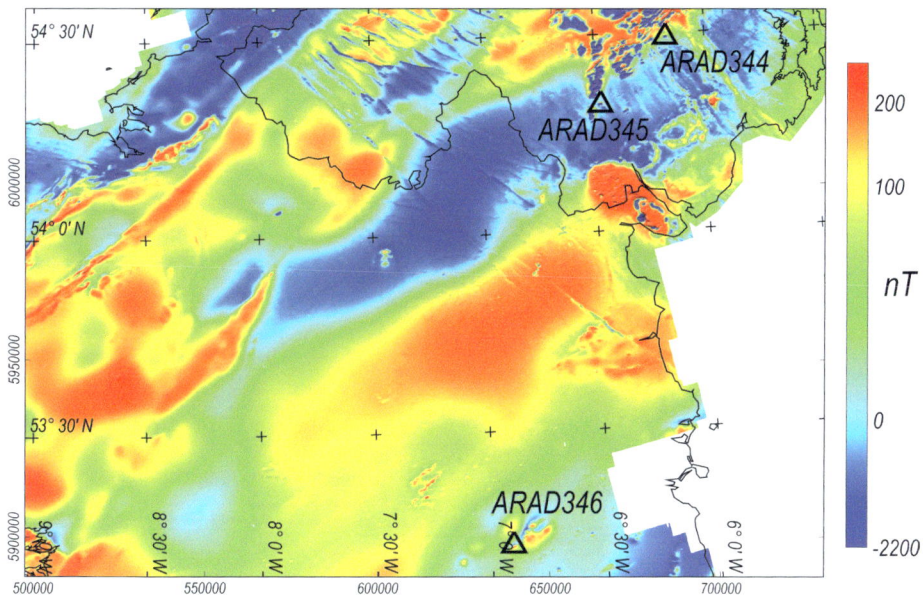

Fig. 15.13. TMI image of a segment of the Tellus Ireland Survey with anomalies ARAD344 to ARAD346.

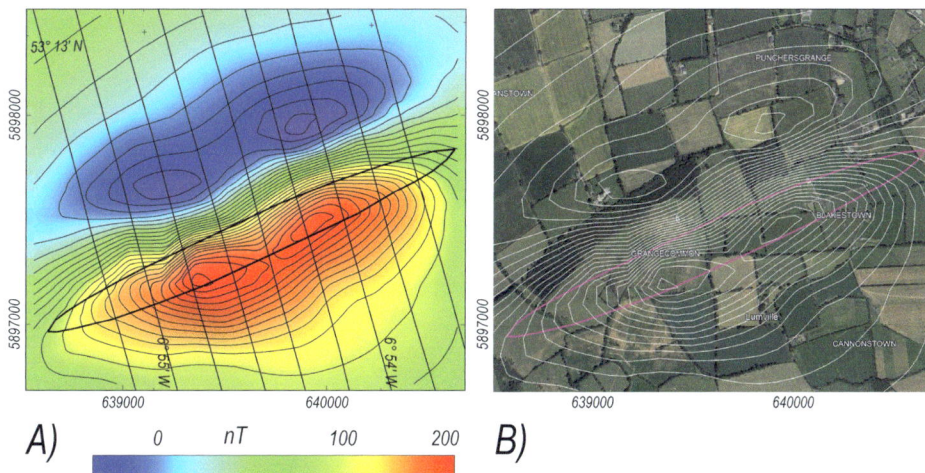

Fig. 15.14. A) TMI anomaly ARAD346 with flightlines and outline of the top surface of the model and B) TMI contours and body outline over a Google Earth image of the area.

of Allen Quarry to the north-east. ARAD346 could be split into two overlapping anomalies but their common pattern suggests that they are (possibly joined) sources of identical or similar magnetisation direction as reported in Table 15.3 (the models can be downloaded from ARAD). A more complex model could improve the match to the input anomaly but would not necessarily increase confidence in the estimated magnetisation direction (the anomaly could also be upward continued to provide data less sensitive to the details of its distribution).

15.5.5 ARAD anomaly 339, Hennock, Devon, UK

Figure 15.15 shows a TMI image from a part of the Tellus SW survey of south-west England (Beamish *et al.* 2014). This is an area with a long and intensive history of metalliferous mining. The survey was flown for the British Geological Survey (BGS) on north–south lines at 200 m spacing and a mean terrain clearance of 91 m. Data can be downloaded from the Tellus South West Project website (https://www.tellusgb.ac.uk/data/ airborneGeophysicalSurvey.html). Regional magnetic field variations are subdued over the magnetically quiet Permian Dartmoor and Cornish granites and over the Devonian and Carboniferous sediments they are intruded into, with higher amplitude magnetic field variations over the contact metamorphic regions around the granites and associated with more localised features, including some associated with mineralisation in and around the granites. These more localised magnetic field variations have a wide range of patterns. Three of the anomalies (of broadly consistent pattern) ARAD339 to ARAD341 are located in Fig. 15.15 and their inversion details are reported in Table 15.3. Figure 15.16A shows the easternmost of these three anomalies to the west of Hennock in the Teign Valley. The anomaly is well defined on four to five flightlines and has opposite polarity to that expected of an induced magnetisation. The anomaly peak and trough amplitudes are more similar than would be expected at this steep geomagnetic inclination, suggesting that the magnetisation has a low inclination. The anomaly is closely matched by a simple steeply plunging pipe of elliptic section and predominantly remanent magnetisation (Table 15.3). The plan of the top of the model is shown over the TMI image in Fig. 15.15A and over a Google Earth image of the area in Fig. 15.15B. A section along the central north– south light-line shown in Fig. 15.17 indicates that the body is narrow, steeply dipping and terminates at or close to surface. The anomaly was modelled with both an ellipsoid and an elliptic-section pipe model with almost identical results in locating the magnetisation and determining its direction. The Teign Valley where the anomaly is located has a long history of base metal mining of deposits associated with intrusion of the Dartmoor Granite, with east–west lodes parallel to apparent elongation of the source bodies for the ARAD339 anomaly. The Great Rock Mine that mined

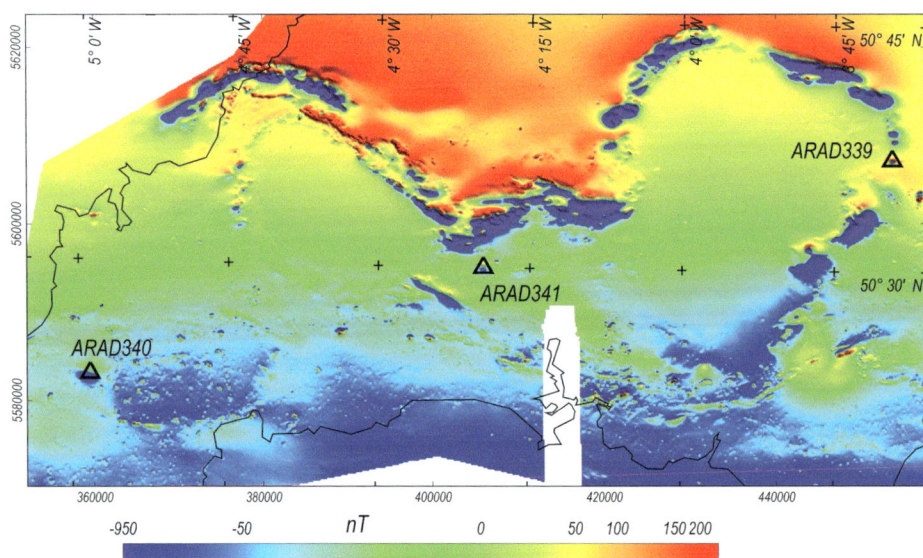

Fig. 15.15. TMI image of a section of the Tellus South-west survey including anomalies ARAD339 to ARAD341.

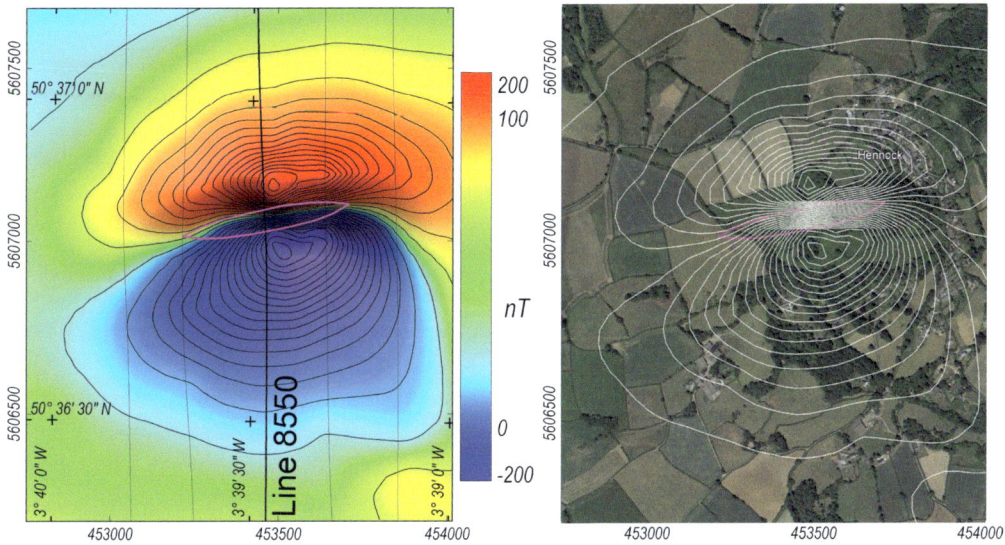

Fig. 15.16. A) TMI anomaly ARAD339 with flightlines and outline of the top surface of the model and B) TMI contours and body outline over a Google Earth image of the area.

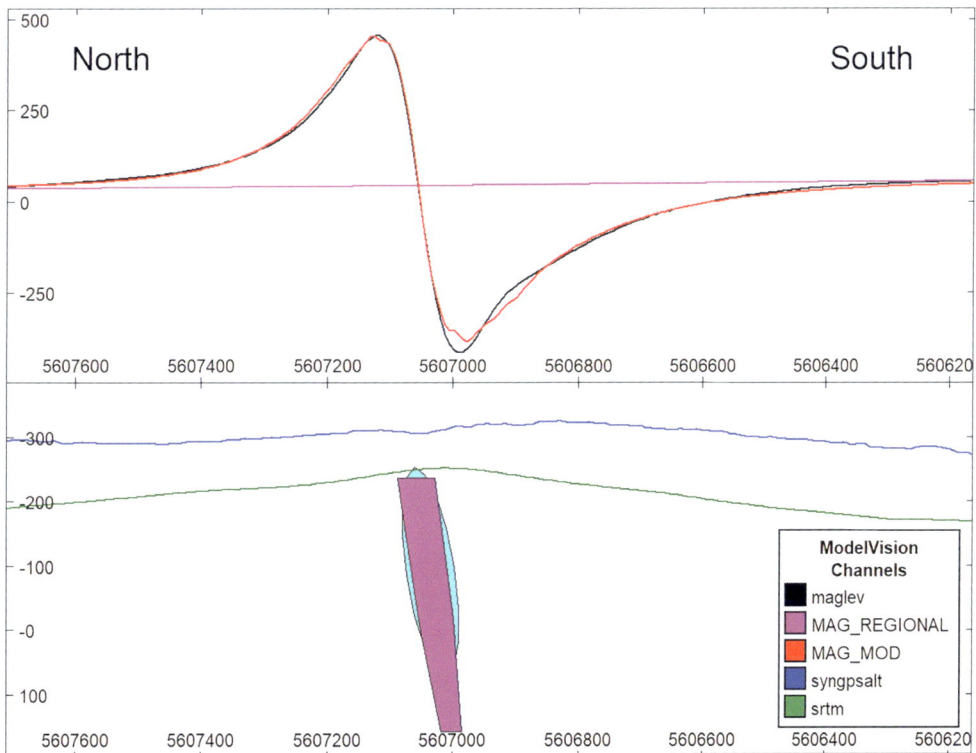

Fig. 15.17. Section along flightline 8550 (for location see Fig. 15.16) with elliptic-section pipe (magenta) and ellipsoid (turquoise) model intersections. The computed field channel is for the elliptic-section pipe model.

micaceous hematite (formerly used to paint the Sydney Harbour Bridge and British warships) is located only 1 km to the north of ARAD339. Two anomalies of similar pattern due to apparently larger and deeper magnetisations, 2 and 3 km to the north of ARAD339 (and visible in Fig. 15.15) might be due to untested mineralisation (but are within the Dartmoor National Park area).

15.5.6 The Haugsbygd anomaly ARAD347, Norway

Figure 15.18 shows a TMI image of Anomaly ARAD347 north of Haugsbygd in the Buskerud Region of Norway, extracted from the 2010 Kroderen heli-magnetic and EM Survey flown by the Geological Survey of Norway on east–west flightlines at 200 m spacing and 75 m terrain clearance. The data can be downloaded from https://www.ngu.no/mins-data/Geofysikk/. The anomaly is well matched by an outcropping or sub-cropping circular-section plunging pipe with a moderate-inclination reverse magnetisation of ARRA 138°. Details of the model magnetisation estimate are listed in Table 15.3 and the model can be downloaded from ARAD. The anomaly and source model are broadly coincident with a topographic high that suggests it may be possible to directly test the magnetisation by sampling of suitable outcrop. The compact anomaly and relatively undeveloped area might make this a suitable site for a lower-level, closely line-spaced drone survey to investigate any complexity of the anomaly unresolved by the currently spare coverage on only four key flightlines.

15.6 EMAG2 DATABASE ENTRIES

Mid- to high-resolution aeromagnetic surveys (line spacings of 500 m and less and terrain clearances of 150 m and less) over shallow magnetisations generally record a population of many small anomalies and progressively fewer large anomalies. The multiple smaller anomalies constitute many of the solutions in the ARAD database but these anomalies attenuate rapidly with height and are either unrepresented or insufficiently sampled by surveys at higher elevation (that generally also have wider line spacing). EMAG2 is a global TMI coverage generated from a composite of many national and regional datasets that are in many cases themselves composites from multiple surveys (https://www.ncei.noaa.gov/products/earth-magnetic-model-anomaly-grid-2). The EMAG2 grid has a cell size of two arc seconds and nominally represents the field at 4 km elevation. It is constructed predominantly from aeromagnetic and marine magnetic ship-track data, with satellite data used for levelling. Validity and resolution of EMAG2 varies with location according to the quality and sufficiency of the input data. At the 4 km elevation the magnetic field is reasonably defined by sparse sampling but the low-elevation input data that are used to derive the high elevation field should be of at least moderate resolution to restrict aliasing before upward continuation. Anomalies in the EMAG2 grid are mostly due to clusters of shallow strong magnetisations (possibly resolved individually in low-elevation magnetic field data) that are broad enough to generate longer wavelength field variations and/or deeper moderate to strong (generally larger volume) magnetisations that in low-elevation magnetic field data are in many cases partially obscured by

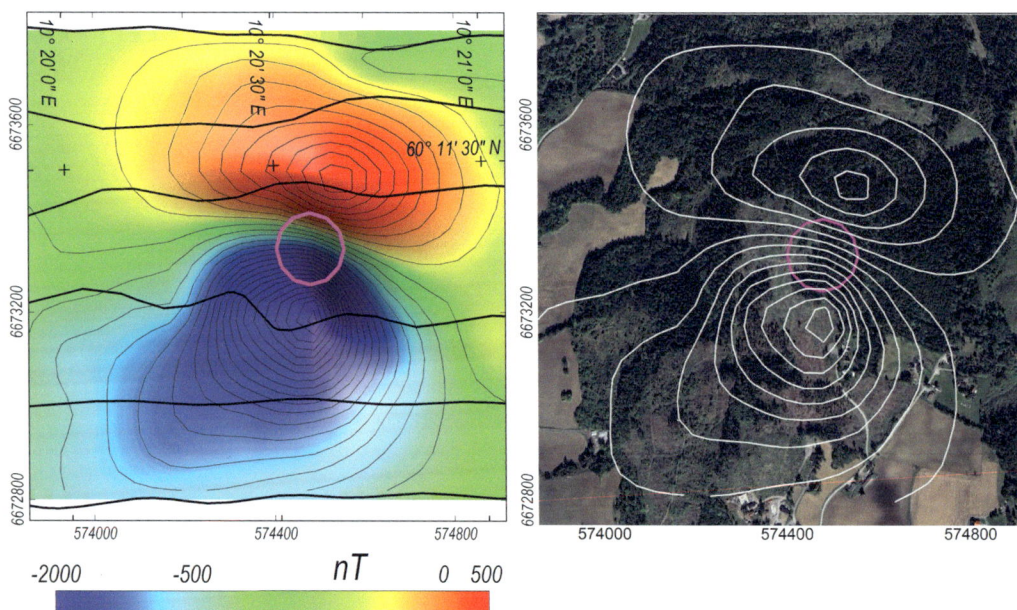

Fig. 15.18. A) ARAD347 TMI anomaly with flightlines and plan of the top of the circular-section pipe model and B) the TMI anomalies and model plan over a Google Earth image.

overprinting short-wavelength field variations of shallow magnetisations. Each anomaly in EMAG2 yields a single magnetisation estimate that is at best representative of the combined magnetisations that generate it. There are undoubtedly also anomalies in EMAG2 that are due to multiple distributed magnetisations for which an estimated magnetisation direction is a factor not just of the contributing (possibly strongly variable) magnetisations but also of their spatial distribution. Nevertheless, EMAG2 provides a valuable resource to estimate large scale magnetisations over regions where input data is of moderate to high quality and suitability but is not easily accessible elsewhere. EMAG2 anomalies in regions where input data are less sufficient or less reliable should be treated with greater caution.

We investigate EMAG2 anomalies in two areas of high-quality input data. We compare magnetisation directions recovered from four prominent EMAG2 anomalies over Mexico with estimates from the anomalies in the North American Magnetic Anomaly Map (https://pubs.usgs.gov/publication/70211067) and in the north-east of Western Australia we compare magnetisation directions derived from EMAG2 anomalies with directions derived from the Australian national TMI grid (https://ecat.ga.gov.au/geonetwork/dashboard/api/records/f2e58161-24cf-42d2-b328-a9039f72113b).

15.6.1 North American Map and EMAG2 anomalies over volcanic intrusions in Mexico

Figure 15.19 shows images of prominent TMI anomalies over volcanic intrusions in Mexico from A) the Magnetic Anomaly Map of North America (MAMNA) at a 1 km elevation and B) from EMAG2 at a 4 km elevation. The EMAG2 data over this region is an upward continuation of the MAMNA grid or of the same input data as for that grid. In the absence of original flightline data it is acceptable to estimate source magnetisation direction from well defined grid anomalies provided there is confidence in the construction of those grids. Derivation of source magnetisation details such as depth to the top of magnetisation is less reliable from grid data (as discussed in Chapter 3). However, as established in Chapters 5 and 6, estimation of mean magnetisation direction from grid data is relatively robust.

A study using data from MAMNA (García-Abdeslem and Calmus 2019) reports a magnetisation direction of declination 5°, inclination 50° for Anomaly ARAD335. Inverting the anomaly for magnetisation direction using elliptic-section pipe and ellipsoid models we obtained directions different to the estimate by García-Abdeslem and Calmus by 20° and 19° respectively and from the equivalent anomaly ARAD348 in the EMAG2 grid (with the same two body types) different by only 4° and 5°.

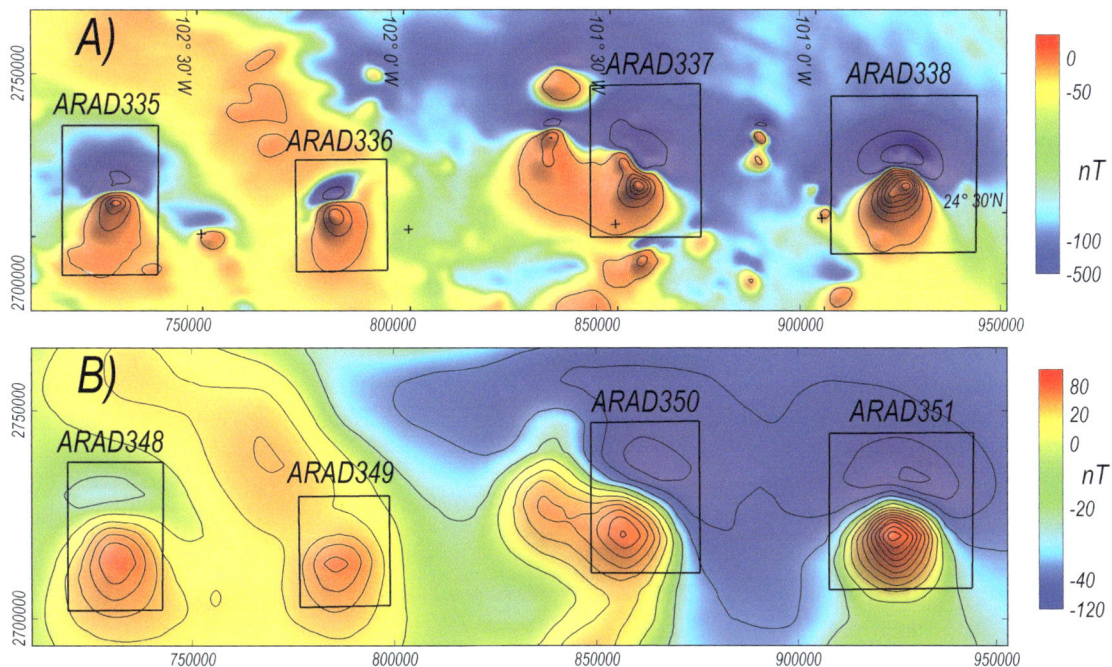

Fig. 15.19. An east–west string of TMI anomalies in Mexico (for location see Fig. 15.4), A) in the North American Magnetic Anomaly Map (elevation 1 km, contour interval 200 nT) and B) in EMAG2 (elevation 4 km, contour interval 20 nT).

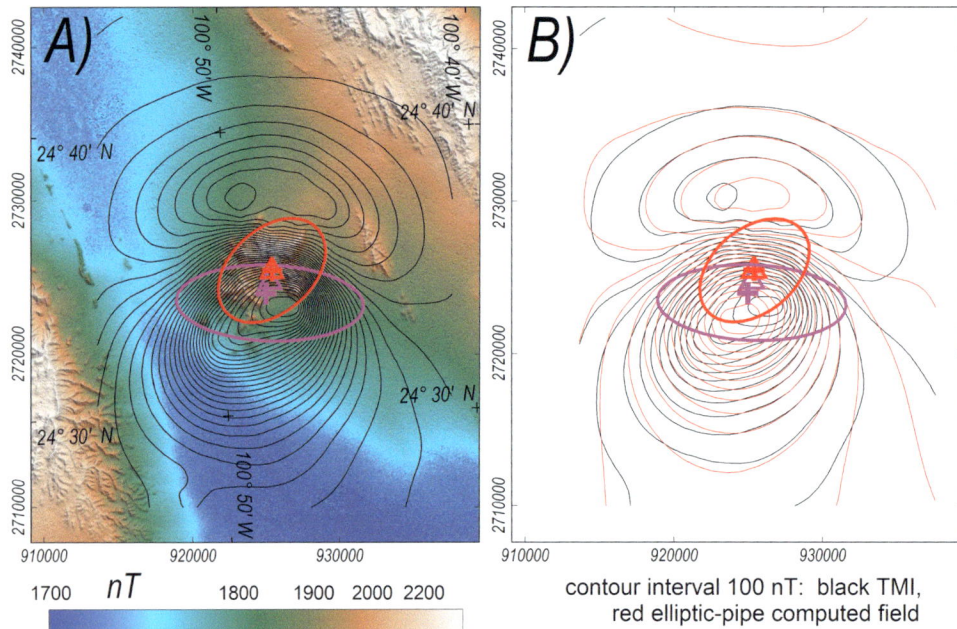

Fig. 15.20. A) contours of ARAD338 from the MAMNA Grid over an image of SRTM and B) those contours (black) and contours of the elliptic-section pipe computed field (red). The top of the elliptic-section pipe models for Anomalies ARAD338 (MAMNA) and ARAD351 (EMAG2) are shown in red and magenta respectively. Crosses mark the centres of the tops of the models and triangles their body centres.

Figure 15.20A shows contours of ARAD338 (the easternmost anomaly in Fig. 15.19B) over an image of SRTM (https://www.usgs.gov/centers/eros/science/usgs-eros-archive-digital-elevation-shuttle-radar-topography-mission-srtm-1). The El Peñuelo intrusive complex (Velasco-Tapia *et al.* 2011) consists of a range of lithologies including monzodiorite and quartz syenite that form a topographic high of 750 m. The range of elevation complicates processing of the MAMNA grid and may persist to influence the EMAG2 grid at higher elevation. The tops of the elliptic-pipe source models from inversion of the MAMNA and EMAG2 data shown in Fig. 15.20 are different but overlap considerably. The centre-points of the tops of the bodies differ by 1,560 m, only one-half of the EMAG2 cell size. The difference between the magnetisation direction of the two models is only eight degrees (the mean difference for the four pairs of anomalies is 9°). For all four pairs of anomalies there is a better fit of the computed fields to the higher and simpler EMAG2 anomalies than to the more complex lower-elevation MAMNA anomalies, but we consider inversion of the lower elevation data to be more reliable because of the higher anomaly amplitudes and more distinct separation from background fields.

15.6.2 EMAG2 anomalies over strong magnetisation in north-east Western Australia

Figure 15.21 shows images of high-amplitude TMI variations in north-east Western Australia: A) from the national Australian TMI grid and B) from EMAG2. Figure 15.22A shows TMI contours in the Australian national grid for the area of Anomaly ARAD354. There is a range of 36,000 nT in a background field of just less than 52,000 (the contour interval in Fig. 15.22A is 1,000 nT). This extreme magnetic field variation causes local rotation of the geomagnetic field that invalidates various transforms of the data. Modelling of the data through computation of orthogonal field components is not susceptible to this problem but the extreme field variations raise a series of challenges in data acquisition and processing. In consequence the models are of lower reliability than models of more moderate magnetisations. These magnetic field anomalies coincide with regions of (sparse) outcrop of the banded and folded Nimingarra Iron Formation, a basal unit of the Archaean Pilbara Supergroup. The extreme magnetisations giving rise to these high-amplitude anomalies include considerable self-demagnetisation effects and most probably significant anisotropy of magnetic

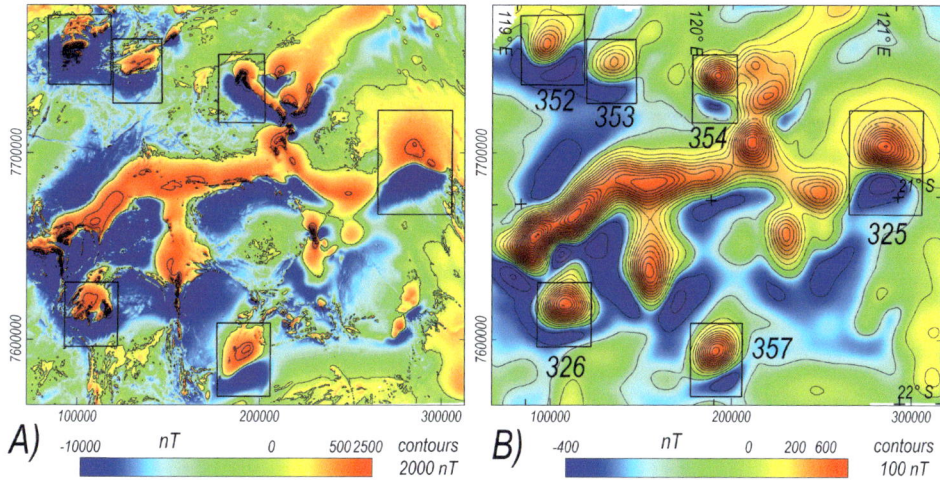

Fig. 15.21. TMI images over north-east Western Australia A) from the national Australian TMI grid and B) from EMAG2.

susceptibility as well as complex remanent magnetisations (magnetisations of the nearby similar Hammersley iron formation are described by Clark and Schmidt 1994 and Guo 2015). In consequence, the effective magnetisation estimated from modelling of the anomalies does not directly represent magnetisations that would be measured on rock samples in the laboratory.

To compare anomalies in the Australian national grid and EMAG2 we directly inverted the Australian national grid data using 14 plunging elliptic-section pipe bodies,

each assigned to explain discrete peaks in TMI (more reliable details could be obtained from inverting the primary survey flightline data). Each body was allowed free resultant magnetisation direction. The field computed from the assemblage of bodies matches the input field closely (Fig. 15.22A) but the overlap of adjacent anomalies reduces confidence in their individual parameter values. The mean magnetisation direction for the model was estimated by vector summation of the magnetisation directions weighted by the product of

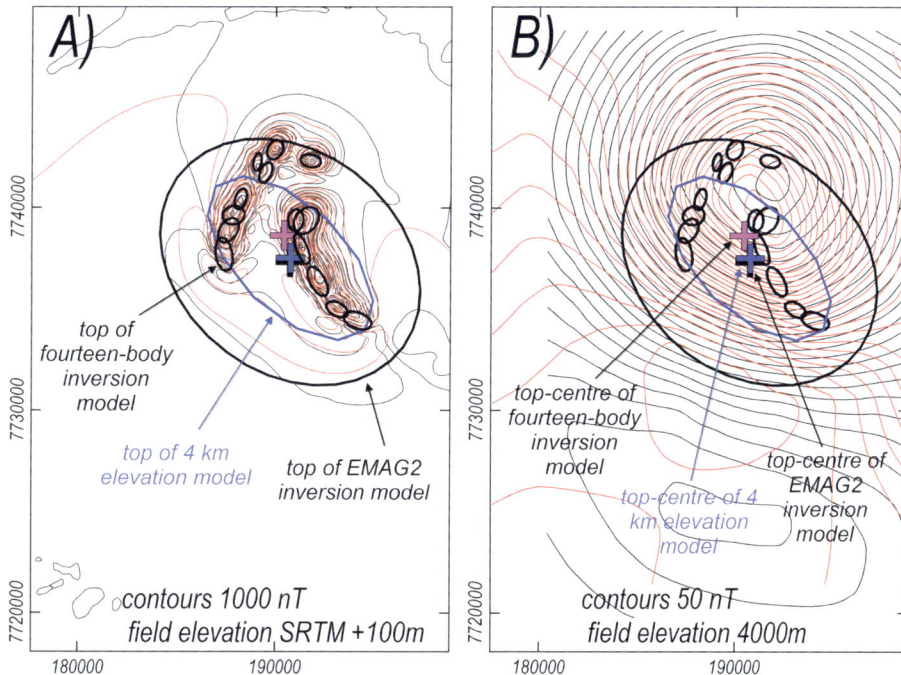

Fig. 15.22. A) contours of the Australian national TMI grid (black) and of the field computed from the 14 elliptic-section pipe model (red) and B) contours of the EMAG2 ARAD354 anomaly (black) and of the field at 4 km elevation computed from the 14-body model (red).

magnetisation-intensity and area of the top of the bodies, and separately by the product of magnetisation-intensity and volume. These two mean directions (listed in Table 15.4) differ by only 2°. The mean difference of the individual body magnetisation directions from the population mean directions was in both cases 13°. We do not know to what extent this range of magnetisation direction represents true variation in magnetisation direction or error in its estimation.

We computed the field from this 14-body model at an elevation of 4 km for comparison with the EMAG2 grid. The EMAG2 contours (black) and model-computed contours (red) are shown in Fig. 15.22B. These two grids, derived in very different ways are broadly similar but differ in detail. We inverted both the EMAG2 anomaly

(Anomaly ARAD354) and the field of the 14-body model computed at the same 4 km elevation, in each case using single plunging elliptic-section pipe models. Magnetisation parameters of these inversion models are reported in Table 15.4. The difference in recovered magnetisation direction estimates from the two inversions is only 4°. The difference in magnetisation direction between the 14-body model and the single-body inversion of its field at 4 km is 13°. The top surface of the elliptic-section pipe model inverted from the 14-body model field is plotted in Fig. 15.22A and Fig. 15.22B. Also plotted in Fig. 15.22B are the outline of the elliptic-section pipe model from inversion of the EMAG2 data and the centre points of both single-body inversion models and the 14-body inversion model. The

Table 15.4. Magnetisation details for models from inversion the EMAG2 ARAD354 anomaly, of the equivalent anomaly in the Australian national TMI grid (the '14-body' model), and of the field of the 14-body model computed at an elevation of 4 km.

Model	Elevation m	Volume km³	Intensity A/m	Magnetic moment A.m².10¹²	Declination degrees	Inclination degrees
Fourteen-body weighted by top surface	100				71	-73
Fourteen-body weighted by volume	100	28.24	74.2	2.1	78	-74
Single-body inversion of 14-body field	4,000	249	9.8	2.4	132	-80
EMAG2	4,000	1,668	9.4	15.7	130	-76

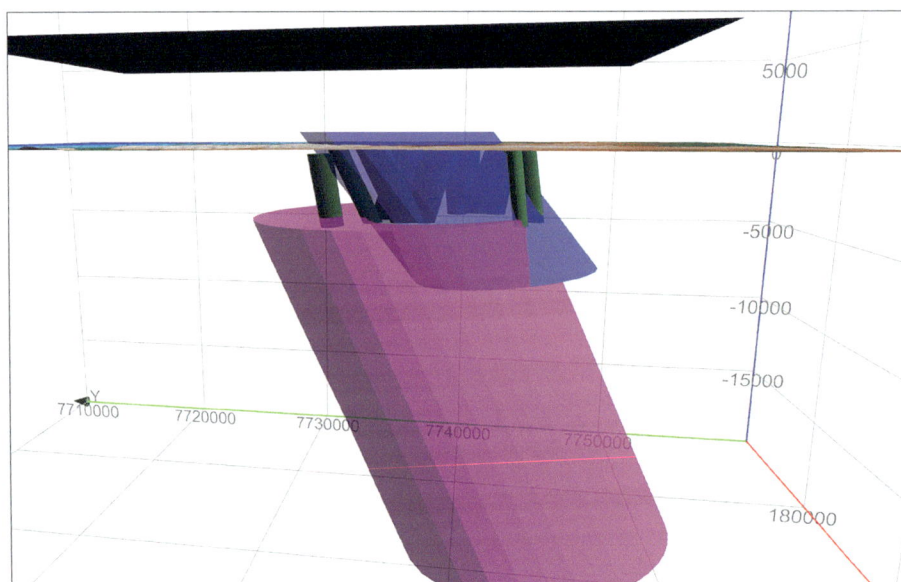

Fig. 15.23. Perspective view of the EMAG2 inversion model (magenta), 14-body Australian national TMI grid inversion model (green) and the body from inversion of the 14-body field computed at 4 km elevation (blue).

two models from inversion of the data at 4 km elevation that have magnetisation differences of only 4° have top surface centres different by only 200 m compared to the separation of 1250 m to the top-centre of the 14-body model that has a difference in magnetisation direction of 13°.

Figure 15.23 shows the three inversion models in perspective view. Although the horizontal centres of the tops of these models and their magnetisation directions are similar, their other spatial details vary considerably. Note that the 14-body model and single-body inversion model, shown in green and blue respectively in Fig. 15.23, produce near-identical fields at the elevation of 4 km although they are very different distributions of magnetisation. The EMAG2 inversion model (in magenta) that produces only a slightly different magnetic field (see Fig. 15.22B) has very different depths to its top and base, with only minor overlaps with the other models.

This study is at a different scale to most magnetic field inversions performed in mineral exploration (and in investigation of most of the much smaller anomalies in ARAD) but it highlights some common features: that spatial detail of the distribution of magnetisation can only be resolved in proximal fields, that inversion of distal fields generally only justifies simple models of homogeneous magnetisation, and that (mean) magnetisation direction and horizontal centre of the top of magnetisation are the more reliable results recovered from magnetic field inversion.

15.7 CONCLUSIONS

We have discussed the design and implementation of the Australian remanent anomalies database and suggest that this database or its blueprint could usefully be applied to establish a global database or linked databases to resolve and integrate magnetisation directions derived from analysis and inversion of magnetic field data with palaeomagnetic and rock magnetic studies at regional, national and global scales. In this publication we present a range of solutions from beyond Australia, similarly derived from FAIR data made available by national geological surveys. We also include study of anomalies from the EMAG2 global magnetic field grid, selected to illustrate the capabilities and limitations of investigating magnetisation at larger scale and greater elevation.

REFERENCES

Beamish D, Howard AS Ward EK White J Young ME (2014) 'Tellus South West airborne geophysical data.' Natural Environmental Research Council, British Geological Survey.

Briggs IC (1974) Machine contouring using minimum curvature. *Geophysics* **39**, 39–48. doi:10.1190/1.1440410

Clark DA (2014) Methods for determining remanent and total magnetisations of magnetic sources – a review. *Exploration Geophysics* **45**, 271–304. doi:10.1071/EG14013

Clark DA, Schmidt PW (1994) Magnetic properties and magnetic signatures of BIFS of the Hamersley Basin and Yilgarn Block, Western Australia. *Exploration Geophysics* **25**, 169. doi:10.1071/EG994169a

Dannemiller N, Li Y (2006) A new method for determination of magnetization direction. *Geophysics* **71**, L69–L73. doi:10.1190/1.2356116

Davis A (2022) Nested anisotropic geostatistical gridding of airborne geophysical data. *Geophysics* **87**, E1–E12. doi:10.1190/geo2021-0169.1

Fedi M, Florio G, Rapolla A (1994) A method to estimate the total magnetization direction from a distortion analysis of magnetic anomalies. *Geophysical Prospecting* **42**, 261–274. doi:10.1111/j.1365-2478.1994.tb00209.x

Foss CA, McKenzie KB (2011) Inversion of anomalies due to remanent magnetisation: an example from the Black Hill Norite of South Australia. *Australian Journal of Earth Sciences* **58**, 391–405. doi:10.1080/08120099.2011.581310

García-Abdeslem J, Calmus T (2019) The monzonitic to dioritic pluton of Cerro Prieto, Durango, Mexico: a Larimide laccolith inferred by nonlinear 3D inversion of aeromagnetic data. *Journal of Applied Geophysics* **160**, 121–130. doi:10.1016/j.jappgeo.2018.10.026

Gerovska D, Araúzo-Bravo MJ, Stavrev P (2009) Estimating the magnetization direction of sources from southeast Bulgaria through correlation between reduced-to-the-pole and total magnitude anomalies. *Geophysical Prospecting* **57**, 491–505. doi:10.1111/j.1365-2478.2008.00761.x

Guo WW (2015) Magnetic mineralogical characteristics of Hamersley iron ores in Western Australia. *Zeitschrift für Angewandte Mathematik und Physik* **3**, 150–155. doi:10.4236/jamp.2015.32023

Helbig K (1963) Some integrals of magnetic anomalies and their relation to the parameters of the disturbing body. *Zeitschrift für Geophysik* **29**, 83–96.

Irving E (1964) 'Palaeomagnetism and its application to geological and geophysical problems' Wiley, 399 pages.

Jarboe NA, Koppers AA, Tauxe L, Minnett R, Constable C (2012) 'The online MagIC Database: data archiving, compilation, and visualization for the geomagnetic, paleomagnetic and rock magnetic communities' In: Abstract GP31A–1063, AGU Fall Meeting.

Li J, Zhang Y, Yin G, Fan H, Li Z (2017) An approach for estimating the magnetization direction of magnetic anomalies. *Journal of Applied Geophysics* **137**, 1–7. doi:10.1016/j.jappgeo.2016.12.009

Liu S, Hu X, Zhang D, Wei B, Geng M, Zuo B, Zhang H, Vatankhah S (2020) The IDQ curve: a tool for evaluating the direction of remanent magnetization from magnetic anomalies. *Geophysics* **85**(5), J85–J98. doi:10.1190/geo2019-0545.1

McElhinny MW (1973) 'Palaeomagnetism and plate tectonics (Earth Science Series)' Cambridge University Press, New York, 368 pages.

Naprstek T, Smith RS (2019) A new method for interpolating linear features in aeromagnetic data. *Geophysics* **84**, JM15–JM24. doi:10.1190/geo2018-0156.1

Phillips JD (2005) Can we estimate total magnetization directions from aeromagnetic data using Helbig's integrals? *Earth, Planets, and Space* **57**, 681–689. doi:10.1186/BF03351848

Pisarevsky SA (2005) New edition of the Global Paleomagnetic Database. *Eos (Washington, D.C.)* **86**(17), 170. doi:10.1029/2005EO170004

Pisarevsky SA, McElhinny MW (2003) Global Paleomagnetic Data Base developed into its visual form. *Eos (Washington, D.C.)* **84**(20), 192. doi:10.1029/2003EO200007

Pisarevsky SA, Li ZX, Tetley MG, Liu Y, Beardmore JP (2022) An updated internet-based global palaeomagnetic database. *Earth-Science Reviews* **235**, 1–14. doi:10.1016/j.earscirev.2022.104258

Pratt DA, White AS, Parfrey KL, McKenzie KB (2020) 'ModelVision User Guide Version 17.0' Tensor Research Pty Ltd. ModelVision 17.5 User Guide (tensor-research.com.au).

Schmidt PW (2014) A review of Precambrian palaeomagnetism of Australia: palaeogeography, supercontinents, glaciations and true polar wander. *Gondwana Research* **25**(3), 1164–1185. doi:10.1016/j.gr.2013.12.007

Schmidt PW, Clark DA (2000) 'Paleomagnetism, apparent polar-wander path, & paleolatitude'. In: *Billion-year Earth History of Australia and Neighbours in Gondwanaland*. (Ed. JJ Veevers) pp. 12–17. Gemoc Press.

Schmidt PW, Powell CM, Li ZX, Thrupp GA (1990) Reliability of Palaeozoic palaeomagnetic poles and APWP of Gondwanaland. *Tectonophysics* **184**(1), 87–100. doi:10.1016/0040-1951(90)90122-O

Staisch LM, Connell DM, Blakely RJ (2024) High-Resolution Airborne magnetic and radiometric survey of the Republic Graben, Okanogan and Kettle Metamorphic core complexes, Kootenay Arc and surrounding regions, Northeastern Washington. U.S. Geological Survey data release. doi:10.5066/P1PKEZOJ

Stavrev P, Gerovska D (2000) Magnetic field transforms with low sensitivity to the direction of source magnetization and high centricity. *Geophysical Prospecting* **48**, 317–340. doi:10.1046/j.1365-2478.2000.00188.x

Velasco-Tapia F, González-Guzmán R, Chávez-Cabello G, Lozano-Serna J, Valencia-Moreno M (2011) Estudio petrográfico y geoquímico del Complejo Plutónico El Peñuelo (Cinturón de Intrusivos de Concepción del Oro), noreste de Mexico. *Bulletin of the Geological Society of Mexico* **63**, 183–199. doi:10.18268/BSGM2011v63n2a4

Index